90 0755799 8

D1765515

The Ecology of Transportation: Managing Mobility for the Environment

ENVIRONMENTAL POLLUTION

VOLUME 10

The titles published in this series are listed at the end of this volume.

The Ecology of Transportation: Managing Mobility for the Environment

edited by

John Davenport

University College Cork, Ireland

and

Julia L. Davenport

University College Cork, Ireland

 Springer

Library of Congress Cataloging-in-Publication Data

ISBN-10 1-4020-4503-4 (HB)
ISBN-13 978-1-4020-4503-5 (HB)
ISBN-10 1-4020-4504-2 (e-book)
ISBN-13 978-1-4020-4504-2 (e-book)

Published by Springer,
P.O. Box 17, 3300 AA Dordrecht, The Netherlands.

www.springer.com

Printed on acid-free paper

CONTENTS

PREFACE

Human transport by land, sea and air has increased exponentially through time in intensity, paralleling rises in population, prosperity and rates of technological change. Transport has considerable ecological effects, many of them detrimental to environmental sustainability. The aim of this volume was to bring together experts from a variety of disciplines to review the ecological effects and their causes in terms of road, rail, ship and aircraft transport. It was also intended that the contributors should have different attitudes and agendas. Some are ecologists, some planners, others social scientists. Focus ranges from identification of threats, through to concentration on amelioration of damaging effects or design of transport systems to minimize environmental degradation. Some chapters consider restricted areas of the globe; others the globe itself. Views encompass deep pessimism and cautious optimism.

Uniquely, the volume considers transport effects in all environments. Normally scientists who are involved in studying, managing or planning land transport systems have little contact with coastal or oceanic scientists and engineers. Ecotoxicologists often talk little with environmental managers. This is the first book that attempts to discuss the relationship between human transport and all ecosystems. Chapters operate at all scales. They consider impacts of ballast water on global biodiversity, and the contribution of motorway underpasses to sustaining mammal biodiversity in The Netherlands. Information on the spread of human disease by aircraft is balanced by accounts of the impact of snowmobiles on national parks.

This book has its origins in international workshops organised at University College Cork in Ireland in 2004 and 2005. These were funded by a grant to the editors from the Higher Education Authority of Ireland as a result of the 2001-2006 National Development Plan. Participants in the workshops decided upon the framework of the book and adopted the sobriquet of TRANSECOS for their group. TRANSECOS recruited additional authors to improve international and discipline coverage. The aim throughout was to write for a general audience of professionals interested in transport and the environment, whether these be scientists, engineers, planners, civil servants or politicians. Parts or the whole of the book should be useful to postgraduate students in a wide variety of disciplines.

Almost all scientists, and the bulk of the world's media and political establishment, have finally accepted global climate change due to human activities as reality. Urbanisation has proceeded to the extent that about half the world's population lives in cities, entirely dependent on complex travel arrangements, and embedded in specialised urban ecosystems. The concept of 'peak oil' and the prospect of continually declining fossil fuel resources over coming decades is now gaining increased acceptance amongst economists. However, there is presently a lack of logical thinking. Politicians express binding commitments to reeling in damaging human activities within the next few decades. However, industries and governments continue to drive forward agendas of enhanced airline activity, increased production of cars and trucks in more and more countries, more road building, extension of tourism and increasingly globalised trade. All of these agendas are incompatible with ecological (as opposed to economic) sustainability, since they inevitably place greater demands on the environment than can be offset by timely technological innovation. Hopefully this volume will help to provide information and ideas to aid in the creation of the necessary integrated thought.

TRANSECOS

MAY, 2005

ACKNOWLEDGMENTS

The editors would like to thank the contributing authors, the referees and the publishers for their invaluable help in bringing together this volume.

The editors acknowledge the funding provided by the Higher Education Authority of Ireland as a result of the 2001-2006 National Development Plan, which provided the impetus for the book.

The editors thank Stig Persson whose photograph of a moose is reproduced on the front cover.

NATIONAL DEVELOPMENT PLAN

Higher Education Authority
An tÚdarás um Ard-Oideachas

CONTRIBUTORS (e mail addresses of first authors)

K.F. Akbar

John Allan

Anthony P. Clevenger

John Davenport   jdavenport@zoology.ucc.ie

Lisa Dolan   l.dolan@student.ucc.ie

Oliver Floerl

Walter Foley

Ruud Foppen

Bella Galil   Bella@ocean.org.il

Edgar A van der Grift   edgar.vandergrift@wur.nl

Jan Olof Helldin

Marcel Huijser   mhuijser@coe.montana.edu

P.J. Keizer

Tom Kelly   T.Kelly@ucc.ie

Roger Mann (E-mail)   rmann@vims.edu

Dan Minchin   minchin@indigo.ie

Richard Moles   Richard.Moles@ul.ie

Anna Occhipinti-Ambrogi

Eugene O'Brien   eugene.obrien@ucd.ie

John O'Halloran

Gerard O'Leary

Vincent O'Malley

Bernadette O'Regan

Sarah O'Reilly

Rogier Pouwels

Rien Reijnen   rien.reijnen@wur.nl

Dario Savini

Andreas Seiler   andreas.seiler@nvb.slu.se

Dagmar Stengel   dagmar.stengel@nuigalway.ie

Adam Switalski

H. Van Bohemen

Padraig Whelan

INTRODUCTION

TRANSECOS

Over the past 100,000 years, humans have increased their movements across the Earth's surface, partly through exponentially rising populations, partly through technological ingenuity. From origins in Africa, they spread to all continents except Antarctica, initially on foot, then, from around 30,000-50,000 years before present (BP), some form of raft or boat, since Native Australians could not have reached Australia by other means. Domestication of beasts of burden (asses, oxen) dates to some 8,000 years BP, with the wheel and yoke appearing around 1,000 years later. Carts, trade, metalled roads and shipping were all evident by 6,500 years BP, though horses were apparently not tamed and used for draught purposes and personal transport until about 4,000 BP. Sailing ships date from around 5,500 BP, so in principle most existing forms of transport have existed for about 5 millennia. The Industrial Revolution of the 18th and 19th Centuries progressively mechanised land and sea transport as well as producing the bicycle; the 20th Century yielded air transport and the conquest of space. A feature of the development of transport has been its exponential nature in terms of both speed of technological change, and intensity of use. Assuming four generations per century, it took humans 200 generations to move from sailing ships to steamers, yet many people born before the first commercial automobile was built lived to see both their great-grandchildren and the exploration of space. Similarly, personal transport capability has rocketed. In 1901 425 cars were sold in the USA, in 1905 6,500 and in 1914 more than 500,000. Today the USA has a car population of 130 million that turns over on a decadal scale; 40% of American families have two or more vehicles, many of them gas-guzzling sports utility vehicles (SUVs) that are not required to meet the USA Corporate Average Fuel Economy (CAFE) fuel consumption targets mandatory for conventional cars, and which use 2-3 times as much fuel as the most economical passenger cars in production. The USA is not alone; European families often own several cars and the major German manufacturers are currently vying with each other in the introduction of ever larger (2-2.5 tons) and more powerful (400-500bhp) saloons and SUVs. Now that US and European car markets are close to saturation, and most major producers have excess production capacity, manufacturers are targeting the billion-strong populations of China and India; it is currently predicted that car production in China alone will exceed that of the USA by 2020-2025. The oil crisis of 1973 led to a brief period when restraint in the production, use and promotion of cars was considered essential. In the subsequent 30 years the praiseworthy increases in fuel efficiency and reduction in emissions of CO, SO_3 and N-oxides achieved by automotive engineers have been undermined by increasing individual car mass (in turn increasing the energy cost of production; roughly 10% of the lifetime energy cost of a car), emphasis on high performance and crash safety rather than economy, plus exponential increases in car populations driven by increased prosperity. Similar statistics for exponential growth can be produced for shipping (which carries 90% of goods), commercial road transport (trucks) and air transport. Although ships have become bigger and individually more efficient, the global use of containerisation over the last 50 years has stimulated international trade and the use of larger trucks on roads throughout the world. After the loss of industrial confidence following the 9/11 attack on the USA in 2001, civil aviation is currently growing dramatically too because of the entrepreneurial zeal of low-cost (and negligible fuel tax) airlines and the consequent enhanced use of aircraft for social and recreational journeys.

Energy usage by transport has increased remorselessly, not only in absolute terms, but also as a proportion of all energy usage by humankind. In 1971 the transport sector accounted for

35% of oil usage by OECD countries when the oil production level was around 40 million barrels per day; in 2005 the figure is close to 60% of 60 million barrels per day, a 150% increase over around 30 years, despite the recognition that rates of oil discovery peaked 30 years ago and have been declining since. In 1999 Dick Cheney, current US Vice-President, then Chief Executive Officer of Halliburton, stated that there was an annual 2% increase in oil demand, but a 3% decrease in production. Despite substantial real oil price increases in 2003-5, the global consumption of oil in 2003-4 rose by 6%, partly because of enhanced demand by China.

Transport on land, by ship and by air has considerable ecological effects. Roads and railways divide habitats, contributing to 'islandification', separating communities and often reducing gene flow. Conversely, fenced-off railway land and road verges can act as refuges and corridors. Road kills by vehicles affect populations in both negative and positive fashion. Land transport creates atmospheric, aqueous and terrestrial pollution that affects biota. Road and rail vehicle construction demands extraction and modification of materials from around the globe, with consequent energy expenditure in those processes, plus transport of the materials themselves. Scrapping and demolition generate waste disposal problems. The existence of roads and railways profoundly influences planning, housing and industrial development, with knock-on ecological consequences. In 2005, urbanisation has proceeded to the extent that about half the world's 6.4 billion population lives in cities, entirely dependent on complex travel arrangements for their existence (especially in terms of food supply), and embedded in specialised urban ecosystems.

The impacts of marine/freshwater transport on ecosystems have attracted wide interest. Oil pollution, the spread of alien species around the world in ballast waters, and problems of antifouling technology (particularly endocrine-disrupting TBT) have generated much media attention. However, disease transmission and involvement of shipping in promoting spread of toxic algal blooms are areas of current medical and aquaculture concern, but less widely publicised. Shipping-driven atmospheric pollution is also an issue as the burning of heavy fuel oil by diesel-powered ships is an unappreciated source of greenhouse gases and carcinogenic hydrocarbon pollutants. Historically, shipping has had dramatic effects on terrestrial ecosystems, particularly of islands, by permitting transport of terrestrial domestic/game animals and pest species that have contributed to loss of species diversity. Several millennia of shipping losses through storms and warfare have created new localized marine habitats – 'artificial reefs' long before the term was thought of. Prehistorically, marine transport allowed hunter-gathering humans to spread to new habitats, with profound effects on the world's ecosystems. In human ecological terms, shipping has also been a major historical cause of spread of pestilence – notoriously of plague via ship rats, but also of measles, rabies and syphilis. As recently as 1991 10,000 people were killed in Peru and 1 million people infected in the Western Hemisphere by cholera, *Vibrio*, carried in ballast water from Bangladesh.

Effective mass transport of material and people by air dates from the 1930s-1940s, initially militarily, then commercially. The speed of air transport increases risks of transport of diseases (as in the recent SARS epidemic) and pest species and this process may be potentiated by global climate change as such organisms become more likely to survive at destinations. Ecological effects of air transport via noise, hydrocarbon usage, generation of greenhouse gases and light-obstructing condensation trails ('global dimming'), plus stratospheric delivery of ozone-depleting chemicals are also globally significant.

All forms of transport (other than the minute fraction utilising biofuels, or electricity derived from renewable sources) are major net contributors of greenhouse gases, particularly CO_2, so are important elements in global climate change processes. Obviously transport is only part of the increasing CO_2 problem, but its relative contribution tends to rise as power generation becomes cleaner and industrial emissions are reduced. Svante Arrhenius identified the basic science of the CO_2-generated 'greenhouse effect' of rising world temperature in the late 19th Century. There has never been disagreement over the fact that CO_2 levels have risen considerably from the onset of the Industrial Revolution. Prior to the industrial age and extensive use of fossil fuels, the concentration of carbon dioxide in the atmosphere stood at about 280 parts per million. It is currently about 380ppm. In the 1950s it rose at around 1ppm per annum and the average rate of rise over the last half century has been about 1.8ppm per annum, but there has been a recent acceleration in concentration rise (in 2003-4 it rose by 3ppm). Predictions suggest atmospheric values of 700-800ppm by the end of the century. Fossil carbon is being released at a million times the rate of its production. However, until relatively recently, it has been difficult to convince politicians, industrialists and the general public that the relatively small rises (ca 0.6°C) in global environmental temperature that have ensued so far are a) due to Man's activities, b) are important, c) presage much greater increases in the next few decades. Non-scientists have interpreted uncertainties about the scale of effects and the timescale of climate change as uncertainty about the reality of the phenomenon itself. At the time of writing the recent (2005) publication of the 2nd Millenium Assessment Report, and a joint statement by the scientific academies of all developed countries in advance of the July 2005 G8 summit meeting at Gleneagles in Scotland, has finally led to a consensus that Man's effects on the climate are established and pose a considerable threat to the world's ecosystems and to human civilisation. However, there is still limited realization that climate change is an accelerating (rather than linear) phenomenon. Ecological effects in terms of altered distributions of aquatic and terrestrial animals and plants have already been repeatedly identified, as have marked changes in phenology (timing of events such as bird breeding), but politicians are still making decisions that will inevitably increase greenhouse emissions. In the justifiable emphasis on climate change, effects of the inexorable rise of CO_2 itself have attracted less attention. CO_2 levels are now at their highest levels for at least 420,000 years, and models show that significant reduction will take millennia even if levels are stabilised at current values. Plant scientists note that the increase in atmospheric carbon dilutes environmental nitrogen with complex effects on quality of forage in natural and agricultural ecosystems. Of perhaps more immediate concern is the developing evidence of the acid buffering capacity of the oceans becoming affected though CO_2 accumulation. The World Ocean Circulation Experiment (WOCE) and the Joint Global Ocean Flux Study (JGOFS) have demonstrated that the surface waters of the oceans have taken up a net 118 billion metric tons of carbon from the atmosphere over the past 200 years. Ocean pH was 8.3 10-12,000 years ago and 8.2 before the industrial age. It is now 8.1 and declining. This has consequences for marine biota reliant on calcification (e.g. corals, coccolithophores), as this process becomes more energy-consuming with decreases in pH.

As transport has burgeoned and environmental effects of transport have increased, transport itself has contributed to our understanding of environmental processes and threats. Between 1872 and 1876 HMS *Challenger*, a sailing ship with a steam engine for manoeuvring on station, sampled 362 stations over 69,000 nautical miles from Arctic to Antarctic. Fifty bible-sized volumes of published data emerged over the next 20 years and the expedition laid the foundation for modern oceanography. Today, aircraft and satellites launched by rocket or Space Shuttle monitor the whole of the world's surface in increasing detail, collecting data ranging from global sea surface productivity and forest cover, to the individual movements of seals, penguins and turtles. Autonomous vehicles collect information from the deepest oceans and from beneath the polar ice. Satellite-based global positioning systems allow sampling precision that would have been unimaginable even a generation ago, while the technology of the world-wide-web can deliver information to users in real-time.

Amongst practitioners in the area of transport and its ecological effects there are a wide range of perspectives, not least because some are ecologists, some environmental planners, and some engineers. Those who have documented steady or increasing levels of environmental degradation, habitat loss and reduced biodiversity (often ecologists) tend towards a pessimistic view, contrasting the scale and accelerating rate of degradation with the slowness and limited effectiveness of societal, regulatory and political responses, plus the resistance to change of most economic vested interests and some politically-powerful religious groups that believe the end of the world is nigh, so restraints on resource use are unnecessary. Pessimists tend to regard technological solutions as too little, too late, since new transport technologies take around 50-100 years to replace existing systems fully. At the other end of the spectrum are those (often engineers or economists) who are optimistic that implementation of green technologies will solve transport-related problems. Such feasible technologies include underground injection of CO_2 derived from fossil fuel burning, plus electric/fuel cell vehicles reliant on the use of biofuels, wind and wave power, and the introduction of novel energy-efficient transport systems such as maglev trains. The optimists believe that these technologies, together with increased global prosperity, will allow humans to arrest a damaging downwards spiral quickly enough to preserve an acceptable environment for future generations. Between these extremes are those whose role is to ameliorate adverse effects of human transport by combinations of biological knowledge, eco-friendly transport engineering and public planning. This volume features chapters by people drawn from this full spectrum; their views and attitudes are diverse and consensus was not an aim. The workshops that have yielded their state-of-the art reviews were conducted in a friendly atmosphere of frequent disagreement! However, the basic pragmatic approach has been to write chapters that identify problems and propose positive remedies wherever possible.

CHAPTER 1: ECOLOGICAL EFFECTS OF AVIATION

TOM KELLY[1] & JOHN ALLAN[2]
[1]Department of Zoology, Ecology and Plant Science, University College Cork, Lee Maltings, Prospect Row, Cork, Ireland
[2]Central Science Laboratory, York, UK

1. Introduction

The technological equilibrium that existed in the middle of the 19th century was punctuated by a series of momentous discoveries and inventions including oil as a lubricant and fuel, the internal combustion engine, Bell's telephone and Edison's electric light bulb. Kerosene and the internal combustion engine were vital contributors to the first powered flight, at Kill Devil Hills, near Kitty Hawk in North Carolina on December 17 1903 (Anderson 2004). Orville and Wilbur Wright's outstanding achievement was quickly followed by a spectacular sequence of inventions in aeronautical engineering leading to, for example, Concorde and commercial supersonic aviation, within 70 years of their inaugural flight (Pascoe 2003). Every step (apart from those which remain classified for military reasons) in the development of aviation is known (e.g. Anderson 1997, 2003). Over this time air transport has become faster, safer, more fuel efficient, less polluting and less noisy. An improving understanding of aerodynamics led to rocketry and ultimately to space exploration and therefore "transport beyond the earth" (Hobsbawm 1994). Satellite imagery of the Earth is now commonplace and is widely used by science to monitor the global landscape, its biodiversity and a wide range of environmental variables including, in particular, the effects of climate change. In a sense the Space Shuttle and associated satellite technology are continuing Darwin's pioneering voyage in the Beagle (1831-1836).

This account aims to describe the positive and negative ecological effects of aviation. These impacts occur on different scales ranging from the global (aviation's contribution of global warming gases to the atmosphere) to the local (airfield habitat and wildlife control measures). It should be stated at the outset that this subject is overwhelmingly dominated by so-called 'grey' i.e. mostly non-peer reviewed literature; relatively little has been published in the mainstream scientific journals. In addition, the stimulus or stimuli that cause animals to respond to aircraft are still, to our knowledge, imprecisely understood.

2. A brief history of aviation and its ecological impacts

The military were the first to make use of the air – despite General Foch of France dismissing aviation as " good sport, but for the army the aeroplane is worthless" (Yergin 1991). In 1915, Britain possessed only 250 planes but by the end of the First World War had produced 55,000 aircraft; broadly similar numbers had been assembled in Germany and France. Initially aircraft were non-combatant and were merely used for reconnaissance but soon these aeroplanes were armed with machine guns and bombs (Yergin 1991; Bilstein 2001; Grant 2004).

However, it was in America that aircraft were put to their first major practical non-military uses, most notably in the delivery of mail, aerial photography and crop spraying (crop dusting) with insecticides against the cotton leaf worm, a noctuid moth caterpillar (*Spodoptera* sp). According to Bilstein (2001) a pilot could "dust" "from 250 to 500 acres per hour". Eventually, the main target became the cotton boll weevil (*Anthromonus grandis*) and the weapon "a swath of lethal white dust (arsenate of lead powder)" (Bilstein *op.cit*) with 500,000

John Davenport and Julia L. Davenport, (eds.) The Ecology of Transportation: Managing Mobility for the Environment, 5–24.
© 2006 *Springer. Printed in the Netherlands*

acres being 'treated' in 1927. Such activities continue throughout the world today, though due to problems of wind drift, as little as 40 per cent of the biocide (including until recently the persistent organochlorines like dichloro-diphenyl-trichloroethane (DDT)) may actually land on the target area. This led to the evolution of insecticide resistance among pest insect species including the *Anopheles* mosquito vectors of malaria and no doubt other species that transmit micro-pathogens to domestic animals and wildlife. It is in this way that aviation began to have a significant negative environmental impact, albeit as an instrument of ever intensifying agricultural production.

2.1 COMMERCIAL CIVIL AVIATION

Gore Vidal (1993) was a passenger on what is claimed to be "the first commercially scheduled airliner" – on July 7 1927, which flew from New York to Los Angeles, less than two months after Charles Lindbergh's pioneering "Spirit of St Louis" non-stop 5,760km flight from New York to Paris. Despite the investment and banking catastrophe incurred during the "Great Crash" of 1929, and the following years of severe economic depression, there were sustained progressive improvements in the airworthiness and design of paying passenger aircraft. Examples include the remarkable DC-3 which first flew in 1935, and the pressurised cabin, in the B-307 "Stratoliner", the first commercial airliner capable of flying at the height of the stratosphere. Competition from the airship industry collapsed with the very public fire, explosion and loss of life, on the *Hindenburg* in May 1937 (Grant 2004). However, it was not until after the end of the Second World War, and especially in the 1950s, that commercial civil aviation began to grow into an alternative form of transport, particularly to travel by ship.

The exponential increase in commercial aviation was largely due to the use of the gas turbine 'jet' engines – first used in military aviation – which substantially increased the range of aircraft and halved the time it took to make these journeys (Bilstein 2001). The first commercial jet airliner was the de Havilland D.H. 106 "Comet" which had its inaugural flight in Britain on July 27 1949 (Gunston 2001). Other designs quickly followed including Boeing's "Dash 8" in 1954 and the highly successful 707 in 1957. In 1955, the French Sud-Est SE 210 Caravelle and the Douglas company's DC-8 had undergone inaugural flights (Gunston 2001). Comparable aircraft were being produced in the USSR.

In what Richard Dawkins (2004) has defined as a progressive process, aviation continued to 'evolve' e.g. in airframe design, electronic transmission systems, hydraulics, and the incurporation of on-board computing leading eventually to the "fly by wire" concept.

Engine design also improved leading to the high-bypass ratio turbofans which are found on the Airbus A320 and more modern Boeing 737 series (e.g. the 300 to 800) (see MacKinnon, 2001). These engines are more fuel efficient, less noisy and less polluting than their predecessors. Overall, air-safety has increased with the notable exception of acts of terrorism. Hijacking of aircraft, the placing of bombs on aircraft and the shooting down of aircraft by terrorists has affected the numbers of people flying, but these reductions are transient and the market appears to recover quite quickly.

Some 1.6-1.8 billion passengers fly on commercial aircraft every year (Robinson 2000; Gunston 2001), and although there was a marked reduction following the September 11 2001 atrocity in New York, growth in the budget travel (e.g. easyJet and Ryanair) market has led to a sharp recovery in the numbers of people travelling by air (Nugent 2005).

A new phase in commercial aviation opened in late April 2005 with the inaugural flight of the Airbus A380 a double-decker "fly by wire" jumbo jet capable of carrying at least 555 passengers. It remains to be seen if this venture will prove to be commercially successful, but all the existing projections indicate that the commercial aviation sector is likely to undergo a sustained expansion-particularly in Africa, South America and Asia (e.g. Robinson, 2000;

Mackinnon 2001). For example, Africa holds only 4% of "world airline aircraft fleet ", and South and Central America 6.8%, compared with 43.7% and 23.7% in the USA and Western Europe respectively (Mackinnon 2001). Therefore, the construction of new airport infrastructure in, for example, Africa and South America, to cater to an expansion in civil aviation is likely to have considerable local ecological effects.

2.2 GENERAL AVIATION

General aviation includes light aircraft, many of which are powered by a single piston engine. It is estimated that there are approximately 339,000 of these aircraft in service, 65% of which are in the USA (Mackinnon 2001). Although light aircraft are the most numerous in civil aviation, they fly for an average of only 135 hours per year (Mackinnon 2001).
As mentioned previously light aircraft were used in aerial spraying and more recently in the destruction of "crops" which are used in the manufacture of drugs.

2.3 MILITARY AVIATION

Although the use of aircraft for military purposes has been the major force in the 'evolution' of aviation generally, less information is available about the number of aircraft and their operations. Remarkable military aircraft include the SR-7IB Blackbird which travels at Mach 3 and exceptionally high altitudes; the eight engined B 52 bomber which is still in service following its launch in 1955, and the stealth bombers Lockheed F-117 Nighthawk and the Northrop B-2 which are effectively invisible to radar surveillance (Grant 2004). Military aviation also includes a variety of helicopters and, increasingly, unmanned reconnaissance aircraft. In addition a vast arsenal of missiles and "smart bombs" have been designed for use in combat.
As with commercial aviation, the air forces of the world require large air bases although these usually lack ancillary infrastructure such as hotels and car parking facilities. However, air forces need large safe areas in which to practice bombing and other military manoeuvres.

2.4 OTHER FORMS OF AVIATION

2.4.1 *Rocketry and space travel*
Following the launch of Sputnik 1 in October 1957, manned space flights were successfully completed in the USSR and the US, in 1961, only 58 years after the Wright brothers' pioneering achievement in North Carolina. The Apollo mission to the moon was launched on July 171969 and the landing took place on July 20th. Thereafter space travel has involved the construction of the MIR space station, the Space Shuttle and missions like the Galileo space probe to Jupiter. This was launched in October 1989, arriving six years later in July 1995 (Grant 2004).

2.4.2 *Helicopters*
Helicopters were flown in Germany in the mid-1930s but were not widely used until after the Second World War. Again there is extensive use of rotary winged aircraft in both military and civil aviation. There are about 33,000 helicopters being used in civil aviation (based on Mackinnon 2001). Helicopters generally fly at low altitudes, and at relatively low speeds. However, this type of aircraft which uses either piston or turboshaft engines, requires relatively little supporting infrastructure, unlike almost all other forms of aviation.

2.4.3 *Powered hang gliding and the microlight*
Hang gliding and paragliding have now become popular sports. In the 1970s powered hang gliders known as microlights were developed, attaching piston engines to a 'pivoting wing' (Grant 2004). These very small aircraft fly at low altitudes and low speeds.

2.4.4 *Gliders*
The use of gliders was developed in Germany in the 1920s (Grant 2004). However, although gliders are still widely flown, the sport does not appear to have attracted large numbers of participants. Gliders generally have to be towed into flight by a light aircraft, but once operating on their own they are relatively noiseless.

3. The ecological effects of air transport

Aviation is likely to have three major ecological effects. First as a polluter of the air, included in which would be the release of the gases which are believed to be responsible for climate change. The scale of this effect will be at both the global and regional level e.g. 72% of all transatlantic air traffic between Europe and North and Central America passes over Ireland (data from the Irish Aviation Authority). This environmental impact is addressed in Chapter 15.

The second ecological impact is noise pollution. This occurs at the regional to local level. For example, the noise from transatlantic aircraft descending in the early morning towards destinations in Britain is clearly audible over the southern counties of Ireland. Likewise, aircraft noise will be obvious in the approach and climb-out zones of most international airports. And finally, airports themselves are generally noisy places (e.g. Burger 1983).

The noise impacts of aviation are controversial as they lead to very negative responses from the public and extensive measures are now being taken to reduce this problem. However, noise and other stimuli produced by the aircraft, are also believed to impact on wildlife (e.g. Kelly *et al.* 2001; Frid & Dill 2002; Komenda-Zehnder & Bruderer 2002; Pepper *et al.* 2003), though the responses of birds in particular are proving difficult to resolve (Kelly *et al.* 2000; Kelly *et al.* 2001; Komenda-Zehnder *et al.* 2003).

Finally, the third major ecological impact occurs at the level of the airfield. This involves the impact of a) the installation of major infrastructural requirements for aviation in the jet age and b) habitat management to reduce the impact to aviation of collisions with wildlife, mainly birds and mammals. Infrastructural requirements for jet and turbo jet aircraft include long concreted runways, wider taxiways, extensive apron space and aircraft parking areas, Instrument Landing Systems (ILS), approach lighting, control and watch towers etc. The new Airbus A380, for example, has a wingspan of 79.8m, and airports endeavouring to receive this aircraft may have to modify existing runways, taxiways and the apron to achieve appropriate stand sizes.

Mammals, birds and reptiles may collide with aircraft causing what is now termed a wildlife strike. This 'strike' may be sufficiently severe to cause the destruction of the aircraft with the loss of life of all of those on board (e.g. Blokpoel 1976; CAA 1998; Mackinnon 2001; Thorpe 2003; Cleary *et al.* 2004). Most wildlife strikes do not have such catastrophic results but nevertheless may reduce the safety margins of flying. Bird strikes are expensive, causing losses of US$1.5 billion per annum (Allan 2002). Therefore, measures must be taken on airfields to prevent or reduce the risk of wildlife strikes. These measures include, albeit rarely, the direct killing of species which pose the most serious risk (e.g. Dolbeer *et al.* 2003), the drainage of wetlands, and the chemical removal of weeds and invertebrates (e.g. earthworms and insects) which attract hazardous birds and mammals. In addition, airfield grasslands are managed to minimise the build up of flocks of species like starlings (*Sturnus vulgaris*), and gulls (*Larus* sp) (CAA 1998; Mackinnon 2001). Finally, a variety of noise emitting devices including shell crackers fired from a pyrotechnic launcher/Very Pistol and re-broadcast distress calls are used to scare or haze birds and mammals from the airfield (e.g. Blokpoel 1976; CAA 1998).

3.1 THE NON-LETHAL EFFECTS OF AIRCRAFT ON WILDLIFE

Non-lethal disturbance of animals by aircraft is an important issue and a source of conflict between the various aviation interests and wildlife conservation bodies (Frid & Dill 2002). This problem has received considerable attention in recent years, and a protocol on how it should be measured, and its effects assessed, is emerging (Efroymson *et al.* 2000; Efroymson *et al.* 2001; Efroymson & Suter 2001; Komenda-Zehnder & Bruderer 2002; Frid & Dill 2002; Frid 2003; Rees *et al.* 2005). Aircraft may cause animals to flee in panic and these behaviours in turn may affect fitness because they are costs in terms of energy expended, energy lost through a reduction in foraging time, decreased time allocated to parental care, increased time allocated to vigilance, and desertion of preferred feeding grounds, territories, and home ranges (e.g. Gill *et al.* 1996; Hill *et al.* 1997; Gill & Sutherland 2000; Frid & Dill 2002; Frid 2003).

However, difficulties remain in overcoming the effects of confounding variables-even where experimental protocols have been carefully prepared (e.g. Komenda-Zehnder *et al.* 2003; Frid, 2003; Rees *et al.* 2005). Thus, while most researchers believe that the noise emitted by the aircraft is the primary source of the disturbance (Efroymson *et al.* 2001; Efroymson & Suter 2001; Pepper *et al.* 2003), others argue forcefully that it is the looming visual stimulus that is the main cause of the panic responses (Frid & Dill 2002). However several studies have also shown that noise looms and sounds that move towards a subject have greater "biological salience than those that move away" (Hall & Moore 2003).

Other studies have mentioned the importance of habitat, location and time of year in explaining the variation in response to sources of human disturbance (e.g. Delaney *et al.* 1999; Frid 2003; Rees *et al.* 2005; Laurson *et al.* 2005). It is also known that there are inter-specific (e.g. Smit & Visser 1993; Pepper *et al.* 2003) and age differences (e.g. Rees *et al.* 2005) in the magnitude of the response to aircraft.

In addition, there is evidence that birds and mammals may habituate to aircraft, particularly if the angle of approach is not direct, and both vertical and horizontal distances are sufficiently large to reduce the probability of a perceived immediate threat (e.g. Conomy *et al.* 1998; Delaney *et al.* 1999; Komenda-Zehnder *et al.* 2003; Pepper *et al.* 2003; Frid 2003). However, Robinson and Pollitt (2002) noted that waterbirds were less likely to habituate to infrequently occurring disturbances. This important observation, echoing those of Burger (1981), has clear implications for the interpretation of results of field experiments where the aircraft-related disturbance is relatively novel in the environment of the studied species.

There is a consensus that helicopters are a greater disturbance to birds and mammals than fixed wing aircraft (e.g. Smit & Visser 1993; Stock 1993; Komenda-Zehnder & Bruderer 2002; Komenda-Zehnder *et al.* 2003).

3.2 NON-LETHAL INTERACTION BETWEEN AIRCRAFT AND ANIMALS AT AIRPORTS

Airports may be a focal point where intense interactions will take place among birds, mammals and aircraft. Some will, as mentioned, involve collisions with aircraft, but these potentially damaging events are relatively rare (generally between 2 and 6 per 10,000 aircraft movements; Kelly & Bolger in prep) given the large number of birds that commute through the at-risk air space. This raises the question – do birds avoid aircraft, and if so, which cue or cues stimulate the avoidance response?

Kelly et al. (1999) showed that at Dublin Airport, Ireland, 88% of birds flying towards the active runway performed obvious avoidance manoeuvres. These were described and classified. Four bird species actively avoided aircraft namely the rook (Corvus frugilegus), black headed gull (Larus ridibundus), woodpigeon (Columba palumbus) and lapwing (Vanellus vanellus). Both the rook and woodpigeon are residents at Dublin Airport, but the black headed gull and lapwing are migrants and winter visitors.

Subsequent research by Kelly et al. (2001) focused on the avoidance behaviour of the rook. Two categories of aircraft are now identified namely those belonging to either Chapter 2 or 'Chapter 3'. 'Chapter 2'-type aircraft (examples include the Boeing 707, 727 and 737 series 200) are much noisier than those in the 'Chapter 3' category (examples of which include the Boeing 737 series 300-800, the Airbus series A300 to A340 and the British Aerospace 146) (Anon 1993). In the study the repertoire of avoidance behaviours shown by the rook to the different categories was analysed. The results showed that significantly fewer rooks responded to the Bae-146, a well-known 'Chapter 3' type aircraft, than to the B737-200, a distinctly more noisy 'Chapter 2' -type aircraft. This suggests that noise may be an important cue for the rooks. However, size differences between these two aircraft may account for the observed pattern of response. The Bae-146 (length 26.3m, height 8.6m and span 26.34m) is shorter and narrower than the B737-200 (length 30.5m, height 11.3m and span 28.4m), but it is questionable whether this contrast in dimensions is sufficient to explain the statistically significant ($P < 0.01$) difference in the response of the rook. Tomlinson et al. (1991) found that birds showed a more marked physiological response to larger aircraft; but others, including Smit and Visser (1993), Koolhaas et al. (1993), Davidson and Rothwell (1993), Conomy et al. (1998), and Ward et al. (1999), have shown that the disturbance caused by aircraft is inversely proportional to their size with, for example, microlights causing the most severe effects.

The results of Tomlinson et al. (1991) suggest that there is a hierarchy in the stimuli to which the birds respond, where visual stimuli elicit the most significant response followed, in order, by a combination of both visual and auditory stimuli and finally purely auditory stimuli.

It is clear that, while noise is potentially an important cue for rooks, it was not possible in the Kelly et al. (2001) study to resolve the confounding effect of the visual stimulus. Kempf and Huppop (1996) refer to this difficulty and state "the noise of aircraft can scarcely be assessed separately from its optical appearance". Further investigations are therefore necessary to establish the relative importance of aircraft noise in the avoidance behaviour of the rook and other bird species found at airports. This may be important as the noisier 'Chapter 2' type aircraft are to be completely phased out from European airports by December 2005.

To summarise, aircraft are a disturbance, sometimes of considerable proportions, to birds and mammals, but research into this problem continues to be hampered by our inability to quantify the relative importance of the noise, versus the visual, stimulus. Birds are well known to respond to multi-component i.e. aural and visual, signals in, for example, their courtship behaviour (Kelly et al. 2000). It is likely, therefore, that their responses to aircraft will prove to be both behaviourally and neurologically complex. However, a greater understanding of the basis of their responses, including the role of learning and habituation, is important, as recreational and military aviation increases in the wilderness of the USA and Canada (see Mackinnon 2001). Paragliding appears to be an increasing problem in Europe (Schnidrig-Petrig & Ingold 2001; Enggst-Dublin & Ingold 2003).

3.3 REDUCTION OF THE NEGATIVE IMPACT OF NON-LETHAL AVIATION

It is generally recommended that over-flying of important conservation areas for birds, marine and terrestrial mammals should be reduced to a minimum. Guidelines proposed by the government of the United Kingdom recommend minimum horizontal and vertical distance of 750m for single engined helicopters, but 1000m for those with twin engines operating close to concentrations of birds in Antarctica. The minimum distances for fixed-winged aircraft with 1 to 2 engines is 450m whereas it is 1000m for those with 4 engines. Raptor species showed no response to aircraft flying at a distance greater than 500m, and Komenda-Zehnder *et al.* (2003) suggest that a minimum altitude of 600m AGL should be sufficient to prevent major disturbances of waterbirds on lakes in Switzerland. Finally, Efroymson *et al.* (2001) and Efroymson and Suter (2001) recommend that "slant distances" be used to estimate the exposure of wildlife to "stressors" emitted by low flying aircraft. However, this method requires the measurements of angles in the field, which appear to be difficult to obtain (Frid & Dill 2002; Frid 2003).

3.4 MILITARY AVIATION AND AERIAL BOMBING CAMPAIGNS

The ecological effects of war have not, for obvious reasons, attracted much detailed study. The bombing of Guernica on April 26 1937 proved to be a portent of the mass destruction of civilian human life in the Second World War up to and including the dropping of the atom bombs on Hiroshima and Nagasaki in 1945. But, even in early January 1940, this scale of devastation appeared a somewhat unlikely eventuality. In the House of Commons when the Secretary of State for Air, Sir Kingsley Wood was pressed to bomb the Black Forest to destroy German timber supplies "the little man replied outraged": "Are you aware it is private property? Why you will be asking me to bomb Essen next!"(Collier 1980).

W.G. Sebald's (2003) book "On the Natural History of Destruction" (a title borrowed from Lord Solly Zuckerman) contains a remarkable but shocking essay entitled "Air War and Literature" which describes the consequences of the Allied bombing campaign against cities in the Reich. On page 35 he quotes Hans Erich Nossack who observed that "Rats and flies ruled the city. The rats bold and fat, frolicked in the streets, but even more disgusting were the flies, huge and iridescent green".

Orians and Pfeiffer (1970) remark that tigers benefited from the Vietnam War and apparently consumed "large numbers of battle casualties". They speculated that the tiger population had "increased much as the wolf population in Poland increased in World War II".

Orians and Pfeiffer (1970), also describe the impact of B-52 bombing raids and particularly the number of craters that were created by the payload, which they estimate to have been 2,600,000 in 1968 alone. Some of these would become fishponds, but others the breeding ground for mosquitoes. Defoliation (notoriously by 'Agent Orange') was one of the main aerial warfare strategies carried out by the USA in Vietnam (See Chapter 15). As this impacted on the mangrove swamps, Orians and Pfeiffer (1970), considered that fish-eating birds including the grey heron (*Ardea cinerea*) were fewer than expected. However, Fitter (1945) noted that during the London Blitz the great heronry at Walthamstow was hardly affected by "a hail of bombs that fell on all sides of it" and in fact "there was an increase of three occupied nests in 1941 compared to 1940". Fitter (1945) also shows that while the black redstart (*Phoenicurus ochrurus*) did breed on bombsites, it had been increasing in the London area before the war. However, Hollom (1975) argues that its "establishment" in areas outside Middlesex and Sussex including the midlands followed " bomb damage of 1940-41".

Fitter (1945) summarises the effects on plant communities resulting from the blitz. Most of the work was undertaken and published by Salisbury (1943). One curious finding was that the London rocket (*Sisymbrium irio*) was not part of the re-colonising plant community after the Blitz, even though it apparently flourished following the Great Fire of 1666.

Rhodes (1986) is one of many who described the devastation caused by the atomic bombs in Japan in 1945, and while no ecological data is presented it must be presumed that most, if not all life, was extinguished within the area of the blast.

4. The wildlife hazard problem

The first human fatality caused by a bird strike occurred in 1912 "when a gull became entangled in the exposed control cables in an aircraft" (Blokpoel 1976). This accident occurred at Long Beach California, and the person killed was Cal Rogers who "had been the first person to fly across America" (Thorpe 2003). Bird hazards to aircraft became a much more serious problem with the arrival of jet propulsion; the greater speed of jet aircraft results in much more severe impact damage resulting from collisions with birds (e.g. Blokpoel 1976; Mackinnon 2001). In addition, jet engines can "ingest" birds and this, albeit rarely, may cause the aircraft to crash (e.g. Blokpoel 1976; CAA 1998; Mackinnon 2001; Sodhi 2002; Thorpe 2003; Cleary et al. 2004).

In an authoritative review, Thorpe (2003) showed that 231 people have been killed in 42 wildlife-related fatal civil aviation accidents. Overall, a total of 80 civil aircraft have been destroyed by collisions with birds. In addition, at least 141 people have been killed as a result of bird strikes to military aircraft (Richardson & West 2000). People on the ground have been killed in bird strike-related fatal accidents involving both civil and military aircraft (Thorpe 2003; Richardson & West 2000). A collision between a mammal and an aircraft has caused one human fatality in the USA (Cleary et al. 2004).

Wildlife strikes are responsible for significant additional (i.e. non-fatal) damage and costs, to both civil and military aircraft (see Allan 2002; Cleary et al. 2004).

The response of the aviation industry to these hazards has been to implement a wide range of control measures to reduce the overall risk and costs of wildlife strikes (e.g. CAA 1998; Mackinnon 2001; Cleary et al. 2004). Managing wildlife hazards is generally confined to the airfield and therefore the effects are mostly local (though see Rees et al. 2005) but necessary planning restrictions to ensure air safety may exert an influence on a wider scale (see Section 5).

4.1 ECOLOGICAL EFFECTS OF AIR TRANSPORT: NUMBERS OF ANIMAL FATALITIES

Most animals that collide with an aircraft are killed. Therefore an increase in mortality is the main impact of air transport on the population dynamics of animals. Culling of hazardous species may also occur on or off the airfield and would therefore constitute an additional source of mortality.

For mortality to exert a negative effect on a population i.e. to cause it to decline, it must be additive, i.e. in excess of natural mortality (Newton 1998). This can be compared with compensatory mortality i.e. where increased losses are "compensated" for by enhanced survival and reproductive success (Newton 1998). It is of interest to know therefore, if the mortality caused by bird and other wildlife strikes is additive, and whether it may, as a consequence, cause the decline of the affected species. It is also relevant to compare aircraft strike induced losses with other anthropogenic sources of collision related mortality. While authorities may disagree over the definition of a bird strike (e.g. Brown et al. 2001), the consequences of differences in interpretation are unlikely to have major implications in calculating the effects on the population dynamics of most, but not all, species (i.e. those in which the bird strikes cause additive mortality). A more substantial problem is the non-reporting of bird strikes. It is suggested that as few as 15-20 per cent are reported (Burger

1985; Linnell *et al.* 1996, 1999; Barras & Dolbeer 2000; Brown *et al.* 2001; Cleary *et al.* 2004). Therefore in calculating the total number of birds and mammals killed by civil aircraft it is necessary to make the appropriate transformations. We have assumed, for the data set compiled by Cleary *et al.* (2004) for the USA, the most comprehensive of its kind in the World, that only 20 per cent of strikes are reported. It is obvious however; that air strikes involving deer are much more likely to be documented than collisions with small animals such as voles, bats and swallows. Therefore, any analysis at this stage is necessarily preliminary, crude and possibly in some cases (e.g. large birds and mammals) an overestimate.

Table 1. Estimates of the numbers of birds killed in collisions with man-made structures. Erickson *et al.* (2001) Percival (2005) Cleary *et al.* (2004) (Data on mammals and reptiles Cleary *et al.* 2004)

Collision with:	Birds (range)	Mammals	Reptiles
Motor vehicles	60-80 million	Not available	Not available
Buildings	98-980 million	Not available	Not available
Power lines	10,000 to 174 million		
Communication towers	4-50 million		
Wind turbines	10,000-40,000	Not available	Not available
Aircraft (civil)	15-30,000***	364	20

*** The data from Cleary *et al.* (2004) refers to the 1990-2003 interval. Therefore the totals for each group were divided by 14 to give an annual average, and then this figure was multiplied by 4, to account for unreported strikes. The range in the number of birds killed by aircraft is based on an average of 1 and 2 respectively, killed per strike.

It is obvious from Table 1, that even when the very major problems of compiling such data, and their obvious limitations, are taken into account (Erickson *et al.* 2001; Percival 2005) collisions with manmade structures are a major source of mortality in birds. However, aviation appears to be responsible for the least amount of birds (and probably mammals and reptiles) being killed in this way. It is unlikely that the level of recorded mortality will prove additive and therefore of conservation concern. While the number of large birds being struck by civil aviation in the USA is increasing, and is itself an expanding threat to air safety (Dolbeer & Eschenfelder 2003), the reason is that their populations are recovering from previous declines. So, while large birds generally have high annual survival rates, and are therefore particularly vulnerable to the effects of additive mortality, there is no evidence as yet that the number of bird strikes to these species poses a threat to their population viability. Most bird strikes involve just one individual being killed, and with improving standards of management the overall losses (per collision) are likely to decrease (Kelly *et al.* 1996).

Culling birds, even when this is designed to improve air safety, is controversial for both ethical and scientific reasons. There is a long history of opposition to the deliberate killing of animals (e.g. Thomas 1983) which precedes the so-called "Bambi effect"(Cartmill 1993), and an increasing number of airports are employing non-lethal methods for reducing wildlife hazards. Some are using dogs, particularly border collies (e.g. Patterson 2000), and others falcons, to scare off birds from the airfield.

However, culling is also being implemented particularly at John F Kennedy Airport in New York. Here gulls are shot as they approach and over-fly the airfield. In total 72,063 gulls were shot between 1991 and 2002 (Dolbeer *et al.* 2003) This procedure has led to a marked and statistically significant reduction in the number of bird strikes caused by gulls (mainly the Laughing Gull *Larus atricilla*) and a 58 per cent reduction in the size of the adjacent gull colony. The numbers of gulls over-flying the airfield has decreased by 14-26 per cent. Nevertheless, Brown *et al.* (2001) recommend alternative procedures including the relocation of the gull colony.

The laysan and black-footed albatrosses posed a problem for air operations on Midway Island. These are a species which would normally have an annual adult survival rate of up to 95 per cent. Some 30,000 had to be killed on the island to ensure air safety, but the actions attracted a lot of publicity and the effects in terms of the viability of the population (900,000 individuals in 1994-1995) were apparently relatively small (Blokpoel 1976; Whittow 1993; Dolbeer *et al.* 1996).

4.2 SPECIES KILLED BY AIRCRAFT

4.2.1 *Birds*
The species of bird involved in a collision with an aircraft is often not expertly established. Consequently it is not possible to provide a definitive list of species that have been killed as a result of bird strikes. However, lists are provided by, among others, Blokpoel (1976), Mackinnon (2001) for Canada; Cleary *et al.* (2004), Rochard and Horton (1980) for the UK, and Owino *et al.* (2004) for Kenya. It is probable that species from most avian orders have been struck.

Damaging bird strikes are caused by species that are heavy (>100g) and that form flocks (e.g. the starling). Thorpe (2003) showed that ingestion of gulls into jet engines was the major cause of fatal bird strikes whereas windscreen penetration by large raptors, including vultures, resulted in the loss of most light aircraft and helicopters.

4.2.2 *Mammals*
The only systematically collected data on mammal strikes that has been published, to our knowledge, is that of Cleary *et al.* (2004). Of the 1272 mammals struck in the USA between 1990 and 2003, 96 (8%) were bats, 88 (7%) were hares and rabbits, 74 (6%) were rodents, 312 were carnivores (24.5%), 643 (51%) were artiodactyls. Of these 574 (42% of all mammal strikes) involved the white tailed deer (*Odocoileus virginianus*) reflecting the remarkable abundance of this species (Cote *et al.* 2004). The white tailed deer was the species involved in the only fatal mammal collision with a civil aircraft between 1990 and 2003 in the United States (Cleary *et al.* 2004).

4.2.3 *Reptiles*
Cleary *et al.* (2004), list 67 records of reptiles being struck by civil aircraft in the United States between 1990 and 2003. Fifty of these (75%) involved turtles of 6 species. In two cases the strikes had a negative effect on the flight. More serious however, were 12 strikes involving the American alligator, three (25 %) of which had a negative effect on the flight.

4.2.4 *Introduced animals*
Although the relevant data does not appear to have been collated, there is evidence that some introduced bird, mammal and reptile species may become a serious hazard to aviation. For example in Great Britain, the Canada goose (*Branta canadinensis*) population has increased exponentially and is posing a serious threat to air safety (Allan *et al.* 1995). In the Netherlands, the Egyptian goose (*Alopochen aegyptiacus*) population has also increased dramatically and has been found at times at Schipol Airport (Schipol Group 2000). The European starling is a problem in the United States where it was involved in the worst bird strike to have been recorded, at Logan Airport in Boston, in October 1960, when 62 people died. In the United States, the introduced rock or feral or racing pigeon (*Columba livia* var *domestica*), was

involved in at least 983 bird strikes between 1990 and 2003, and 245 (25%) of these caused some damage to the aircraft (Cleary *et al*. 2004).

In Holland, some intending passengers release their pets including cats, racoons and mink on the outskirts of Schipol Airport (Schipol Group 2000). The main threat these animals pose is that while hunting, they disturb flocks of birds, and therefore have to be trapped or culled ((Schipol Group 2000).

Finally, the introduced green iguana has become a problem in Puerto Rica (Quinones & Engeman, 2004). This 1.5-2m lizard has become well established in and around Luis Munoz International Airport, where hundreds of iguanas invade the area of the runway to breed. Determined attempts to exclude these reptiles from the aircraft manoeuvring areas, included the construction of a 300m three strand electric fence. But "rather than being repelled by the electric charges", "the iguanas charged straight forward through the fences"! (Quinones & Engeman 2004). In the end a 30cm high chicken wire fence led to a reduction in the "intrusions" by the iguanas. Cleary *et al*. (2004) mention five green iguana strikes of which three caused some damage to the aircraft involved.

5. Airports in the environment

The vast majority of discussions surrounding the impact of aviation on the environment centres on noise and atmospheric pollution. What is less often considered is the direct impact of the airport and surrounding infrastructure on the nearby environment in terms of land take and the control that needs to be applied to developments on and around the airport that may impact upon flight safety. As well as controlling the most obviously hazardous developments, such as buildings that may obstruct the approach paths of aircraft or interfere with radio communications, aviation regulators are required to control the hazards posed to aircraft by collisions with wildlife, especially birds (birdstrikes). Birdstrikes are conservatively estimated to cost the world civil aviation industry US$1.2 billion per year (Allan 2002). They have also resulted in the loss of 80 civil aircraft and 231 lives (Thorpe 2003). In order to control this risk the International Civil Aviation Authority (ICAO) introduced a new standard for controlling bird-attracting sites near airports in November 2003. The standard states *inter alia* that: Garbage disposal dumps or any such other source attracting bird activity on or in the vicinity of an aerodrome shall be eliminated or their establishment prevented unless an appropriate aeronautical study indicates that they are unlikely to create conditions conducive to a bird hazard problem.

In small densely populated countries, such as the United Kingdom, the rigorous application of this standard could have profound effects on the environment. ICAO defines a distance of 13km around aerodromes as the radius within which the presence of birds may be considered a hazard. The UK Civil Aviation Authority has interpreted 'in the vicinity of an aerodrome' to mean the ICAO 13km circle. If all UK licensed civil airports implemented this standard, and the Ministry of Defence were to do the same at its military airfields, then approximately 40% of the land surface of England would be subject to these controls. There is clearly scope for conflict between flight safety requirements and national and local commitments to nature conservation and biodiversity enhancement. Similarly, local and national regulations concerning planning and economic development may conflict with flight safety. For example, the need to extract minerals such as sand and gravel for construction work frequently leaves large voids, which have often been landscaped to form wetlands for biodiversity benefits and/or recreational use for local communities. Such sites often attract hazardous birds and pose significant problems to the aviation industry (CAA 1998). The fact that many aerodromes are sited on large areas of flat ground such as river floodplains, which often also have rich deposits of sand and gravel further complicates the situation.

The projected growth of civil aviation, which will result in the need for more and larger airport facilities at least over the medium term future, combined with increasing pressure for sustainable economic growth that takes due account of environmental issues such as biodiversity and climate change, means that increasing attention will be paid to the impact of aviation on the environment. Whilst issues such as noise and pollution are closely monitored, the other impacts of airports on the environment are less well understood and very poorly measured. The following sections attempt to describe some of these impacts and to estimate, where possible, their effects on the environment.

5.1 DIRECT IMPACT OF AIRPORT INFRASTRUCTURE

There are around 13,860 airports with paved runway surfaces in the world plus a large number of smaller facilities without paved surfaces (CIA 2002). Airports vary considerably in size depending both on the length, number and alignment of the runways, and on the quantity of ancillary infrastructure that surrounds them. An airport with a single runway and one or two terminal buildings may have a relatively low land take compared to an airport with two or three runways set at different angles to allow aircraft to take-off and land into the wind. Where land-take is not a major constraint, airports may be constructed with numerous long runways and multiple terminal facilities. Like runway length, the passenger throughput of an airport is also not a good guide to the physical size of the facility. Former military airfields converted to civil use may be very large, but with relatively low passenger numbers, whilst purpose built civil airports can accommodate large passenger throughputs with relatively small airfields. Although the length of the runways at airports is documented and information on passenger and flight numbers can be obtained, it is impossible to make an accurate determination of the land take of airports around the world. Notwithstanding this, the actual land take of airports is relatively small compared to other forms of transport. Civil and military airports combined constitute around 1% of the total transportation land-coverage in the EU (European Environment Agency 2001). A study in Germany estimated that air travel utilised 0.4ha of land per 1,000,000 passenger miles compared to 2.6ha for rail travel (UK Commission For Integrated Transport 2001).

5.2 IMPACT OF ANCILLARY STRUCTURES

As with the size of the airport itself, estimating the impact of the ancillary structures such as hotels, car parks, road and rail links, office and industrial buildings etc. associated with an airport is difficult. It is often unclear which buildings are linked directly with the airport and which would be in that location anyway. Some data are available on car parking as a component of airport size (see Table 2), but this does not include car parks associated with other buildings around the airport.

Table 2. The total area and area of car parking at 3 major UK airports. Numbers from Whitelegg (1994)

	Total site area (Ha)	Car park land take (Ha)
Heathrow	1197	58.19
Gatwick	759	68.57
Stansted	975	57.6

5.3 ECOSYSTEM MANAGEMENT FOR AIR SAFETY

The majority of aerodromes around the world operate some form of programme to deter hazardous wildlife from the property. The techniques used can be broadly categorised as habitat management and bird deterrence. The techniques used vary depending upon the habitat on the airfield, the bird species involved and on the manpower and resources available to the aerodrome manager to carry them out.

5.3.1 *Aerodrome habitat management*
Birds exploit the habitat provided by aerodromes for a variety of reasons, but these can broadly be categorised as feeding, breeding and roosting opportunities. Different bird species have adaptations that permit them to exploit airports in different ways. Grassland feeders, such as gulls (*Larus sp.*), plovers (*Vanellus sp.*) starlings (*Sturnus sp.*) corvids (*Corvus sp.*) pigeons (*Columba sp.*) and egrets (*Egretta sp.*) are all able to exploit the vegetated areas between runways and taxiways as feeding sites, but all do so in slightly differing ways, some relying on soil surface invertebrates, some on subsoil invertebrates, others on plant species in the sward. The basic principle of aerodrome habitat management is either to remove the resources that the birds are exploiting, or to deny birds access to these resources. The choice of which resource to concentrate on can be a difficulty. For example, in Western Europe the practice of growing the grassed areas between the runways as dense monocultures with a sward height of 15-20cm is commonly employed to deter species such as Lapwing (*Vanellus vanellus*), Rook (*Corvus frugilegus*) and Gulls (Brough & Brigeman 1980, Mead & Carter 1973). The swards thus created provide a haven for voles (*Microtus sp.*), which in turn attract birds of prey such as Kestrels (*Falco tinnunculus*) and owls (*Strigidae*). The final choice of habitat management technique therefore depends on the ecology of the bird species that constitute the greatest risk at the airport concerned. As well as impacting directly on the wildlife that is deterred from using the airfield, wildlife management can have more general environmental impacts. The long grass swards described above require regular inputs of chemical fertilisers to maintain good grass growth and periodic application of selective herbicides to remove broad-leaved weeds. Insecticides and lumbricides may also occasionally be used to remove invertebrates from the sward that are attracting birds, but their use is not generally encouraged because of the impact on predatory invertebrates which can result in a superabundance of prey invertebrates that requires further chemical applications to correct (Allan 1990).
Risk assessment, followed by the formulation of an appropriate habitat management plan is now becoming increasingly common as part of safety management systems at civil airports (Allan *et al.* 2003). Whatever the techniques used, the outcome of most airport habitat management programmes is to minimise the resources available to hazardous bird species and to lower the diversity of the aerodrome as a habitat for birds, thus simplifying the task of deterring any birds that remain. Because this process is restricted to the aerodrome, it has relatively little impact on surrounding ecosystems, but the presence of the airport may act as an interruption to wildlife corridors. Some species of conservation concern can benefit from airfield habitat management. The Skylark (*Alauda arvensis*) thrives in the long grass swards used to deter other species from airfields. Its small size means that it poses little threat to aircraft, but precautionary landings following strikes with small birds may still result in significant financial losses (Allan 2002). In recent years, attempts have been made to develop habitat management techniques that remove the attraction offered by aerodromes to hazardous birds whilst retaining a diverse flora and fauna (Dekker 2003). If successful, this approach has the potential to provide significant ecological benefits, especially for plant and invertebrate species, but controlled scientific evaluations have yet to be completed.

5.3.2 Management of the surrounding area

The birdstrike reduction standards imposed by ICAO do not apply directly to airports, but are instead enacted by the various countries that are signatories to the Chicago Protocol. The methods by which birdstrike hazards that arise off the airfield are controlled are therefore highly variable. Some countries employ prohibitions on particular land uses, such as landfills, within a specified distance of an aerodrome, some require consultation on a case-by-case basis, while others have little or no effective control in place. The impact of planning controls on the environment is thus impossible to quantify, but it has the potential to be significant in densely populated countries with large numbers of aerodromes and a relatively small land area. In the UK, for example, every application for a new development within 13km of a licensed civil or military airport is assessed for its potential to attract birds. The civil airport manager or the Ministry of Defence is given the opportunity to object to the development on the grounds of flight safety. It is often possible to reach a compromise solution involving a re-design of problematic applications that satisfies both the developers and aviation interests, but where developments such as nature reserves or restoration of industrial sites to areas of amenity value for the local community are concerned, biodiversity benefits are often given a high profile in the applicant's submission. Compromises to ensure flight safety will almost always result in a reduction in the biodiversity value of the site. For example, removal of wetlands to avoid attracting hazardous waterfowl can substantially reduce the habitat diversity on a site and diminish the botanical and invertebrate interest as well as affecting the aesthetics of the design. Experience has shown that if a compromise is to be reached, an early consultation with flight safety interests is vital (Jackson & Allan 2000). It is frequently difficult to persuade applicants to change designs that have been developed over a long period and at considerable cost, whereas offering a consultant the challenge of developing a biodiverse proposal that does not attract hazardous birds from the outset of the process can produce results that are acceptable to both sides. One possible solution is to develop sites that are targeted at particular species of conservation interest rather than at a general increase in biodiversity. If a site is designed with a particular non-hazardous species in mind then it is far easier to implement design changes or management plans specifically aimed at deterring hazardous wildlife providing that it does not adversely affect the species that the site was designed to protect.

Further to this, designing for habitat rather than particular species will often satisfy the needs of the conservation advocates. In addition, if a management plan is designed to promote or protect a particular species part of the challenge presented to the designer can be to establish a target of limited numbers.

5.3.3 Managing existing attractions

The control of risks arising from sites around airports that already exist is far more problematic. Aside from the conflicting demands of commitments of national and local regulators to biodiversity on the one hand and flight safety on the other, most countries lack any sort of legislative framework that allows airports or aviation regulators to compel landowners to manage flight safety risks that arise from wildlife on their property.

In many instances it may be legally impossible to undertake actions to reduce numbers of a protected species or to degrade the habitat in order to discourage them from areas close to an airport where their presence causes a flight safety risk. In the EU 7% of RAMSAR sites have aviation infrastructure within 5km of their centre, and in the UK, 15% of special bird areas are within 5km of an airfield (UK Commission For Integrated Transport 2001). These areas have statutory legal protection and attempting to manage habitat or otherwise control hazardous bird populations may meet with practical and legal difficulties that make effective risk management impossible. One such situation has occurred at JFK International Airport, New York where Laughing Gulls (*Larus atricilla*) nest in large numbers in the adjacent Jamaica Bay Wildlife Refuge. The Port Authority, which operates the airport was unable to reach an agreement with the US Fish and Wildlife Service, which manages the wildlife refuge, to permit them to prevent nesting by the gulls in an effort to relocate the colony. This has resulted in staff from a third government agency (the US Department of Agriculture) deploying staff along the airport boundary during the gull nesting season to shoot any Laughing Gull that flies over the airport. This has proved highly successful in reducing the number of birdstrikes with Laughing Gulls by 62%.

In some circumstances it may be possible to utilise other regulatory processes to enhance flight safety. For example, all landfill sites in the UK have their operating permits periodically renewed. The Environment Agency, which is responsible for this process, is now requiring bird control to be implemented at any landfill site within 13km of a licensed aerodrome if the aerodrome operator determines that there is a birdstrike risk arising from birds feeding at the landfill.

Where there is no conflict with biodiversity issues, the main problem in achieving wildlife management objectives off the aerodrome is often one of cost. Many landowners would be only too happy to have a colony of roof-nesting gulls removed or to have flocks of geese dispersed from their fields, but they are unwilling to bear the cost of something that they do not see as their problem.

5.4 SUMMARY OF ENVIRONMENTAL IMPACTS

The direct environmental impact of aerodromes on the environment as a result of land take and the need for wildlife management on the airport to preserve flight safety is localised and small compared to other forms of transportation. The cumulative impact of controlling developments close to aerodromes to prevent wildlife attractions may be significant over longer time periods, especially in those countries where there are high densities of population. Opportunities for biodiversity enhancement on aerodromes themselves are limited, but may be possible if new developments are carefully designed to deter hazardous birds and other wild-life. Outside the airport perimeter there is considerable scope for biodiversity enhancement providing that an imaginative approach is taken at an early stage in the design process.

6. Aviation and the transport of alien species

It is well known that all transport systems are capable of transferring alien species over great distances and facilitating their entry to new geographical areas (Pimental 2002; Kelly 2004). These introductions may be extremely damaging to the receiving ecological systems and have lead to the extinction of endemic species. Alien species often become pests of agriculture, horticulture and forestry, leading to huge economic losses. These in turn require the implementation of expensive control programmes some of which are themselves ecologically destructive (reviewed by Pimental *et al.* 2001).

Parasites especially viruses and other directly transmitted (i.e. person to person) micro-pathogens constitute a major threat to public health but also to economic stability. For example, the SARS virus epidemic which ran from March to July 2003, involved 8439 cases of which 812 died (9.6%) (Anon 2003). But the epidemic also cost "nearly" US$100 billion, and caused massive disruption to air travel, and tourism, particularly in China, Hong Kong and

Canada. (and almost forced the cancellation of Bird Strike 2003 in Toronto!) The Millennium Ecosystem Synthesis Report (2005) observes that "An event similar to the 1918 Spanish Flu pandemic, which is thought to have killed 20 to 40 million people worldwide could now result in over 100 million deaths in a single year. Such a catastrophic event, the possibility of which is being seriously considered by the epidemiological community, would lead to severe economic disruption and possibly even rapid collapse in a World economy dependent on fast global exchange of goods and services". It is also likely that modern Health Services in the developed world would be unable to cope with such a crisis.

Aviation provides ideal opportunities for the rapid dissemination of a highly infectious pathogen to all the major cities of the world, within a very short time (Grais *et al.* 2003; Enserink 2004). While considerable attention is being given to an avian flu epidemic, other zoonotic diseases are also emerging (Kelly 2004). The Millennium Ecosystem Synthesis Report (2005) warns of the growing bush meat "economy" in Africa and Asia which is "harvesting species closely related to man from which could emerge a hitherto unknown pathogen". This might enter a city (bush meat has often been confiscated by airport customs authorities) where there is little or no Herd Immunity, and if the pathogen is highly infectious i.e. has a high R_0 value and is highly pathogenic then a major epidemic would follow.

Aviation is also being associated with the spread of pathogens which have evolved resistance to various drugs (Witte 2004; Roper *et al.* 2004). It has been suggested by Roper *et al.* (2004), in the context of the intercontinental spread of Pyrimethamine-resistant malaria, that careful thought should be given to preventing the further import of resistant parasites, perhaps by screening and treatment of passengers travelling from southeast Asia or South America to Africa.

Imported malaria is now relatively common in Europe, with up to 7000 cases occurring every year. However, the great majority of these cases are due to people becoming infected while travelling – mostly in Africa. Imported malaria in most temperate regions is unlikely, at present, to be transmissible via local mosquitoes i.e. become authochthonous (Gratz *et al.* 2000; Snow 2000), although this does appear to have occurred in both Germany (Kruger *et al.* 2001) and Italy (Romi *et al.* 2001) (see also Marchant *et al.* 1998 and Snow 2000).

Airport malaria, though not common or involving many cases, is regarded as being serious because of possible delays in diagnosis. Airport malaria is caused by the long distance transport (mostly from West Africa) of mosquito vectors – particularly species belonging to the *Anopheles gambiae* complex. These may escape from the aircraft after its arrival and bite people living nearby. These people may become infected with malaria, though this, at least initially, might not be considered in the differential diagnosis. Another, and more worrying, scenario is where the mosquitoes are in the luggage and get transported over greater distances from the airport (Reviewed by Gratz *et al.* 2000). This is known as baggage malaria and diagnosis may be compromised by the apparent 'missing link' with a transmitting organism. It is now widely recognised that special precautions need to be taken where aircraft have departed from tropical airfields, especially in West Africa where in some areas malaria is hyperendemic and mosquitoes are very abundant. The techniques which can be employed to 'deinsect' aircraft are outlined by Gratz *et al.* (2000).

Modern aviation has an obvious potential for the rapid dissemination of infectious disease agents. Since no two points on the globe are now more than 24 hours from each other, there are unprecedented risks of transferring pathogens from remote isolated disease foci to distant densely populated, cities. The efficiency and decreasing costs of international air travel inevitably increases the risks of a major pandemic, since the receiving urban populations may have no immunity to the novel pathogen(s). The severity of such disease outbreaks (particularly those caused deliberately in biological warfare) will depend on the R_0 values associated with the pathogen, its incubation period and lethality (Ferguson *et al.* 2003; May *et al.* 2001; Antia *et al.* 2003).

In the case of directly transmitted pathogens – of which SARS would be an example -air transport poses the greatest transport-mediated risk as it enables the rapid, widespread and multiple initiation of "chain reaction" type infection processes following contact between infectious and susceptible hosts (see Ferguson *et al.* 2003).

In contrast, it appears that vector borne diseases, and particularly the vectors themselves, are more effectively transported via shipping. Mosquitoes are most likely to become established in a distant receiving environment if they are dispersed by ships as larvae and pupae rather than by air as adults (Lounibos 2002).

References

Allan JR. 1990; The impact of a lumbricide treatment on airfield grassland. Proceedings of the 20th Meeting Birdstrike Committee Europe, Helsinki. 531-542.

Allan JR. 2002; The costs of birdstrikes and birdstrike prevention. In: Clarke L. (ed.) 147-153 Human Conflicts With Wildlife: Economic Considerations. US Department of Agriculture, Fort Collins.

Allan JR, Kirby JS, Feare CJ. 1995; The biology of Canada geese (*Branta Canadensis*) in relation to the management of feral populations. Wildlife Biology 1:129-143.

Allan JR, Orosz A, Badham A, Bell JC. 2003; Development of birdstrike risk assessment procedures, their use on airports and the potential benefits to the aviation industry. Proceedings of the 26th International birdstrike Committee, Warsaw, 73-80.

Anderson JD Jr. 1997; A History of Aerodynamics. Cambridge University Press New York.

Anderson JD Jr. 2003; The Airplane: A History of its Technology. Reston, VA: American Institute of Aeronautics and Astronautics.

Anderson JD Jr. 2004; Inventing Flight- The Wright Brothers and Their Predecessors. The Johns Hopkins University Press. Baltimore and London.

Antia R, Regoes RR, Koella JC, Bergstrom CT. 2003; The role of evolution in the emergence of infectious diseases. Nature 426: 658-661.

Anonymous. 1993; Volume 1, Aircraft noise, Annex 16 International standards and recommended practices environmental protection. International Civil Aviation Organisation Montreal 3.8.

Anonymous. 2003 SARS What have we learned? Nature 424:121-126.

Barras SC, Dolbeer RA. 2000; Reporting bias in bird strikes at John K Kennedy International Airport, New York Proceedings of the International Bird Strike Committee 25 (1): 99-112.

Bilstein RE. 2001; Flight in America: from the Wrights to the Astronauts. The Johns Hopkins University Press. Baltimore and London.

Blokpoel H. 1976; Bird Hazards to Aircraft. Clarke, Irwin & Co. Ltd., Canada.

Brough T, Bridgman CJ. 1980; An evaluation of long grass as a bird deterrent on British airfields. Journal of Applied Ecology 17:243-253.

Brown KM, Erwin RM, Richmond ME, Buckley PA, Tanacredi JT, Avrin D. 2001; Managing birds and controlling aircraft in the Kennedy Airport Jamaica Bay Wildlife Refuge complex: The need for hard data and soft options. Environmental Management 28: 207-224.

Burger J. 1981; Behavioral responses of herring gulls (Larus argentatus) to aircraft noise. Environmental Pollution (Series A) 24:177-184.

Burger J. 1983; Jet aircraft noise and bird strikes: why more birds are being hit. Environmental Pollution (Series A) 30:143-152.

Burger J. 1985; Factors affecting bird strikes on aircraft at a coastal airport. Biological Conservation 33:1-28.

CAA. 1998; CAP 680 Bird control on aerodromes. Civil Aviation Authority, London.

CIA. 2002; The World Fact Book Central Intelligence Agency. http://www.umsl.edu/services/govdocs/wofact2002/fields/2053.html Accessed 04-05-2005.

Cartmill M.1993; A View to Death in the Morning - Hunting and Nature through History. Harvard University Press. Cambridge Massachusetts.

Cleary EC, Dolbeer RA, Wright SE. 2004; Wildlife hazard management at airports: Wildlife strikes to civil aircraft in the United States, 1990-2003. Federal Aviation Administration, National Wildlife Database Serial Report Number 10. Washington DC.

Collier R, 1980; 1940 The World in Flames. Penguin Books. London.

Conomy JT, Dubovsky JA, Collazo JA, Fleming WJ. 1998; Do black ducks and wood ducks habituate to aircraft disturbance? Journal of Wildlife Management 62:1135-1142.

Cote D, Rooney TP, Tremblay JP, Dussault C, Waller DM. 2004; Ecological impacts of deer overabundance. Annual Review of Ecology Evolution and Systematics 35: 113-147.
Davidson NC, Rothwell PI. 1993; Human disturbance to waterfowl on estuaries: Conservation and coastal management implications of current knowledge. Wader Study Bulletin 68 (Special Issue): 97-105.
Dawkins R. 2004; The Ancestors Tale: A Pilgrimage to the Dawn of Life. Weidenfeld & Nicolson. London.
Dekker A. 2003; Taking habitat management one step further. Proceedings of the International Birdstrike Committee 26:265-272.
Delany DK, Grubb TG, Beier P, Pater LL, Reiser MH. 1999; Effects of helicopter noise on Mexican spotted owls. Journal of Wildlife Management 63: 60-76.
Dolbeer RA, Arrington DP, LeBoeuf E, Atkins C. 1996; Can albatrosses and aircraft coexist on Midway Atoll? Proceedings of Birdstrike Committee Europe 23: 327-335.
http://www.themes.eea.eu.int/.../indicators/supply/TERM18,2001/Capacity_of_infrastructure_networks___TERM_20 01.pdf.
Dolbeer RA, Chipman RB, Gosser AL, Barras SC. 2003; Does shooting alter flight patterns of gulls: case study at John F. Kennedy International Airport. Proceedings of the International birdstrike Committee 26:547-562.
Dolbeer RA, Eschenfelder P. 2003; Amplified bird-strike risks related to population increases of large birds in North America Proceedings of the International birdstrike Committee 26:49-67.
Efroymson RA, Suter II, GW, Rose WH, Nemeth S. 2000; Ecological risk assessment framework for low altitude overflights by fixed wing and rotary-wing military aircraft (ORNI/TM-2000/289). Oak Ridge TN: Oak Ridge National Laboratory http://www.esd.ornl.gov/programs/ecorisk/documents/overflight-e1.pdf. Accessed 06-05-2005.
Efroymson RA, Suter II GW. 2001; Ecological risk assessment framework for low altitude aircraft overflights: II. Estimating effects on wildlife. Risk Analysis 21: 263-274.
Efroymson RA, Suter II GW, Rose WH, Nemeth S. 2001; Ecological risk assessment framework for low altitude aircraft overflights:1 Planning analysis and estimating exposure. Risk Analysis 21: 251-262.
Enggist-Dublin P, Ingold P. 2003; Modelling the impact of different forms of wildlife harassment, exemplified by a quantitative comparison of the effects of hikers and paragliders on feeding and space use of chamois *Rupicapra rupicapra*. Wildlife Biology 9: 37-45.
Enserink M. 2004; Looking the pandemic in the eye. Science 306: 392-394.
Erickson WP, Johnson GD, Strickland MD, Young DPJ, Sernka KJ, Good RE. 2001; Avian Collisions with Wind Turbines: a summary of existing studies and comparisons to other sources of avian mortality in the United States. National Wind Coordinating Committee (NWCC) Resource Document.
European Environment Agency. 2001; http://www.themes.eea.eu.int/.../transport/indicators/consequences/TERM08,2001/Land_take_TERM_2001.doc.pdf Accessed 06-05-2005.
Ferguson NM, Keeling MJ, Edmunds WJ, Gani R, Grenfell B, Anderson RM, Leach S. 2003; Planning for smallpox outbreaks. Nature 425:681- 685.
Fitter RSR. 1945; London's Natural History. Collins New Naturalist Series. Collins. London.
Frid A. 2003; Dall's sheep responses to overflights by helicopter and fixed wing aircraft. Biological Conservation 110: 387-399.
Frid A, Dill L. 2002; Human caused disturbance stimuli as a form of predation risk. Conservation Ecology 6 (1): 11.
Gill JA, Sutherland WJ. 2000; Predicting the consequences of human disturbance from behavioural decisions. In: Gosling LM & Sutherland WJ (eds) Behaviour and Conservation, 51-64. Cambridge University Press, Cambridge, UK.
Gill JA, Sutherland WJ, Watkinson AR.1996; A method to quantify the effects of human disturbance on animal populations. Journal of Applied Ecology 33: 786-792.
Grais R, Ellis HJ, Glass G. 2003; Assessing the impact of airline travel on the geographic spread of pandemic influenza. European Journal of Epidemiology 18:1065-1072.
Grant RG. 2004; Flight: 100 Years of Aviation. DK. London & New York.
Gratz NG, Steffan R, Cocksedge W. 2000; Why aircraft disinfection? Bulletin of the World Health Organisation 78: 995-1004.
Gunston B. 2001; Aviation Year by Year. DK. London & New York.
Hall DA, Moore DR. 2003; Auditory neuroscience: The salience of looming sounds. Current Biology 13: 91-93.
Hill D, Hockin D, Price D, Tucker G, Morris R, Treweek J. 1997; Bird disturbance: improving the quality and utility of disturbance research. Journal of Applied Ecology 34: 275-288.
Hobsbawm E. 1994; Age of Extremes - The Short Twentieth Century 1914-1991. Michael Joseph. London.
Hollom PAD. 1975; The popular handbook of British birds. Witherby Ltd. London.
Jackson VS, Allan JR. 2000; Nature reserves and aerodromes – resolving conflicts. Proceedings of the International birdstrike Committee 25: 339-344.
Kelly TC. 2004; Globetrotting Germs-and the Transport of their Carriers. In: Davenport J & Davenport JL (eds.) The Effects of Human Transport on Ecosystems: Cars and Planes, Boats and Trains, 227-243. Dublin: Royal Irish Academy.
Kelly TC, Bolger R, O'Callaghan MJA. 1999; The behavioural responses of birds to commercial aircraft. Proceedings of Bird Strike 1999, Vancouver International Airport Authority, Vancouver: 77-82.

Kelly TC,.Buurma L, O'Callaghan MJA, Bolger R. 2000; Why do birds collide with aircraft-a behavioural perspective. Proceedings of the International Bird Strike Committee 25 (1): 47-48.

Kelly TC, O'Callaghan MJA, Bolger R. 2001; The avoidance behaviour shown by the rook (*Corvus frugilegus*) to commercial aircraft. Advances in Vertebrate Pest Management 2: 291-299.

Kelly TC, Murphy J, Bolger R. 1996; Quantitative methods in bird hazard control; preliminary results Proceedings of the International Bird Strike Committee 23: 227-233.

Kempf N, Huppop O. 1996; The effects of aircraft noise on wildlife: A review and comment. Journal Fur Ornithologie 137:101-113.

Komenda-Zehnder S, Bruderer B. 2002; Einflus des Flugverkehrs auf die Avifauna- Literaturstudie. Schriftenreihe Umvelt Nr 344. Bundesamt fur Umwelt und Landschaft. Bern.

Komenda-Zehnder S, Cevallos M, Bruderer B. 2003; Effects of disturbance by aircraft overflight on waterbirds-an experimental approach. Proceedings of the International Bird Strike Committee 26 (1): 157-168.

Koolhaas A, Dekinga A, Piersma T. 1993; Disturbance of foraging Knots by aircraft in the Dutch Wadden Sea in August – October 1992. Wader Study Group Bulletin 68: 20-22.

Kruger A, Rech A, Su XZ, Tannich E. 2001; Two cases of autochthonous Plasmodium falciparum malaria in Germany with evidence for local transmission by indigenous *Anopheles plumbeus*. Tropical Medicine and International Health 6: 983-985.

Larsen K, Kahlert J, Frikke J. 2005; Factors affecting escape distances of staging waterbirds. Wildlife Biology 11: 13-19.

Linnell MA, Conover MR, Ohashi TJ. 1996; Analysis of bird strikes at a tropical airport. Journal of Wildlife Management 60: 935-945.

Linnell MA, Conover MR, Ohashi TJ. 1999; Biases in bird strike statistics based on pilot reports. Journal of Wildlife Management 63: 997-1003.

Lounibos LP. 2002; Invasions by insect vectors of human disease. Annual Review of Entomology 47: 233-266.

Mackinnon B. 2001 (ed) Sharing The Skies An Aviation Industry Guide to the Management of Wildlife Hazards. Transport Canada http://www.tc.gc.ca/CivilAviation/Aerodrome/WildlifeControl/tp13549/menu.htm (06-05-2005).

Marchant P, Ebling G, Van Gemert G-J, Leake CJ, Curtis GF. 1998; Could British mosquitoes transmit falciparum malaria? Parasitology Today 14: 344-345.

May RM, Gupta S, McLean AR. 2001; Infectious disease dynamics: What characterises a successful invader? Philosophical Transactions of the Royal Society London B 356: 901-910.

Mead H, Carter AW. 1973; The management of long grass as a bird repellent on airfields. Journal of the British Grassland Society. 28: 219-221.

Millennium ecosystem synthesis report. 2005; Island press, Washington D.C. http://www.millenniumassessment.org/en/Products.Synthesis.aspx.

Newton I. 1998; Population Limitation in Birds. Academic Press. London.

Nugent H. 2005; Low–cost airlines lead flying resurgence. The Times, April 14th 2005.

Orians GH, Pfeiffer EW. 1970; Ecological effects of the war in Vietnam. Science 168: 544-554.

Owina A, Biwott N, Amutete G. 2004; Bird strike incidents involving Kenya Airways flights at three Kenyan airports, 1991-2001. African Journal of Ecology 42: 122-128.

Pascoe D. 2003; Aircraft. Reaktion Books. London.

Patterson B. 2000; Wildlife control at Vancouver International Airport ; Introducing border collies. Proceedings of the International Bird Strike Committee 25: 81-87.

Pepper CB, Nascarella MA, Kendall RA. 2003; A review of the effects of aircraft noise on wildlife and humans, current control mechanisms, and the need for further study. Environmental Management 32: 418-432.

Percival S. 2005; Birds and Windfarms: What are the real issues? British Birds 98: 194-204.

Pimental D. (ed) 2002; Biological Invasions Economic and Environmental Costs of Alien Plant, Animal and Microbe species. CRC Press London and New York.

Pimental D, McNair J, Janecka J, Wightman J, Simmonds C, O'Connell C, Wong E, Russel J, Zern J, Aquino T, Tsomondo T. 2001; Economic and environmental threats of alien plant, animal, and microbe invasions. Agriculture, Ecosystems and Environment 84: 1-20.

Quinones PF, Engeman RM. 2004; Use of fencing to deter iguanas from airport runways at Luis Munoz Marin International Airport. Abstract No.10, 6th Joint Meeting Bird Strike Committee USA/Canada September 13-17th 2004.

Rees EC, Bruse JH, White GT. 2005; Factors affecting the behavioural responses of whooper swans (Cygnus c.cygnus) to various human activities. Biological Conservation 121: 369-382.

Rhodes R. 1986; The Making of the Atomic Bomb. Penguin Books. London.

Richardson WJ, West T. 2000; Serious birdstrike accidents to military aircraft: updated list and summary. Proceedings of the International Birdstrike Committee 25: 67-97.

Robinson M. 2000; Is the possibility of costly bird strike growing? Proceedings of the International Bird Strike Committee25 (1): 169-178.

Robinson JA, Pollitt MS. 2002; Sources and extent of human disturbance to waterbirds in the UK: an analysis of Wetland Bird Survey data, 1995/96 to 1998/99. Bird Study 49: 205-211.

Rochard JBA, Horton N. 1980; Birds killed by aircraft in the United Kingdom. Bird Study 27: 227-234.
Romi R, Boccolini D, Majori G. 2001; Malaria incidence and mortality in Italy in 1999-2000. Eurosurveillance 8: 143-147.
Roper C, Pearce R, Nair S, Sharp B, Nosten F, Anderson T. 2004; Intercontinental Spread of Pyrimethamine-Resistant Malaria. *Science* 305 (Issue 5687), 1124 http://www.sciencemag.org/cgi/content/full/sci;305/5687/1124, Accessed 20-04-05.
Salisbury EJ. 1943; The flora of bombed areas. Nature 151: 462-466.
Schipol Group. 2000 Fauna management Amsterdam Airport Schipol. Proceedings of the International Bird Strike Committee25 (2): 243-262.
Schnidrig-Petrig R, Ingold P. 2001; Effects of paragliding on alpine chamois *Rupicapra rupicapra rupicapra*. Wildlife Biology 7: 285-294.
Sebald WG. 2003; On the Natural History of Destruction. Penguin Books. London.
Smit CJ, Visser GJN. 1993; Effects of disturbance on shorebirds: A summary of existing knowledge from the Dutch Wadden Sea and Delta area. Wader Study Group Bulletin 68: 6-19.
Snow K. 2000; Could malaria return to Britain? Biologist 47: 176-180.
Sodhi N. 2002; Competition in the air: Birds versus Aircraft. The Auk 119 (3): 587-595.
Stock M. 1993; Studies on the effects of disturbances on staging Brent Geese: a progress report. Wader Study Group Bulletin 68: 29-34.
Thomas K. 1983; Man and the Natural World Changing Attitudes in England 1500-1800. Pengiun Books. London.
Thorpe J. 2003; Fatalities and destroyed civil aircraft due to birdstrikes. Proceedings of International Birdstrike Committee 26: 86-113.
Tomlinson S, Buckingham J, Erwin R, Carpenter B, Wilhote S. 1991; Physiological response of birds to approaching aircraft. DOT/FAA/CT-91/14, Springfield, Virginia, USA. 98 pages.
UK Commission For Integrated Transport. 2001; http://www.cfit.gov.uk/reports/racomp/a3.htm Accessed 06-05-2005.
Vidal G. 1993; Love of Flying. In: Silvers RB & Epstein B (eds) Selected essays from the first 30 years of The New York Review of Books, 263-288. New York Review of Books.
Ward DH, Stehn RA, Erickson WP, Derksen DV. 1999; Response of fall-staging brant and Canada geese to aircraft over-flights in south-western Alaska. Journal of Wildlife Management 63: 373-381.
Whitelegg J. 1994; Transport and Land Take. Report to Council for the Preservation of Rural England.
Whittow GC. 1993; Laysan Albatross (*Diomedea immutabilis*) In: Poole A. & Gill F (eds) The Birds of North America No 66 Philadelphia: The Academy of Natural Sciences, Washington, D.C.: The American Ornithologists Union.
Witte W. 2004; International dissemination of antibiotic resistant strains of bacterial pathogens. Infection, Genetics and Evolution 4(3) 187-191.
Yergin D. 1991; The Prize – The epic quest for oil, money and power. Simon and Schuster, London and New York.

CHAPTER 2: THE LOCAL COSTS TO ECOLOGICAL SERVICES ASSOCIATED WITH HIGH SEAS GLOBAL TRANSPORT

ROGER MANN

Virginia Institute of Marine Science, School of Marine Science, College of William and Mary, Gloucester Point, VA 23062, USA
"If you build it they will come." Kevin Costner - Field of Dreams.

1. Introduction

Human mobility is the engine that has fueled the development of global society as we observe it today. That society reflects human history over very recent geological time and the ability of humans to drastically alter the global ecosystem, a situation accelerated by population growth and the globalisation of economies. In contrast, the elements of the global ecosystem are a reflection of the interplay of evolution, planetary geology and climate change over a time frame of tens of millions of years. In recent human decades we have begun to appreciate the cumulative and continuing destructive impacts of societal development on the global ecosystem.

Human settlement has always been closely related to aquatic resources. The majority of current human populations reflect this historical trend in residing in close proximity to rivers, estuaries and continental coastlines. These remain avenues of and termini for current global transport systems. Commercial shipping remains the dominant mechanism for global transportation of commodities, and the scale of economy offered by current commercial shipping continue to mould national economies and the location of manufacturing of consumer goods on a global basis. Shipping and its support needs have acted as both a disturbance to local ecosystems and a vector for biological invasions, thus influencing global biogeography for the entirety of human history. Coastal and estuarine ecosystem integrity and function have suffered cumulative abuse as a result of this process.

In the face of continuing environmental degradation we are faced with challenge of how to address restoration of ecosystem function through, optimally, native species management, or, more frequently, manage ecosystem services as either an analogue of former native ecosystems or in societal service (e.g. agriculture) using a mix of native and non-native animal and plant species. This chapter focuses on temperate estuarine ecosystems and examines the contrasting time frames of individual species and community evolution over geological time (millions of years), with desired restorative action of ecosystem services in societaly-relevant time frames of years to decades. Examples are given from the estuaries of the Atlantic coastline of the United States, but are generically applicable on a wider geographical scale. Temperate estuarine ecosystems serve as examples of wider cumulative impacts, particularly on rivers and coastal ecosystems, associated with human transport, settlement and societal development. Scaled, practical application of ecological engineering as the tool to effect restoration is discussed. Ecological engineering is defined in this context as the product of collaboration between biologists, who frame questions in terms of ecosystem function, engineers who provide tractable approaches of developing and implementing technical solutions, and politicians who provide a legal framework to regulate the process.

John Davenport and Julia L. Davenport, (eds.) The Ecology of Transportation: Managing Mobility for the Environment, 25–38,

2. Time frames of evolution of species assemblages in coastal environments with emphasis on keystone species

Species assemblages in coastal and estuarine environments are spatially diverse and complex. Ecological function and services are maintained and structured by this diversity along physical and biological gradients. In the aquatic realm these are multifaceted gradients of depth, light, energy and turbidity as the shoreline is approached, with salinity dominating along the transition from seawater to fresh water. What are the seminal elements in temperate estuaries and coastal marine communities, and over what time frame did these evolve in the temperate estuarine of the globe? Temperate estuaries may be viewed with permanence in the time frame of recent human history, but they are ephemeral in geological time frames, being the transient product of continental drift and sea level fluctuation in concert with climate change. Resident benthic communities are complex patchworks related to sediment and substrate type, in turn products of geological erosion and deposition processes. They effect bentho-pelagic coupling and form the base of food webs serving apex predators, many of which are seasonally anadromous fishes that use estuaries as breeding grounds. The disproportionate productivity of these regions, in comparison with oligotrophic oceans that cover the majority of the Earth's surface, underscore their equally disproportionate importance in maintaining aquatic communities over wider geographical scales as they export juveniles of mobile species, predominantly fishes, to the coastal ocean and beyond (see Constanza *et al.* 1997). Their roles as depositional sinks for sediments of upstream watershed origin also distinguishes them as repositories for pollutants from industry and agriculture that have potentially long term deleterious impacts on estuarine food webs. Last, but not least in this context, they are the repositories for anthropogenic compounds associated with transport processes and their usually deleterious impacts – examples of which include tributyltin compounds used extensively as hull antifoulants which have toxic impacts on ecologically important marine invertebrate species (Bryan *et al.* 1988, 1989; Gibbs *et al.* 1987, 1988; Minchin *et al.* 1996; Stickle *et al.* 1990; Burnell & Davenport 2003).

The fossil record illustrates that the invertebrate species contributing to temperate estuarine benthic communities have representative lineages that often extends back for tens of millions of years, suggesting stability in the complexity of the community over geological time frames, that complexity being accommodated by the lethargic pace of sea level fluctuation. Oysters of the genus *Crassostrea*, chosen for this discussion because they are widely recognized as keystone species in temperate estuaries and coastal regions on a worldwide basis (see contributions in Kennedy *et al.* 1996), provide a suitable example having a lineage dating to the Cretaceous. Many invertebrates with such representation exhibit complex life histories with planktonic larvae providing dispersal capabilities where adult forms are sessile or attached. The origin of the planktotrophic larval form of marine invertebrates has been recently reviewed (Havenhand 1995; Levin & Bridges 1995; Wray 1995). This is generally considered to be the primitive form from which other early life history forms are derived. This primitive form is still employed by a vast number and variety of invertebrates, including *Crassostrea*, with obvious success over evolutionary time scales. Pelagic larvae are the life history stages that facilitate invasion and occupation of ephemeral environments. Keystone species such as reef forming oysters become seminal biological as well as physical features in estuarine biology. Oyster larvae initially settle on available hard substrate during the colonization phase. Gregarious settlement emphasises the recruitment signal to the benthos. The juveniles grow, their collective filter feeding forming a vital link between pelagic primary production and benthic food chains that, in turn provide trophic structure supporting both benthic and pelagic apex predators. In soft bottom communities the diverse infauna, notably polychaetes and clams, form analogous benthic-pelagic couplers through both phytoplankton and detrital feeding with subsequent connections to higher trophic level predators.

Keystone species are central to the maintenance of diverse community structure in coastal and estuarine environments – again the oyster provides an example. The adult oyster makes an irreversible commitment to estuarine life. Unlike the majority of the approximately 8,000 extant bivalve mollusc species the adult oyster has forsaken a foot and one of its two adductors to adopt a primitive monomyarian (single adductor muscle) form that cements itself to other adults. If it were not for this form, which by its nature provides sanctuary to newly settling generations from both predation and physical stress, oysters would arguably have become extinct; however, the combination of longevity and gregariousness slowly gives rise to oyster reefs, virtual biological condominiums for a staggering variety of other invertebrates and refuges for reef dwelling fishes in estuarine and coastal environments over a large latitudinal range (see Harding & Mann 2001). In undisturbed situations such reef systems serve to dominate the ecology of temperate estuaries for thousands of years until gradual changes in climate and/or sea level fluctuation again hasten local extinction as the environment becomes less commensurate with the plasticity of the form. The fact that the longevity of the oyster *Crassostrea* in the geological record far exceeds that of the current coastal and estuarine features that delineate their present distribution is very important. The disparity illustrates that oysters, like so many other representatives of the complex communities of estuaries, have, over geological time exhibited a distribution that has been a moving mosaic migrating back and forth across the subsumed coastal margins and associated estuaries with a long time frame driven by sea level rise and fall. Speciation has occurred over time frames of millions of years. The rates of occupation of new environments are much quicker, arguably just hundreds to thousands of years. Unfortunately, even the shortest of these time frames is not consistent with our expectancy of restoration of ecological services in ecosystems degraded by human action.

Benthic communities are not alone in providing essential linkage elements in estuarine and coastal ecosystems. Indeed, the land-water edge may appear as a convenient visual boundary but biological processes function seamlessly across it. Sea grasses, salt marshes and wetlands form a continuum at the land-water margin that serve to process nutrients of land origin with accompanying high primary productivity, control sediment runoff and provide complex shallow water habitat for early life history stages of many resident and anadromous species. Like the benthic communities they exhibit spatial complexity associated with relief. Modest changes in height relative to tidal excursion, and hence inundation/exposure, drive the dominant flora and the associated faunal assemblages. Each floral component presents a typical relationship of optimal height with respect to mean water level, soil/sediment type, root penetration, aerial canopy, and more. Disturbance in the aquatic zone, a characteristic of port development and operation, cascades to the shoreline and vice versa. For example, fragmentation of vertical structure in the aquatic zones serves to remove the natural barriers to the effect of wind on fetch over extensive but shallow bodies of water. The following sequence of events includes increased turbidity, decreased light transmission and failure of submerged aquatic vegetation (Moore *et al.* 1997; Moore & Wetzel 2000) with subsequent impact on physical refugia for early life history stages, with eventual impacts on communities. Open fetch can also result in spatial regrading of the suspended sediment with increasing energy – the larger waves sort and exhume the finer sediments leaving sandy sediments and thus alter the infaunal community. In lower energy regions the finer material settles and mud accumulates. Open fetch can combine with degradation of stabilising shoreline communities, such as fringing reef or vegetation structures, in accelerating physical shoreline destruction or change of shoreline topography with associated reduction in species diversity. The impacts are

4. Rates of change of communities in response to changing volume and types of vectors over human history

Successful colonisation of new environments has driven positive feedback loops with increased transport services and premeditated local environmental modification. There have, however, been accompanying unintentional environmental consequences associated with accidental transfer of aquatic species beyond their native ranges. Carlton (1999) provides a recent review of the scope and magnitude of this cause of ecosystem degradation in coastal and estuarine aquatic systems. The progression from ancient to medieval wooden ships provided a period of several thousand years when hull-attached organisms were relocated globally by early explorers. The advent of iron and steel hulls simply provided larger hulls and decreased times of passage, and it was not until anti-fouling paints were widely adopted that curtailment, although not elimination of hull attached transfer was possible. Modern antifoulants, as mentioned earlier, are not without their own deleterious, often cumulative impacts on estuarine and coastal biota. The emergence of oil as the fuel of choice for developed countries hastened the development of ever-larger bulk carriers, and the emergence of ballast water as the major vector for continuing redistribution of species across ocean basins. Current ballast water management practices are inadequate to prevent this continuing assault on coastal ecosystems worldwide, and the rates and volumes of ballast water transfer continue to increase thus ensuring the perpetuation of the problem. The zebra mussel (*Dreissena polymorpha*) invasion of the freshwater systems of the eastern United States remains as the poster child illustrating both the ecological and economic consequences of the failure to appreciate and act in regulating ballast water management, and the magnitude of the continuing threat[1].

---

[1]The impact of ballast water mediated invasions is addressed in this volume by Minchin (Chapter 5). Regulation of ballast water discharge to moderate and eventually eliminate pelagic forms presents a difficult compromise between vessel safety on one side, and biological control on the other. Through such bodies as the International Council for the Exploration of the Seas (ICES) there has been extensive examination of high seas ballast water exchange as a mechanism to eliminate pelagic forms of coastal and estuarine species replacement by oceanic forms that, it is presumed, will not survive in receptor locations. Mandatory ballast water exchange is now a U.S. Coast Guard requirement for inbound vessels in international trade. Unfortunately ballast water exchange is neither effective in ensuring the elimination of pelagic forms, nor commensurate with safe operation of bulk carriers in bad weather. Bulk carriers in transit operate with large holds either empty or full of seawater ballast. Partial filling provides mobile surfaces that can reduce stability. Ballast exchange requires either periods of open mobile surface, which decreases stability, or exchange by overflow and dilution, which is inefficient and ineffective. Both the International Maritime Organization (IMO 2000-2005) of the United Nations and the U.S. Congress, through reauthorisation of the National Invasive Species Act (NISA), are seeking to develop enforceable standards for discharge of ballast water in units of included viable organisms of specified size ranges per unit volume, for more detail see IMO (2004). Agreement on such standards will undoubtedly fuel technological development of very large-scale water treatment for shipboard application, with the goal of treating all ballast water to acceptable standards before discharge, without recourse to ballast exchange.

5. Restoration options: native community structure and function versus ecological function in isolation

Restoration is an exercise in scaled landscape management, and must start with stabilising the situation, or even reinserting keystone species back into the ecosystem under stress. This is predicated on our emergent understanding of metapopulation dynamics. This is essentially concerned with how fragmentation of environment and species distribution at various scales dictates how species reproduce with some spatial regions acting as net sources of juveniles that are exported throughout the range, while other regions act as net sinks of juveniles, where import exceeds export (Hanski 1994; Dias 1996). In the aquatic environment fragmentation of benthic structure disrupts physical corridors that are optimal in maintaining community structure. Ecologists must provide quantification of populations to sustain long-term equilibrium in birth, death, immigration and emigration rates of the keystone organisms (BIDE models: **B**irth, **I**mmigration, **D**eath, **E**migration; see Pulliam 1988). Such restoration includes both generic and site specific challenges; their resolution requires iterative discussion between ecologists and environmental engineers.

The first choice in restoration must be to focus on native community structure. This is usually well documented, or can be researched to provide a framework for target species actions. The presumption follows that restoration of species inherently rebuilds ecosystem function. However restoration of native species may not be tractable. There is growing understanding that human population growth, urban development, modern agriculture and short-term economic policies have accelerated destruction of local community structure with disassembly of native ecosystem function. With this has come the realisation that we cannot restore native ecosystems because we have irretrievably altered their physical and biological environment. The question arises can we restore or at least mimic ecological function of the extirpated species? If restoration options involve an element of mimicry, then we must identify the ecological services provided by the extirpated species in either a singular or combined value, and use a multidisciplinary approach to replacing them with analogous capabilities. Why multidisciplinary? Because reversing the trends of ecosystem damage is not simply a matter of reversing previous actions. Society has placed irreversible demands on the environment, so restoration must balance societal needs with ecological services, relying on ecologists and engineers to develop and implement practical and rational strategies to attain well-defined goals while functioning within an established regulatory environment. The dimensions of this problem cannot be understated. Current world population exceeds seven billion people; all are stakeholders in international trade and global ecology. Within this framework we seek to maintain the avenues of transport from personal mobility to the routes of international commerce. The role of the ecologist is to define the nature of the impacted ecological services paying particular attention to scales of time and space. Ecologists should take the highest ground possible and present the most holistic problem to engineers with no compromise. The engineering answer must not compromise ecological function. These are long-term challenges, typically decadal in prospect, and our perspectives cannot be limited by current technology.

Successful examples of managed terrestrial ecosystems serving societal needs are, in fact, globally abundant in modern agriculture (e.g. wheat, corn and cereal crops, rice paddies and cattle range lands) and silviculture (e.g. northern hemisphere softwoods for both structural lumber and pulp, plus an increasing area of tropical hardwoods such as teak and mahogany to reduce pressure on natural stands of these species), but they stand in marked contrast to the complexity of estuarine and coastal ecosystems that are the focus of this contribution. Such managed ecosystems have generally been intentionally fragmented to spatially and temporally maintained monocultures. Also, modern agriculture typically involves replacement of native with non-native species with accompanying reduction in species diversity – it is relevant to note that 90% of the worlds agricultural production is based on 20 plant and 6 animal species with global distribution from intentional introductions. Despite this reduction in diversity it can be argued that agriculturally managed ecosystem function has been so central to colonisation success and development of economically and politically stable human society that without it the settlement of North America and the eventual emergence of the United States as a world economic superpower would probably not have occurred. Species that are now central to manipulated ecosystem management cannot be discarded to attempt retrofit of native species that either did not survive manipulation, or cannot contribute to the economic value of the maintained ecosystem. The extension of spatially explicit monoculture to provide ecosystem function in the estuarine and coastal regions is, however, precluded by the steep environmental gradients that occur in these target zones and the recruitment processes of aquatic species that utilise mobile larval and/or adult life history stages. Estuarine and coastal regions cannot be served by a single species management approach as used in agriculture because of their spatial complexity and the associated diversity of niches and trophic structure. Thus they must be managed as a multi-species complex.

6. The regulatory environment

We live in ecosystems that, by design or default, are managed to varying degrees. Management tools are typically regulated. Extant regulation must serve a wide range of goals from reduction and eventual curtailment of the impacts of destructive local practices. It is necessary to incorporate appropriate mitigation, compensation and restorative actions with defined and attainable end points, to limit continuing vectors of external stresses. These include unintentional species introductions, which offer the prospect of no cure in that established invaders are rarely amenable to extermination. Thus the regulatory environment addresses, where applicable, local, regional, national and international issues and includes legally 'grand fathered' (i.e. changes compliant with the law at that time) and culturally sensitive, notably those relating to native cultures, components. The balance of federal versus state regulation abundantly illustrates this complexity in U.S. law, as does the evolving national versus community level regulation within the European Economic Union. The complex structure of regulation at varying levels is a result of decades of societal action, yet the need for ecological restoration is a recent addition to this process and is struggling to maintain pace in this complex regulatory mix. In the following sections examples are provided of the ecological challenges of restoration, and comments made on the regulatory environment that guides, or in some instances, confuses such action.

7. Approaches to restoration of native species

Recent work with oysters is presented as an example of scaled, practical application of ecological engineering as a restoration tool. Methods used in Chesapeake Bay over the past decade provide specific details. Chesapeake Bay is approximately 10,000 years old and was formed with rising sea level by inundation of a wide but shallow coastal river valley. The native oyster *Crassostrea virginica* has a current distribution from Prince Edward Island in the Canadian Maritime Province, to the Yucatan peninsula in the Gulf of Mexico. Significant recent fossil deposits of oyster shell on the inner continental shelf of the eastern seaboard of the United States illustrate the longevity of the species over this latitudinal range. The precolonial disposition of the Chesapeake Bay population was as a complex reef system both within and between subestuaries. That the patchwork nature of this system developed in concert with estuarine circulation and sedimentary deposition is illustrated in bay wide surveys of habitat by Stevenson (1894) in Maryland, and by Baylor (1894) in Virginia. A summary of the development of this species and its habitat in the Chesapeake Bay is given by Hargis (2000). A summary of the current status of oyster habitat is given by Berman *et al.* (2002) and in recent revision at VIMS (2004a). The demise of the resource and a commentary on the causative agents is given by de Broca (1865), Ingersoll (1881), Brooks (1890) and Hargis and Haven (1988). The most obvious among these is over a century of harvest at levels in excess of maximum sustainable yield. Less obvious, although undoubtedly important, is the gradual change in sediment associated with deforestation, and more recently both point source (urban origin) and non point source (agricultural) run off of excess nutrients resulting in eutrophication and local hypoxia or even anoxia events.

Management of the oyster fisheries of the Chesapeake Bay have involved elements of habitat maintenance for approximately 100 years in the form of shell replenishment on known oyster reefs and cultivated hard substrata. There has been a continuing transformation of the original three dimensional habitat structure, that represents the end point of ecosystem development in shallow estuarine waters, to a two dimensional veneer of shell on the footprints of former reef habitat. The first significant efforts at restoration of three dimensional structure to restore long term ecological services, as opposed to supply a fishery product, occurred in 1993 with construction of a one hectare reef in the Piankatank River, a small sub-estuary on the western shore of the Chesapeake Bay (Bartol & Mann 1997; Mann 2001). A smaller reef followed this in 1996 in the Great Wicomico River and the supplementation of that reef with addition of brood stock oysters (Southworth & Mann 1998). Over one hundred reef sanctuaries have been constructed in the past decade and much has been learned to support future efforts. Modest efforts in otherwise degraded ecosystems face little chance of long-term positive impact. Scaling issues are critical since mosaicism in the environment was central to the stability of reef systems prior to colonial times. In a patchwork benthic environment, edge effects from complex edges providing shelter or microniches, while vertical texture from mature reef structure fosters an increase in benthic epifauna and provides refugia for juvenile stages of sessile or attached species. Such observations prompt the question "what is the minimum size at which restoration can provide a sustainable impact?"

Early efforts at oyster restoration, and through this the restoration of the estuarine community itself, lacked both a suitable model for population rebuilding (analogous to that of fishery management models with biological reference points) and an appreciation of minimal population size to support rebuilding. In effect, modest sanctuaries (hectare scale) seeded with oyster brood stock in estuaries with areas of tens to hundreds of hectares provided recruitment signals that waned in successive years. Estuarine circulation ensured dispersal of larval forms from point sources, but limited immigration was accomplished because of the lack of reciprocal larval supply from other sanctuaries. The inability of restoration ecologists to provide several sanctuaries simultaneously in a 'leaky' estuarine system ensured that continual dilution of the output from single sources would thwart stable recruitment over many years. The role of engineers in resolution of this problem is seminal. Systems dynamics modeling is the commonly used tool to address diverse problems that have common elements. The same approach used in inter- and intra-settlement transport models for roads can be used in estimating genetic flow and propagule exchange between populations in a metapopulation mosaic. In restoration ecology the generic approach to exchange must, however, be sensitive to geomorphology of the restoration site. Building upon this approach a restoration effort focused on a single estuary, the Great Wicomico River, is currently underway using multiple reef sanctuaries stocked with selectively bred, disease-tolerant oysters with known genetic markers in a location where the entire historical habitat of oyster shell relief has been rebuilt to a pre-1900 landscape. The continuous monitoring of this system provides for hypothesis testing on scaling of simple population models in real world applications. Significant gaps exist in the ability of ecologists to build stable metapopulation models to drive restoration in aquatic species, a situation that is even more acute when models attempt to incorporate both terrestrial and aquatic realms in a coordinated effort.

The regulatory environment surrounding restoration of estuarine and coastal aquatic and shoreline resources is diverse. In the U.S.A. maintenance of navigable waterways is driven by federal law while associated dredging and fill are generally addressed by both federal and state regulations. Fishery management may be a federal domain for anadromous species effected through a federal-state partnership, yet benthic target species such as oysters and clams are predominantly state-regulated within state waters, including leasing of bottom and/or water column for aquaculture purposes. State jurisdiction generally presides on issues of individual riparian rights (those of waterfront landowners for use of adjacent waters) wetland preservation for ecological services, and sediment runoff which impact on both total sediment load and possible coliform introduction. Federal regulations drive state actions on various water quality parameters including nutrient loading. While ecological sanctuaries, of which oyster broodstock regions are an example, are generally supported by public opinion, they currently represent a modest proportion of exploited bottom area and little inconvenience to boat traffic of all sizes. Much of the exploited bottom area may be held in public trust for the common good of the state's citizens, so accompanying legislation may be particularly difficult to amend to incorporate exclusion from public access. Significant increases in sanctuary size and number are a likely need to be considered as our quantitative understanding of large-scale efforts evolve, but both present obvious obstacles for public opinion to consider. Public education on the necessity for ecological services is the most powerful asset available to promote a continuing atmosphere of public support, see for an example VIMS (2004b). The role of non-governmental organisations cannot be understated here as amply demonstrated, for example, by the contribution of the Chesapeake Bay Foundation (CBF 2005).

8. Approaches supporting use of non-native species as a restoration tool

Non-native species provide a tool for restoration of ecological function; however, their use is a societal decision and it is regulated by society to both promote rational progress, while preserving ecological function. Their use should be treated with utmost caution. The Convention on Biological Diversity generally considers alien invasive species as the second most important threat, after habitat destruction, to indigenous biodiversity. When alien introducts and habitat destruction are considered together, their aggregate impact drives the argument that there are no longer any intact native ecosystems in the estuarine and coastal zones of most, if not all, of the globe. However, global biodiversity provides a huge catalogue of species that may be candidates to provide ecological services in novel environments, notably in drastically altered ecosystems that are common in coastal and estuarine regions. An example of the latter may be a situation where a native species degradation is associated with the presence of introduced parasites and disease to which the natives are susceptible with catastrophic results. Both ecological services and economic vitality have arguably been restored in many instances through the careful introduction of non-native species to degraded situations. The rebuilding of the French oyster industry through introduction of *Crassostrea gigas* after the decimation of the native species by diseases is such an example. The broader challenge to restoration ecologists is to use that non-native catalogue wisely, while admitting to our ignorance of niche definition and possible interaction of species in novel environments. The choice of a candidate species for application in ecological engineering in estuarine and coastal applications is assisted only by relatively crude guidelines. Terrestrial agriculture typically involves a large component of environmental structure or physical limitation. In contrast, aquatic releases are generally beyond control once affected. Ploidy manipulation offers some elements of control in certain instances, but universal application is not practical. ICES has been actively involved in the responsible guidance for use of non-native species in fishery restoration for several decades, and has developed guidelines for considering proposed introductions. Essentially identical guidelines have also been developed by the American Fisheries Society (AFS 1989) and European Inland Fisheries Advisory Committee (EIFAC 1989). Here the ICES guidelines are used for discussion. The guidelines recommend observation of the candidate species in its native (donor) environment and comparison of those observations with the native species in the receptor environment. Observation in this context should attempt to describe the niche of the species in terms of both its physical environment and its biological interactions. The ecologists' ability to define the niche of either the native or the candidate introduced species in their respective environments is very limited by the historical changes in both locations associated with human activity. The track record of prediction of impacts of aquatic introductions is limited. So what is an ecologist to do? A poorly examined option is that of consideration of the evolution of the genus of the candidate species over recent geological time, the conditions that caused its range extension or retraction, and the stability of the species over that time. In comparison with some terrestrial situations where rate of change of communities may be rapid – for example the eastern Mediterranean rim or Northern Africa where whole ecosystems have changed in <10,000 years – the diversity and composition of coastal and estuarine communities is arguably fairly stable, it simply migrates with sea level rise and fall. This suggests that the individual species in the candidate pool is reasonably stable over extended time – the restoration ecologists' challenge is to decipher the protocol to identify optimal match of the presumed niche over evolutionary time with the changing status of the restoration site and the ecological service to be fulfilled.

The regulation of intentional use of non-native species in ecological restoration (and other services) is covered by a plethora of local, state, national and international jurisdictions or advisory agreements that have diplomatic but little legal weight. Within this maelstrom the presiding jurisdiction is often unclear and awaits legal challenge to provide clarity, especially at the U.S. federal-state junction. The body of law addressing non-native species in agriculture appears better developed than laws relating to other ecological settings, notably the aquatic realm. For example, the intentional movement of non-native species across state lines within the USA is controlled by the Lacey Act, originally enacted in 1900 and amended in 1998 (P.L. 97-79, 16 U.S.C. 3371-3378). The Act makes it unlawful to engage in interstate or foreign commerce involving any fish, wildlife or plant material taken, possessed, transported or sold in violation of state or foreign law. However, the Act also contains a clause that delegates authority to a state where the laws of the state specifically address this subject and vest such authority in the state. The Commonwealth of Virginia has such authority in its state Code (Article 2, Chapter 8, Title 28.2, section 825) with respect to aquatic organisms. But proposed introduction of a non-native species of aquatic organisms with mobile life history stages has potential impact on neighbouring states. How can such states address their concerns? The National Environmental Policy Act of 1970 (NEPA, P.L. 91-190, as amended; 42 U.S.C. 4321 et seq.), Endangered Species Act (ESA, P.L. 93-205 as amended, 16 U.S.C. 1531-1543) and the Nonindigenous Aquatic Nuisance Prevention and Control Act of 1990 (NANPCA; Title I of P. No.101-646, 16 U.S.C. 4701 et seq., currently under reauthorization) all provide some limited avenues for recourse, but explicit authority may be lacking (see discussion in Corn et al. 1999). Above this debate, compliance with international guidelines on introduction (ICES, AFS, EIFAC as cited earlier) may be accorded in terms of practice, but the comments of neighbouring or potentially impacted national jurisdictions may be ignored.

9. Future prospects

Port location over historical time has, by necessity, been concentrated in estuarine and coastal ecosystems that are exceedingly complex in both physical and biological function. Disturbance of these ecosystems has ramifications over extended spatial regions in that they provide ecosystem services to both associated terrestrial and oceanic regions. The complexities of these interactions are subjects of intense research, and quantitative descriptions are beginning to emerge that identify critical targeting for restoration activity. Port development and operations are central to the global economy, and will increase in future years. Indeed the pace of projected increase in container services alone provides alarming prospects for destruction of both aquatic and terrestrial communities in the vicinity of port facilities. While detailed economic development plans accompany port development, consideration of the maintenance of ecological services through mitigation or restoration measures has enjoyed minimal attention, shows little prospect for keeping pace with port development, exists in a complex and occasionally untested regulatory environment, and suggests a continuing scenario in which many port regions will suffer increasingly impaired surrounding ecosystems.

Acknowledgments

I thank Prof. John Davenport for his invitation to participate in the writing of this volume. My colleague Mr. Lyle Varnell (VIMS) and fellow volume contributors to this volume shared valuable discussions. The contributing work on oysters in ecosystem function was supported by NOAA Chesapeake Bay Stock Assessment Committee grant numbers NA66FU0487 and NA17FU2888, the NOAA Office of Sea Grant under grant number NA56RGO141, and EPA Chesapeake Bay Program (CB983649-01-0). This is Contribution Number 2665 from the School of Marine Science, Virginia Institute of Marine Science. This chapter is dedicated to the memory of Michael Castagna and Jay D. Andrews and their contributions to shellfish biology and restoration.

References

AFS 1989; American Fisheries Society www.afsifs.vt.edu.

Bartol I, Mann R. 1997; Small-scale Settlement Patterns of the Oyster *Crassostrea virginica* on a Constructed Intertidal Reef. Bulletin of Marine Science 61(3): 881-897.

Baylor JB. 1894; Method of defining and locating natural oyster beds, rocks and shoals. Oyster Records (pamphlets, one for each Tidewater, Virginia county, that listed the precise boundaries of the Baylor Survey). Board of Fisheries of Virginia.

Berman M Killeen S, Mann R, Wesson J. 2002; Virginia Oyster Reef Restoration Map Atlas. Comprehensive Coastal Inventory, Center for Coastal Resources Management, Virginia Institute of Marine Science, Gloucester Point, VA 23062, USA.

de Broca P. 1865; Etude sur l'industrie huitriere des Etats-Unis. (English translation: On the oyster industries of the United States. Report Comm. U.S. Comm. Fish and Fisheries. 1873-1875: 271-319 {1876}).

Brooks WK. 1891; The oyster, re-issued, 1996 Edition with a foreword by K.T. Paynter, Jr. Johns Hopkins University Press, Baltimore, Maryland.

Bryan G, Gibbs P, Burt G. 1988; A comparison of the effectiveness of tri-n-butyltin chloride and five other organotin compounds in promoting the development of imposex in the dog whelk *Nucella lapillus*. Journal of the Marine Biological Association U.K. 68: 733-744.

Bryan G, Gibbs P, Huggett R, Curtis L, Bailey D, Dauer D. 1989; Effects of tributyltin pollution on the mud snail, *Ilyanassa obsolete*, from the York River and Sarah Creek, Chesapeake Bay. Marine Pollution Bulletin. 20(9): 458-462.

Burnell G, Davenport JL. 2003; A review of the use of Tributyltin (TBT) in marine treansport and its impact on coastal environments. In: Davenport J & Davenport JL (eds) The effects of Human Transport on Ecosystems, 266-276. Royal Irish Academy Publications, Dublin.

Carlton JT. 1999;. Molluscan Invasions in Marine and Estuarine Communities. Malacologia. 41(2): 439-454.

CBF 2005; Chesapeake Bay Foundation www.cbf.org (accessed 04-04-2005).

Constanza, R , d'Arge R, de Groot R, Farber S, Grasso M, Hannan B, Limburg K, Naeem S, O'Neill RV, Paruelo J, Raskin RG, Sutton P, van der Belt M. 1997; The value of the world's ecosytem services and natural capital. Nature 387: 253-260.

Corn ML, Buck EH, Rawson J, Fischer E. 1999; Harmful Non-Native Species: Issues for Congress. Congressional Research Service Issue Brief available from: The National Council for Science and the Environment, 1725 K Street, NW, Suite 212, Washington, D.C. 20006. cnie@cnie.org (accessed 04-04-2005).

Dias PC. 1996; Sources and sinks in population biology. Trends in Ecological Evolution. 11:326-330.

EIFAC 1989; European Inland Fisheries Advisory Commission. Codes of practice and manual of procedures for consideration of introductions and transfers of marine and freshwater organisms. European Inland Fisheries Advisory Commission Occasional Papers – EIFAC/OP23. EIFAC Secretariat; Viale delle Terme di Caracalla, 00100 Rome, Italy.

Gibbs P, Bryan G, Pascoe P, Burt G. 1987; The use of the dog whelk *Nucella lapillus* as an indicator of tributyltin (TBT) contamination. Journal of the Marine Biological Association U.K. 67:507-523.

Gibbs P, Pascoe P, Burt G. 1988; Sex change in the female dog whelk *Nucella lapillus*, induced by tributyltin from antifouling paints. Journal of the Marine Biological Association U.K. 68:715-731.

Hanski I. 1994; Patch-occupancy dynamics in fragmented landscapes. Trends in Ecological Evolution.Ecol. Evol. 9:131-135.

Harding JM, Mann R. 2001; Oyster reefs as fish habitat: opportunistic use of restored reefs by transient fishes. Journal of Shellfish Research, 20: 951-959.

Hargis WJ Jr. 2000; The Evolution of the Chesapeake oyster reef system During the Holocene Epoch. In: Luckenbach, Mann MR & Wesson J E (eds) 2000. Oyster Reef Habitat Restoration: A Synopsis of Approaches, 5-23. Virginia Institute of Marine Science, Gloucester Point, Virginia 23062, USA.

Hargis WJ Jr., Haven DS. 1988; The imperiled oyster industry of Virginia. VIMS Special Report (Number 290 in Applied Marine Science and Ocean Engineering. 130p. Virginia Institute of Marine Science, Gloucester Point, Virginia 23062, USA.

Havenhand JN. 1995; Evolutionary Ecology of Larval Types. In: McEdward, L.(ed) Ecology of Marine Invertebrate

IMO 2000-2005; International Maritime Organization www.imo.org (accessed 05-05-2005).

IMO 2004; International Maritime Organization. http://globallast.imo.org/ (accessed 05-05-2005).

Ingersoll E. 1881; The oyster industry. In the history and present condition of the fishery industries: Tenth Census of the United States, Department of the Interior, Washington, D.C. 251pp.

Kennedy VS, Newell RIE, Eble AF. (Eds.) 1996; "The Eastern Oyster, *Crassostrea virginica*." University of Maryland Sea Grant Press, College Park, MD. 734p.

Kurlansky M. 1997; Cod: A Biography of the Fish That Changed the World. Penguin Books, 294 pp.

Levin L, Bridges TS. 1995; Pattern and Diversity in Reproduction and Development. In: McEdward L (ed). Ecology of Marine Invertebrate Larvae, 1-48. CRC Press, Boca Raton.

Louisiana Coastal Wetlands Restoration Plan 1993; http://www.lacoast.gov/reports/cwcrp/1993 (accessed 18-05-2005).

Mann R. 2001; Restoration of the oyster resource in the Chesapeake Bay. Bulletin of the Aquaculture Association of Canada. 101(1): 38-42.

Minchin D, Stroben E, Oehlmann J, Bauer B, Duggan C, Keatinge M. 1996; Biological indicators used to map organotin contamination in Cork Harbour, Ireland. Marine Pollution Bulletin. 32(2): 188-195.

Moore KA, Wetzel RL, Orth RJ. 1997; Seasonal pulses of turbidity and their relations to eelgrass survival in an estuary. Journal of Experimental Marine Biology and Ecology 215: 115-134.

Moore KA, Wetzel RL. 2000; Seasonal variations in eelgrass responses to nutrient enrichment and reduced light availability in experimental ecosystems. Journal of Experimental Marine Biology and Ecology 244: 1-28.

Port of Hampton Roads. 2001; http://www.odu.edu/bpa/forecasting/2001chapter3.pdf (accessed 18-05-2005).

Pulliam HR. 1988; Sources, sinks and population regulation. American Naturalist. 132: 652-661.

Reyes E, White ML, Martin JF, Kemp GP, Day JW, Aravamuthand V. 2000; Landscape modeling of coastal habitat change in the Mississippi Delta. Ecology. 81(8): 2331-2349.

Southworth M, Mann R. 1998; Oyster Reef Broodstock Enhancement in the Great Wicomico River, Virginia. Journal of Shellfish Research. 17(4): 1101-1114.

Stevenson CH. 1894; The oyster industry of Maryland. Bulletin U.S. Fish Commission for 1892: 205-297.

Stickle W, Sharp-Dahl J, Rice S, Short J. 1990; Imposex induction in *Nucella lima* (Gmelin) via mode of exposure to tributyltin. Journal of Experimental Marine Biology and Ecology. 143: 165-180.

R. MANN

VIMS 2004a; Virginia Institute of Marine Science 2004 Virginia Oyster Reef Restoration Map Atlas.
http://www.vims.edu/mollusc/oyrestatlas/index.htm (accessed 05-05-2005).
VIMS 2004b; Virginia Institute of Marine Science 2004 Molluscan Ecology Program Educational Programs and
Materials www.vims.edu/mollusc/education/meeducate.htm (accessed 05-05-2005).
Wray GA. 1995; Evolution of Larvae and Developmental Modes. In: McEdward L(ed) Ecology of Marine
Invertebrate Larvae, 413-441. CRC Press, Boca Raton.

CHAPTER 3: SHIPWRECKED – SHIPPING IMPACTS ON THE BIOTA OF THE MEDITERRANEAN SEA

BELLA S. GALIL

National Institute of Oceanography, Israel Oceanographic & Limnological Research, POB 8030, Haifa 31080, Israel

1. Introduction

Seafaring in the Mediterranean Sea may have begun as early as 9000 years ago. Archaeological finds, texts and iconographic evidence, all reveal the important role of maritime shipping in the ancient Mediterranean. In fact, maritime transport of people and goods allowed the early spread of the 'Neolithic revolution' that shaped the history of the entire region. The remains of harbours and imported goods hint at extensive maritime activity throughout the Mediterranean since the early Bronze Age. Maritime power and maritime trade have played a significant role in the emergence of the great Bronze Age mercantile seafaring empires – the Minoans, the Mycenaeans and the Phoenicians, with their complex networks of long-distance trade. The Roman, Byzantine and Ottoman naval and merchant fleets ruled over, what each Empire in its turn considered, *Mare Nostrum*. The Roman seaborne trade extended to West Africa, the British Isles, and through the Red Sea, to Arabia and India, but was mostly concerned with supplying the metropolitan with grain, oil and wine. The rupture of the economic and cultural ties that followed the breakup of Roman hegemony and the division of the sea between the Christian and Muslim domains, brought on the contraction of large-scale maritime trade. The mercantile horizons that opened with the geographical discoveries of the 15th century, together with political upheavals in the lands bordering the Mediterranean, further marginalised the intra-Mediterranean trade. But following the opening of the Suez Canal in 1869, the Mediterranean regained its prominence as a hub of commercial shipping, and ever more so since the development of the Middle Eastern oil fields, and the ascendance of the southeast Asian economies. It is estimated that about 220,000 vessels of more than 100 tonnes cross the Mediterranean annually, carrying 30% of the international seaborne trade volume, and 20% of the petroleum. With some 2000 merchant ships plying the Mediterranean at all times, accidental pollution as results of collisions or operational mishaps, and pollution stemming from the regular operation of ships, is significant.

This chapter presents an overview of the main biotic impacts of shipping-related pollution in the Mediterranean Sea. The extent and distribution of that pollution, together with past trends, environmental impacts and the relevant existing legal framework are discussed.

2. Shipping-related petroleum hydrocarbons in the Mediterranean Sea

2.1 THE DISTRIBUTION AND ACCUMULATION OF SHIPPING-RELATED PETROLEUM HYDROCARBONS IN THE MEDITERRANEAN SEA

Estimates in the 1970s placed the amount of oil that was spilled or discharged into the Mediterranean at between half a million and a million tonnes annually (Le Lourd 1977). Following ratification of the International Convention on Prevention of Pollution from Ships (MARPOL 73/78) and its more stringent protection measures for the Mediterranean Sea, plus the Convention for the Protection of the Mediterranean Sea against Pollution (Barcelona Convention) and its Protocols (see below), estimates were lowered in the 1990s to 600,000

John Davenport and Julia L. Davenport, (eds.) The Ecology of Transportation: Managing Mobility for the Environment, 39–69,
© 2006 *Springer. Printed in the Netherlands*

tonnes. This is still an amount that marks the sea as one of the most oil-polluted regions in the world. Public attention is mostly drawn by dramatic images of oil tanker accidents and their gruesome aftermaths, as proven by the long-lasting notoriety of the *Prestige, Erika, Haven, Braer, Amoco Cadiz* and *Torrey Canyon*, but most of the hydrocarbon emissions from ships results from routine operations such as loading, unloading and bunkering, and from illegal discharges of bilge waters, used oils and tank washing on the high seas. Though deliberate petroleum discharge from ships is banned in the Mediterranean and all 'operational' oily waste emissions are illegal, there is ample evidence for numerous, repeated, ongoing offences: analysis of satellite photos identified over 1,600 oil slicks in 1999 (Pavlakis *et al.* 2001), and 2350 oil slicks were observed in 2000 (EC 2002) in the Mediterranean.

Though merchant shipping, naval vessels, fishing and recreational boats contaminate the sea with hydrocarbons, it is estimated that three-quarters of the shipping-derived petroleum hydrocarbons in the Mediterranean result from the transport of crude oil and its refined products (EC 1997). The annual oil trade between countries of the north and south Mediterranean is estimated at 360 million tonnes (MT) (REMPEC 1998). Scores of oil-related sites (i.e. pipeline terminals, refineries, off shore platforms, etc.) are spread along the Mediterranean coastline, from and to which crude oil and petroleum products are loaded, unloaded and transported by 250-300 oil tankers daily. The major shipping axis runs from east to west, from the Middle East (about 150 MT going through the Suez Canal and the Sumed pipeline), passing between Sicily and Malta and following closely the coasts of Tunisia, Algeria and Morocco. Traffic on that axis attenuates as it moves westwards and branches off towards unloading terminals near Piraeus, Greece, the northern Adriatic, the Gulf of Genoa and near Marseilles; it is intersected by tanker routes connecting the Algerian and Libyan loading terminals (about 100 MT) with the northern Mediterranean oil ports. The second important route (only partially used in the past decade due to the Iraqi embargo and the post war disruptions in Iraqi oil production) connects crude oil terminals in the Gulf of Iskenderun and on the Syrian coast with Gibraltar and the northern Mediterranean ports. A third route, from the Black Sea through the Istanbul Straits (about 70 MT) leads westwards to join the main axis. Satellite images of the Mediterranean taken in 1999 reveal oil spill concentrations along those routes, in the vicinity of the Egyptian coast, the Adriatic, the straits of Sicily, the Ligurian Sea and the Gulf of Lion covering an area of 17,141 km^2 (Pavlakis *et al.* 2001). Oil transport to and through the Mediterranean is expected to rise with the full lifting of economic sanctions from Libya, and completion of pipelines from the Caspian Sea oil fields: Baku-Ceyhan (capacity 1 million bbl/d, by 2005), Baku-Supsa (115,000 bbl/d in 2001, proposed upgrade to 600,000 bbl/d), Baku-Novorossiisk (50,000 bbl/d in 2001, 100,000 bbl/d capacity), Baku-Novorossiisk (Chechnya bypass) (120,000 bbl/d current, 360,000 bbl/d, by 2005), Kazakhstan- Novorossiisk (400,000 bbl/d in 2002, 565,000 bbl/d capacity, 1.34 million bbl/d, by 2015), and several 'Bosphorus Bypass' pipelines planned with termini at Omisalj (Croatia), Trieste (Italy) and Alexandropoulis (Greece) (EIA 2004). The implementation of the new "motorways of the sea" component of the EU "Trans-European Transport Network" initiative, set for 2010, will increase further the volume of maritime traffic in the Mediterranean Sea.

Acute oil pollution usually results from accidents, such as foundering, grounding, fire, explosion, collision at sea or ramming into quay, pier or bridge. Because of the dense traffic, the risk of accidents is high: in the 1980s nearly 15% of the world shipping accidents occurred in the Mediterranean, Black Sea and Suez Canal. Of the 268 accidents listed by the Regional Marine Pollution Emergency Response Centre for the Mediterranean (REMPEC) for the 1977-1995 period, more than three-quarters involved oil. As a result of shipping accidents an estimated 22,223 tonnes of oil were spilled into the sea between 1987 and 1996. The number of reported accidents is rising, with 94 events reported in 2000-2003, 24 of which resulted in oil spills (Table 1) (REMPEC 2004).

Table 1. Accidental Oil Spills in the Mediterranean 2000-2003 (Source of Data: REMPEC2004)

Date	Location	Spilled	Remarks
21/2/2000	Vatika bay, Neapolis Voion, Greece	Bunkers	The ship sank with 20T of diesel and 5 barrels of lube oil on board. Leakage of bunkers and oily mixtures reported
9/3/2000	Salamis Island	Bunkers	Slight leakage of gas oil reported after the grounding
15/6/2000	Kynosoura, Salamis Island, Greece	Oily waste	Black iridescent oil slick spotted by A.HCG. Surveillance aircraft after the explosion of the "Slops" which is used as a storage and treatment vessel for oil residues
29/8/2000	Kithira Island, Greece	Fuel oil	The ship had 250T of fuel oil, 25T of gas oil and 7T of lubricating oil on board, part of which was spilled and polluted the shoreline
1/9/2000	Khalkis Port, Greece	Bunkers	The ship broke in two during loading cement and eventually sank. She had 670T of fuel oil, 25T of diesel oil and 775L of lube on board. Serious shoreline pollution reported
8/9/2000	Porto Vesme, Italy	Bunkers	Following grounding, a leak of bunker oil was reported. She had a total of 170T of fuel oil and 35T of diesel oil on board
1/10/2000	Psara Island, Greece	Other oily wastes	After grounding a small spill noted around the ship by HCG surveillance aircraft. The ship had 289T of fuel oil, 38T of gas oil and 167T of fuel sludge
10/11/2000	Pachi, Megara, Greece	Crude oil unspecified	A leak from the vessel at Greek petroleum new quay
24/11/2000	Eleusis Bay, Greece	Other oily wastes	The vessel, which was laid up, developed a sudden list and sank. Small pollution reported
23/1/2001	South of Fontvielle, Monaco	Bunkers	The source of pollution reported unknown
27/3/2001	Near Tripoli, Lebanon	Bunkers	An oil spill of unknown origin and described as fuel oil was spotted near Tripoli, covering "few kilometres"
5/5/2001	Southwest of Tsoungria Island, Greece	Bunkers	The ship grounded with 239T of fuel oil and 24T of diesel on board. A private contractor appointed to undertake response operations. 3 tag boats, 3 patrol boats and an antipollution vessel
12/6/2001	Agioi Theodoroi, Greece	Oily waste	Minor pollution caused when gasket on hose coupling burst during loading of lube oil base by the tanker (according to: Hazardous Cargo Bulletin)
18/6/2001	West of Kavo Maleas, Greece	Fuel oil, diesel oil	The tanker grounded under "unidentified conditions" and after cargo transfer operation was completed on 22-06, refloated and towed towards Piraeus for repairs

Date	Location	Spilled	Remarks
11/1/2002	Thessaloniki, Greece	Bunkers	The incident occurred during bunkering of passenger/ro-Ro Express Appolon
4/8/2002	Algeciras, Spain	Bunkers	Small quantity of oil (150L) spilled during bunkering of the salvage tug
7/9/2002	Thessaloniki, Greece	Bunkers	Pollution was caused by a leak from the terminal's underwater pipeline during discharge operations
21/12/2002	Algeciras, Spain	Bunkers	The vessel which spilled fuel oil during refueling operation
3/1/2003	Karystos, Greece	Bilges	Slick 30 N/miles long and 10m wide
7/1/2003	2 n/m from Coero Cape-Ancona, Italy	Bunkers	Slick pushed back out to the sea by winds
29/1/2003	Off Tipaza, West of Algiers, Algeria	Bunkers	Cougar sank quickly under the weight and uneven distribution of its cargo of red potters clay
5/3/2003	Kalymnos, Greece	Bunkers	Oil slick successfully dispersed by Matsas Star
11/7/2003	Eleusis Port, Greece	Bunkers	Medoil III sustained 1.5m fracture above waterline causing 200 m^2 diesel oil pollution
22/10/2003	Agioi Theodoroi, Greece	Crude oil	Incident due to a disconnected discharging arm

Studies of dissolved/dispersed petroleum hydrocarbons (DDPH) conducted in the 1970s show a wide variation, due, no doubt to the patchiness of the hydrocarbon as well as to different methodologies. Generally, the findings indicated "… considerable petroleum contamination", with the highest subsurface concentrations occurring in the Alboran Sea (25.8µg/l), off the Libyan coast (24.9µg/l), and southeast of Crete (20.5µg/l) (Zsolnay 1979). Subsurface seawater samples collected in the Levantine Basin between 1977 and 1979 showed high levels of DDPH (>40.0µg/l) between Cyprus and Crete, an area where "deballasting, release of wash waters and bilges from tankers and other ships was permitted in the past and is probably still practiced"; but even "clean" areas showed more than a trace of hydrocarbons (0.9-4.3µg/l) (Ravid et al. 1985). Similarly, a 1983 survey of DDPH in coastal and offshore waters off the northeastern Levantine Basin found 3.8µg/l offshore, and levels as high as 33µg/l in the bay of Iskenderun, with its two oil terminals and extensive tanker and ship traffic (Saydam et al. 1985). DDPH values in samples collected in the Aegean and Ionian seas in 1993 ranged mostly from 1.3 to 4.3 µg/l, with some higher contamination levels in the Ionian Sea (6.4, 37.2 µg/l), the latter possibly a spill (Sakellariadou et al. 1994).

A compilation of measurements in the entire Mediterranean (dating to 1982) gave an average of 2.0µg/l. More than half the samples were above 0.4µg/l with an average of 3.7µg/l. The average Mediterranean concentration is 100 times that of the North Sea value, with highest values in the Levantine Basin (De Walle et al. 1993).

2.2 THE DISTRIBUTION AND ACCUMULATION OF TAR IN THE MEDITERRANEAN

Maritime tar in the Mediterranean results from the transportation and usage of petroleum products: the deballasting or the flushing of oily waters from tankers; fuel, lubricating oils and grease residues; engine-room bilges; and accidental oil spills. The weathering of an oil slick, the evaporation of its volatile components and its emulsification, produces a highly viscous layer that adsorbs suspended mineral and organic particles. The gradually condensed and compacted mass is broken into the familiar tar lumps by water movements. Approximately 30% of the oil discharged intentionally or unintentionally into the sea forms tar. Tar particles occur near shipping lanes where deballasting or flushing of oily waters takes place, or more frequently, near oil-loading terminals. Fingerprinting pelagic tar particles collected in the western Mediterranean by gas chromatography, Albaiges *et al.* (1979) determined that over 80% were discharged oil tanker washings, originating in vessels carrying Middle Eastern crude oils. Tar is a persistent contaminant, very resistant to biodegradation, and in a small, enclosed sea such as the Mediterranean, pelagic tar is likely to be washed up along the coast.

Tar contamination is a Mediterranean-wide problem: 80% of the 120 neuston samples collected from 40 stations during a 1974 cruise that traversed the Mediterranean contained tar pellets (Benzhitskiy & Polikarpov 1976); as had more than 97% of the 143 neuston samples collected between 1975 and 1977 in the western Mediterranean (Ros & Faraco 1978). Tar lumps floating on the sea surface have been noted in the Mediterranean since the mid 1950s, but the first study of their distribution and abundance there was carried out by Horn *et al.* (1970). Their extensive survey, from the Levantine Basin to the Alboran Sea, showed that tar particles were found in 75% of the 734 samples taken, and that the most tar-polluted areas were the Ionian Sea, the Libyan coast and the Levantine Basin with 360, 125 and 46 mg/m^2 respectively. Analysis of pelagic tar particles collected in 1970 in 38 sites in the eastern Mediterranean, marked large concentrations of tar off the Libyan coast, in the Levantine basin between Syria and Cyprus, and in the northern Ionian Sea, and it was proposed that the tar concentrations are "....situated off the main oil ports of the Levant where large quantities of crude and refined oils are loaded, and where tankers unload ballast waters" (Oren 1970). The accumulation of pelagic tar off the Egyptian coast was sampled in 1971: average concentrations between El Sallum and Marsa Matruh were 45mg/m^2, with a maximal value of 405mg/m^2 (El Hehyawi 1979). The highest concentration of pelagic tar (33mg/m^2) in the northeastern Levantine waters was sampled in the Gulf of Iskenderun, home to two pipeline terminals and extensive tanker and ship traffic (Saydam *et al.* 1985). At the other end of the Mediterranean tar was found in each of 34 samples taken at the Alboran Sea in 1981-82, where the highest contamination (19.8 - 25.6mg/m^2) was found along the crowded tanker lanes off the Spanish and Algerian coasts (De Armas 1985). The chromatographical analysis indicated the tar particles originated from oil discharged with tank washings and deballasting by crude oil tankers. Samples collected in 1974 showed that tar contamination in the Ionian and Tyrrhenian seas, and along the North African coast averaged 10, 7 and 6.8mg/m^2, respectively, with maximal concentration in the Ionian Sea reaching 28mg/m^2 (Benzhitskiy & Polikarpov 1976). 101 neuston samples were collected during a series of nearly simultaneous cruises in the Eastern Mediterranean and the Alboran Sea in 1987 (Golik *et al.* 1988a). Analysis of their tar contents showed that the highest concentrations were in the Gulf of Sirte, off the Libyan coast (9.7-19.9mg/m^2), in the central Levantine Basin (4.1-16mg/m^2), and along the southern coast of Turkey (4.7-12.9mg/m^2). The distributional shift in pelagic tar particles may stem from political and economical events in the Middle East: these include the suspension of Iranian oil shipments through the Ashkelon terminal (Israel) following the Iranian regime change, the decline in Lebanese and Syrian oil terminals' activities following regional disputes, and the increased capacity of both the Kirkuk-Ceyhan oil terminal in the

Gulf of Iskenderun (Turkey), and of the Sumed pipeline (121 MT/y) at Sidi Kerir (Egypt). The excessive tar contamination in the Gulf of Sirte is attributed to the "intense activity of the oil terminals on the Libyan coast", as well as to flagrant disregard for environmental issues (four of the five oil terminals at the time lacked reception facilities for oil residues) (UNEP 1986). A comparison of the data collected in 1969, 1974 and 1987 (Horn *et al.* 1970; Golik *et al.* 1988b; Loizides 1994) reveals a steep reduction in (mean) concentrations of pelagic tar, from 37mg/m² in 1969 to 9.7mg/m² in 1974 and 1.2mg/m² in 1987. Though overall concentrations of pelagic tar may be declining, samples taken in Antikithira Strait (between the Ionian and Cretan Seas) in 1997 revealed high densities of tar – up to 4.3mg/m². Analysis of surface water hydrographic data indicates that the tar may have originated in the southern Ionian Sea, due to expansion in Libyan oil shipments, and/or illegal deballasting along the Ionian Sea tanker lanes (Kornilios *et al.* 1998).

Beach-stranded tar is derived from pelagic oil slicks, and its landfall is dependent on the circulation of surface water masses and wind patterns. Tarry beaches were the bane of the Mediterranean in the 1960s and 1970s: "on all beaches are flat fragments of tar, mostly between 0.5cm and 5cm in diameter, concentrated in swash marks near the upper limit of storm and ordinary waves" (Oren 1970). Chromatographic analyses of beached tar collected along the Israeli coastline showed it was derived from Middle Eastern crudes deballasted or flushed offshore, and the closer the beach to an oil terminal or shipping lane, the heavier its contamination (Golik 1982). Tar accumulation (380g/m²/15 days) on Egyptian beaches near Alexandria, was blamed on passage of tankers through the Suez Canal, the Sumed pipeline loading terminal, and accidental oil pollution from Alexandria harbour (Wahby 1979). Aybulatov *et al.* (1981) described smaller pellets accumulating in the supralittoral, and forming "… beach ramparts with a height of several tens of centimeters and 3m to 7m wide" along a North African beach, where, following a northeastern storm, the accumulation of tar pellets were estimated at 12,300kg/km. The 1980s saw a significant reduction in beach-stranded tar along much of the Mediterranean coastline. Beach-stranded tar contamination at Lara, Cyprus, declined from 6156g/m² in 1978 to 121g/m² in 1989, and in Paphos from a mean of 268g/m² in 1976-78 to 67g/m² in 1983, and at Haifa, Israel, it was reduced from 3600g/m² in 1975-76, to 12g/m² in 1984 (Golik *et al.* 1988b; Loizides 1994).

2.3 IMPACT OF PETROLEUM HYDROCARBONS ON THE MEDITERRANEAN BIOTA

Studies have shown that seabirds and sea turtles ingest tar pellets (van Franeker 1985; Carr 1987), and floating tar balls may serve as rafts aiding the dispersal of their associated epibionts (Minchin 1996).

Seventeen of the 92 loggerhead turtles inadvertently caught by Maltese fishermen in 1986 were contaminated with crude oil and tar (Gramentz 1988). Most of the impacted turtles had oil and tar covering their mouth and pharynx, one had a lump of tar attached to its palate, a couple defecated 40-60 tar balls, and others had tar smeared on their body. Tar was identified in the digestive tract of 14 of the 54 loggerhead turtles captured off the Spanish Mediterranean coast. The effects of ingested oil and tar are unknown, but heavy contamination "could immobilise and cause death due to exhaustion in younger and smaller sea turtles" (Gramentz 1988). Epipelagic fish too are likely to ingest tar: large amounts of tar were discovered in the stomachs of several specimens of the saury, *Scomberesox saurus*, an important forage fish for larger predators such as porpoises (Horn *et al.* 1970).

Members of the pelagic fouling biota such as the isopod crustaceans *Idothea mettalica* and *Eurydice truncata* and the goose barnacles *Lepas pectinata* and *Lepas anatifera* have been found on pelagic tar particles, with some lumps carrying scores of barnacles: Horn *et al.* (1970) collected as many as 150 individuals from four tar lumps. The age of the largest attached barnacles, inferred from their length, and based on the growth rate of the species, was estimated at 2-3.5 months (Horn *et al*. 1970; Benzhitskiy & Polikarpov 1976). It is assumed that once the thriving biota overgrows the positive buoyancy of the tar lump, it sinks its tarry raft.

Oil spill impacts arise from smothering and toxicity. Lethal spills that cause wide mortality and habitat destruction are rare but, in plain sight, imprint themselves on our collective memory. The chronic toxic effects of oil include interference with reproduction and feeding, abnormal growth and behaviour, and susceptibility to disease, these may lead to changes in species abundance and diversity in the affected area (UNEP 1982). Eastern Mediterranean neuston was found highly contaminated with petroleum hydrocarbons, suggesting that the near-surface plankton store and concentrate the hydrocarbons from the badly polluted surface films (Morris 1974; Youssef *et al*. 1999). A study of DDPH in seawater and in fish in the Gulf of Iskenderun, Turkey, has demonstrated a correlation between levels of DDPH in seawater and fish liver (Salihoglu *et al*. 1987). A mussel watch survey conducted in 1988-89 along the Mediterranean coast of France and Italy showed that concentrations of total aliphatics (average 470μg/g dry wt) and PAHs (average 233μg/g dry wt) were higher near the harbours of Marseille, Toulon and Genoa (Villeneuve *et al*. 1999).

Experimental pollution of sediments in a shallow cove in the Gulf of Fos, France, with crude oil, caused a "massive and almost complete destruction of macrofauna" due to the toxic action of the aromatic fraction of hydrocarbons (Plante-Cuny *et al*. 1993). The annihilation was followed within three months by mass colonisation of juvenile individuals of opportunistic taxa, and consistent with succession in organically-enriched sediments, reached diversity and density similar to native populations after nine months. The *Agip Abruzzo* and *Haven* accidents occurred within a day and 150km of each other along the Ligurian coast in April 1991. Following the *Agip Abruzzo* spill that dumped 30,000tonne of crude oil, a decline was observed in meiofaunal density due to oil-induced mortality (Danovaro *et al*. 1995). But though analysis of the nematode fauna at depth of 10m showed a disturbed assemblage immediately following the spill, within a month diversity patterns were similar to pre-spill values. Similarly, a study of the impacts of tar resulting from the *Haven* accident that spilled 36000 tonne of crude oil, concluded that there were no "appreciable detrimental effects on the soft-bottom macrobenthos, which appeared to have recovered towards natural pristine conditions" (Guidetti *et al*. 2000). However, nine years after the *Haven* spill specimens of a flatfish, *Lepidorhombus boscii*, collected at the site of the spill, displayed both high incidence of micronuclei in hepatic cells and hepatic lesions, indicative of chronic exposure to petroleum-contaminated sediments (Pietrapiana *et al*. 2002).

2.4 THE POLICY AND MANAGEMENT OF SHIPPING-RELATED OIL POLLUTION IN THE MEDITERRANEAN SEA

Estimates based on satellite images of oil spills taken in 1999 put the volume of illegal operational oil discharges at "four times greater than the average amount spilled in the region by ship accidents" (Pavlakis *et al*. 2001). However, oil tanker accidents resulting in major spills in coastal waters receive most media attention. Their histrionic presentation and economic implications drive popular concern and move policy makers to legislate measures for preventing future occurrence. Though the policy and management of transport-related oil pollution in the Mediterranean is concerned both with accidental oil pollution due to collisions or groundings in the highly trafficked and navigationally difficult sea, and routine shipping operations such as fuel and bilge oil discharges, tank-washing and deballasting, it is the highly visible threat of the former that have been emphasised, regardless of ample evidence that most shipping-related oil pollution stems from operational discharges.

The most important regulations for preventing shipping-associated oil pollution are contained in Annex I of the International Convention for the Prevention of Pollution from Ships, 1973, as modified by the Protocol of 1978 (MARPOL 73/78). The provisions of MARPOL prescribe strict standards and conditions for discharge of oily wastes and residues into the sea, with even more stringent requirements for the seas designated as 'Specials Areas'. The land-enclosed Mediterranean Sea, crowded with maritime traffic, is considered particularly sensitive to chronic pollution, and is recognised in MARPOL as a 'Special Area' deserving more rigorous protection.

The Convention for the Protection of the Mediterranean Sea against Pollution (Barcelona Convention) was adopted in 1976 and amended in 1995. The Protocol Concerning Cooperation in Combating Pollution of the Mediterranean Sea by Oil and Other Harmful Substances in Cases of Emergency (Emergency Protocol), an instrument of the Convention to manage a specific issue, had been adopted in 1976. It came into force in 1978, and by 1991 was ratified by 18 Mediterranean countries and the EEC. Before the adoption of the Barcelona Convention and its Emergency Protocol, only four of the countries bordering the sea (France, Greece, Israel, Italy) had oil spills contingency plans. The Regional Oil Combating Centre (ROCC), administered by the International Maritime Organization (IMO), was created in 1976 to strengthen the capabilities of the countries bordering the Mediterranean to develop both national and regional oil spill emergency plans, and to establish a communications network to alert national authorities in case of oil-spill emergencies. The focus shifted in the 1980s from response to prevention of shipping generated pollution, and the Centre was renamed in 1990 'Regional Marine Pollution Emergency Response Centre (REMPEC)'.

The overall decline in tar contamination in the 1980s can be credited to the implementation of the above Conventions, improvement in national enforcement of the rules and regulations, installation of coastal facilities for reception of oily wastes and to technological innovations in shipping and handling oil.

The spillage of 10,000 tonnes of heavy fuel oil from the infamous oil tanker *Erika* in December 1999, prompted the European Commission "to bring about change in the prevailing mentality in the seaborne oil trade". That accident brought to the fore the risks inherent in old, poorly maintained ships, and the multiform national rules on maritime safety issues. Three months later, with uncharacteristic alacrity, the Commission adopted a series of proposals, named the '*Erika* I package', to increase controls on safety-deficient vessels and to speed up the timetable for eliminating single-hull tankers. By the end of 2000 a second set of measures was adopted, the '*Erika* II package', to deal with compensation for oil spills, and the creation of a European maritime safety agency. Following the 'Prestige' accident, the Commission proposed an immediate ban on transport of heavy fuel oil by single hull tankers, and an earlier withdrawal of single-hull tankers. The *Erika* I legislative package came into force in July 2003, but of the EU member states that border the Mediterranean Sea only France and Spain – stricken by the recent oil spills – have adopted the requisite laws, regulations and administrative provisions, and transposed them into their national laws.

The absence of adequate compliance and verification mechanisms governing the application of the above rules, regulations, directives and protocols, limits their effectiveness – the widespread and persistent occurrence of oil slicks from illegal discharges offers persuasive evidence of the continuous violation of the law in the Mediterranean. A recent survey sponsored by REMPEC of reception facilities for oily wastes in ports along the eastern and southern rims of the sea, has shown that only Cyprus, Israel and Malta have adequate facilities in all their ports (Wolterink *et al.* 2004), and recommended building new facilities, or improving existing ones in 19 other ports. But even proper facilities will not ensure the end of illegal oily discharges, as Wolterink *et al.* (2004) remark "Ship owners are relatively reluctant to pay for their ship-generated waste treatment. ...While environmental consciousness is increasing in the shipping industry, illegal discharges at sea are still quite common". Many of those discharges occur, presumably deliberately, beyond the 12 nautical mile limit of the coastal states' control and jurisdiction. Effective monitoring and prosecution is crucial to end this common practice.

Recognising those shortcomings, The Contracting Parties in 2003 adopted the 'Catania Declaration' that proposed "to ensure the enforcement of national legislation related to prosecution of offenders illicitly discharging polluting substances…. to establish a comprehensive regional network for the monitoring, detection and reporting of illicit discharges from ships; … to strengthen the level of enforcement and the prosecution of illicit discharge offenders; … to provide all major ports … with adequate reception facilities for wastes generated on board ships" (UNEP/MAP 2003, Annex V, 17b,d-f). Also, "Recognising that grave pollution of the sea by oil and hazardous and noxious substances" the Contracting Parties replaced the 1976 'Emergency Protocol' with the 'Protocol Concerning Cooperation in Preventing Pollution from Ships and, in Cases of Emergency, Combating Pollution of the Mediterranean Sea' (Prevention and Emergency Protocol) in 2004. It is hoped that the surveillance capabilities of the newer satellites can be harnessed to safeguard the environment by improving compliance monitoring and enforcement of existing and future regulations.

3. Ship-generated marine litter in the Mediterranean Sea

3. 1 DISTRIBUTION AND ACCUMULATION OF SHIP-GENERATED LITTER IN THE MEDITERRANEAN

Ship-generated litter is a persistent but overlooked problem for marine ecosystems worldwide, and its potential as a hazard for marine biota has been acknowledged only in recent decades (Pruter 1987; Ryan & Moloney 1993; Derraik 2002). It is of even greater importance in the land-enclosed Mediterranean Sea with its intensive shipping activity. Horsman (1982) calculated that merchant ships dump 639,000 plastic containers each day into the ocean, and estimates of vessel-generated refuse discarded into the Mediterranean, published by the National Academy of Sciences (USA), based on 1964 shipping data, were close to 325,000 tonnes (NAS 1975). In the decades since, marine mercantile activity and the size of naval and fishing fleets plying the waters of the Mediterranean, have increased dramatically. It is reasonable to suppose that litter input from vessels have increased as well. Yet, despite the frequently expressed concern that the Mediterranean Sea is a singularly sensitive ecosystem (Turley 1999), little is known about the extent of problem. Studies of marine litter conducted in the Mediterranean included surveys of coastal litter (Shiber 1979, 1982, 1987; Shiber & Barrales-Rienda 1991; Gabrielides et al. 1991; Golik & Gertner 1992; Bowman et al. 1998), floating litter (Morris 1980; Saydam et al. 1985; McCoy 1988; Kornilios et al. 1998; Aliani et al. 2003), and seabed debris on the continental shelf, slope and bathyal plain (Bingel et al. 1987; Galil et al. 1995; Galgani et al. 1995, 1996, 2000; Stefatos et al. 1999). In most studies plastic items accounted for much of the debris, sometimes as much as 90% or more of the total, due to their ubiquitous usage and poor degradability (Galgani et al. 1996). The composition, abundance and distribution of the litter have been discussed, but few studies have investigated its sources, or its impact on the biota.

Studies of beach stranded litter (Gabrielides et al. 1991; Golik & Gertner 1992; Bowman et al. 1998) in the Mediterranean have shown that most component items (food and beverages containers, suntan lotion bottles and sundry items) are consistent with land-based recreational usage, and their abundance related to intensity of use. However, debris on beaches located down current from a port may have a sizable component of shipping-related items such as crates and pallets drifting from the nearby harbour (Golik, 1997). The distribution and abundance of seabed plastic debris along the southeastern coast of Turkey was highest near Iskenderun and Mersin, both ports serving "considerable foreign traffic"; indeed, where labels were present, many attested to the foreign origin of the items (Bingel *et al.* 1987). Seabed debris in the heavily trafficked Gulf of Patras, at southeastern Ionian Sea, was estimated at 240 items/km². Much of the debris collected consisted of beverage packaging and general packaging, attributed to shipping: "The most frequent tendency of passengers aboard ships is to consume beverages while on the open deck during their trips. However, not only passengers are the culprits as, it is known that some vessels, both cargo and passenger deliberately dump their litter in mid ocean" (Stefatos *et al.* 1999). Mapping the concentrations of debris in the Gulf of Lion, Galgani *et al.* (1996, 2000) found large amounts of debris off Marseilles and Nice and related it to urban activity. However, Marseilles and Nice are port cities and some of the litter may have its origin in shipping. This may be supported by the fact that at the central and southern parts of the Gulf the debris consisted mainly of non-buoyant objects such as glass and metal that probably sank *in situ*. A great amount of floating debris (2000 items/km²), much of it plastic, was observed southwest of Malta near major shipping lanes (Morris 1980). While land-based litter may be blown and washed into coastal waters during storms, with river drainage and sewers, and some debris is obviously introduced by the fishing industry, a considerable amount of marine litter is vessel-generated. Most offshore floating and seabed debris in the Mediterranean is probably ship-based. A month-long cruise to study the deep sea benthic biota of the Eastern Mediterranean, from the Gulf of Taranto to the southeastern Levant, provided an opportunity to examine the extent and potential fate of marine litter in that region (Galil *et al.* 1995). Seventeen trawls were taken, at depths ranging from 194m to 4,614m, of these twelve contained litter. The most common litter of the 13 categories (glass bottles, glass, metal beverage cans, metal wires, metal, plastic bags, plastic sheeting, plastics, nylon rope, paper-cardboard, cloth rags, paint chips and miscellaneous) were paint chips (44%) and plastics (36%). The presence of paint chips in half of the sites surveyed indicates that much of the litter collected in that study originated from shipping activity.

3.2 IMPACT OF SHIP-GENERATED LITTER ON THE MEDITERRANEAN BIOTA

The impact of marine debris on marine ecosystems is considerable. It injures or kills organisms that become entangled in or swallow litter; when settled on the seafloor it alters the habitat, either by furnishing hard substrate where none was available before, or by overlaying the sediment, inhibiting gas exchange and interfering with life on the seabed. Floating debris may serve as rafts aiding the dispersal of epibionts – a potential vector for alien organisms.

The entanglement of larger marine organisms, particularly turtles (Carr 1987; Duguy *et al.* 1998), but also mammals and birds (Laist, 1997), has frequently been ascribed to discarded fishing nets and lines. But, near the Maltese islands, where high concentrations of floating debris were found near busy shipping lanes, Morris (1980) noticed "a small turtle was seen at the surface attempting to swim with a large piece of what appeared to be plastic sheet wrapped around its shell". A study of the loggerhead sea turtle, *Caretta caretta*, found that 8 of the 92 turtles inadvertently caught by Maltese fishermen were affected by marine debris (Gramentz 1988). The turtles examined had ingested transparent, milky white or light blue plastic, styroform and nylon debris, possibly mistaking them for jellyfish. At low ingestion levels debris may cause dietary dilution by replacing food with non-nutritive material that, if persistent, may reduce growth (McCauley & Bjorndal 1999). If large enough, these items can lodge in the turtle's gastrointestinal tract, injuring the animal or causing its death. Indeed, one of the turtles examined had a plastic piece 16cm² removed from its intestines during necropsy. Pelagic loggerhead juveniles are indiscriminate generalist predators frequently found in convergence zones where surface marine debris is likely to accumulate. Forty-one of 54 juvenile loggerhead turtles illegally captured off the northeastern Spanish Mediterranean coast and examined had plastic debris lodged in their intestines, though most of the debris was small enough to pass through the digestive tract without causing obstruction (Tomás *et al.* 2002). However, the high frequency of occurrence of debris in the loggerhead guts in the Mediterranean is reason for concern. The loggerhead, widely considered one of the emblematic animals of the Mediterranean, is classified as 'vulnerable' by the International Union for the Conservation of Nature (IUCN), and so attracts much conservation efforts. The impacts of maritime litter at the surface and seabed of the Mediterranean on the 'non-emblematic' biota are poorly documented beyond anecdotal finds of fish and larger invertebrates 'necklaced' with debris (Figure 1).

Figure 1. Fish with plastic 'necklace' (B.S. Galil)

During its pelagic phase floating debris furnishes a new, anthropogenic substratum for settlement of encrusting biota and other epibionts (Winston 1982; Minchin 1996). Drifting debris, driven by winds and currents, provides opportunities for long-range transport of fouling assemblages, increasing their abundance and distribution with the proliferation of maritime flotsam. The resident community of plastic debris in temperate and warm seas may include representatives of encrusting bryozoans, algae, hydroids, molluscs, serpulid polychaetes and barnacles. The most common of the 14 macrobenthic species identified from floating debris collected off the Ligurian coast were the lepadomorph barnacle *Lepas pectinata* and the isopod crustacean *Idotea metallica*, frequently found in the offshore fouling community (Aliani & Molcard, 2003). However, individuals of species rarely found floating offshore, like seagrass epibionts, were identified as well, 'rafting' on plastic debris. Microscopic examination of plastic flotsam off the Catalan coast (northwestern Mediterranean) revealed the presence of benthic diatoms and potentially harmful dinoflagellates including *Alexandrium taylori*, a species notorious for forming blooms. Masó *et al.* (2003) suggest that "drifting plastic debris as a potential vector for microalgae dispersal". The sheer volume of floating debris aids the dispersal of epibionts and multiplies chances of introducing them into new regions: two cases of alien-bearing plastic debris had been documented, though none yet from the Mediterranean Sea (Gregory 1978; Winston 1982).

3.3 POLICY AND MANAGEMENT OF SHIP-GENERATED LITTER

Concern over the impact of persistent maritime litter on the marine environment has led to attempts at regulation through international and regional conventions. At the international level, mitigation of the problem of ship-based litter was sought through the adoption of three global conventions: the 'Law of the Sea Convention' (LOS) (United Nations 1982), the 'Convention on the Prevention of Pollution by Dumping of Wastes and Other Matter' (LDC), and 'Protocol to the International Convention for the Prevention of Pollution from Ships' (MARPOL). The LOS Convention prohibits vessel source pollution and, importantly, establishes a relationship with other issue-specific environmental agreements, such as the Dumping Convention (LDC) and Regional Seas conventions. The LDC, written in 1972, is concerned solely with disposal of wastes in the sea by dumping including "persistent plastics and other persistent synthetic materials, for example, netting and ropes", recognising them as an environmental hazard. However, dumping as defined under LDC does not pertain to vessel-generated wastes, only to discharges of land-based waste loaded onto vessels. MARPOL, ratified in 1987, recognises that vessels are a significant and controllable source of marine environmental pollution (International Maritime Organization, 1982). Five categories of ship-based pollution (oil, chemicals, hazardous substances in packaged form, sewage and garbage) are regulated under MARPOL, through its five annexes. Annex V prohibits "the disposal into the sea of all plastics, including but not limited to synthetic ropes, synthetic fishing nets and plastic garbage bags". Regulation 5(2)(a)(1) bans disposal of plastics within 'special areas' such as the Mediterranean Sea. In fact, the disposal of all litter except food waste is prohibited in the Mediterranean, and even that should be disposed of at least 12 nautical miles from the nearest shore.

In 1975 the Mediterranean countries and the EEC adopted the 'Mediterranean Action Plan' (MAP) and in 1976, the 'Convention for the protection of the Mediterranean Sea against pollution' (Barcelona Convention). Article 6 of the Convention declares that "The Contracting Parties shall take all measures in conformity with international law to prevent, abate and combat pollution of the Mediterranean Sea Area caused by discharges from ships and to ensure the effective implementation in that area of the rules which are generally recognised at the international level relating to the control of this type of pollution" (UNEP 1992). Protocols are the instruments of the Barcelona Convention to manage specific issues. The first protocol, for the prevention of pollution by dumping from ships and aircraft entered into force in 1978, and was amended in 1995 to include provisions prohibiting incineration at sea. However, Article 3.4(a) excludes "The disposal at sea of wastes or other matter incidental to, or derived from, the normal operations of vessels or aircraft and their equipment".

Recent decades have seen the increased utilisation of synthetic materials and their consequent accumulation in the marine environment. Although the present seabed coverage of plastics and plastics sheeting in the Mediterranean is low (Galil et al. 1995), owning to their persistence, these materials pose a severe problem, with ever increasing amounts of non-degradable debris littering the sea surface and seabed. The density of litter washed up on the shores of a remote island in the South Atlantic Ocean has increased exponentially throughout the 1980s, with the entry into force of Annex V of MARPOL having "no discernible impact on the rate of increase" (Ryan & Moloney 1993). Incidental evidence from the Mediterranean is in concurrence: the dates stamped on some recovered items collected by Galil et al. (1995) and Stefatos et al. (1999) are evidence that vessel-generated refuse is a major source of litter into the marine environment even after regulations that prohibit disposal of all litter except food waste went into force.

Concern over the problem has abated following the ratification in 1987 of Annex V of MARPOL, yet neither international, regional or national regulations provide specific measures for disposal at sea of ship-based litter, and at present there are no practical solutions for effective implementation and enforcement of the ban.

4. Ship-generated noise in the Mediterranean Sea

4.1 DISTRIBUTION OF SHIP-GENERATED NOISE IN THE MEDITERRANEAN

Marine noise levels have been increasing since the mid 19th century, with ships considered the single most important unintentional anthropogenic source. Comparisons with surveys taken in the middle of the 20th century indicate that ambient noise has increased in many of the world's oceans by 3-5dB per decade, amounting to an increase over the past sixty years of somewhere between one and two orders of magnitude (Andrew *et al.* 2002, NRC 2003). Ships generate noise through their engines, machinery, propeller cavitation (the collapse of bubbles formed by propeller movement) and hull vibration. Ship-generated noise is the greatest, most pervasive source of anthropogenic low frequency (5-500 Hz) noise, though characteristic ship noise consists of narrow- and broad-band sounds over a wide range of frequencies. Shipping noise affects large areas of the world ocean, but is more intense near well-travelled shipping lanes, straits and canals, and busy ports: the ambient noise in areas of heavy shipping could range between 85-95dB (1Hz bandwidth), peaking at 100Hz (Ross 1976; Curtis *et al.* 1999). Fast ferries emit sound peaking at 130dB around 500Hz (at 900m distance) (Browning & Harland 1999), supertankers, large bulk carriers, container ships and cargo vessels fill the frequency band below 500Hz with an incessant sonic output reaching 190dB (at source) (Richardson *et al.* 1995; Ross 1976; NRC 2003). The Mediterranean is one of the busiest sea routes in the world, with high-volume ports and crowded shipping lanes; indeed, ambient low frequency sound (LFS) levels are two to three orders of magnitude higher in the Ligurian Sea than in the Gulf of California.

4.2 IMPACT OF SHIP-GENERATED NOISE ON THE MEDITERRANEAN BIOTA

Most of the concern regarding the possible effects of LFS focuses on whales and dolphins, species that produce and use sound to communicate, to feed, to navigate and sense their environment (NRC 2000). But LFS is also the primary range of hearing of most marine fish and turtles (Tyack 1998; Bartol *et al.* 1999) therefore it is possible that human-generated LFS interferes with their natural behaviour as well. The cumulative impact of the rise in the spatial and temporal prevalence of shipping-generated noise in the Mediterranean Sea is unknown. However, it is considered of important concern as it affects the acoustically sensitive and potentially vulnerable cetaceans and turtles. Reports of possible impacts include masking, stress, auditory damage, alteration or disruption of behavioural patterns such as vocalisations, feeding, breathing and diving, and spatial avoidance (Richardson *et al.* 1995; McCarthy 2004). Other studies suggest that marine mammals may become habituated to vessel noise (Browning & Harland 1999), or even attracted to them (Angardi *et al.* 1993; Diaz Lopez *et al.* 2000). A visual and acoustic survey of small cetaceans and ships in the Alboran Sea allowed that though the cetaceans "do not completely avoid contact with passing vessels" (Perez *et al.* 2000), a negative correlation existed between the sounds they produce and ship noise. The authors concluded that the acoustic impacts of maritime traffic on dolphins and pilot whales would be to reduce "their ability to explore their environment through sound production and reception".

4.3 POLICY AND MANAGEMENT OF SHIP-GENERATED NOISE

The stranding of 12 Cuvier's beaked whales in Kyparissiakos Gulf, Greece, in 1996, associated with NATO tests of active sonar, has defined the issue of marine noise pollution in the Mediterranean (Frantzis 1998). Though high volume shipping in the Mediterranean is acknowledged as the preeminent source of noise pollution, and as such "a very important cause of concern" (Roussel 2002), the scientific uncertainty concerning noise effects on the biota precluded attempts to regulate sound produced by commercial vessels. An IMO resolution 'Guidelines for the designation of special areas and guidelines for the identification and designation of particularly sensitive sea areas' (IMO 2002) recognises that "in the course of normal operations, ships may release a wide variety of substances... such pollutants include ... sewage, noxious solid substances, anti-fouling paints, foreign organisms and even noise." However, even though the Mediterranean is recognised as a 'Special Area' deserving a higher level of protection from pollutants, existing treaties and regional agreements such as MARPOL and the Barcelona Convention do not consider the problem of ship-generated noise pollution. Neither is the issue of ship-generated noise pollution addressed in 'The Agreement on the Conservation of Cetaceans of the Black Sea, Mediterranean Sea and contiguous Atlantic Area' (ACCOBAMS) that entered into force in 2001. Though a report presented the following year to the ACCOBAMS Secretariat included a section on 'Traffic noise: shipping, pleasure boats and whale watching' that proposes mitigation measures (Roussel 2002). The parties (France, Italy, Monaco) to the 'Agreement concerning the creation of a Marine Mammal Sanctuary in the Mediterranean' (INTFISH 2002) that entered into force in 2002, undertook to protect marine mammals and their habitat "from any negative direct or indirect impacts resulting from human activities" (Article 4), and to "fight against any form of pollution" (Article 6), but only at the very last sentence of the Declaration concluding the agreement does it mention the need to examine "the question of noise" produced by vessels.

5. Shipping-transported alien biota in the Mediterranean

5.1 DISTRIBUTION OF SHIPPING-TRANSPORTED ALIEN BIOTA IN THE MEDITERRANEAN

Shipping has been implicated in the dispersal of numerous neritic organisms, from protists and macrophytes to fish (Carlton 1985). It is seldom possible to ascertain the precise means of transmission, as one species may be transported by a variety of vectors, yet it is assumed that port and port-proximate aliens are dispersed primarily by shipping. The transport on the hulls of ships of boring, fouling, crevicolous or adherent species is certainly the most ancient vector of aquatic species introduction. Fouling generally concerns small-sized sedentary, burrow-dwelling or clinging species, though large species whose life history includes an appropriate life stage may be disseminated as well (Zibrowius 1979). Ballast (formerly solid, but for the past 130 years aqueous) is usually taken into dedicated ballast tanks or into empty cargo holds when offloading cargo, and discharged when loading cargo or bunkering (fuelling). Ballast water therefore consists mostly of port or near port waters. Water and sediment carried in ballast tanks, even after voyages of several weeks' duration, have been found to contain many viable organisms. Since the volume of ballast water may be as much as a third of the vessel's deadweight tonnage, it engenders considerable anxiety as a vector of introduction. Slower-moving and frequently moored vessels, such as drilling platforms employed in offshore exploration of oil and gas in the Mediterranean, serve as large artificial reefs and therefore pose a high risk of alien species transmission. The oil platform 'Discovery II' arriving in

Genoa, Italy, in 1977 from the Indian Ocean carried "una ricca fauna tropicale vivente tra cui Teleosti Blennidi e Scorpenidi e Decapodi" (Relini Orsi & Mori 1979). The oil drilling platform 'Southern Cross' originating in Australia was brought to Haifa Bay, Israel, in 2003 for maintenance work including in-water scraping of its extensive fouling. The local divers employed described unfamiliar fish and crustaceans among the dense fauna, and from the shells that had been collected by the divers twelve species of molluscs were identified as new records for the Mediterranean (Mienis 2004).

The Mediterranean Sea, a hub of commercial shipping lanes and encircled by major ports, is susceptible to ship-borne aliens, whether they occur in fouling communities or in ballast. The global maritime trade connections of Mediterranean ports sustain a large-scale dispersal process of both inbound and outbound biota. Shipping is also an important vector for secondary introduction – the dispersal of an alien beyond its primary location of introduction. The widely invasive algae *Sargassum muticum, Caulerpa taxifolia* and *Caulerpa cylindracea* spread across the Mediterranean by ships, fishing boats and recreational craft (Knoepffler-Péguy *et al.* 1985; Meinesz 1992; Verlaque *et al.* 2003). A small Erythrean mytilid, *Brachidontes pharaonis*, settling in dense clusters on midlittoral and infralittoral rocks, piers and debris in the Levantine Basin, has spread as far west as Sicily in ship fouling. Trade patterns ensure that the Mediterranean exports biota as well as imports: The Indo-West Pacific portunid crab *Charybdis hellerii*, an alien present in the eastern Mediterranean since the 1920s, was collected in 1987 in Cuba (Gómez & Martinez-Iglesias 1990), and in rapid succession in Venezuela, Colombia, Florida, and Brazil (Mantelatto & Dias 1999). Transport in ballast tanks is the most probable mode of dispersal since the crab's arrival corresponds with increased coal shipping from Port Drummond, Colombia, to Israel. The presence of two Erythrean aliens, *Alepes djedaba* and *Stephanolepis diaspros*, identified along with four other fish species in a survey of biota in floodable cargo holds and dedicated ballast tanks arriving in Baltimore, U.S.A., from Israel, attest that this is a major pathway for transoceanic dispersal (Wonham *et al.* 2000). The movement of ballast water also provides opportunities for the transfer of microorganisms, including pathogens, which exceed concentrations of other taxonomic groups by several orders of magnitudes (Galil & Hülsmann 1997). Of special concern are possible human pathogens such as the bacteria *Vibrio cholerae* 01 and 0139, agents of human cholera. *Vibrio cholerae* is endemic in the Mediterranean and indeed, a survey of plankton arriving in ballast water in Chesapeake Bay, U.S.A., from the Mediterranean revealed viable *Vibrio* bacteria (Ruiz *et al.* 2000). The risk of invasion of a new strain is of grave concern given the proximity of some ports to aquaculture facilities and to bathing shores (Drake 2002).

The increase in shipping-related invasions was noted in a recent series of Atlases that summarised the extant knowledge on 'Exotic species in the Mediterranean' (CIESM 2004). The increase may be attributed to the increase in shipping volume throughout the region, changing trade patterns that result in new shipping routes, improved water quality in port environments, augmented opportunities for overlap with other introduction vectors, and rising awareness and research effort. However, the choice of taxa treated in the Atlases – fishes, decapod crustaceans, and molluscs – emphasises the precept that taxonomic and biogeographic data are biased in favour of larger taxa of economic importance. Indeed, a basin-wide targeted effort to survey the presence and abundance of the shipping-transported species is warranted as most of the records stem from fortuitous finds. Often reports of new records depend upon intensity of research effort, and since the latter vary greatly along the coasts of the Mediterranean, and even the better studied locales suffer temporal and taxonomical lacunae, there are some doubts concerning the actual spatial and temporal patterns of vessel-transported invasions. The records of vessel-transported alien species in the Mediterranean (Table 2) were culled from the CIESM Atlases, research papers, biota surveys and conference abstracts. The list is doubtlessly an underestimation due to lack of knowledge concerning some taxa, the presence of cryptogenic species and lack of concerted efforts to survey port environments for alien biota. Since the likelihood of encountering a stray incursion is diminishingly small, most recorded alien species are considered as 'established' species that have self-maintaining populations of some duration in the Mediterranean Sea. It is recognised that some alien species may fail to maintain populations over time and thus a single record dating back several decades may be considered an ephemeral entry. The distinction between the 'established' and 'ephemeral' aliens can vary spatially and temporally, and is sometimes difficult to discern and circumscribed in large part by our ignorance. Many 'cosmopolitan' members of the fouling community are quite possibly older introductions into the Mediterranean (Boudouresque & Ribera 1994). Serpulid polychaete worms of the genus *Hydroides* are frequently found in tropical fouling communities and are well established in ports and lagoons throughout the Mediterranean, where they cause major fouling problems on artificial substrata, but are absent from natural marine habitats. They are among the earliest documented invaders in the Mediterranean: *Hydroides dianthus* was documented in Izmir as early as 1865, and *H. dirampha* and *H. elegans* were recorded in the harbour of Naples in 1870 and 1888, respectively.

Table 2. Vessel-transported organisms in the Mediterranean Sea.

Date - date of first observation, or in its absence, first record. Origin - the probable area of origin of the species. Vector - probable dispersal mechanism: Ac- Aquaculture, Aqr - Aquarium trade, B- Ballast, F - Fouling, S- Suez Canal. Distribution in the Mediterranean: Adr - Adriatic Sea, Alg - Algeria, Blr - Balearic Is., CS - Corsica and Sardinia, Fr - France, Gr - Greece, It - Italy, Lby - Libya, Lvt – Levant (Cyprus, Syria, Lebanon, Israel, Egypt), Mrc - Morocco, Sc - Sicily, Sp - Spain, Tn - Tunisia, Trk - Turkey.
Sources: Atta 1991; Boudouresque & Verlaque 2002; Brunetti & Mastrototaro 2004; Golani 2004; Occhipinti-Ambrogi 2002; Rezig 1978; Rosso 1994; Zibrowius 1979, 1992.

Taxa	Date	Origin	Vector	Distribution in the Mediterranean
Macrophyta				
Acetabularia calyculus	1968	Pantropical	F?S?	Blr, Mrc, Lvt
Acrothamnion preissii	1969	Indo Pacific	F	Blr, Fr, wIt, Sc
Aglaothamnion feldmanniae	1976	N Atlantic	F	Fr, wIt
Antithamnionella amphigeneum	1989	Tropical Pacific	F	Sp, Fr, wIt, Alg, Mrc
Antithamnionella ternifolia	1926	Southern Ocean	F	Fr
Apoglossum gregarium	1997	Indo Pacific	F	Sp, Blr, Fr, wIt, Sc
Asparagopsis armata	1923	Cosmopolitan	F	Sp, Blr, Fr, CS, It, Alg, Mrc, Gr, Trk
Audouinella codicola	1952	Subcosmopolitan	F	Sp, Fr, Sc, Tn
Audouinella robusta	1950	Subcosmopolitan	F	Egypt
Audouinella subseriata	1950	Indo Pacific	F	Egypt
Bonnemaisonia hamifera	1910	Circumboreal	F	Sp, Fr, Adr, Alg, Mrc, Tn
Botryocladia madagascariensis	1997	Indian Ocean	F?S?	Sc, Mlt, Trk
Caulerpa mexicana	1941	Pantropical	F?B?S?	Lvt
Caulerpa cylindracea	1990	SW Australia	F?	Sp, Blr, Fr, Cs, It, Sc, Al, Cr, Tn, Lby, Gr, Trk
Ceramium strobiliforme	1992	Tropical Atlantic	F	Sc, Adr
Chondria curvilineata	1983	Pantropical	F	Fr, Egypt
Chondria polyrhiza	1987	Pantropical	F	It, Gr
Halothrix lumbricalis	1979	Circumboreal	F?B?	Fr, It, Trk
Hypnea cornuta	1948	Pantropical	F?S?	sIt, Lvt
Hypnea spinella	1977	Pantropical	F	Sp, Blr, CS, Sc, Alg, Tn, Gr, Trk, Lvt
Hypnea valentiae	1898	Indo Pacific	F?S?	Fr, Gr, Lvt
Lophocladia lallemandii	1938	Indo Pacific	F?S?	Sp, Blr, CS, It, Alg, Tn, Lby, Gr, Trk, Lvt
Neosiphonia sphaerocarpa	1970	Pantropical	F	Sp, Fr, Tn, Gr
Pleonosporium caribaeum	1974	Pantropical	F	Sp, Fr
Plocamium secundatum	1991	Southern Ocean	F?	Sc
Rhodymenia erythraea	1948	Indian Ocean	F?S?	Lvt

Taxa	Date	Origin	Vector	Distribution in the Mediterranean
Stypopodium schimperi	1982	Indian Ocean	F?S?	Lby, Trk, Lvt
Symphyocladia marchantioides	1985	Indo Pacific	F	It
Wormersleyella setacea	1987	Pantropical	F	Sp, Blr, Fr, CS, It, Mlt, Gr
Plantae				
Cladophora cf *patentiramea*	1992	Indo Pacific	F/S	Lvt
Codium fragile subsp. *tomentosoides*	1950	N Pacific	F/Ac	Sp, Blr, Fr, CS, It, Alg, Mrc, Tn, Trk
Codium tailori	1980	Atlantic	F?	Alg, Lvt
Cnidaria				
Clytia hummelinki	1996	Atlantic	F	It
Diadumene cincta	1993	Atlantic	F?	Adr
Garveia franciscana	1978	Indo Pacific	F	Adr
Haliplanella lineata	1971	W Pacific	F	Cs, Fr
Oculina patagonica	1965	SW Atlantic	F	Sp, wIt, Lvt
Ctenophora				
Mnemiopsis leidyi	1992	NW Pacific	B/?	Gr, Trk
Polychaeta				
Branchiomma luctuosum	1983	Red Sea	F	It
Ficopomatus enigmaticus	1922	Southern Ocean	F	Sp, Blr, CS, Adr, Tn
Hydroides dianthus	1865	NW Atlantic	F	Fr, wIt, Adr, Tn, Trk
Hydroides dirampha	1870	W Atlantic	F	Sp, wIt, Tn, Lvt
Hydroides elegans	1888	Pantropic	F	Sp, Fr, wIt, Adr, Lvt
Pileolaria berkeleyana	1979	E Pacific	F	Fr
Spirorbis marioni	1970s	E Pacific	F	Sp, Fr, CS, wIt, Mrc, Trk, Lvt
Bryozoa				
Arachnoidea protecta	1998	W Pacific	F	Sc
Celleporella carolinensis	1994	W Atlantic	F	Adr
Electra tenella	1990	?	F	Sc
Tricellaria inopinata	1982	Indo Pacific	F	Adr
Crustaca, Decapoda				
Callinectes danae	1981	W Atlantic	B	Adr
Callinectes sapidus	1949	W Atlantic	B	Fr, wIt, Adr, Gr, Trk, Lvt
Dyspanopeus sayi	1992	W Atlantic	B/Ac	Adr
Eriocheir sinensis	1959	NW Pacific	B?	Fr
Hemigrapsus sanguineus	2001	W Pacific	B	Adr
Herbestia nitida	2002	E Atlantic	F?	Adr
Libinia dubia	1997	W Atlantic	B	Tn

Taxa	Date	Origin	Vector	Distribution in the Mediterranean
Menaethius monoceros	1978	Indo Pacific	F?	wIt
Percnon gibbesi	1999	E Pacific, Atlantic	F	Blr, CS, It, Malta
Plagusia tuberculata	1981	Indo Pacific	F	Lvt
Rhithropanopeus harrisii	1994	W Atlantic	B/Ac	Fr, Adr
Scyllarus caparti	1977	E Atlantic	F?B? Aqr?	Adr
Thalamita gloriensis	1977	Indo Pacific		Blr, wIt
Crustaca, Cirripedia				
Balanus reticulatus	1958	Indian Ocean	F	Fr, Lvt
Crustaca, Isopoda				
Paracerceis sculpta	1978	EP		distributed in wmed ports
Paradella dianae	1991	P		Alexandria, Egypt
Sphaeroma walkeri	1977	IO		distributed in wmed ports
Crustaca, Amphipoda				
Elasmopus pectenicrus	1982	Indian Ocean	F	Adr
Pycnogonida				
Ammothea hilgendorfi	1981	Indo Pacific	F	Adr
Mollusca				
Aeolidiella indica	1968	Pantropic	F	Wit, Malta
Anadara demiri	1972	Indian Ocean?	F	Adr, Gr
Anadara inaequivalvis	1969	Indo Pacific	F	It
Chlamys lischkei	1985	W Atlantic	F/B?	Sp, Sc
Conus fumigatus	1986	Red Sea	F/B?	Lby
Crepidula aculeata	1973	Pantropic	F	Sp
Cuthona perca	1977	Pantropic	F	Adr
Dendrostrea frons	1998	Indo Pacific	F	Trk
Melibe viridis	1970	Indo Pacific	B?	Sc, Tn, Adr, Gr
Musculista senhousia	1960	W Pacific	F	F, It,
Petricola pholadiformis	1994	W Atlantic	F/B?	Gr
Polycera hedgpethi	1986	NE Pacific	F	wIt
Polycerella emertoni	1964	Atlantic	F	wIt, Gr, Malta
Rapana venosa	1974	NW Pacific	B/Ac	wIt, Adr, Gr
Saccostrea cucullata	1999	Indo Pacific	F	Trk, Lvt
Spondylus cf. *multisetosus*	1992	Indo Pacific	F	Trk
Thais sacellum	2000	Indian Ocean	F	Lvt
Xenostrobus securis	1992	Pacific	F/Ac	Fr, Adr

Taxa	Date	Origin	Vector	Distribution in the Mediterranean
Echinodermata				
Asterias rubens	1993	NE Atlantic	F/B/?	Trk
Tunicata				
Botrylloides violaceus	1993	Indo Pacific	F	Adr
Microcosmus squamiger	1963	Australia	F	It, Tn, Lvt
Polyandrocarpa zorritensis	1974	E Pacific	F	Sp, It
Fishes				
Abudefduf vaigiensis	1959	Indo-Pacific	F	wIt, Lvt
Omobranchus punctatus	2003	Indo-Pacific	F/B	Lvt
Pinguipes brasilianus	1990	W Atlantic	B?	wIt, Sc

5.2 IMPACT OF SHIPPING TRANSPORTED ALIENS ON THE MEDITERRANEAN BIOTA

Alien biota may have ecological, economic and human health impacts. The latter, naturally, draw the most attention: the introduction of the toxic dinoflagellate *Alexandrium catanella* to Thau Lagoon, France, is of concern for shell fish mariculture, as blooms may render the molluscs unfit for human consumption on account of paralytic shellfish poisoning toxins (Lilly *et al.* 2002); while the transport of *Vibrio cholerae* in ballast water may pose a serious health hazard (see above). Though little is known of the impacts of even the many high-abundance invaders, yet their presence cannot fail to have impacted on the native biota: it was noted that native species are outcompeted wholly or partially displaced by the invaders. Though there is scarce documentation of direct competition between shipping-transported and indigenous species, there are many instances of sudden changes in abundance; competition is one explanation.

The presence of the alien filamentous, turf-forming rhodophytes *Acrothamnion preissii* and *Womersleyella setacea* reduces species number and diversity in the affected area by trapping sediments which prevent the development of other species (Piazzi & Cinelli 2001). The latter species may establish an almost monospecific stratum suffocating the underlying coralligenous communities (Boudouresque 1994). In the Tuscan Archipelago *A. preissii* often overgrows rocks, other macrophytes and seagrasses, overwhelming *Posidonia oceanica* rhizomes. The toxic secondary metabolites produced by another alien macrophyte, *Asparagopsis armata*, induce grazers such as the sea urchin *Paracentrotus lividus* and the sea bream *Sarpa salpa*, to shun it, and frees it to form dense stands in the northwestern Mediterranean (Ribera & Boudouresque 1995).

Caulerpa cylindracea, misidentified earlier as a particularly invasive variety of *C. racemosa*, is a recently introduced invasive chlorophyte algae endemic to south-west Australia. It has spread rapidly throughout the Mediterranean, from Cyprus and Turkey to Spain, and all the larger islands. Its occurrence in the Mediterranean close to harbours "provides compelling evidence in favour of its secondary dispersal via shipping" (Verlaque *et al.* 2003). *Caulerpa cylindracea* colonises both hard and soft bottoms to depth of 60m, grows rapidly (up to 2cm/d), forms a dense canopy that overgrows native algae, and significantly decreases diversity and cover of native macrophytes (Piazzi *et al.* 2001). The invasive alga has replaced, between 1992 and 1997, the native sea grass meadows of *Posidonia oceanica* in Moni Bay,

Cyprus, and prompted significant change in the benthic macrofauna: the abundance of gastropods and crustaceans decreased, whereas that of polychaetes, bivalves and echinoderms increased (Argyrou *et al.* 1999).

The encrusting colonies of the South American hermatypic coral *Oculina patagonica* are widely spread along the Mediterranean coast of Spain, Ligurian coast of Italy and the southern Levantine coast (Zibrowius 2002). With its high growth rate, endurance, and ability to reproduce both sexually and asexually, it inevitably forms substantial patches both in natural habitats and in polluted and disturbed sites, displacing the native mytillids.

The calcareous tube worm *Ficopomatus enigmaticus* builds reefs in western Mediterranean lagoons, altering the environment by forming in the Albufera de Menorca, Balearic Islands, for instance, a continuous layer up to 3m thick, encrusting molluscs and small rocks (Fornos *et al.* 1997).

The veined rapa whelk, *Rapana venosa*, native to the Sea of Japan, was first recorded in 1947 from the oil-exporting port of Novorossiysk in the Black Sea. It later spread to the Aegean and Adriatic Seas, possibly mediated by ballast transport of larvae. In the Black Sea *R. venosa* has expanded rapidly and nearly eliminated the commercially valuable *Mytillus galloprovincialis*, but no damage has been observed to mussel or clam beds in the Mediterranean (Occhipinti-Ambrogi 2002). The bivalve *Anadara inaequivalvis* has replaced the native olive cockle, *Cerastoderma glaucum*, on soft bottoms in the northern Adriatic lagoons (Occhipinti-Ambrogi 2000). The xanthid crabs *Dyspanopaeus sayi* and *Rhithropanopeus harrisii* too are well established in the Adriatic lagoons, the former species is at present the most common crab there, far exceeding in abundance the native xanthids (Mizzan 1995). The native bryozoan populations in the invasion-prone lagoon of Venice suffered severe decline, some species to the point of disappearance, following the spread of the Indo-Pacific bryozoan *Tricellaria inopinata* in the late 1980s, when it dominated the bryozoan fauna (Occhipinti Ambrogi 2000).

5.3 POLICY AND MANAGEMENT OF SHIPPING-TRANSPORTED BIOTA

Shipping has universally been regarded as the single largest vector for the movement of aquatic alien species. Though modern anti-fouling technology has decreased the chances of dispersal of fouling species, and rapid port turn-around times and fast ships are less conducive to settlement and survival of transported biota, recent inventories of alien biota in the Mediterranean Sea have shown a growing trend in the number of ship-transported species, increasing concerns among researchers and policy and management personnel dealing with anthropogenic changes in the marine environment.

International conferences on the environment in the past decade, including the United Nations Conference on Sustainable development in 1992, the Conference of Parties to the Convention on Biological Diversity, the World Summit on Sustainable development in 2002 and the 5th World Congress on Protected Areas in 2003, have invariably highlighted the issue of invasive species, and called upon governments to act 'to prevent the introduction of, control or eradicate those alien species which threaten ecosystems, habitats or species' (CBD 1992, Article 8 h).

Since ballast-mediated bioinvasions into freshwater, estuarine and marine habitats have caused significant economic losses in the past two decades, the International Maritime Organization (IMO) and the shipping industry have concentrated their attention on ways to address that issue. In February 2004 the new 'International Convention on the Control and Management of Ship's Ballast Water and Sediments' was adopted by a Diplomatic Conference (IMO 2004). This Convention, a significant environmental achievement, provides a uniform international instrument to regulate ballast water management, though to be effective, the parties to the

Convention have to implement it through appropriate national legislation and enforcement. Like an earlier IMO resolution (A.868(20), November 1997) it relies on Ballast Water Exchange (BWE, the replacement of coastal water with open ocean water) to reduce the risk of inoculation. A review of recent studies of biota entrained in ballast sediments following BWE raised questions as to the reliability of the procedure as an effective control measure, as fine particles were detected in ballast tanks even after a million-fold dilution of the ballast water (Galil & Hülsmann 2002; Forsberg *et al.* 2005), though it is effective in reducing waterborne organisms (Ruiz *et al.* 2004).

Hull fouling, an important vector in the Mediterranean for the dispersal of both macrophytes and invertebrates, was held in check since the 1970s by the widespread use of biocidal antifouling paints. However, the adoption of an IMO Convention prohibiting the application of tributyltin (TBT)-based antifouling paints as of January 2003, may lead to an increase in fouled hulls, and consequently, hull-transferred biota (Mineur *et al.* 2004).

The Barcelona Convention 1976 and its relevant protocols, initially aimed at reducing pollution (see above), have been updated with the adoption of new protocols. The Protocol concerning Specially Protected Areas (SPA), that had been adopted in 1982 and came into force in 1986, prohibits "the introduction of exotic species" (Article 7 e). In 2003 the Mediterranean Action Plan (MAP), United Nations Environment Programme (UNEP 2003), drafted an "Action Plan concerning species introductions and invasive species in the Mediterranean Sea". Both versions of the disputed Article 7 recognise that shipping is a major vector of introduction into the Mediterranean Sea. Article 22 of the Action Plan strongly recommended that "Given the importance of shipping-mediated introductions of non-indigenous species in to the Mediterranean ... a regional project be developed to overcome gaps for the Mediterranean countries, and strengthen the capacities of the countries to reduce the transfer of aquatic organisms via ships' ballast water and sediments and hull fouling".

6. Shipping-derived antifouling biocides in the Mediterranean Sea

6.1 THE DISTRIBUTION AND ACCUMULATION OF BIOCIDAL ANTIFOULANTS IN THE MEDITERRANEAN SEA

It is universally acknowledged that antifouling paints are the most important contributors of organotin compounds to the marine environment. Organotin-based antifouling paints were introduced in the mid 1960s and, as they proved to be highly effective biocides, their use increased unchecked for two decades. The deleterious impacts of organotin contamination were first noticed in the late 1970s when reproductive failure and shell deformations affected shellfish farms on the Atlantic coast of France (Alzieu *et al.* 1980). Since then, TBT and its degradation products, mono- (MBT) and di- butyltin (DBT), and triphenyltin (TPT), were recognised as the most toxic materials intentionally introduced into the sea, and confirmed as harming a wide range of organisms: their ecotoxicological impacts have been amply documented. Studies revealed that organotin compounds degrade slowly: TBT half-life in shelf seawater is estimated at 1-3 weeks (Seligman *et al.* 1986, 1988), and in the sediment at 1-5 years (Adelman *et al.* 1990), and is predicated on microbial degradation, UV photolysis and temperature. However, it is possible that the half-life for TBT in the open sea is considerably longer, at least for the oligotrophic waters of the Mediterranean, with their low kinetic biodegradation (Michel & Averty 1999). The prevalence and persistence of organotin compounds in the marine environment, and the damage they cause, are of grave concern to scientists, policy and management personnel.

The first coordinated survey in the Mediterranean Sea of TBT and its degradation derivatives was conducted in 1988. One hundred and thirteen water samples were collected along the French Mediterranean coast, the Tyrrhenian coast of Italy, the southern coast of Turkey, and 35 sediment samples were taken off Alexandria, Egypt. At most sites examined the concentrations of TBT in seawater exceeded 20ng/l. The harbours of Mersin, Turkey (936ng/l), and Livorno, Italy (810ng/l) displayed the highest level in contamination among the sampled harbours, but TBT levels inside recreational marinas generally exceeded contamination levels at commercial shipping ports, with particularly high levels at Cecina and Punta Ala (Italy) and the old port of Marseille (France) (3930, 960 and 736 ng/l respectively). All the sediment samples from Alexandria contained TBT; highest concentrations were detected in the western and eastern harbours and in the Bay of Abu Kir (975, 260 and 252 ng/l respectively) (Gabrielides *et al.* 1990). Subsurface water samples were taken that same year at several additional locations in the western Mediterranean: the Ebro delta, the port of Barcelona and El Masnou marina, on the Spanish Mediterranean coast, along the Midi coast and along the French and Italian rivieras. Substantial contamination was reported for the entire region with elevated levels of TBT in all samples, with highest records at Toulon harbour, and at the Beaulieu and San Remo marinas (Alzieu *et al.* 1991).

Regulations concerning the use of organotin-based antifouling paints in the Mediterranean Sea were introduced in 1991 (see below). Yet, organotin compounds were detected in all subsurface water samples taken in 1995 from ports and marinas along the Côte d'Azur, with high concentrations persisting in the ports of Antibes, Golfe Juan, Cannes and Nice (459, 348, 142 and 138 ng/l respectively), though levels in recreational marinas were substantially less than those recorded in 1988. Measurable levels of TBT and DBT were noted in samples taken from the bathing beaches of Eze, Nice, Cannes and Villefranche (<0.6-5.2ng/l) (Tolosa *et al.* 1996). Subsurface water samples taken in 1999 at 14 sites along the Corsican coast proved that contamination levels at the commercial harbours of Bastia, Porto-Vecchio and Ajaccio (200, 169 and 88 ng/l respectively), as well as in the marinas of Ajaccio, Porto-Vecchio and Propriano (189, 169 and 161 ng/l respectively) were "quite excessive" (Michel *et al.* 2001). More discouraging was the presence of contamination in the immediate vicinity of Scandola nature reserve (7.2ng/l), and in the Lavezzi Islands nature reserve (2.0ng/l), far from maritime shipping, when TBT concentrations of 1-2ng/l have been shown to induce deleterious effects (Alzieu, 2000). Sediments from harbours and marinas along the Catalan and Alboran seas were sampled in 1995 and 1999-2000, respectively. The highest TBT concentrations were associated with large vessel input like the Barcelona commercial harbour (maximum 18722, average 4487 ng/g dry wt), and Almería commercial harbour (2135ng/g dry wt), though high values were noted also in fishing and recreational ports such as the harbour of Sant Carles (maximum 5226, average 1617 ng/g dry wt), and the Sotogrande recreational marina (3868ng/g dry wt) (Díez *et al.* 2002). High TBT concentrations, in excess of 10,000ng/g dry wt, were also found also in sediment sample taken from Piraeus Harbour, Greece (Tselentis *et al.* 1999). None of the 14 locations sampled along the Israeli coast in 2003 were free from contamination, but the highest concentrations of TBT were recorded in the sediments of the commercial harbours of Haifa and Ashdod (770 and 730 ng/g dry wt, respectively); high levels of contamination (>100ng/l) were detected in the waters of four recreational marinas in addition to Haifa harbour (Herut *et al.* 2004). However, organotin contamination is not limited to port and port-proximate environments. Samples collected in 1998 in the northwestern Mediterranean along vertical profiles offshore, between 25m and 2500m depth, have shown that contamination of surface waters was as high as 0.47 ng/l 20 km offshore, and 0.08ng/l midway between Toulon and Corsica; contamination of abyssal water reached a maximum of 0.04ng/l at 1200m (Michel & Averty 1999). TBT compounds may reach great depth possibly with winter cooling and descent of the surface mass, or with chipped and discarded paint fragments (Galil *et al.* 1995). Deep sea fishes collected at depths of 1000-1800m in the Gulf of Lion carried as much as 175ng/g wet wt total butyltin residues in their tissues, comparable with contamination levels in coastal fish collected along the Catalan coast, attesting to exposure of deep sea biota to TBT (Borghi & Porte 2002).

With the restrictions on the use of organotin-based compounds in antifouling paints (see below), they were replaced by a number of alternative biocidal treatments, mostly copper and zinc compounds combined with organic booster herbicides such as the triazine compound Irgasol (Evans *et al.* 2000a). Few recent data are available regarding the spread and accumulation of the alternative biocides in the Mediterranean. Substantial levels of Irgasol 1051 were present in water samples collected in 1992 and again in 1995 along the French Riviera, with higher concentrations recorded from recreational marinas (Fontvielle 1700ng/l, St Laurent 640ng/l) than in the harbour of Antibes (264ng/l in 1995) (Readman *et al.* 1993; Tolosa *et al.* 1996), confirming its use primarily on small boats at the time. Irgasol 1051 was the main pollutant (along with another herbicide, Diuron) among the recently introduced antifouling pesticides detected in recreational marinas, fishing ports and harbours along the Mediterranean coast of Spain between 1996 and 2000, with concentrations as high as 330ng/l (Martinez *et al.* 2001). Similarly, the presence of Irgasol 1051 and two other 'booster bio-cides' was recently confirmed in sediments collected from Greek harbours and recreational marinas, with concentration being highest (690ng/g dry wt) in marinas (Albanis *et al.* 2002).

6.2 IMPACT OF ANTIFOULING BIOCIDES ON THE MEDITERRANEAN BIOTA

TBT is "... probably the most toxic substance ever introduced deliberately into the marine environment" (Mee & Fowler, 1991). An effective long-action antifoulant, TBT also has an impact on non-target biota, especially in harbours and marinas with high vessel density and restricted water circulation. Marine molluscs are notably sensitive to the substance with well-documented sublethal impacts such as the superimposition of male sex characters in female gonochoristic prosobranch gastropods (imposex) at TBT concentrations as low as 1ng/l (Smith 1981), and shell malformation and reproductive failure in bivalves at concentrations of 20ng/l (Alzieu 2000). As the severity of imposex characteristics in a population has been correlated with concentrations of TBT in the environment, it has served as a widely used and sensitive indicator for monitoring TBT contamination, though it has been shown that TBT is not the sole causative agent of imposex (Evans *et al.* 2000b).

The first study in the Mediterranean to relate levels of TBT in the sediment to vas deferens and penis development in females of the common muricid gastropod *Hexaplex trunculus* was conducted in Malta in 1992 (Axiak *et al.* 1995). All female gastropods collected near major recreational marinas and within the commercial harbours of Marsamxett, Rinella, Marsaxlokk and Marsascala were impacted, and the severity of the phenomenon was correlated with the levels of organotins in their digestive glands and the gonads, and the amounts of TBT in the superficial sediments. Most females in the highly contaminated harbours exhibited split capsule glands, and might have been sterile. Females of *Hexaplex trunculus* sampled at 15 yachting, fishing and commercial harbours along the Italian coast in 1995-96, exhibited nearly 100% sterility in all but Linosa and Lampedusa islands where yachting activity was limited to the summer months, though even in heavily impacted populations no evidence was found of decrease in abundance (Terlizzi *et al.* 1998). Very high levels of imposex were found in the harbour of Napoli (66.7% sterile females, Relative Penis Size Index 77.2, Vas Deferens Sequence Index 4.8). All female *H. trunculus* collected in the canal connecting the lagoon of Bizerte, Tunisia, to the sea showed external male characteristics (Lahbib *et al.* 2004). Another common muricid, *Bolinus brandaris*, has been used in monitoring TBT along the Catalan coast. At five of the six locations sampled in 1996-97 imposex affected all the female specimens collected (Solé *et al.* 1998), whereas nearly all the females in the samples collected between 1996 and 2000 displayed advanced imposex characteristics, though population dynamics were unaffected (Ramón & Amor 2001). TBT and its degradation products accumulate within tissues of marine organisms and move up the food chain. Very high concentrations have been found in top predators such as the bottlenose dolphin, bluefin tuna and blue shark collected off the Italy, with total butyltin in dolphin's liver tissues reaching 1200-2200 ng/g wet wt (Kannan *et al.* 1996).

There is little published data on the toxicity and possible environmental impacts of many of the alternative biocidal compounds used in antifoulants. Irgasol 1051 inhibits photosynthetic electron transport in chloroplasts, and though not harming marine organisms it affects non-target algae. It is feared that if accumulated in high enough levels, it may damage periphyton, algae and seagrasses, and thus affect primary productivity (Thomas *et al.* 2001).

6.3 POLICY AND MANAGEMENT OF TBT IN THE MEDITERRANEAN SEA

France pioneered regulations restricting the use of organotin-based antifoulants: as early as 1982 the use of organotin paint on boats longer than 25m was prohibited (with exemption for aluminium hulls). The legislation reduced contamination within shellfish culture areas on the French Atlantic coast, but "the efficacy of the legislation does not extend to the Mediterranean coast" (Alzieu *et al.* 1991, see also Michel & Averty 1999).

The Mediterranean countries were the first to take action to restrict the use of organotins on a region-wide basis. The Protocol of the Barcelona Convention for the Protection of the Mediterranean Sea against Pollution from Land-Based Sources, signed in 1980, listed organotin compounds (Annex I, A.5) among substances for which legal measures should be proposed and adopted. UNEP's Mediterranean Action Plan (MAP), with the cooperation of international agencies, conducted a pilot study of organotin contamination in 1988, which recorded "high and potentially toxic concentrations of TBT ... in the vicinity of harbours and marinas" (Gabrielides *et al.* 1990). These data led to the adoption, in 1989, by the Contracting Parties to Barcelona Convention, of measures limiting the use of TBT antifouling paints in the Mediterranean that entered into effect 1991, including a ban on organotin-based antifouling paints "On hulls of boats having an overall length... of less than 25m". A recommendation was made that "a code of practice be developed in minimising the contamination of the marine environment in the vicinity of boat-yards, dry docks, etc., where ships are cleaned of old anti-fouling paint and subsequently repainted" (UNEP 1989). In 1990 the International Maritime Organization (IMO) adopted a resolution recommending governments to adopt measures to eliminate anti-fouling paints containing TBT (MEPC 46(30). However, despite regulations, concentrations of TBT in sediments and water in ports, coastal regions and offshore, have failed to decline, and in some cases have increased in the 1990s. In 2001 the IMO, long concerned with the effects of organotin compounds on the marine environment, adopted the International Convention on the control of harmful anti-fouling systems on ships that calls for a global prohibition of the application of organotin compounds. Annex I attached to the Convention and adopted by the Diplomatic Conference states that by 1.1.2003 all ships shall not apply or re-apply organotin compounds, and that by 1.1.2008 ship hulls will either be free from organotin compounds, or those will be coated over to prevent leaching (IMO 2001).

7. Coda

The Mediterranean seemed "Infinite and of unmeasured depth" (Oppian 1987) for millennia, and was treated as limitless and immutable until the late 20th century. Only then have the nations clustered around its shores woken to the sea's parlous state and have adopted the

'Barcelona Convention' and its protocols, some dealing with shipping-related pollution. The past decade saw the adoption of several important global and regional initiatives to protect the oceans from shipping-related damage, and some laudable successes. Yet, the accelerating globalisation and greater economical interdependence between distant markets mean an increase in the number of vessels plying the sea. The absence of Mediterranean-wide mechanism for detection, enforcement and prosecution of offenders will result in continuing degradation of the environment.

One wishes that the words of another ancient poet would guide the shipping industry, policy decisions, and management programmes: *They that go down to the sea in ships, that do business in great waters; these see the works of the Lord, and his wonders in the deep.* Psalms 107:22, 23.

Acknowledgments
I am grateful to Elisabeth Apt, the IOLR librarian, for bibliographical help, and to D. C. Moore for kindly remarks on an earlier version of the manuscript.

References

Adelman D, Hinga KR, Pilson MEQ. 1990; Biogeochemistry of butyltins in an enclosed marine ecosystem. Environmental Science and Technology 24: 1027-1032.

Albaiges J, Borbon J, Ros J. 1979; Source identification of tar balls from the western Mediterranean. Workshop on pollution of the Mediterranean, CIESM & UNEP, pp. 103-109.

Albanis TA, Lambropoulou DA, Sakkas VA, Konstantinou IK. 2002; Antifouling paint booster biocide contamination in Greek marine sediments. Chemosphere 48(5): 475-485.

Aliani S, Griffa A, Molcard A, 2003; Floating debris in the Ligurian Sea, north-western Mediterranean. Marine Pollution Bulletin 46(9): 1142-1149.

Aliani S, Molcard A. 2003; Hitch-hiking on floating marine debris: macrobenthic species in the Western Mediterranean sea. Hydrobiologia 503: 59-67.

Alzieu C. 2000; Impact of tributyltin on marine invertebrates. Ecotoxicology 9 (1-2): 71-76.

Alzieu C, Michel P, Tolosa I, Bacci E, Mee LD, Readman JW. 1991; Organotin compounds in the Mediterranean: continuing cause for concern. Marine Environmental Research 32: 261-270.

Alzieu C, Thibaud Y, Heral M, Boutier B. 1980; Evaluation des risques des à l'emploi des peintures antisalissures dans les zones conchylicolis. Revue des Travaux de l'Institut des Peches maritimes 44: 301-348.

Andrew RK, Howe BM, Mercer JA. 2002; Ocean ambient sound: comparing the 1960s with the 1990s for a receiver off the California coast. Acoustic Research Letters Online 3: 65-70.

Angradi AM, Consiglio C, Marini L, 1993; Behaviour of striped dolphins (*Stenella coeruleoalba*) in the central Tyrrhenian Sea (Mediterranean Sea) in relation to commercial ships. Proceedings of the 7th annual conference of the European Cetacean Society 1993: 77-79.

Argyrou M, Demetropoulos A, Hadjichristophorou M. 1999; Expansion of the macroalga *Caulerpa racemosa* and changes in soft bottom macrofaunal assemblages in Moni Bay, Cyprus. Oceanologica Acta 22 (5): 517-528.

Atta MM. 1991; The occurrence of *Paradella dianae* (Menzies, 1962) (Isopoda, Flabellifera, Sphaeromatidae) in Mediterranean waters of Alexandria. Crustaceana 60(2): 213-217.

Axiak V, Vella AJ, Micallef D, Chircop P, Mintoff, B. 1995. Imposex in *Hexaplex trunculus* (Gastropoda: Muricidae): first results from biomonitoring of tributyltin contamination in the Mediterranean. Marine Biology 121: 685-691.

Aybulatov NA, Nemirovskaya IA, Nesterova MP. 1981; Oil pollution of the North African shelf of the Mediterranean Sea. Oceanology 21(5): 589-592.

Bartol SM, Musick JA, Lenhardt ML, 1999. Auditory evoked potentials of the loggerhead sea turtle (*Caretta caretta*). Copeia 1999(3): 836-840.

Benzhitskiy AG, Polikarpov GG. 1976. Distribution of petroleum aggregates in the hypneustal zone of the Mediterranean Sea in April-June 1974. Oceanology 16: 45-47.

Bingel F, Avsar D, Unsal M. 1987; A note on plastic materials in trawl catches in the north eastern Mediterranean. Meeresforschung 31: 227-233.

Borghi V, Porte C. 2002; Organotin pollution in deep-sea fish from the northwestern Mediterranean. Environmental Science and Technology 36: 4224-4228.

Boudouresque CF. 1994; Les espèces introduites dans les eaux côtières d'Europe et de Méditerranée: Etat de la question et consequences. In: Boudouresque CF, Briand F, Nolan C (eds) Introduced species in European coastal waters. CEC Ecosystem Research Report 8: 8-27.

Boudouresque CF, Ribera MA. 1994; Les introductions d'espèces végétales et animales en milieu marin – conséquences écologiques et économiques et problèmes législatifs. In: Boudouresque CF, Meinesz A & Gravez V (eds) First International workshop on *Caulerpa taxifolia,* 29-102. GIS Posidonie Marseille.

Boudouresque CF, Verlaque M. 2002; Assessing scale and impact of ship-transported alien macrophytes in the Mediterranean Sea. Alien marine organisms introduced by ships in the Mediterranean and Black Seas. CIESM Workshop Monographs, 20: 53-62.

Bowman D, Mano-Samsonov N, Golik A. 1998; Dynamics of litter pollution on Israeli Mediterranean beaches: a budgetary, litter flux approach. Journal of Coastal Research 14 (2): 418-432.

Browning LJ, Harland EJ, 1999; Are bottlenose dolphins disturbed by fast ferries? Proceedings of the 13th annual conference of the European Cetacean Society 1999: 92-98.

Brunetti R, Mastrototaro F. 2004; The non-indigenous stolidobranch ascidian *Polyandrocarpa zorritensis* in the Mediterranean: description, larval morphology and pattern of vascular budding. Zootaxa 528: 1-8.

Carlton, JT., 1985. Transoceanic and interoceanic dispersal of coastal marine organisms: the biology of ballast water. Oceanography and Marine Biology annual Review 23: 313-371.

Carr A. 1987; Impact of nondegradable marine debris on the ecology and survival outlook of sea turtles. Marine Pollution Bulletin 18 6(B): 352-357.

CBD 1992; The Convention on Biological Diversity, UN Conference on Environment and Development, Rio de Janeiro. www.biodiv.org (accessed 04-04-2005).

CIESM 2004; Commission Internationale pour l'Exploration Scientifique de la Mer Méditerranée 2004: www.ciesm.org/atlas (accessed 04-04-2005).

Cormaci M, Furnari G, Giaccone G, Serio D. 2004; Alien macrophytes in the Mediterranean Sea: a review. Recent Research Developments in Environmental Biology 1(1):1-202.

Curtis KR, Howe BM, Mercer JA, 1999; Low-frequency ambient sound in the North Pacific: long time series observations. Journal of the Acoustical Society of America 106: 3189-3200.

Danovaro R, Fabiano M, Vincx M. 1995; Meiofauna response to the *Agip Abruzzo* oil spill in subtidal sediments of the Ligurian Sea. Marine Pollution Bulletin 30(2): 133-145.

De Armas JD. 1985; Pelagic tar in the western Mediterranean 1981-82. Workshop on pollution of the Mediterranean, 555-559. CIESM & UNEP.

De Walle FB, Lomme JJ, Nikolopoulou-Tamvakli M. 1993; General overview of the environmental quality of the Mediterranean Sea. In: De Walle FB, Nikolopoulou-Tamvakli M & Heinen WJ (eds) Environmental condition of the Mediterranean Sea, 34-179. Kluwer, Dordrecht.

Derraik JGB. 2002; The pollution of the marine environment by plastic debris: a review. Marine Pollution Bulletin 44: 842-852.

Díaz Lopez B, Mussi B, Miragliuolo A, Chiota D, Valerio L. 2000; Respiration patterns of fin whales (*Balaenoptera physalus*) off Ischia island (southern Tyrrhenian Sea, Italy). Proceedings of the 14th annual conference of the European Cetacean Society 2000: 125-129.

Díez S, Ábalos M, Bayona JM. 2002; Organotin contamination in sediments from the western Mediterranean enclosures following 10 years of TBT regulation. Water Research 36: 905-918.

Drake LA. 2002; Ship-transported virio- and bacterio- plankton. In: Alien marine organisms introduced by ships in the Mediterranean and Black Seas. CIESM Workshop Monographs 20: 35-39. www.ciesm.org/publications/ (accessed 04-04-2005).

Duguy R, Moriniere P, Le Milinaire C. 1998; Faceurs de mortalité observés chez les tortues marines dans le golfe de Gascogne. Oceanologica Acta 21(2): 383-388.

EC, European Commission. 1997; The EMARC Project. MARPOL rules and ship generated waste. European Commission. Directorate-General for Transport Directorate Development of Transport Policy; Research and Development VV-E. Project Funded by the European Commission under the Transport RTD Programme of the 4th Framework Programme. Project WA-95-SC.097.

EC, European Commission. 2002; Marine oil pollution. Technologies and methodologies for detection and early warning. European Commission. ISPRA, Italy, EUR 20231.

EIA. 2004; Energy Information Administration www.eia.doe.gov (accessed 04-04-2005).

El Hehyawi ML. 1979; New data on the distribution of pollutants and their effect on some hypneuston constituents in the S.E. Mediterranean. Workshop on pollution of the Mediterranean, CIESM & UNEP, 301-305.

Evans SM, Birchenough AC, Brancato MS. 2000a; The TBT ban: out of the frying pan into the fire? Marine Pollution Bulletin 40(3): 204-211.

Evans SM, Kerrigan E, Palmer N. 2000b; Causes of imposex in the dogwhelk *Nucella lapillus* (L.) and its use as a biological indicator of tributyltin contamination. Marine Pollution Bulletin 40(3): 212-219.

Fornos JJ, Forteza V, Martinez-Taberner. 1997; Modern polychaete reefs in western Mediterranean lagoons: *Ficopomatus enigmaticus* (Fauvel) in the Albufera de Menorca, Balearic Islands. Palaeogeography, Palaeoclimatology, Palaeoecology 128 (1-4): 175-186.

Forsberg R, Baier R, Meyer A, Doblin M, Strom M. 2005; Fine particle persistence in ballast water sediments and ballast tank biofilms. The Adhesion Society Meeting, Mobile Alabama.

Frantzis A. 1998; Does acoustic testing strand whales? Nature 392: 29.

Gabrielides GP, Alzieu C, Readman JW, Bacci E, Aboul Dahab O, Salighoglu I. 1990; MED POL survey of organotins in the Mediterranean. Marine Pollution Bulletin 21 (5): 233-237.

Gabrielides GP, Golik A, Loizides L, Marino MG, Bingel F, Torregrossa MV. 1991; Man-made garbage pollution on the Mediterranean coastline. Marine Pollution Bulletin 23: 437-441.

Galgani F, Jaunet S, Campillo A, Guenegan X, His E. 1995; Distribution and abundance of debris on the continental shelf of the north-western Mediterranean Sea. Marine Pollution Bulletin 30 (11): 713-717.

Galgani F, Leaute JP, Moguedet P, Souplet A, Verin Y, Carpentier A, Goraguer H, Latrouite D, Andral B, Cadiou Y, Mahe JC, Poulard JC, Nerisson P. 2000; litter on the sea floor Along European coasts. . Marine Pollution Bulletin 40 (6): 516-527.

Galgani F, Souplet A, Cadiou Y. 1996; Accumulation of debris on the deep sea floor off the French Mediterranean coasy. Marine Ecology Progress Series 142: 225-234.

Galil BS, Golik A, Türkay M. 1995; Litter at the bottom of the sea: a sea bed survey in the Eastern Mediterranean. Marine Pollution Bulletin 30 (1): 22-24.

Galil BS, Hülsmann N. 1997; Protist transport via ballast water – biological classification of ballast tanks by food web interactions. European Journal of Protistology 33: 244-253.

Galil BS, Hülsmann N. 2002; The biological efficacy of Open Ocean Exchange – implications for Ballast Water Management. In: Leppäkoski E, Gollasch S, Olenin S (eds) Invasive aquatic species of Europe. Distribution, Impacts and Management, 508-510. Kluwer, Dordrecht.

Golani D. 2004; First record of the muzzled blenny (Osteichthyes: Blenniidae: *Omobranchus punctatus*) from the Mediterranean, with remarks on ship-mediated fish introduction. Journal of the Marine Biological Association of the United Kingdom 84: 851-852.

Golik A. 1982; The distribution and behaviour of tar balls along the Israeli coast. Estuarine, Coastal and Shelf Science 15: 267-276.

Golik A. 1997; Debris in the Mediterranean Sea: types, quantities and behavior. In: Coe JM & Rogers DB (eds.) Marine debris: sources, impacts, and solutions, 7-14. Springer series on environmental management.

Golik A, Gertner Y. 1992; Litter on the Israeli coastline. Marine Environmental Research 33: 1-5.

Golik A, Weber K, Salihoglu I, Yilmaz K, Loizides L. 1988a; Pelagic tar in the Mediterranean Sea. Marine Pollution Bulletin 19 (11): 567-572.

Golik A, Weber K, Salihoglu I, Yilmaz K, Loizides L. 1988b. Decline in tar pollution in the Mediterranean Sea. Rapports et Proces-Verbaux des Reunions Commission Internationale pour l'Exploration de la Mer Méditerranée 31:164.

Gómez O, Martinez-Iglesias JC. 1990; Reciente hallazgo de la especie indopacifica *Charybdis helleri* (A.Milne Edwards, 1867) (Crustacea: Decapoda: Portunidae) en aguas cubanas. Caribbean Journal of Science 26 (1-2): 70-72.

Gramentz D. 1988. Involvement of loggerhead turtle with the plastic, metal and hydrocarbon pollution in the Central Mediterranean. Marine Pollution Bulletin 19 (1): 11-13.

Gregory MR. 1978. Accumulation and distribution of virgin plastic granules on New Zealand beaches. New Zealand Journal of Marine and Freshwater Research 12: 399-414.

Guidetti P, Modena M, La Mesa G, Vacchi M. 2000; Composition, abundance and stratification of macrobenthos in the marine area impacted by tar aggregates derived from the Haven oil spill (Ligurian Sea, Italy). Marine Pollution Bulletin 40 (12): 1161-1166.

Herut B, Shefer E, Cohen Y. 2004; Quality of coastal waters along the Mediterranean coast of Israel, 2003. IOLR Report H23/2004.

Horn MH, Teal JM, Backus RH. 1970; Petroleum lumps on the surface of the sea. Science 168: 245-246.

Horsman PV. 1982; The amount of garbage pollution from merchant ships. Marine Pollution Bulletin 13 (5): 167-169.

IMO. 1982; International Maritime Organization 1982; Inter-governmental Conference on the Convention on the Dumping of Wastes at Sea, London, 30 October – 13 November, 1972. Final Act of the Conference with attachments including the Convention for the Prevention of Marine Pollution by Dumping of Wastes and Other Matter. (1982 Edition) IMO, London.

IMO 2001; International Maritime Organisation 2001. International Convention on the Control of Harmful Anti-fouling Systems on Ships. http://www.imo.org/Conventions/mainframe.asp?topic_id=529 (accessed 04-04-2005).

IMO. 2002; International Maritime Organisation 2002. Guidelines for the designation of special areas under Marpol 73/78 and guidelines for the identification and designation of particularly sensitive sea areas. www.imo.org/Environment (accessed 16-05-05).

IMO. 2004; International Maritime Organisation 2004. International Convention on the Control and Management of Ship's Ballast Water and Sediments. www.globallast.imo.org (accessed 04-04-2005).

INTFISH 2002 Internet Guide to International Fisheries Law. www.intfish.net/treaties/sanctuary.htm (accessed 04-04-2005).

Kannan K, Corsolini S, Focardi S, Tanabe S, Tatsukawa R. 1996; Accumulation pattern of butyltin compounds in dolphin, tuna, and shark collected from Italian coastal waters. Archives of Environmental Contamination and Toxicology 31 (1); 19-33.

Knoepffler-Peguy M, Belsher T, Boudouresque CF, Lauret M. 1985; *Sargassum muticum* begins to invade the Mediterranean. Aquatic Botany 23(3): 291-295.

Kornilios S, Drakopoulos PG, Dounas C. 1998; Pelagic tar, Dissolved/dispersed petroleum hydrocarbons and plastic distribution in the Cretan Sea, Greece. Marine Pollution Bulletin 36 (12): 989-993.

Lahbib Y, Trigui El-Menif N, Le Pennec M, Bou Maiza M. 2004; Le phénomène d'imposex observe pour la première fois en Tunisie chez un mollusque gasteropode, *Murex trunculus*. Rapports et Proces-Verbaux des Reunions Commission Internationale pour l'Exploration de la Mer Méditerranée 37: 218.

Laist DW. 1997; Impacts of marine debris: entanglement of marine life in marine debris including a comprehensive list of species with entanglement and ingestion records. In: Coe JM & DB Rogers (eds.) Marine debris: sources, impacts, and solutions, 99-139. Springer series on environmental management.

Le Lourd P. 1977; Oil Pollution in the Mediterranean Sea. Ambio 6: 317-320.

Lilly EL, Kulis DM, Gentien P, Anderson DM. 2002; Paralytic shellfish toxins in France linked to a human-introduced strain of *Alexandrium catanella* from the western Pacific: evidence from DNA and toxin analysis. Journal of Plankton Research 24: 443-452.

Loizides L. 1994. Oil pollution of the sea waters and coastal areas in Cyprus. International Symposium on Pollution in the Mediterranean Sea. WTSAC-IAWQ, Cyprus. pp. 131-140.

Mantelatto FLM & Dias LL (1999) Extension of the known distribution of *Charybdis hellerii* (A. Milne-Edwards, 1867) (Decapoda, Portunidae) along the western tropical South Atlantic. Crustaceana 72 (6): 617- 620.

Martinez K, Ferrer I, Hernando MD, Fernandez-Alba AR, Marce RM, Borrull F, Barcelo D. 2001. Occurrence of antifouling biocides in the Spanish Mediterranean marine environment. Environmental Technology 22 (5): 543-552.

Masó M, Garcés E, Pagès F, Camp J. 2003; Drifting plastic debris as a potential vector for dispersing Harmful Algal Bloom (HAB) species. Scientia Marina 67 (1): 107-111.

McCarthy E, 2004; International regulation of underwater sound: establishing rules and standards to address ocean noise pollution. Kluwer Academic Publishers.

McCauley SJ, Bjorndal KA. 1999; Conservation implications of dietary dilution from debris ingestion: sublethal effects in post hatchling loggerhead sea turtles. Conservation Biology 13 (4): 925-929.

McCoy FW. 1988; Floating megalitter in the Eastern Mediterranean. Marine Pollution Bulletin 19 (1): 25-28.

Mee LD, Fowler SW. 1991; Editorial In: Roesijadi G, Spies RB, Widdows J. (eds) Special Issue of Organotin. Marine Environmental Research 32 (1-4): 1-5.

Meinesz A. 1992; Modes de disseminations de l'algue *Caulerpa taxifolia* introduite en Méditerranée. Rapports et Proces-Verbaux des Reunions Commission Internationale pour l'Exploration de la Mer Méditerranée 33: 44.

Michel P, Averty B. 1999; Distribution and fate of tributyltin in surface and deep waters of the northwestern Mediterranean. Environmental Science and Technology 33: 2524-2528.

Michel P, Averty B, Andral B, Chiffoleau JF, Galgani F. 2001; Tributyltin along the coasts of Corsica (Western Mediterranean): a persistent problem. Marine Pollution Bulletin 42 (11): 1128-1132.

Mienis HK. 2004; New data concerning the presence of Lessepsian and other Indo-Pacific migrants among the molluscs in the Mediterranean Sea with emphasize on the situation in Israel. In Öztük & Salman A. (eds): Proceedings 1st National Malacology Congress, Izmir. Turkish Journal of Aquatic Life 2 (2): 117-131.

Minchin D. 1996. Tar pellets and plastics as attachment surfaces for lepadid cirripeds in the North Atlantic ocean. Marine Pollution Bulletin 32 (12): 855-859.

Mineur F, Johnson M, Maggs C. 2004; Vectors for introduction of alien macroalgae in Europe: hull fouling. Abstracts of the 13th International Conference on Aquatic Invasive Species, 16.

Mizzan L. 1995; Notes on the presence and diffusion of *Dyspanopeus say* (Smith, 1869) (Crustacea, Decapoda, Xanthidae) in the Venitian lagoon. Bollettino del Museo civico di Storia naturale di Venezia 44 [1993]: 121-129.

Morris RJ. 1974; Lipid composition of surface films and zooplankton from the Eastern Mediterranean. Marine Pollution Bulletin 5(7): 105-109.

Morris RJ. 1980; Floating plastic debris in the Mediterranean. Marine Pollution Bulletin 11: 125.

NAS, National Academy of Sciences. 1975; Marine litter. In Assessing Potential Ocean Pollutants, 405-438. A report of the Study Panel on Assessing Potentail Ocean Pollutants to the Ocean Research Council, Washington DC.

NRC (National Research Council). 2000; Marine Mammals and Low-Frequency Sound. National Academy Press, Washington, D.C.

NRC (National Research Council). 2003; Ocean noise and marine mammals. National Academy Press, Washington, D.C.

Occhipinti Ambrogi A. 2000; Biotic invasions in a Mediterranean Lagoon. Biological Invasions 2: 165-176.

Occhipinti Ambrogi A. 2002; Susceptibility to invasion: assessing scale and impact of alien biota in the northern Adriatic. In: Alien marine organisms introduced by ships in the Mediterranean and Black Seas. CIESM Workshop Series 20: 69-73.

Oppian 1987. Halieutica. The Loeb Classical Library. 219: 200-515.

Oren OH. 1970; Tar pollutes the Levant Basin. Marine Pollution Bulletin 11: 149-150.

Pavlakis P, Tarchi D, Sieber AJ, Ferraro G, Vincent G. 2001; On the monitoring of illicit vessel discharges – a reconnaissance study in the Mediterranean Sea. ECDG Joint Research Centre, 20 pp.

Perez A, Canadas AM, Sagarminaga R, San Martin G. 2000. The effects of acoustic pollution on the cetaceans of the Alboran Sea (Spain). www.geocities.com (accessed 5-05-2005).

Piazzi L, Ceccherelli G, Cinelli F. 2001; Threat to macroalgal diversity: effects of the introduced green alga *Caulerpa racemosa* in the Mediterranean. Marine Ecology Progress Series 210: 149-150.

Piazzi L, Cinelli F. 2001; Distribution and dominance of two introduced turf-forming macroalgae on the coast of Tuscany, Italy, Northwestern Mediterranean Sea in relation to differenct habitats and sedimentation. Botanica Marina 44 (5): 509-520.

Pietrapiana D, Modena M, Guidetti P, Falugi C, Vacchi M. 2002; Evaluating the genotoxic damage and hepatic tissue alterations in demersal fish species: a case study in the Ligurian Sea (NW-Mediterranean). Marine Pollution Bulletin 44: 238-243.

Plante-Cuny MR, Salen-Picard C, Grenz C, Plante R, Alliot E, Barranguet C. 1993; Experimental field study of the effects of crude oil, drill cuttings and natural biodeposits on microphyto- and macrozoobenthic communities in the Mediterranean area. Marine Biology 117: 355-366.

Pruter AT. 1987; Sources, quantities and distribution of persistent plastics in the marine environment. Marine Pollution Bulletin 18 (6B): 305-310.

Ramón M, Amor MJ. 2001; Increasing imposex in populations of *Bolinus brandais* (Gastropoda: Muricidae) in the northwestern Mediterranean. Marine Environmental Research 52: 463-475.

Ravid R, Oren OH, Ben-Yosef J, Hornung H. 1985; Oil pollution in the Eastern Mediterranean. Marine Pollution Bulletin 16 (2); 81- 84.

Readman JW, Kwong LLW, Grondlin D, Bartocci J, Vielleneuve JP. Mee LD. 1993; Coastal water contamination from a triazine herbicide used in antifouling paints. Environmental Science and Technology 27: 1940-1942.

Relini Orsi L, Mori M. 1979; Due reperti Mediterranei di *Thalamita gloriensis* Crosnier, 1962 (Crustacea Decapoda: Portunidae). Oebalia 1: 7-13.

REMPEC 1998; Regional Marine Pollution Emergency Response Centre for the Mediterranean Sea, Regional Information System, March 1998. www.rempec.org (accessed 04-04-2005).

REMPEC 2004; Regional Marine Pollution Emergency Response Centre for the Mediterranean Sea, Database. www.rempec.org/databases.asp (accessed 04-04-2005).

Rezig M. 1978; Occurrence of Paracerceis sculpta (Crustacea, Isopoda, Flabelllifera) in the Lake of Tunis. Bull. Off. Natl. Peches (Tunisia) 2(1-2): 175-191.

Ribera MA, Boudouresque CF. 1995; Introduced marine plants with special reference to macroalgae: mechanisms and impacts. Progress in Phycological Research 11: 187-268.

Richardson WJ, Greene CR Jr, Malme CI, Thompson DH. 1995; Marine mammals and noise. Academic Press, San Diego.

Ros J, Faraco F. 1978; Pollution par les hydrocarbures des eaux superficielles de la Méditerranée Occidentale – Première partic: boules de goudron. IV Journées Étude. Pollutions, CIESM, pp. 111-115.

Ross D. 1976; Mechanics of underwater noise. Pergamon Press, New York.

Rosso A. 1994; Segnalazione di Electra tenella (Hincks) (Bryozoa) lungo le coste sud-orientali della Sicilia. Bolletino Accademia Gioenia di Scienze Naturali, Catania 27(346): 241-251.

Roussel E. 2002; Disturbance to Mediterranean cetaceans caused by noise. In: Notarbartolo di Sciara (ed) Cetaceans of the Mediterranean and Black Seas: state of knowledge and conservation strategies, 18 p. A report to ACCBAMS Secretariat, Monaco.

Ruiz G, Rawlings TK, Dobbs FC, Drake LA, Mullady T, Huq A, Colwell RR. 2000; Global spread of microorganisms by ships. Nature 408: 49-50.

Ruiz G, Smith G, Verling E, Chaves S. 2004; Ballast water management: toward understanding treatment efficacy. 13th International Conference on Aquatic Invasive Species, Ireland. pp. 18-19.

Ryan PG, Moloney CL. 1993; Marine litter keeps increasing. Nature 361:23.

Sakellariadou F, Tselentis V, Tzannatos E. 1994; Dissolved/Dispersed Petroleum Hydrocarbon content in Greek Seas. International Symposium on Pollution in the Mediterranean Sea. WTSAC-IAWQ, Cyprus. pp. 151-155.

Salihoglu I, Saydam C, Yilmaz A. 1987; Long term impact of dissolved/dispersed petroleum hydrocarbons (DDPH) in Gulf of Iskenderun. Chemosphere 16: 381-394.

Saydam C, Salihhoglu I, Sakarya M, Yilmas A. 1985. Dissolved/dispersed petroleum hydrocarbons, suspended sediment, plastic, pelagic tar and other litter in the north-eastern Mediterranean. 7th Workshop on Pollution in the Mediterranean, Lucerne, International Commission for the scientific exploration of the Mediterranean Sea, Monaco, 509-518.

Seligman PF, Valkirs AO, Lee RF. 1986; Degradation of tributyltin in marine and estuarine waters. Oceans '86 Conference, 4:1189-1195.

Seligman PF, Valkirs AO, Stang PM, Lee RF. 1988; Evidence for rapid degradation of tributyltin in a marina. Marine Pollution Bulletin 19 (10): 531-53.

Shiber JG. 1979; Plastic pellets on the coast of Lebanon. Marine Pollution Bulletin 10: 28-30.

Shiber JG. 1982; Plastic pellets on Spain's 'Costa del Sol' beaches. Marine Pollution Bulletin 13: 409-412.

Shiber JG. 1987; Plastic pellets and tar on Spain's Mediterranean beaches. Marine Pollution Bulletin 18: 84-86.

Shiber JG, Barrales-Rienda JM. 1991; Plastic pellets, tar, and megalitter on Beirut beaches, 1977-1988. Environmental Pollution 71: 17-30.

Smith BS. 1981; Male characteristics on female mud snails caused by antifouling bottom paints. Journal of Applied Toxicology 1: 22-25.

Solé M, Morcillo Y, Porte C. 1998; Imposex in the commercial snail *Bolinus brandais* in the northwestern Mediterranean. Environmental Pollution 99: 241-246.

Stefatos A, Charalambakis M, Papatheodorou G, Ferentinos G. 1999; Marine debris on the seafloor of the Mediterranean Sea: examples from two enclosed gulfs in Western Greece. Marine Pollution Bulletin 36 (5): 389-393.

Terlizzi A, Geraci S, Minganti V. 1998; Tributyltin (TBT) pollution in the coastal waters of Italy as indicated by *Hexaplex trunculus* (Gastropoda: Muricidae). Marine Pollution Bulletin 36 (9): 749-752.

Thomas KV, Fileman TW, Readman JW, Waldock MJ. 2001; Antifouling paint booster biocides in the UK coastal environment and potential risks of biological effects. Marine Pollution Bulletin 42 (8): 677-688.

Tolosa I, Readman JW, Blaevoet A, Ghilini S, Bartocci J, Horvat M. 1996; Contamination of Mediterranean (Côte d'Azur) coastal waters by organotins and Irgarol 1051 used in antifouling paints. Marine Pollution Bulletin 32 (4): 335-341.

Tomás J, Guitart R, Mateo R, Raga JA. 2002; Marine debris ingestion in loggerhead sea turtles, *Caretta caretta*, from the western Mediterranean. Marine Pollution Bulletin 44: 211-216.

Tselentis BS, Maroulakou M, Lascourreges JF, Szpunar J, Smith V, Donard, OFX. 1999; Organotins in sediments and biological tissues from Greek coastal areas: preliminary results. Marine Pollution Bulletin 38 (2): 146-153.

Turley CM. 1999; The changing Mediterranean Sea – a sensitive ecosystem? Progress in Oceanography 44: 387-400.

Tyack P, 1998; Acoustic communication under the sea. In: Hoop SL, Owren MJ & Evans CS (eds.), Animal acoustic communications: Recent technical advances, 163-220. Springer-Verlag, Heidelberg.

United Nations. 1982; United Nations Convention on the Law of the Sea. UN DOC. A/CONF.62/122, December 10, 1982.

UNEP 1982; United Nations Environment Programme. 1982; The health of the oceans. UNEP Regional Seas Reports and Studies. 16.

UNEP 1986; United Nations Environment Programme. 1986; Assessment of the present state of pollution by petroleum hydrocarbons in the Mediterranean Sea. UNEP/WG. 144/9.

UNEP 1989; United Nations Environment Programme. 1989. Report of the Sixth Ordinary Meeting of the Contracting Parties to the Convention for the Protection of the Mediterranean Sea against Pollution and its related Protocols, Athens, 3-6 October, 1989. UNEP (OCA)/MED.IG.1/5. UNEP, Athens, 146 pp.

UNEP 1992; United Nations Environment Programme. 1992; Mediterranean Action Plan and Convention for the Protection of the Mediterranean Sea against pollution and its related protocols. UNEP, Athens.

UNEP 2003; United Nations Environment Programme. 2003; Action Plan concerning species introductions and invasive species in the Mediterranean Sea. UNEP(DEC)MED WG.232/6, Tunis.

UNEP/MAP 2003; United Nations Environment Programme/Mediterranean Action Plan 2003; Report on the thirteenth ordinary meeting of the contracting parties to the convention for the protection of the Mediterranean Sea against pollution and its protocols. UNEP(DEC)/MED IG.15/11, Athens.

van Franeker JA. 1985; Plastic ingestion in the North Atlantic fulmar. Marine Pollution Bulletin 16: 367-369.

Verlaque M, Durand C, Huisman JM, Boudouresque CF, Le Parco Y. 2003; On the identity and origin of the Mediterranean invasive *Caulerpa racemosa* (Caulerpales, Chlorophyta). *European* Journal of Phycology 38: 325-339.

Villeneuve JP, Carvalho FP, Fowler SW, Cattini C. 1999. Levels and trends of PCBs, chlorinated pesticides and petroleum hydrocarbons in mussels from the NW Mediterranean coast: comparison of concentrations in 1973/1974 and 1988/1989. Science of the Total Environment 237-238 (1-3): 57-65.

Wahby, SD. 1979; Pollution by petroleum hydrocarbons along Alexandria coast. Workshop on pollution of the Mediterranean, CIESM & UNEP, pp. 93-97.

Winston JE. 1982; Drift plastic – an expanding niche for a marine invertebrate? Marine Pollution Bulletin 13 (10): 348-351.

Wolterink JWK, Hess M, Schoof LAA, Wijnen JW. 2004; Port reception facilities for collecting ship-generated garbage, bilge waters and oily wastes. Activity B. Optimum solutions for collecting, treatment and disposal of relevant ship-generated solid and liquid wastes. Final Report. REMPEC project MED.B4.4100.97.0415.8.

Wonham, MJ, Carlton JT, Ruiz GM, Smith LD. 2000; Fish and ships: relating dispersal frequency to success in biological invasions. Marine Biology 136:1111-1121.

Youssef AK, Durgham H, Baker M, Noureddin S. 1999; Accumulation of petroleum hydrocarbons in zooplankton of Banyas coastal waters (Syria). Marine Pollution 427-428.

Zibrowius H. 1979; Serpulidae (Annelida Polychaeta) de l'Océand Indien arrives sur des coques de bateaux à Toulon (France, Méditerranée). Rapports et Proces-Verbaux des Reunions Commission Internationale pour l'Exploration de la Mer Méditerranée 25/26 (4): 133-134.

Zibrowius H. 1992; Ongoing modification of the Mediterranean fauna and flora by the establishment of exotic species. Mésogée 51: 83-107.

Zibrowius H. 2002; Assessing scale and impact of ship-transported alien fauna in the Mediterranean? In: Alien marine organisms introduced by ships in the Mediterranean and Black Seas. CIESM Workshop Series 20: 63-68.

Zsolnay A. 1979. Hydrocarbons in the Mediterranean Sea, 1974-1975. Marine Chemistry 7: 343-352.

CHAPTER 4: SNAKES AND LADDERS: NAVIGABLE WATERWAYS AS INVASION CORRIDORS

BELLA S. GALIL[1] & DAN MINCHIN[2]

[1]*National Institute of Oceanography, Israel Oceanographic and Limnological Research, P.O.B. 8030, Haifa 31080, Israel*
[2]*Marine Organism Investigations, 3, Marine Village Ballina Killaloe, Co Clare, Ireland*

1. Introduction

Herodotus, the Greek geographer and historian, supplied us with the earliest reference to a navigable canal – it was constructed in the 6th century BCE and joined the easternmost arm of the Nile with the northern Red Sea. Schooled early in drainage and irrigation engineering, and trained in colossal construction projects, the nilotic civilization built a canal "four day's voyage in length, and it was dug wide enough for two triremes to move in it rowed abreast. It is fed by the Nile, and … it issues into the Red Sea" (in Godley 1975, II: 158). The idea of digging through the Isthmus of Corinth also dates to the 6th century BCE, and was considered, successively, by Julius Caesar, Caligula and Nero, but each met an untimely death that prevented completion of the canal (Werner 1997). The Grand Canal in China, constructed in the 4th century BCE, connected Peking to Hangchow, a distance of almost 1000km, linking the Huang-ho and Yangtze rivers, was one of the great aquatic engineering projects of the ancient world. However, not until the technological innovations of the 18th century, were processes set in motion that led to a proliferation of canal building, and an expansion of the network of navigable inland waterways, first in Europe and then worldwide.

River-borne transport had been increasing in Europe since the 16th century, following demographic and economic growth. The wider usage of the steam engine in the 19th century facilitated waterborne commerce by powering ships and dredges. The removal of navigational obstacles by deepening river channels, reinforcing riverbanks and connecting river systems allowed for expansion of the navigable network. The early navigable waterways were developed to transport coal, timber and ores to the manufacturing centres and to improve market links. Rapid industrialisation led to an increase of waterborne transport to accommodate the enlargement in trade volume: the length of navigable inland waterways in Germany alone doubled between 1873 and 1914 (Ville 1990). The interconnection of the watersheds of the North American continent, east of the Rocky mountains, by a complex array of canals and canalised rivers from the Laurentian Great Lakes to the Gulf of Mexico, was achieved mainly during the 19th century and the early part of the 20th century. Two interoceanic canals were products of the same period: the Suez Canal (1869) opened a direct route from the Mediterranean Sea to the Indo-Pacific Ocean, and the Panama Canal (1914) afforded passage between the Atlantic and the Eastern Pacific.

Following a period of decline in waterborne transport and conversion of many smaller canals to recreational usage, the last decades of the 20th century saw a revival in inland waterways' expansion and recognition of their economic importance. A European agreement on the Main Inland Waterways of International Importance was established in the framework of the United Nations Economic Commission for Europe (UNECE), and was joined by Russia in 2002. This Pan-European inland waterway network comprises 28,000km of navigable rivers and canals and connects about 350 ports of international importance. The White Sea and the North Sea are now connected over vast distances across Russia and Europe via a dense network of inland waterways, with the Mediterranean, the Black and the Caspian Seas. These cross-continental

John Davenport and Julia L. Davenport, (eds.) The Ecology of Transportation: Managing Mobility for the Environment, 71–75,

systems of rivers, canals, lakes and inland seas are used by a large number of vessels transporting a significant volume of cargo.

- Through the 280,000km of navigable rivers, lakes and canals in East Asia, more than 1 billion tons of cargo and 500 million passengers are transported each year; the inland waterway fleet consists of 446,000 vessels with combined capacity of 27.5 million tons.
- 11,725 ocean-going vessels passed through the Panama Canal in 2003, transporting 242.5 million tons of cargo.
- 14,000 commercial vessels transit the Suez Canal annually.
- Though closed to commercial traffic between November and April, the Volga-Don Canal has been traversed by 400,000 ships since its opening 50 years ago.
- The volume of cargo expected to pass through the Mittellandkanal in 2010 is 42 million tons.

Inland waterways transport is more energy efficient than overland transport and so produces a lower emission of pollutants per ton of cargo transported. For these reasons water transport is considered to be more environmentally friendly. In seeking to limit the increasingly destructive impact of transport upon the environment, UNECE has promoted inland waterway transport as a more pro-ecological transport system. However, the development of inland waterways entails enlargement of existing canals to allow passage of larger vessels, construction of dams, locks and levees, and reinforcement of riverbanks to withstand the vessels' increased speed. The hydrological changes threaten floodplains, water meadows and wetlands. The canalisation of rivers, and the prevalent aquatic pollution and eutrophication, tend to homogenise their water quality. The increasing depth and width of the canals, and the creation of reservoirs mean a larger volume of water that buffers temperature and salinity fluctuations, and so provides a more uniform environment, leading to a decrease in habitat diversity and to diminishing biodiversity (Tittizer & Banning 2000). Similarly, the chronic physical disturbance of the river beds leads to habitat loss. Apart from providing transport routes, the cross-continental systems of rivers, canals, lakes and inland seas serve an ever increasing fleet of sea-river vessels that may carry alien species or provide them with many opportunities for natural dispersal. Estuarine ports servicing both inland waterways and overseas shipping, are prone to inoculations of trans-oceanic biota, and may provide occasions for secondary spread of alien biota upstream.

2. Keystone invasive species

Invasive organisms have been dispersed through canal and river systems, examples include: in North America, the alewife, *Alosa pseudoharengus*, and the sea lamprey, *Petromyzon marinus*; both entered Lake Ontario through the Erie Canal. The alewife became abundant in its new environment and competed with native fishes, and the lamprey, a predatory fish, reduced the native salmonid populations and whitefish species (GLFC 2004). The construction of the Welland Canal, bypassing the Niagara Falls, also allowed for their dispersal throughout the Laurentian Lakes (Aron & Smith 1971). The zebra mussel, *Dreissena polymorpha*, formerly a Ponto-Caspian endemic, began its global spread in the early 19th century (Köppen 1883) by reaching the Baltic Sea through canals linking the Dnieper with the Vistula and the Neman, and through canals linking the Volga and the Neva. This mussel also spread through the central European waterways to the Rhine, and with the timber trade to western European ports (Minchin *et al.* 2002). The zebra mussel was first sighted in Lake St. Clair in 1988, and spread within two years throughout the Laurentian Great Lakes, and subsequently dispersed through many of the inland waterways of the eastern United States with profound economic and ecological consequences (Vanderploeg *et al.* 2002). Being an effective filter feeder, it outcompeted native filter feeders, to alter total benthic biomass and species composition, thereby radically changed the trophic structure by decreasing planktonic abundance, and shifting the trophic structure from the water column to the benthos (Karatayev *et al.* 1997). The zebra mussel is the most prominent of a suite of Ponto-Caspian taxa, from amphipods to fish that have spread through the Eurasian inland waterways to western and northern European estuarine ports, and in some cases onward to North America.

The freshwater Asiatic clam, *Corbicula fluminea*, native to south-east Asia, is a further example of an invasive mollusc to both North America and Europe. The clam was recorded at Vancouver Island, Canada, in 1924 and within 20 years it was found throughout North America south of the Great Lakes. It is believed that the species spread via pelagic larvae carried in rivers, canals and irrigation systems (McMahon 1983). In Europe it was found in 1989 at the confluence of the Rhine and Meuse rivers near the port of Rotterdam, and spread rapidly: by 1990 it was collected upstream of Bonn on the Rhine (Bij de Vaate 1991). In 1999 it was already reported from the Danube, in 2000 from the drainage basin of the Seine, in 2001 from the Elbe, and in 2003 from the Saône and Rhone rivers and in the Canal du Midi. It has been suggested that navigational waterways played a major role in its dispersal (Brancotte & Vincent 2002). Its recent appearance in the Broads of Britain may be due to leisure craft movements and subsequent spread via the British canal system is expected.

The Chinese mitten crab, *Eriocheir sinensis*, introduced to Germany in the early 1900s, dispersed in the 1920s and 1930s to several European rivers by active migration, carried in ballast or attached to boat hulls. Its present distribution ranges from Finland to Portugal and the Mediterranean coast of France, and from the Czech Republic to Great Britain, where its population has increased in recent years. In 1965 it was found in the Detroit River, and in 1973 it reached the Laurentian Great Lakes. Since it was first reported in 1992 from San Francisco Bay, it rapidly expanded range and density in the Sacramento-San Joaquin Delta and watershed. In Europe, the burrows excavated by the mitten crab accelerated bank erosion, thereby reducing the stability of dykes, riverbanks and levees, where crabs occurred in high densities. Damage to fishing nets and catch has been reported in Europe and California. In California, the invasion of the mitten crab clogged pumps and blocked water systems at several power stations.

The Atlantic comb jelly *Mnemiopsis leidyi* has been linked to changes in the ecology and productivity of the Black Sea, since its introduction, probably in ballast water. It was first noted in the 1980s during which it attained abundance estimated to be hundreds of million of metric tons. Its presence was linked to a 90% decline in zooplankton biomass with concomitant decreases in commercial catches of anchovy and sprat. In 1999 *M. leidyi* was recorded from the Caspian Sea, having been transported through the Volga-Don Canal. In the following summer they attained an abundance of up to 170 ind/m². The immediate impact of the outbreak was again a serious decline in the biomass and abundance of zooplankton, and deterioration in the size/weight ratio of the planktivorous anchovy-like kilka (*Clupeonella spp.*) that forms the bulk of the Caspian fisheries. Within two years the kilka catch was halved. The cascading ecological effect of the invasion of the comb jelly has raised concern for the endemic Caspian seal, *Phoca caspica*, which feeds mainly on kilka (Ivanov *et al.* 2000). Nineteen alien species, believed to be ship-transported, have already established viable populations in the Caspian Sea since the opening of the Volga-Don Canal half a century ago (Aladin *et al.* 2002).

The Suez Canal serves as a conduit for Red Sea and Indo-Pacific biota into the Mediterranean. Despite impediments such as the canal's length, shallowness, turbidity, temperature and salinity variations, hundreds of Red Sea species have become established along the Levantine coasts of the Mediterranean Sea, with some extending their range westwards to Tunis, Malta and Sicily (Galil 2000). The Suez Canal has provided access for over 80% of all records of alien fish, decapod crustaceans and molluscs in the Mediterranean Sea (CIESM 2004). This will have resulted in competition, and replacement of some local populations of native species. Some alien taxa are considered as pests or cause nuisance, whereas other invaders are of commercial value – Red Sea prawns and fish presently constitute nearly half of the trawl catches along the Levantine coast. In contrast, the fresh waters of Lake Gatun form an effective barrier to the dispersal of marine biota through the Panama Canal. Only seven Atlantic decapod crustaceans were collected from the Pacific drainage, and a single Pacific crab from the Atlantic drainage. None of these are known to have established populations outside the canal (Abele & Kim 1989), apart from the tarpon, *Megalops atlanticus*, tolerant of reduced salinities, and now established near the Pacific terminus of the Panama Canal and around Coiba Island. Eight other species, predominantly blennies and gobies, that entered the Canal, failed to breed (Gunter 1979).

3. Future trends

The expanding global trade engenders greater volume of shipping, and economic development of new markets brings about changes to shipping routes. The navigable inland waterways may be facing a significant increase of bioinvasions. The confluence of the following political and economical events enhances a higher potential inoculation and spread:
• International trade with the eight Baltic and central European countries that have joined the EU will increase.
• Expansion of the Caspian Basin oil production will result in an increase in shipping traffic along the Volga-Don route.
• The Russian Ministry of Transport, The European Bank for reconstruction and Development and the European Commission have begun financing the modernisation of Russia's navigable waterways.
• The development of a transportation corridor between the Persian Gulf and the Baltic Sea is expected to increase shipping along the Volga-Baltic waterways.
• The Trans-European Network for Transport (TEN-T) plans a 1,980 km long navigational canal linking the Danube, Oder and Elbe rivers.
• China has embarked on a major improvement of the inland waterborne transport infrastructure, including the expansion of the 1,794 km long ancient Beijing-Hangzhou Grand Canal, and construction of 20 inland river channels totalling 15,000 km to facilitate north-south and east-west transport.
• A canal across the Kra Isthmus in Thailand to facilitate the movement of oil tankers to Japan is under consideration.

4. Conclusions

The economic arguments in favour of the development of inland waterways infrastructure should not obscure their significant long-term cost to the environment. Most inland waterways are no longer pristine: watershed engineering, discharge of agricultural, industrial and domestic wastes, and power plant cooling water, have contributed to varying impairments of the environment and harmed the native biota, leaving it vulnerable to invasion. Reconstruction of existing navigable waterways and constructions of new canal systems expedite range expansion of taxa within the interconnected watersheds, promoting both homogenisation of faunas and secondary and tertiary invasions. The rise in sea-river transport in the inland waterways transport market will enhance the spread of alien taxa, arriving in transoceanic shipping. With ample evidence that inland waterways serve as major invasion corridors, environmentally-considerate waterway engineering should include barriers that might preclude future invasions.

References

Abele LG, Kim W. 1989; The decapod crustaceans of the Panama Canal. Smithsonian Contributions to Zoology 482: 1-50.
Aladin NV, Plotnikov IS, Filippov AA. 2002; Invaders in the Caspian Sea. In: E. Leppäkoski, S. Gollasch & S. Olenin (eds). Invasive Aquatic Species of Europe: Distribution, Impacts and Management, 351-359. Kluwer Academic Publishers.
Aron WI, Smith SH. 1971; Ship canals and aquatic ecosystems. Science 174: 13-20.
Bij de Vaate A. 1991; Colonization of the German part of the river Rhine by the Asiatic clam, *Corbicula fluminea* Müller, 1774 (Pelecypoda, Corbiculidae*).* Bulletin Zoölogisch Museum, Universiteit van Amsterdam. 13(2): 13-16.
Brancotte V, Vincent T. 2002; L'invasion du reseau hydrographique Francais par les mollusques *Corbicula* spp. Modalite de colonization et role preponderant des canaux de navigation. Bulletin de la Peche et de la Pisciculture 365-366: 325-337.
CIESM 2004; Commission Internationale pour l'Exploration Scientifique de la Mer Méditerranée 2004: www.ciesm.org/atlas (accessed 05-04-2005).
Galil BS. 2000; A sea under siege - alien species in the Mediterranean. Biological Invasions 2: 177-186.
GLFC 2004; Great Lakes Fishery Commission: www.glfc.org/sealamp/how.asp (accessed 05-04-2005).
Godley AD. 1975; Herodotus. In: G.P. Goold (ed.) The Loeb Classical Library. Harvard University Press, William Heinemann. I: 1-504.
Gunter G. 1979; Marine fishes of Panama as related to the Canal. Gulf Research Reports 6(3): 267-273.
Ivanov VP, Kamakin AM, Ushivtzev VB, Shiganova T, Zhukova O, Aladin N, Wilson SI, Harbison GR, Dumont HJ. 2000; Invasion of the Caspian Sea by the comb jellyfish *Mnemiopsis leidyi* (Ctenophora). Biological invasions 2(3): 255-258.
Karatayev AY, Burlakova LE, Padilla DK. 1997; The effects of *Dreissena polymorpha* (Pallas) invasion on aquatic communities in Eastern Europe. Journal of Shellfish Research 16(1): 187-203.
Köppen FT. 1883; Notiz über die Rückwanderung der *Dreissena polymorpha*. Beiträge zur Kenntnis des Russischen Reich und der angrenzenden Länder Asiens. St. Petersburg. 6:269-281.
McMahon RF. 1983; Ecology of an invasive pest bivalve, *Corbicula*. In: W.D. Russell-Hunter (ed). The Mollusca., 505-561. Vol 6. Ecology, Special Edition. Academic Press, Orlando, Florida.
Minchin D, Lucy F, Sullivan M. 2002; Zebra mussel: impacts and spread. In: Leppäkoski E, Gollasch S & Olenin S. (eds). Invasive Aquatic species of Europe: Distribution, Impacts and Management, 135-146. Kluwer Academic Publishers.
Tittizer T, Banning M. 2001; Biological assessment in the Danube catchment area: Indications of shifts in species composition induced by human activities. European Water Management 3(2): 35-45.
Vanderploeg HA, Nalepa TF, Jude DJ, Mills, EL, Holeck KT, Liebig JR, Grigorovich IA, Ojaveer H. 2002; Dispersal and emerging ecological impacts of Ponto-Caspian species in the Laurentian Great Lakes. Canadian Journal of Fisheries and Aquatic Science 59: 1209-1228.
Ville SP. 1990; Transport and the development of the European Economy 1750-1918. MacMillan Press, London.
Werner W. 1997; The largest ship trackway in ancient times: the Diolkos of the Isthmus of Corinth, Greece, and early attempts to build a canal. The International Journal of Nautical Archaeology, 26(2): 98-119.

CHAPTER 5: THE TRANSPORT AND THE SPREAD OF LIVING AQUATIC SPECIES

DAN MINCHIN

Marine Organism Investigations, 3 Marina Village, Ballina, Killaloe, Co Clare, Ireland

1. Introduction

The distribution of aquatic species was brought about by natural processes but the development of human transport has meant that species can be transported around the world, both deliberately and accidentally. Dissemination of species by rapid modern transport results in a greater survival rate of transported organisms. Fast modern transport is also essential for the much of the present day worldwide trade in living aquatic species (e.g. fish, shellfish). This trade may also result in accidental transfers of other species, either with the designated cargo, or by direct association with the transport vector (Carlton 2001; Williamson *et al.* 2002; Minchin & Gollasch 2002).

Increased world trade particularly over the past 50 years, together with increases in human populations, improvements to transport and increases in prosperity, has led to an increasing number of organisms being inadvertently transmitted (Carlton 1988; Gollasch 1996; Hewitt 2002). These include many aquatic organisms, for example, species used in culture for angling bait, ornamental species and those intended for direct human consumption (Warren-Rhodes *et al.* 2003). Most of the international trade passes through one or more hubs, usually ports and airports, which then act as centres for further distribution. This distribution may involve one or more forms of transport. Such trade fluctuates in accordance with economies, political events and changing environments. Redistribution of products from main ports of entry to smaller hubs at local level is increasing. The development of new marinas, garden-centres, aquarium shops as well as outlets for speciality foods (Padilla & Williams 2004) has led to transport networks that have the potential to disperse species rapidly over a wider geographical area.

Open markets and general trade seldom consider the consequences arising from the 'leakage' of biota between different biological provinces. Improved codes of practice and legal enforcement are likely to reduce deliberate introductions into to the wild. However, a general awareness of how transport acts in the spread of species is also required, so that legislation can be put into place to prevent the accidental movement of species.

Unlike most terrestrial species, aquatic species, apart from large unusual species, are most often spread beyond an area feasible for eradication measures when first found. Where management is possible the action taken will very much depend on the life history stages and mode of life of the species as well as the vectors (both natural and otherwise) involved in their spread. In this account some examples of species carried by different transport modes with options for management are discussed and some future scenarios are considered.

2. Ships and other floating craft

All vessels have the capability of transmitting species including small craft (see Chapter 6). Usually this is by attachment to the hull or by the uptake of water to ballast tanks (to stabilise the ship while in transit). Either type of transmission can lead to species with very different life-histories, from virus to fish, becoming broadcast (Gollasch *et al.* 2002). Recent studies

John Davenport and Julia L. Davenport, (eds.) The Ecology of Transportation: Managing Mobility for the Environment, 77–97,

have shown that mobile species are not only transported in ballast water, but can also be carried with the assemblages attached to ships' hulls (Minchin & Gollasch 2003). One feature of ships is that on leaving a dry-dock they have opportunity to accumulate organisms by progressive fouling, or in ballast water, with each port visited.

Ports are the principal regions through which trade passes. Shipping has been a basis for this trade over some thousands of years. However, early trade was based on small vessels making occasional, often seasonal journeys, to specific areas where there was shelter. Conditions in port regions were not greatly modified. From such times the size of ships and their frequency of transit have greatly increased with specialised ports evolving, changes to the geography of ports and increased urbanisation and waste discharges. The ports have evolved different trading practices according to the nature of the products in trade. Vessels themselves have become specialised according to their cargo. The routes of ships have also become specialised and range from regular routes with exports of oil between industrial regions to infrequent seasonal routes, often to remote areas as in the case of some cruise ships. Rare events (including warfare) may also be important in species movements. For example, transmission of the slipper limpet *Crepidula fornicata* took place in Normandy in June 1944 with the movement of Mulberry Harbours (Blanchard 1995) during World War II, while the ascidian *Styela clava* arrived on the south coast of Britain with returning warships following the Korean War (Minchin & Duggan 1988).

The great majority of trade of islands and between continents is based on shipping, while ships exclusively transport bulk products. The number of ships in transit is increasing and, with port modernisation, the turn-around times are shortening. These circumstances ensure that there are greater volumes of ballast water being released in ports and a greater range of opportunities provided to hull fouling species. Increases in world trade have been possible with more ships in service and with the development of non-traditional trading routes. The majority of cosmopolitan species spread worldwide by shipping could have been transmitted either as hull fouling, including sea-chests (spaces in the ships hull from where ballast water is taken-up) or within ballast water.

2.1 TRANSPORT AND HULL FOULING

Management of hull fouling has been of great concern to ship builders for centuries. Reconstructions of old sailing ships can provide insights into the species carried by them in previous centuries when studied using attached settlement plates (Carlton & Hodder 1995). Wood has a limited life span when introduced to the water, and under marine conditions becomes colonised by those molluscs and crustaceans that erode timber by boring into it forming galleries that gradually weaken the hull. In some world regions where wooden craft are still in regular use, these borers reduce the working life of the craft (Nagabhusanam & Sarojini 1997). Over time, various applications were developed to preserve the wood; these included aerial exposure and periods of immersion in freshwater. Boring biota and fungal growths were controlled with varying degrees of success. Later developments included the exclusion of settling organisms by the tacking of copper sheets or iron cladding to the wooden hull. As most ships over the past century, have been constructed of steel, boring is no longer a significant problem.

The main issue for steel ships is the management of the fouling that gradually accumulates on the hull. Fouling increases drag, causing more fuel to be used and journeys to take longer. Even a small film of 1mm thickness can result in a significant increase in drag. Managers that control hull-fouling do so to reduce this drag, not to reduce the spread of invasive species. However, any reduction of hull fouling deals with both issues at the same time.

2.1.1 *Hull fouling prevention using paint coatings*

Antifouling paints provide a toxic surface to settling organisms thus reducing settlement and development of fouling growths on the hull. The active ingredients contained in the paints become less effective over time and growths are more likely to develop with paint ageing. Repainting of the hull takes place when the ship undergoes dry-dock inspection and repairs. Copper-based paints were used for over the 50 years until organotin paints became available in the early 1970s (Champ & Pugh 1987). Since then the majority of ships have used paints containing organotins, in particular tri-butyl-tin (TBT). This substance not only prevented settlements on the hulls but also leached into the environment to have a serious impact on the populations of some species. Some snail populations declined and became locally extinct (Bryan *et al.* 1986). Pacific oysters became unmarketable as a result of mortalities, distorted shells and poor meat yields (Alzieu *et al.* 1982). Several experimental studies since have shown that a wide range of organisms including phytoplankton could be affected at very low TBT concentrations (Lee 1991). The larvae of many animal species were found to be intolerant of the toxic conditions. Improvements made to these paints following the bonding the active ingredient to the paint matrix – copolymer paints – led to a reduced leaching of TBT into the environment. However, although the degree of contamination has declined, some ports still remained heavily contaminated (Burnell & Davenport 2004). The IMO made a resolution to improve water quality by banning the application of organotins to all hulls from January 2003. By 2008 no vessels are expected to have such paints on their hulls. It is anticipated that there will be a decline in contamination in port water. In addition, following implementation of the European Union Water Framework Directive, it may be expected that ports in coastal, estuarine and freshwater regions will have notable improvements in water quality. The likely consequence of this is that arrivals of species attached to ship hulls or carried in discharged ballast water will then have fewer challenges and so will have a greater opportunity of becoming established. Some of these may be robust invaders.

2.1.2 *Specialised ships and structures*

Some vessels or structures, on account of their size or type of function they were designed for, have a high probability of transmitting species. For example, oil platforms, normally operating at fixed positions for long periods, can develop extensive fouling (Lambert 2002). These may then transferred over long distances for service elsewhere, towed by ocean-going tugs at slow speeds. From time to time, platforms may be brought inshore for servicing where the progeny of hard fouling species may be capable of attachment to coastal rock substrate. Other vessels towed by ocean-going tugs include barges. As barges travel more slowly, their hull maintenance is not as carefully managed as for ships. Even local movements of barges have been implicated in transmission of oyster parasites on the south coast of Britain (Howard 1994) and of tunicates in New Zealand (Coutts 1999).

Ships traversing oceans carry estuarine and marine species on their hulls but some highly specialised vessels may be capable of transmitting freshwater hull fouling species. Lighter Aboard SHip Vessels (LASH) are large containers carried on a mother ship and unloaded from the stern into the water. A mother ship will carry some hundreds of these floating modules that are towed upstream in some European and Atlantic North American rivers (Minchin & Gollasch 2003). Idle containers can develop hull-fouling assemblages. These could subsequently become transported across the Atlantic. Aerial exposure under cool and damp conditions of some tolerant species may allow some fresh water species to survive to establish new populations. The zebra mussel is one such species that can survive aerial exposure of up to 18 days under such conditions (Minchin pers.ob.). This travelled from Europe to the Great Lakes of North America in about 1986 and is now well established in the

Mississippi River (Allen *et al.* 1999) following downstream dispersal. This river is used as a route for floating modules and could easily distribute zebra mussels elsewhere, or return a differently challenged form of the zebra mussel to Europe. Should such vessels trade with China it is likely the zebra mussel could spread there too.

2.1.3 *Decommissioned vessels*

Some vessels retired from service may be held in storage in an idle state for several years before their final journey to a breakers yard. Even after arrival at a yard a ship/structure may be held for long periods before being broken up. Such idle periods can lead to a build-up of mature fouling communities composed of short to long-lived biota and can include mobile species (Gollasch 2002). When the ship/structure is moved over large distances these communities may be carried to regions beyond their natural range. These ships undertake journeys under-tow by ocean-going tugs at slow speeds.

On the east coast of the United States in Chesapeake Bay, within the estuary of the James River, is an area where military service vessels have been held following decommissioning for up to twelve years (www.hazegray.org/worldnav/usa/decom.htm). Four vessels were towed to a North Sea estuary at Teeside in 2004 and may have carried fouling species from the James River, an area where the American oyster *Crassostrea virginica* is fished. This movement carries a risk of transmitting the vertically-transmitted protozoan parasite of oysters *Perkensis marinus* (Bower *et al.* 1994). An autumn transfer enhances this possibility (R. Mann pers. comm.). Other transmissions from North American estuaries are possible that could include the predatory snail, *Rapana venosa* an avid feeder of infaunal bivalves in Chesapeake Bay (Savini *et al.* 2002) first recorded there in 1998 (Harding & Mann 1999), and a North Pacific shore crab *Hemigrapsus sanguineus*, a predator and avid consumer of shore organisms (McDermott 1991). Should any of these organisms establish themselves after transfer they will almost certainly spread their range.

In advance of moving the battleship *USS Missouri* to Pearl Harbour, Hawaii, to act as a floating museum, it was brought to an estuarine site to purge the living biota attached to the hull by exposure to freshwater. However, mussels, *Mytilus galloprovincialis,* survived on the lower hull below 5m where the water was more saline. These mussels spawned on arrival at Pearl Harbour (Apte *et al.* 2000). Transmission of historically important ships or ships of special design, such as light vessels (light-ships), carry similar risks.

It is unclear whether species populating the hull can be taken-up in the ballast water. At least theoretically, species settling or moving into the sea-chests could produce propagules that might be released and drawn in with the periodic uptake of ballast water. These could subsequently be released with a ballast water discharge. Short journeys are likely to enhance survival rates and are more likely to transmit species with short pelagic phases.

Sea stars are able to remain attached to subsurface rocks even on exposed coasts. It is therefore unsurprising that small individuals can adhere directly to hull surfaces or to the fouling community. The transmission of *Asterias amurensis* from Japan to Tasmania may have been on a hull surface with detachment on arrival possibly following exposure to surface freshwater in the Derwent Estuary. Detached starfish would have survived the deeper marine conditions. However, the more general assumption is that the species arrived with ballast water (Buttermore *et al.* 1994). The related sea-star *Marthasterias glacialis* has a range that extends from the Norwegian coast south to the Canary Islands and includes the Mediterranean Sea (Mortensen 1927). The disjunct population that occurs on the cool South African coast (Penney & Griffiths 1984) is difficult to explain but could reflect carriage there as hull fouling or from ballast water should it have arrived after about 1880. Some small specimens have been found attached to the hull of a fishing boat in Ireland (Minchin unpublished).

The oyster *Crassostrea angulata* is now accepted as being the same species as *C. gigas* first recognised by Menzel (1974) and was already present in Portugal by 1819. As oysters are commonly found on the hulls of modern ships, it is likely that sailing ships introduced this species from Taiwan at a time of trading between the early 1500s and the 1700s (Wolff & Reise 2002). Had the oysters been carried as cargo they would have required periodic freshening of the stock with sea water, whereas their transmittal as hull fouling would not have required special conditions except avoidance of prolonged exposure to freshwater.

Increases to fuel costs are likely to lead to a more regular hull maintenance, because even small amounts of fouling lead to significant increases of fuel consumption. Dry-docking of a ship, depending on its design and age, normally takes place every five years. Cleaning during the intervening period is likely to be needed particularly if the ship has been in service in warmer seas. This can be achieved using divers and underwater robots to scrub the hull surface. The dislodged material falls to the sea floor and may survive. Serpulid worms when crushed will often release gametes and some may be detached alive (T. Moen pers. comm). Any mobile species associated with the hull may swim or float away during cleaning. Codes of practice for hull cleaning are not generally applied. However, the possibility of cleaning stations being sited in areas over anoxic water or in freshwater areas could be considered.

2.2 TRANSPORT OF ORGANISMS WITH SHIPS' BALLAST

The stability of a ship often depends on the carriage of additional weight within the ships hull so that it's seated in the water at an immersion level that allows for manoeuvrability by providing rudder bite, fan thrust and stability. The amount of ballast necessary will depend on the sea-state conditions expected, and there are levels marked on the hull to which a voyaging ship must be immersed according to latitude and season. All vessels carry some ballast and some of this may not even be moved during the life of the vessel. Vessels loaded with cargo will have a smaller requirement for carrying ballast because the cargo will provide much of the stability required. However, when the cargo is off-loaded, in the absence of further cargo, ballast must be taken on board.

2.2.1 *Ship solid ballast*
Sailing ships required large amounts of ballast to balance the large sail areas they carried. The ballast was made up of stone, or more rarely sands and gravels, and placed low within the hull. When the ship carried cargo, stone would be removed. The practice of discharging stone ballast to navigation channels was generally legislated against and areas for the storage of this stone were set up on the shore near the docking area. For convenience the areas were positioned at different shore levels to facilitate the use of the stones at all stages of the tide. The loading and unloading was labour intensive and time consuming. Stone that had been stored from mid- to low-shore could have resulted in the transmission of the periwinkle *Littorina littorea* to the Atlantic coast of North America from Europe (Carlton 1982). Many other species were probably transmitted, but generally these events would have gone unrecorded.

2.2.2 *Water ballast*
The use of water for ballast was made possible with the building of steel ships, where compartments for storing pumped water were incorporated. The use of water for ballast provided distinct economic and practical advantages as uptake or discharge could take place at any time, even in advance of arrival at a port. Many of the first steel vessels were not designed to take-up water for ballast, therefore both stone and water ballast were in general use over two to three decades from the 1880s (Carlton 1996).

Ballast tanks on ships vary in size and design according to ship class and size, and their position on the ship. Most tanks have ladders and platforms for access and include baffles and transverse sections to strengthen the ship. The position and configuration of these tanks provide different environmental conditions within the ballast water. The forepeak tank, near the bow, is normally more agitated than tanks further astern, with consequent differences in ballast water circulation, temperature and quantities of dissolved gases. When water is taken on board it is distributed to tanks in accordance with the positioning of the loaded cargo. Thus uptake and discharge of ballast results in different histories of the contained water in any one tank.

Throughout the year almost 3km^3 of ballast water is distributed, and an estimated 3,000 to 4,000 species carried in transit every day (Gollasch *et al*. 2002). Species include those with planktonic phases and mobile stages. The arrival of the alga *Biddulphia sinensis* in 1908 is probably one of the earliest records of a ballast water introduction to Europe (Boalch 1993), and was closely followed by the arrival of the Chinese mitten crab *Eriochir sinensis* in Germany in 1912 (Gollasch 1999). It was not until the 1980s that there was a general recognition that a wide range of different species in transit represents a major environmental problem. The incidence of invasive species appearing in different ports worldwide at an apparently increased frequency has heightened this concern (Carlton & Geller 1993). The International Maritime Organisation (IMO) attempted to manage the spread of such species with new controls, but only one method, the mid-ocean exchange of ballast water, is in general practice. However, this exercise can only be undertaken under certain weather conditions; under poor sea-state conditions such activities could endanger the safety of the ship. This is because the process can result in a series of stresses on the hull not considered in the original design. It is such poor sea-state conditions that also result in an increased risk of sediments in the bottom of the tanks being re-suspended and discharged in the port following arrival. It is now known that sediments carry an extraordinary number of biota, ranging from viruses to larger components of the benthos (Drake *et al*. 2001; Hulsmann & Galil 2002; Gollasch *et al*. 2003; Lockett & Gomon 2001). Of particular current concern are the resting stages found in different taxonomic groups, in particular dinoflagellates. There is evidence that some dinoflagellates cysts survive years in ballast sediments (Nehring 1997). This finding may explain the apparent increase in toxic algal events worldwide (Hallegraeff & Bolch 1992; Hallegraeff 1993). This contention is supported by the appearance of dinoflagellate blooms in areas where they were previously unknown, and in one case was strengthened by a stratigraphic sampling of sediments that demonstrated a recent appearance of *Gymnodinium catenatum* in Australia (McMinn *et al*. 2000). The occurrence of disease organisms in ballast water is a more recent cause for concern (Drake *et al*. 2001). It is clear that bacteria and viruses are transmitted and are not easily identified. Noting the problem of ballast water mediated biological invasions the IMO developed the Convention on Ballast Water and Ships´ Sediments. This instrument sets standards for allowable organism concentrations in discharged ballast water. This standard will be phased in according to ships' ballast water capacity and age. Ballast water exchange is seen as an interim solution until ballast water treatment options are available.

Sampling water in ballast tanks is difficult. Collections of water are limited to manhole covers or sounding pipes (used to verify the depth of water within a tank) as tanks are otherwise sealed. Such sampling is unlikely to be fully representative. Extraction of biota also has variability. Calibration of the different methods of sampling used worldwide has shown significant differences according to method used, net design and mesh size, thereby hindering the study of organisms in transit (Gollasch *et al*. 2003). Nevertheless there are some general trends. It is thought that some organisms are killed and injured when ballast water is taken on. There is often an apparent increase in numbers of organisms during the day following the take-up of ballast water. This could be explained by there having been a sufficient length of time for the recovery of some organisms, enabling them to re-enter the water column. Apart from this feature, declines in the number of taxa and abundances within a species occur with each subsequent day of transit (Taylor *et al.* 2002). Organisms that die during ballast water uptake or during a voyage will provide a resource for bacteria. The lack of light within a ballast tank will compromise species that lack energy reserves or are dependent on photo-synthetic energy. These conditions provide opportunities for decomposers and could create conditions that become inhospitable for other biota.

Long journeys result in lower survival rates except for a few species that have robust resting stages. On one long journey the abundance of an harpacticoid crustacean increased (Gollasch *et al.* 2000), as the species is a scavenger it was presumably able to increase population size by consuming organisms that died during the voyage.

In order to reduce the numbers of species transmitted by ballast water, standards have been implemented in order to minimise the numbers of organisms that are present in the water (IMO 2004). The general assumption is that the lower the number of organisms carried, the less risk there is of establishing a new population at the receiving site. However, the numbers needed to develop a new population remain unknown for virtually all species. Theoretically these numbers can be very low and for some species a new population could be formed from a single individual as is the case in the spread of the Asiatic clam *Corbicula fluminea* or dinoflagellate *Alexandrium tamarense*. Regulations being developed to control invasive species have evolved more from a management perspective because little scientific progress in reducing biota in ballast tanks has been made. For some countries this perspective has resulted in unilateral actions that may have some consequences for future trade. Most ballast water control techniques are still being developed with varying effects (see Box 1). Intercalibration of the efficacy of the different techniques is needed so that appropriate measures to best reduce overall risk of transmissions can be determined for different tank designs and sizes, different ship classes and journey length. To-date this has not been done. Some control measures involving chemicals may be unsuitable should discharges contain elements of the toxin or of its degraded elements that cause harm following release. Indeed any releases in European ports will need to accede to requirements under the Water Framework Directive (EU 2000) as well as regulations concerning discharges of contaminated sediments (Borja *et al.* 2004).

Development of a technique to control organisms in ballast water should be practical and of low cost, must be undertaken safely in all weather states, should treat both ballast water and its accumulated sediments, and should be undertaken on all conventional vessels by their crew. The method should process large volumes of ballast water of varying condition, sediment loading, chemical composition, salinity and temperature and operate to a reliable quality standard. It would be of advantage if the procedures involved low maintenance and treatments that could be used on entering and leaving ports.

2.2.3 *Ballast water sediments*

Ballast sediments contain large numbers of protozoa (Hulsman & Galil 2002), bacteria and viruses (Drake *et al.* 2001) and this is also the case for estuarine sediments (Middelbeau & Glud 2003). In port regions sediments are often enriched with organics as a result of food processing and human and animal and other wastes (e.g. discards of by-catch in fishing ports). This input enhances the abundance of micro-organism populations. As these sediments can be disturbed by tidal scouring and weather events, as well as by port activities, e.g. dredging, any uptake of ballast water will result in their entry into ships' tanks. The sedimentation that follows will occur within the ballast tank in areas of least energy. Such accumulations vary according to their position in the ship and with ship class. These in turn will be exposed to different levels of agitation. For example, wing tanks on the side of a sun-exposed ship passing through the sub-tropics and tropics may develop higher temperatures.

Sediments vary greatly between ballast tanks and between ships, but any large vessel can carry 10s to 100s of tonnes of sediment (Hamer 2002).

Box 1. Treatment measures for ballast water under study or with potential

Mid-ocean ballast water exchange: Developed originally for vessels undergoing long voyages across oceans. The water taken-up in port is exchanged with oceanic water in depths of over 2000m. The concept is that the oceanic species taken-up are less likely to survive in coastal waters where the ballast water will be released. This method may eliminate 95% of the organisms taken up at the point of departure.

Continuous flow through ballast tanks: This method involves the flushing of water through full ballast tanks while underway. At least three times the contained volume needs to be flushed to achieve a 95% removal of organisms taken-up at the point of departure.

Dilution of ballast water: Selective flushing with water entering the top of ballast tanks and removal via a pipe from the bottom of ballast tanks.

Filtration: The removal of organisms by filtration can be achieved while taking ballast water on-board and any backwash used to clean the filters can be disposed of directly. The process can filter up to $>5000m^3$ per hour and remove >90% of larger planktonic organisms and >75% of zooplankton and phytoplankton. However, bacteria and viruses are not controlled.

Cyclonic separation: This involves the removal of organisms heavier than water by centrifugal action. Some sediments may also be removed using this system when used with flow rates of up to $3000m^3$ per hour.

High velocity pumping: Fast water movement during pumping could disrupt animal tissues and result in higher mortalities. There may be limits to the practicality of the method as it puts strain on the pipework of the ballast water delivery network.

Ultra-violet (UV) irradiation: UV light is often used in the sterilisation of large volumes of water and the efficacy depends on the size of the organisms present. Larger organisms may be unaffected and can shadow other organisms from the UV light. The method has some effect on bacteria and viruses but many cysts can be resistant to the treatment.

Heat treatment: Most aquatic species will be eliminated at sustained temperatures that exceed 40ºC. These include the cysts of dinoflagellates that may occur in ballast sediments. Heating of water to these temperatures can only be achieved on long journeys in tropical areas. It is not known whether the increased temperatures would increase corrosion or would affect ships' stability. Should temperatures not reach the required killing temperatures microbiota may thrive.

Ballast water-cooling: Reduction of temperature will cause mortalities to warm water aquatic species. However, many organisms are able to survive low temperatures by means of resting stages.

Ultrasound: High levels of vibration can result in cavitation, tissue damage and may remove dissolved gases from solution, resulting in mortality. It is unclear whether such vibrations would result in damage to the ship and its efficacy on different biota remains untested.

Electrical shock: The generation of electrical currents can have an impact on biota, indeed dinoflagellate cysts succumb at 5 seconds exposure to 100v. A likely place for this treatment is where the ballast water enters the ship in the sea-chest cavity.

Electrochemical treatment: Electrodes result in the production of chemical oxidising agents that can be supplemented with other oxidising chemicals. The technique results in an effective system for sterilising water containing microbiota.

Gas super-saturation: Injection of gases such as nitrogen to create bubble 'mists' can result in the formation of gas bubbles in animal tissues and result in floating organisms to the surface of the ballast water.

Biocides: Additions of toxic chemicals will result in mortality, the chemicals used will have different levels of efficacy depending on the taxonomic group and life-history stage, some resting stages may survive. Chemical additions may be easily injected to the ballast water on intake so that all of the water taken-up is treated. The breakdown products of the biocide will need to be environmentally neutral if they are to be in general use.

Permanent ballast water: Water remains on the ship at all times. This approach can only have limited usage for vessels that operate with very little changes in loading, such as cruise vessels.

Changing salinity: Alterations to salinity by uptake of freshwater where a discharge to a marine port is to take place and vice versa. Provision of freshwater in a port may be possible in some regions.

Shore ballast-water reception facilities: Where ballast water can be delivered to and from the shore for treatment.

(See Taylor *et al.* 2002 and Matheickal & Raaymakers 2004 for reviews).

2.3 AQUATIC PRODUCTS AS CARGO

One of the earliest of the aquatic products transported by ships on a regular basis was the oyster. For some decades from the mid-1800s, consignments of naturally-settled spat of native oysters were imported to Britain and Ireland from areas in France and the Netherlands that had intense natural oyster-settlements. The shorter ocean-transit times of steam ships compared with sailing ships across the Atlantic meant that from the 1870s deck cargo of half-grown American oyster *Crassostrea virginica*, contained in wooden barrels, were imported from Long Island Sound, initially to Britain then to Ireland. The oysters were laid on the shore for a full growing season before sale. Until 1939, millions of American oysters were imported annually and distributed by rail to oyster growing areas and then after a growing season sent to market. As this oyster was unable to spawn at these latitudes these imports continued. The arrival of the American oyster in Britain was not without consequence for Northern Europe, since other species accompanied them on their journey. One species in particular, the slipper limpet, *Crepidula fornicata*, became established ca 1887 on the south-east coast of Britain. From there it spread around the south coast of Britain but was spread by many vectors including the movement of Mulberry Harbours,. Today the slipper limpet ranges from Norway to Spain. It causes significant trophic competition. For example in France it has a biomass of millions of tonnes with 700,000Mt in the Baie de Granville alone (Blanchard 1995).

Oysters were also moved in cases as deck cargo during the 1920s and 1930s to the north-east Pacific from Miyagi Prefecture, Japan to the Georgia Strait and Puget Sound in the north-west of North America, a crossing taking 10-12 days. Pests imported in these consignments included a predatory snail *Ocenebra japonica*, the brown seaweed *Sargassum muticum*, a parasitic copepod *Mytilicola orientalis* and the flatworm *Pseudostylochus ostreophagus* (Quayle, 1969).

3. Overland transport

Movement overland was initially impeded by soft ground, forest, parched lands and altitude barriers. The development of transport routes through such terrain provided new opportunities for the spread of exotic aquatic communities away from major rivers and sheltered areas of coastline. Roads were the first to be developed from tracks providing coach roads but when motor vehicles were in general use these gradually became paved and allowed for movement in all seasons and with increasingly heavier loads. Today some bridges that originally had been designed for a coach and four horses bear loads of vehicles of over 40 tonnes.

3.1 RAIL AND ROAD

Steam power allowed for the development of rail for overland transport of goods. Rail networks became highly evolved and stole much of the trade from canal traffic during the 1800s by providing a more rapid delivery and by being able to cross rugged terrain. Railroads spread species across continents at a time for which few records survive; but such consignments involved fish and shellfish movements for stocking aquatic systems. In the early years railheads often terminated at fishing ports and so would have distributed commercial molluscs and crustaceans from unpolluted bays to the cities. Towards the end of the 1800s, once the techniques of fish-hatchery rearing had been developed there were movements of fertilised salmonid eggs for stocking lakes. One notable introduction, in 1879, was the Atlantic striped bass *Morone saxatilus* to California. Initially this seemed unsuccessful until bass appeared regularly at local fish markets and today in San Francisco Bay it is an important sport fish (Cohen & Carlton 1995). At this time there would also have been movements of oysters within North America and Europe. Train transport undertaken during the cooler parts of the year resulted in distributions over large distances and in large quantities. Indeed this led to a high exploitation of oysters and declines in many wild fisheries, especially in those areas where recruitment was variable or poor. These circumstances led to an interest in shellfish cultivation; but the biology of oysters in those early days was not fully known, so many attempts at cultivation were unsuccessful. Production was sustained with the distribution of small oysters from areas with significant annual settlements, as well as movements of half-grown oysters distributed by ship and rail. It was not until the development of pond culture on the west coast of Ireland in the early 1900s, and later in Wales in the 1920s-1930s that new ideas began to evolve to enhance local resources. These approaches generally led to a better understanding and development of hatchery culture techniques (Loosanoff 1969) reducing the need for more extensive overland transmissions from natural settlement areas.

Political unrest has led to the inadvertent transport of some aquatic species. In 1919, during the Russian civil war of 1917-1922, rail transport of small boats with attached mussels, *Mytilaster lineatus,* on the 800km rail connection between Butamy in the Black Sea and Baku in the Caspian Sea (Zaitev & Öztürk, 2001) resulted in the introduction of the mussel to the Caspian Sea. Transport during damp and cool conditions presumably made such an introduction possible. They now form a biomass of about 1kg/m^2 in parts of the western Caspian Sea and weigh over 10 million tonnes (41% of the biomass of the benthos). They compete with the native zebra mussel (Zaitev & Öztürk 2001).

The introduction of the red king crab *Paralithodes camtchaticus* from Peter the Great Bay in the Pacific Ocean to the Barents Sea took place from 1961 to 1969 with movements of 1.5 million larvae, 10,000 juveniles and 2,500 adults. Much of the stock was transferred by ship and plane in insulated boxes involving 128 to 135 hours of travel but during bad weather conditions some consignments were transferred by train (Orlov & Ivanov 1978). The introduction of this valuable food product is not without consequence since the species has become well established and has expanded its range to northern Norway where there have been notable changes to the benthos on which it avidly feeds (Petryashov *et al.* 2002).

Much overland transport involves short distance movements of stock between different bays e.g. the local distribution of oyster spat, scallops, clams and mussels. Fish stocks are transported on trucks carrying tanks containing re-circulated water. Re-circulated water may contain exotic species which may also be distributed. For example, salmon fingerlings that had been sourced from the rivers that drain to the Baltic Sea, were moved in water-tanks by truck from Swedish hatcheries to Norwegian rivers draining to the Atlantic. The seemingly harmless helmith parasite *Gyrodactylus salaris,* to which the Swedish stock had become adapted was transferred in the water and caused serious salmon mortalities in Norwegian river catchments from 1974 (Johnsen & Jensen 1991).

Road transport in water tanks of products intended for culture and subsequent human consumption may also have resulted in the spread of some serious parasites. The nematode worm *Anguillicola crassus* was first inadvertently transported by aircraft to Germany from Japan in consignments of its host, the eel *Anguilla japonica,* imported for culture in the 1980s (Jansson 1994). It is likely this parasite spread to the wild with the release of water containing young life-history stages. These can reside in a wide range of paratenic hosts and can also complete their life history within the eel *Anguilla anguilla.* The early infective life-history stage does not occur in seawater and adult eels are not known to make local movements between islands. It is therefore suspected that the further spread to the islands of Britain and Ireland was via tanks used to collect consignments of adult eels from fisheries. Such trucks collect small consignments from different fisheries in Britain and then in Ireland. As fish are crowded, the water in the tanks is replenished en route. This action could have resulted in the parasites' release. The nematode is associated with extensive bleeding and thickening of the air bladder wall and a loss in condition of other organs (Würtz & Taraschewski 2000). Its prevalence in lakes and rivers may have consequences for spawning migrations of the Atlantic stock.

Regularly each season there are road consignments of the Japanese scallop *Patinopecten yessoensis* transmitted to different regions of Honshu from Mutsu Bay in Japan. These are carried within insulated boxes between layers of damp sacking and usually transported overnight by truck for distances of up to 300km (Ventilla 1982). These scallops are unable to survive in some of the more southern areas unless they are cultivated in deeper cooler water. Similarly the New Zealand mussel industry *Perna canaliculus* in the North of the South Island receives its seed mussels from the northern end of the North Island where they are stranded on the shores of the hundred mile beach then delivered by truck.

Temperate fishes and aquatic plants for garden ponds and aquaria are transported by road, and are often exchanged privately. Fish such as the sterlet *Acipenser ruthenus,* pumpkinseed *Lepomis gibbosus* and various forms of ornamental carp are generally available and often sold in garden centres. Aquatic plants that include *Crassula helmsii, Azolla filiculoides, Elodea canadensis* and *E. nutallii, Hydrocotyle ranunculoides* are available at most garden centres, yet these plants are known to cause problems on release to the wild (Wallentinus 2002).

Road consignments also have unexpected capabilities for potential inoculations. For example, molluscs may be imported for processing from distant regions. This can happen once monitoring for dinoflagellate toxins in molluscs reveal concentrations above a standard level causing a discontinuation to local harvesting. This deficit in production can cause processors to supplement their orders with molluscs from unaffected areas. In one case mussels were imported to Ireland from the Venice Lagoon. These had spawned during the journey and were not accepted with the result that the entire consignment was dumped on the shore. Such acts could have consequences for introducing species.

An importation of the European flat oyster, *Ostrea edulis*, from the United States (it had been deliberately introduced to California at an earlier time) to France, carried with it the then unknown protozoan blood disease *Bonamia ostreae.* This caused a decline in the native flat oyster production. The Pacific oyster *C. gigas* was flown from Japan to France to replace the decline in production of the flat oyster in the mid-1970s. These large imports were accompanied by other species that passed brine-dip controls following arrival. These included pests and parasites (Gruet *et al.* 1976) and almost certainly some diseases and a syndrome 'summer mortality' (Cheney *et al.* 2000). Almost twenty years later this stock was transported to other maritime states by truck within the European Union following the open trade agreement in January 1993. This included a movement to areas where disease-free stock had been in culture following quarantine and for the first time losses from 'summer mortality' occurred. This syndrome is known to be carried by stocks of Pacific oysters where quarantine precautions were not exercised (Cheyney *et al.* 2000).

Recreational activities such as watersports that involve widespread road transport of small craft (see Chapter 6), plus the hobbies of gardeners and fish fanciers, may all result in deliberate and inadvertent releases of non-native aquatic species.

4. Transport by aircraft

Movements of live air-cargo did not generally take place until the 1960s. The development of plastics greatly aided this trade, as these are used for hygienic insulation and containment. For many transmissions the organisms need to be held close to the airport prior to departure where back-up facilities are available should flights not run on schedule.

4.1 LONG DISTANCE JOURNEYS

Most aquatic animals require attention when they are in transit over several days, or they may lose condition and weaken. Aircraft can transmit consignments rapidly and reduce this risk. Some of these movements can be over 10,000km and undertaken within a day. Consignments of Atlantic eel elvers *Anguilla anguilla* have been regularly exported to Taiwan and Japan from Europe by plane for the cultivation of eels for the Asian markets (Aoyama *et al.* 2000). Organisms can be held in water within sealed bags inflated with oxygen and carried within insulated boxes (e.g. aquarium products including tropical fish), or carried under damp conditions in coolboxes (e.g. live angling bait including nereid polychaetes).

Tropical fish are well suited to the air-conditioned and centrally heated homes of many western nations in cooler climates. Such aquarium species are in demand in North America, Japan and Europe and with the network of airline routes they are easily distributed from the Indo-Pacific, Caribbean, South America and Africa. In 2000 in the United States there were approximately 11 million aquarium enthusiasts and many were able to order material for their aquaria which include 'living rocks' which consist of corals, worms, snails, bivalves, crustaceans, tunicates etc. The industry worldwide at the beginning of this century was worth, annually, more than 25 billion US$ (Padilla & Williams 2004). Tropical species are normally cultured or gathered and brought to holding areas near departure points before transmission. A matter of concern for some years (Adams *et al.* 1970) has been the fact that infections can be spread within holding areas by stressed and dying fish, and consignments of apparently healthy fish have revealed harmful pathogens in the water associated with them (Shotts *et al.* 1976). Future consignments could include a further spread of the Epizootic Ulcerative Syndrome that has reduced fish-farm production in the Indo-Pacific. The World Aquarium Council has a code of practice that ensures that trade between members reduces risks from mortality and disease and that the source areas are not overexploited, but not all traders follow this code.

Aquaculture stock movements include, for example, the transmissions of fertilised salmonid eggs that form an important part of production in some world regions. Fertilised eggs are easily transported and have been distributed worldwide. Rainbow trout *Oncorhynchus mykiss*, originally from the Northeast Pacific region, have been introduced to temperate regions, including mountain areas in tropical regions (MacCrimmon 1971) and remote islands. Consignments may consist of triploid fish or of special stock selections. Transmission of some aquatic organisms by aircraft has developed into a regular trade. Pacific oysters are exported from North America at the eyed-larval stage and are settled on shells on arrival. Unexpectedly the larvae survive in this state while in transit without undue harm. In the past, stock movements of shrimp, salmon and oyster led to the transmission of serious diseases (Sinderman 1993) but, by reducing the biomass of a consignment and by moving species at a smaller size, the risk of transporting pathogens and pests is considerably reduced.

In the 1970s, following serious declines in production of the native oyster, *Ostrea edulis*, the Japanese oyster was imported from Japan to Europe. Subsequently, the unexpected importa-tion of a sporozoan blood disease *Bonamia ostreae* of *O. edulis*, imported directly by plane from trial cultures in the north-east Pacific led to a fall in native oyster production in France (Goulletquer *et al.* 2002). Imports of half-grown Pacific oysters were flown to France to supplement this deficit. On arrival these oysters were dipped in brine to kill species attached to the shell. However, several Japanese plants and invertebrates were introduced and became established (Goulletquer *et al.* 2002). History repeats itself, because these included species that had been introduced to the north-east Pacific from Japan some thirty years earlier.

Stocks of micro-algae are held at institutions to provide sterile cultures for hatcheries (e.g. in the rearing of filter feeding molluscs). The micro-algae are frequently dispatched by air for use in areas outside their natural range. In Ireland such marine algae of distant origin have been used in shellfish cultivation (Table 1). The algae used are considered beneficial, and their deliberate or accidental release to the wild has not been considered harmful.

North American lobsters *Homarus americanus* compete favourably in price with the almost identical European lobsters. Their airfreight from the east coast of North America to Europe is part of a regular though seasonal trade. Lobsters are packed in insulated boxes and following arrival are held in tanks. Should the lobsters be held in shore tanks pathogens (e.g. the bacterium *Gaffkaemia* that causes a wasting disease) may escape to the wild. Other as yet undescribed diseases may also be spread with unforeseen consequences (Dove *et al.* 2004). Although releases of American lobsters are legally not permitted, some have taken place, hybrids of *H. americanus* and *H. gammarus* having been found in coastal areas in Europe (Hopkins 2002).

4.2 SHORT DISTANCE JOURNEYS

Aircraft are also responsible for local movements and can provide an efficient means of, for example, getting fish fingerlings to remote angling lakes, as occurs in North America. Planes distributing fish must gain altitude and descend slowly to avoid gas disease due to pressure changes. Helicopters are frequently used for short transmissions of fish, such as the movements of smoltified salmon from freshwater to marine cages. Some species such as scallops must be transmitted within twelve hours of removal from water to water (Minchin *et al.* 2000) and have been successfully transported from Ireland to France and Spain.

Reef fish, mainly composed of serrainids, labrids and lutjanids, are transmitted by air from the Indo-Pacific region to markets in southern Asia; Hong Kong restaurants receive about 25% of this trade. In 1995 trade in reef fish was valued at about 1 billion US$. There are concerns that this trade, which began in the 1960s, is now unsustainable (Warren-Rhodes *et al.* 2003).

5. Discussion

Over the past 200 years transport processes, and their frequency of operation, have changed due to advances in technology. The great majority of exotic species have been spread during this time and, with the development of new transport routes and expansion of recreational activities and hobbies, this will continue. New technology has meant that some forms of transport no longer operate. For example, transatlantic seaplanes were used in passenger services. These carried bilge-water and plants entangled on mooring lines or on canvas sea anchors. Eno *et al.* (1997) suggested that the red alga *Pikea californica* may have arrived in the south of Britain in this way, but this is disputed (Maggs & Ward 1996). Carlton (1979) did suggest that this transport mode may have been important in the transmission of marine species into the Salton Sea in California. Some products are no longer distributed in the same way. In the 1820s zebra mussels arrived in several North European ports. They had been attached to floating logs which had been transported by ship (Kinzelbach 1992). Now timber is sawn and bailed before export, therefore, this vector no longer operates.

Table 1. Sources of stock of algal cultures used in Ireland for the purpose of cultivation in aquaculture. Species underlined are in regular usage. Source: B. Ottway pers. comm.

Organism	Source of Irish cultures	Where isolated	Occurrence of species in Irish waters
DIATOMA			
Skeletonema costatum	Dunstaffnage, Scotland & Algarve, Portugal	Rhode Island, Long Island USA & Algarve	Yes
Thalassiosira pseudonana	Maine, USA & Toronto, Canada	Long Island, USA	Probably
Chaetoceros muelleri	Dunstaffnage, Scotland	Hawaii, USA ?	Not known
Chaetoceros simplex var. *calcitrans*	Dunstaffnage, Scotland	Japan	Probably
Nitzschia sp	Eilat, Israel	Eilat, Israel	Not known
Phaeodactylum tricornutum	Plymouth, England	Plymouth, England	Yes
CHLOROPHYTA			
Nannochloris atomus	Dunstaffnage, Scotland	England	Yes
Nephroselmis pyriformis	Plymouth, England	Plymouth, England	Yes
Micromonas pusilla	Plymouth, England	Plymouth, England	Yes
Pyramimonas sp	Plymouth, England?	English Channel?	Probably
Tetraselmis tetratheala	Dunstaffnage, Scotland	Probably UK	Yes
Tetraselmis suecica	Dunstaffnage, Scotland & Plymouth, England	Italy	Yes
Tetraselmis chui	Dunstaffnage, Scotland	Millport, Scotland	Yes
Chlorella sp.	-	-	-
EUSTIGMATOPHYTA			
Nannochloropsis spp	Austin, USA & Dunstaffnage, Scotland	Japan & Scotland	Probably
PRYMNESIOPHYTA			
Isochrysis galbana	Plymouth, England	Isle of Man	Yes
Isochrysis sp (=Tahitian isochrysis)	Dunstaffnage, Scotland	Tahiti	Probably not
Pseudoisochrysis paradoxa	?	Chesapeake Bay, USA	?
Pavlova lutheri	Dunstaffnage, Scotland	Millport, Scotland	Yes
Pleurochrysis carterae	Dunstaffnage, Scotland	UK	Yes
Hymenomonas elongata	Dunstaffnage, Scotland	Millport, Scotland	Yes
CRYPTOPHYTA			
Rhodomonas baltica	Menai Bridge, Wales	Channel Islands, UK	Probably
Rhinomonas reticultata	Milford, USA	Milford	Probably

The majority of those exotics that are unintentionally spread worldwide are carried in water ballast or on hulls of ships, whereas airfreight currently poses the greatest threat from deliberate movements of products worldwide. Airfreight ensures that aquatic species, together with their associated biota, are favourably treated in transit and are seldom challenged beyond their normal physiological limits. It is because of these controlled conditions that the risk of future introductions will continue. Airfreight transport risks can be more easily managed at the point of departure rather than at the point of arrival, from which it may be rapidly distributed. The management of the unintentional spread of species by shipping is far more difficult to control. Ports that have a history of primary inoculations need to consider the routes and volumes of ships coming from different world regions and identify those species that need to be monitored. Countries that normally receive species from neighbouring regions should monitor ports for those species that may yet arrive.

The main change in trends of non-bulk type products is from a ship and rail dominated transport in the 1800s to an air and road-dominated transport since the early 1960s. Airports act as hubs for the immediate collection and radial dissemination to receptor regions from where smaller consignments spread. Aquatic species are rapidly disseminated because of the stress that can be posed by travel. Some operators provide quarantine facilities, most usually re-circulation systems, for revival on arrival and the removal of moribund individuals. Subsequent transport is seldom undertaken by rail because road transport provides a more flexible, extensive and direct network for distribution without restriction of time schedules, thus enabling rapid distribution. Currently, the belief held by management that there is no danger posed by the introduction of any species unable to survive a winter period, because this lies outside their normal physiological range, needs to be reviewed. If the species is released, their parasites or diseases may be capable of survival and cross to native species during the warmer seasonal periods, or the species itself may find warm-water outflows (e.g. from power stations) that could provide over-wintering refuges.

The great majority of biota moved by ships is unintended. As ships have destinations in ports, these are the regions where monitoring for recent arrivals should take place. Unless an early knowledge of an arrival takes place it is seldom possible to eliminate an unwanted species. Full port surveys have been devised in Australia (Hewitt & Martin 1996) and it was such a survey that identified the presence of *Mytilopsis sallei*, and a direct eradication programme led to its elimination (Thresher 1999). However, such surveys are expensive and time consuming. An alternative approach is to monitor for a specific suite of species known to be in transit worldwide. The 'next pest' list approach is a rational and practical approach, but is based on harmful biota known to already exist in some world regions; but this does not take into account those species that have apparently benign behaviours (Hayes *et al.* 2002). Another problem is that micro-biota, from virus to protozoa, are usually overlooked because of a lack of knowledge and difficulties posed by identification. Genetic markers would enable a better understanding of the risks involved. Unless rapid genetic markers are identified, this area is likely to remain under-studied. A rapid diagnostic PCR kit to reveal the presence of white spot syndrome (WSSV) a serious virus disease of shrimp is now available (UPM 2005). The virus has spread between shrimp farms most probably with movements of contaminated larval shrimp; but other wild and farmed marine and freshwater crustaceans may act as carriers. This disease was recorded from Taiwan in 1993 and has since spread through much of the Indo-Pacific, and by 1999 was found in the Americas. It is possible the virus could be carried by other crustaceans such as copepods, frequent in ballast water, or barnacles that foul ship hulls. There is now the possibility of evaluating whether the virus can be carried in ships ballast water or as hull fouling. It may even be possible for small craft to distribute the virus. The knowledge gap of such transmissions urgently requires attention.

The positioning of aquaculture operations in ports where large volumes of ships ballast water are released may lead to serious future difficulties from releases of pathogens that could have consequences for human and animal health. Indeed in Mobile Bay in the United States cessation of oyster harvesting was necessary following the accumulation of the bacterium *Vibrio cholerae* virulent strain 01. The strain appeared in the ballast water of five of twelve ships examined (McCarthy & Khambaty 1994) and it is probable that a cholera outbreak in south-east Asia was caused by the same strain as an epidemic in Peru which had started a short time previously. Ships may also carry diseases and parasites that could lead to declines in production of aquaculture species. Once established the organism may easily be spread elsewhere in the course of normal culture operations. Should the organism be species-specific its spread could have serious consequences for the production of those species cultured worldwide. Some, such as the Pacific oyster *C. gigas* are frequently carried on the hulls of ships and to date there has been no investigation undertaken on the risk this poses. As aquaculture and shipping activities both normally have requirements for sheltered bays it is difficult to see how this issue can be fully resolved.

Managing ballast water is a complex issue and one major problem is the treatment of ballast water over short journeys because it is unlikely that treatment can be effective for voyages that only take a few days. In the future new vessels will, no doubt, be designed to reduce traveling time even further thereby increasing the risk, unless ships in the future are designed in such a way that they do not require 'conventional' ballasting systems. Nevertheless the great majority of conventional craft will be in service for many more years and the issue needs to be managed now. Short voyages result in a higher survival of species and of different taxa leading to a greater probability of secondary spread. Invasive species on arrival normally expand their range locally and then spread to nearby areas so that, over time, more source regions for further expansion are developed. Unless these smaller scale movements of species can be controlled, many species that already exist in a biological province will expand their ranges.

Some of the unexpected and unexplained events worldwide could perhaps be associated with releases of ships' ballast water, particularly where microbiota may be involved. There have been several notable events worldwide since ballast water has been in general usage. These include the high mortality of the pilchard *Sardinops sagax* off the Australian coast (Hyatt *et al.* 1997), the die-off of the sea urchin *Diadema* in the Caribbean, the appearance of turtle tumours in the Atlantic Ocean (Williams & Sinderman 1992), the white spot disease of corals (Raymundo *et al.* 2003), the haplosporidian parasite of cultured bay scallop *Argopecten irradians* in China (Chu *et al.* 1996), the withering syndrome of black abalone *Haliotus cracherodii* in California (Gardner *et al.* 1995), the summer mortality of the mussel *Mytilus edulis* in the Gulf of St Lawrence (Myrand & Gaudereault 1995). The eel-grass wasting disease occurred in many widely separated areas of the world is difficult to explain otherwise. Indeed the causative organism *Labrynthula macrocystis* has several congeners known from ballast tanks (Hulsman & Galil, 2002).

A majority of metazoa carried by ships will be from sheltered estuaries. Because of this they are likely to be seasonally abundant and tolerant of fluctuations of temperature and salinity and of turbid water. Such robust species may cope easily with the challenges of transmission through different environments to new destinations. Consequently estuarine species that are abundant in world ports are at high risk of transmittal. The species most likely to create problems in other world regions are best known within the home range, so information on such species could be gathered and exchanged as risk donor species with other world regions. Such a process could greatly aid in early identifications and management.

Very often ballast water has been ascribed as the causative mode of transmission, but in some cases it is equally as plausible that hull fouling may be responsible. Incorrect deductions of a vector process could lead to targeting the wrong ship class in a transmission. It is important, if management is to succeed in managing the spread of a species; that actual means of transmission is accurately identified. For example, gobies are widely distributed by ships. The extended ranges of these fish may be due to their crevice dwelling behaviour. They could take up residence within sea chests and be transmitted as adults, or eggs could be laid attached to the hull surface. The development time within the egg-capsule could result in them being carried over large distances before hatching. It has generally been assumed that the round goby *Gobius melanostomus* arrived on the Polish coast of the Baltic Sea from the Ponto-Caspian region in ballast water (Skóra 1999). However, the spread as hull fouling was not fully considered. Because this goby has benthic larval and adult stages it is unlikely sufficient numbers could be taken-up in ballast water. However, should eggs have been carried on a hull, their hatching could have resulted in a release of a large number of individuals and, as there are weak tidal currents in the southern Baltic ports, they would be unlikely to be dispersed far, so a breeding population could have been established.

Over the centuries there have been great changes in trading routes between ports. These changes often involved political events. Indeed new trade routes continue to evolve and some species as a result are likely to develop cosmopolitan distributions. Increased trade within the European Union is likely to result in a continued spread of Ponto-Caspian species. There will be an expansion of trade with China and south-east Asia as economies in that region continue to grow. However, one of the most significant events likely to occur within a few decades is the development of a direct route between the North Atlantic and the Pacific boreal and temperate regions. Hitherto transmissions between these two oceans involved passage through warm tropical water conditions *via* routes through the Suez and Panama canals or south of Africa or South America. With the continuing retreat of the Arctic ice-sheet, seasonal shipping routes between the Pacific and Atlantic oceans will provide the first cold-water route and be of considerable consequence in the transmission of cold-water biota, opening up a new invasion corridor.

While it is most unlikely that the spread of all unwanted species can be controlled or curbed those species that are deliberately distributed in trade from specific regions are susceptible to effective management control. Codes of practice for many industries do not always take into account their role in the spread of exotic species. Improved codes of practice for different sectors and general education towards a shared responsibility can reduce much of the release of species to the wild. The awareness of exotic species as being an environmental threat is difficult for many to realise when much of the current prosperity in aquaculture, as in agriculture, is based on non-native species farming. The International Council for the Exploration of the Sea has a code of practice that is used can considerably reduce the risks of spreading unwanted biota (ICES 2003).

Historically, large numbers of unwanted species have been distributed along with value products. For trade to prosper there must be guidelines to prevent transmissions of harmful species that are likely to cause serious impacts. However, too many restrictions on the movement of a commercial species may erode the viability of that trade. In the case of species used in trade over a long period of time, there may be incomplete checks, so presenting windows of opportunity for spread. This may only need to take place on a single occasion, to compromise an entire industry. Should the responsible organism be recognised at an early stage, so that it remains confined to a small area, control may be possible. However, once it disperses from such a hub its elimination is unlikely to succeed. Disease restrictions on traded species are based on known diseases, yet new diseases continue to be described. As a result any movement of a species should be considered capable of creating some risk and this is the basis of a risk assessment and is a likely future component in environmental audits.

In the management of transmissions of exotic species awareness by the public is critically important in reducing environmental impacts. This awareness, as well as the appropriate actions to be taken, need to be considered by those involved in managing the different forms of transport. Educational outreach to these industries needs to become a standard means of reducing overall risk and could lead to a positive outcome for each transport industry.

Acknowledgments

I thank Rick Boelens, Stephan Gollasch, Roger Mann, Toril Moen, Sergej Olenin, Brian Ottway and Mohamed Shariff whose assistance and discussions contributed to this work. Without the encouragement from John and Julia Davenport this account would have not been written. This work was supported, in part, under the EU 6th Framework Programme 'Assessing Large Scale environmental risks for biodiversity with tested methods (ALARM)', contract GOCE-CT-2003-506675).

References

Adams RM, Remington JS, Steinberg J, Seibert J. 1970; Tropical fish aquariums. A source of *Mycobacterium marinum* infections resembling sporotrichosis. Journal of the American Medical Association 211: 457-461.

Allen YC, Thompson BA, Ramcharan CW. 1999; Growth and mortality rates of the zebra mussel, *Dreissena polymorpha*, in the lower Mississippi River. Canadian Journal of Fisheries and Aquatic Sciences 56: 748-759.

Alzieu C, Heral M, Thiabud Y, Dardignac KJ, Feuillet M. 1982 ; Influences des peintures antisalissures a base d'organostanniques sur la calcification de la coquille de l'huître. Revue Travailles Institute de Pêche Maritimes 45: 101-106.

Aoyama J, Wantanbe S, Miyai T, Sesai S, Nishida M, Tsukamoto K. 2000 ; The European eel, *Anguilla anguilla* (L.), in Japanese waters. Dana 12: 1-5.

Apte S, Holland BS, Godwin LS, Gardner PA. 2000; Jumping ship: a stepping stone event mediating transfer of a non-indigenous species via a potentially unsuitable environment Biological Invasions 2: 75-79.

Blanchard M. 1995 ; Origine et état de la population de *Crepidula fornicata* (Gastropoda Prosobranchia) sur le littoral français. Haliotis 24: 75-86.

Boalch GT. 1993; The introduction of non-indigenous marine species to Europe: Planktonic species. In: Bourderesque CF, Briand F & C Nolan (eds) Introduced species in European Coastal Waters, 28-31. Ecosystems Research Report 8.

Borja A, Valencia V, Franco J, Muxika I, Bald J, Belzunce MJ, Solaun O. 2004; The Water Framework Directive: water alone, or in association with sediment and biota, in determining quality standards. Marine Pollution Bulletin 49: 8-11.

Bower SM, McGladdery SE, Price IM. 1994; Synopsis of infectious diseases and parasites of commercially exploited shellfish. Annual Review of Fish Diseases 4: 1-199.

Bryan GW, Gibbs PE, Hummerstone GL, Burt GR. 1986; The decline in the gastropod Nucella lapillus around south-west England: evidence for the effect of tributyltin from antifouling paints. Journal of the Marine Biological Assocciation of the United Kingdom 66: 611-640.

Burnell G, Davenport JL. 2004; A review of the use of tributyltin (TBT) in marine transport and its impacts on caostal environments. In: Davenport J & Davenport JL (eds) The effects of human transport on ecosystems: cars and planes, boats and trains, 267-276. Royal Irish Academy, Dublin.

Buttermore RE, Turner E, Morrice MG. 1994; The introduced northern Pacific seastar Asterias amurensis in Tasmania. Memoirs of the Queensland Museum 36(1): 21-25.

Carlton JT. 1979; History, biogeography and ecology of the introduced marine and estuarine invertebrates of the Pacific Coast of North America. PhD Dissertation. University of California: Davis, California, USA.

Carlton JT. 1982; The historical biogeography of Littorina littorea on the Atlantic coast of North America and implications for the interpretation of the structure of New England communities. Malacological Review 15: 146.

Carlton JT. 1988; Changes in the sea: the mechanisms of dispersal of marine and aquatic organisms by human agency. Journal of Shellfish Research 7(3): 552.

Carlton JT. 1996; Marine bioinvasions: the alteration of marine ecosystems by non-indigenous species. Oceanography 9 (1): 1-8.

Carlton JT. 2001; Introduced species into US coastal waters: environmental impacts and management priorities. Pew Oceans Commission, Arlington, Virginia.

Carlton JT, Geller JB. 1993; Ecological roulette: the global transport of nonindigenous marine organisms. Science 261: 78-82.

Carlton JT, Hodder J. 1995 Biogeography and dispersal of coastal marine organisms: experimental studies on a replica of a 16th –century sailing vessel. Marine Biology 121: 721-730.

Champ MA, Pugh WL. 1987; Tributyltin antifouling paints: introduction and overview. In: The Ocean ' An international Workplace' Proceedings International Organotin Symposium 4: 1296-1308.

Cheyney DP, MacDonald BF, Elston RA. 2000 Summer mortality of Pacific oysters, *Crassostrea gigas* (Thunberg): initial findings on multiple environmental stressors in Puget Sound, Washington, 1998. Journal of Shellfish Research 19(1): 353-359.

Chu F-LE, Burreson EM, Zhang F, Chew KK. 1996; An unidentified haplosporidian parasite of bay scallop *Argopecten irradians* cultured in the Shandong and Liaoning provinces of China. Diseases of Aquatic Organisms 25:155-158.

Cohen AN, Carlton JT. 1995; Nonindigenous aquatic species in a United States estuary: a case study of the biological invasions of the San Francisco Bay and Delta. Report to United States Fish and Wildlife Service, Washington D.C.

Coutts ADM. 1999; Hull fouling as a factor in species transport across the Tasman Sea, MSc Thesis, Faculty of Fisheries and Marine Environment, Australian Maritime College, Tasmania, Australia.

Dove ADM, LoBue C, Bowser P, Powell M. 2004; Excretory calcinosis: a new fatal disease of wild American lobsters *Homarus americanus*. Diseases of Aquatic Organisms 58:215-221.

Drake LA, Choi K-H, Ruiz GM, Dobbs FC. 2001; Global redistribution of bacterioplankton and viroplankton communities. Biological Invasions 3: 193-199.

EU 2000; Directive 2000/60/EC of the European Parliament and of the Council 23 October 2000.

Gardner GR, Harshbarger JC, Lake JL, Sawyer TK, Price KL, Stephenson MD, Haaker PL, Togstad HA. 1995; Association of prokaryotes with symptomatic appearance of withering syndrome in Black abalone *Haliotus cracherodii*. Journal of Invertebrate Pathology 66: 111-120.

Gollasch S. 1996; Untersuchungen des Arteintrages durch den internationalen Schiffsverkehr unter besonderer Berücksichtigung nichtheimischer Arten. Dissertation University of Hamburg; Verlag Dr. Kovac, Hamburg.

Gollasch S. 1999; *Eriocheir sinensis*. In: Gollasch S, Minchin D, Rosenthal H & Voigt M (eds) Exotics Across the Ocean. Case Histories on Introduced Species, 55-60. Logos Verlag, Berlin.

Gollasch S. 2002; The importance of ship hull fouling as a vector of species introductions into the North Sea. Biofouling 18: 105-121.

Gollasch S, Lenz J, Andres H-G. Dammer M. 2000; Survival of tropical ballast water organisms during a cruise from the Indian Ocean to the North Sea. Journal of Plankton Research 22: 923-937.

Gollasch S, Macdonald E, Belson S, Botnen H, Christensen JT, Hamer JP, Houvenaghel G, Jelmert A, Lucas I, Masson D, McCollin T, Olenin S, Persson A, Wallentinus I, Wetsteyn LPMJ, Wittling T. 2002; Life in ballast tanks. In: Leppäkoski E, Gollasch S, Olenin S (eds). Invasive aquatic species of Europe: Distribution, Impact and Management, 217-231. Kluwer, Dordrecht, the Netherlands.

Gollasch S, Olenin S (eds). Invasive aquatic species of Europe: Distribution, Impact and Management, 240-252. Kluwer, Dordrecht, the Netherlands.

Gollasch S, Rosenthal H, Botnan H, Crncevic M, Gilbert M, Hamer J, Hülsmann N, Mauro C, McCann L, Minchin D, Öztürk B, Robertson M, Sutton C, Villac MC. 2003; Species richness and invasion vectors: sampling techniques and biases. Biological Invasions 5: 365-377.

Goulletquer P, Bachelet G, Sauriau P-G. Noel P. 2002; Open Atlantic coast of Europe – a century of introduced species into French waters. In: Leppäkoski E, Gollasch S, Olenin S (eds). Invasive aquatic species of Europe: Distribution, Impact and Management, 276-290. Kluwer, Dordrecht, the Netherlands.

Gruet Y, Heral M. Robert J-M. 1976; Premières observations sur l'introduction de la faune associée au naissain d'huîtres Japonaises *Crassostrea gigas* (Thunberg), importé sur la côte Atlantique Française. Cahiers de Biologie Marine 17: 173-184.

Hallegraeff GM. 1993; A review of harmful algal blooms and their apparent global increase. Phycologia 32: 79-99.

Hallegraeff GM. Bolch CJ. 1992; Transport of diatom and dinoflagellate resting spores in ships' ballast water: implications for plankton biogeography and aquaculture. Journal of Plankton Research 14(8): 1067-1084.

Hamer JP. 2002 Ballast tank sediments In: Leppäkoski E, Gollasch S, Olenin S (eds). Invasive aquatic species of Europe: Distribution, Impact and Management, 232-234. Kluwer, Dordrecht, the Netherlands.

Harding JM, Mann R. 1999; Observations on the biology of the veined rapa whelk, Rapana venosa (Valenciennes, 1846) in the Chesapeake Bay. Journal of Shellfish Research 18: 9-17.

Hayes KR, McEnnulty FR, Sliwa C. 2002; Indentifying potential marine pests – an industive approach. Final Report to Environment Australia, National Priority Pests Project.

Hewitt CL. 2002; The distribution and diversity of tropical Australian marine bio-invasions. Pacific Science 56(2): 213-222.

Hewitt CL, Martin RB. 1996; Port surveys for introduced marine species – background considerations and sampling protocols. Centre for Research on Introduced Marine Pests, CSIRO Division of Fisheries, Hobart, Tasmania Technical Report 4.

Hopkins CCE. 2002; Introduced marine organisms in Norwegian waters, including Svalbard. In: Leppäkoski E, Eno NC, Clark RA, Sanderson WG. 1997; Non-native marine species in British waters: a review and directory. Joint Nature Conservation Committee, Peterborough, UK.

Hopkins CCE, 2002. Introduced marine organisms in Norwegian waters including Svalbard. In: Leppakoski E, Gollasch S, & Olenin S (eds), Invasive aquatic species of Europe, distribution, impacts and management, 240-252. Kluwer Academic Publishers, Dordrecht.

Howard AE. 1994; The possibility of long distance transmission of *Bonamia* by fouling on boat hulls. Bulletin of the European Association of Fish Pathologists 14(6): 211-212.

Hülsmann N, Galil BS. 2002; Protists – a dominant component of the ballast – transported biota. In: Leppäkoski E, Gollasch S & Olenin S (eds). Invasive aquatic species of Europe: Distribution, Impact and Management, 20-26. Kluwer, Dordrecht, the Netherlands.

Hyatt AD, Hine PM, Jones JB, Whittington RJ, Kearns,C, Wise, TG, Crane MS, Williams LM. 1997; Epizootic mortality in the pilchsrd *Sardinops sagax neopilchardus* in Australia and New Zealand in 1995. II. Identification of a herpesvirus within the gill epithelium. Diseases of Aquatic Organisms 28: 17-29.

ICES 2003; ICES Code of Practice on the Introductions and Transfers of Marine Organisms (www.ices.dk) (accessed 4-04-2005).

IMO. 2004; International Maritime Organisation 2004. International Convention on the Control and Management of Ship's Ballast Water and Sediments. www.globallast.imo.org (accessed 04-04-2005).

Jansson K. 1994; Alien species in the marine environment. Introductions to the Baltic Sea and Swedish west coast. Swedish Environmental Protection Agency Report 4357: 1-68.

Johnsen BO, Jensen AJ. 1991 The *Gyrodactylus* story in Norway. Aquaculture 98: 289-302.

Kinzelbach R. 1992; The main features of the phylogeny and dispersal of the zebra mussel *Dreissena Polymorpha*. In: Neumann D & Jenner HA (eds) The Zebra Mussel *Dreissena Polymorpha*, 5-17. Gustav Fisher Verlag, New York.

Lambert G. 2002; Non-indigenous ascidians in tropical waters. Pacific Science 56(3): 291-298.

Lee RF. 1991; Metabolism of tributyltin by marine animals and possible linkages to effects. Marine Environmental Research 32: 29-35.

Lockett MM, Gomon MF. 2001; Ship mediated fish invasions in Australia: two new introductions and a consideration of two previous invasions. Biological Invasions 3: 187-192.

Loosanoff VL. 1969; Development of shellfish culture techniques. Proceedings of the Conference on Artificial Propagation of Commercially Valuable Shellfish, October 22-23, 1969 pp9-40. College of Marine Studies, University of Delaware.

MacCrimmon HR. 1971; World distribution of rainbow trout (*Salmo gairdneri*). Journal of the Fisheries Research Board of Canada 28: 663-704.

Maggs CA, Ward BA. 1996; The genus *Pikea* (Dumontiaceae, Rhodophyta) in England and the North Pacific: Comparative morphological, life history, and molecular studies. Journal of Phycology 32: 176-193.

Matheickal JT, Raaymakers S. (eds) 2004; Proceedings of the 2nd International Ballast Water Treatment Symposium, IMO, London 2003. GloBallast Monograph Series No. 15.

McCarthy SA, Khambaty FM. 1994; International dissemination of epidemic *Vibrio cholerae* by cargo ship ballast and other nonpotable waters. Applied and Environmental Microbiology 60(7): 2597-2601.

McDermott JJ. 1991; A breeding population of the western Pacific crab *Hemigrapsus sanguineus* (Crustacea: Decapoda: Grapsidae) established on the Atlantic coast of North America. Biological Bulletin 181: 195-198.

McMinn A, Hallegraeff G, Smith J, Lovell A, Jenkinson A, Heijnis H. 2000; Recent appearance of *Gymnodinium catenatum* at Port Lincoln, South Australia In: 9th International Conference on harmful Algal Blooms. Conference Abstracts. Intergovernmental Oceanographic Commission of UNESCO, Paris, France.

Menzel RW. 1974; Portugese and Japanese oysters are the same species. Journal of the Fishereis Research Board of Canada 31: 453-456.

Middelboe M, Glud RN. 2003 Distribution of viruses and bacteria in relation to diagenic activity in an estuarine sediment. Limnology and Oceanography 48(4): 1447-1456.

Minchin D, Duggan CBD. 1988; The distribution of the exotic ascidian Styela clava in Cork Harbour. Irish Naturalists' Journal 22: 388-393.

Minchin D, Gollasch S. 2002; Vectors – how exotics get around. In: Leppäkoski E, Gollasch S, & Olenin S (eds). Invasive aquatic species of Europe: Distribution, Impact and Management, 183-192. Kluwer, Dordrecht, the Netherlands.

Minchin D, Gollasch S. 2003; Fouling and ships' hulls: how changing circumstances and spawning events may result in the spread of exotic species. Biofouling (Supplement) 19: 111-122.

Minchin D, Haugum G, Skjaeggestad H, Strand O 2000; Effect of air exposure on scallop behaviour, and the implications for subsequent survival in culture. Aquaculture International 8: 169-182.

Mortensen Th. 1927; Handbook of the echinoderms of the British Isles. Oxford University Press.

Myrand B, Gaudreault J. 1995; Summer mortality of blue mussels (*Mytilus edulis* Linnaeus, 1758) in the Magdelan Islands (Southern Gulf of St Lawrence, Canada) Journal of Shellfish Research 14(2): 395-404.

Nagabhusanam R, Sarojini R. 1997; An overview of Indian research efforts on marine wood-boring and fouling organisms. In: Nagabhushanam R & Thompson M-F (eds) Fouling organisms of the Indian Ocean, biology and control technology, 1-30. A.A. Balkema, Rotterdam.

Nehring S. 1997; Dinoflagellate resting cysts from recent German coastal sediments Botanica Marina 40: 307-324.

Orlov YuI, Ivanov BG. 1978; On the introduction of the Kamchatka king crab *Paralithodes camtschatica* (Decapoda :Anomura : Lithodidae) intoi the Barents Sea. Marine Biology 48: 373-375.

Padilla DK, Williams SL. 2004; Beyond ballast water: aquarium and ornamental trades as sources of invasive species in aquatic systems. Frontiers in Ecology and Environment 2(3): 131-138.

Penney AJ, Griffiths AB. 1984; Prey selection and the impact of the starfish *Marthasterias glacialis* (L.) and other predators on the mussel *Chloromytilus meridionalis* (Krauss). Journal of Experimental Marine Biology and Ecology 75: 19-36.

Petryashov VV, Chernova NV, Denisenko SG, Sundet JH. 2002; Red king crab (*Paralithodes camtschaticus*) and pink salmon (*Oncorhynchus gorbuscha*) in the Barents Sea. In: Leppäkoski E, Gollasch S & Olenin S (eds) Invasive aquatic species of Europe: Distribution, Impact and Management, 147-152. Kluwer, Dordrecht, the Netherlands.

Quayle DB. 1969; Pacific oyster culture in British Columbia. Bulletin of the Fisheries Research Board of Canada 169: 1-192.

Raymundo LJH, Harvell CD, Reynolds TL. 2003; *Porites* ulcerative white spot disease: description, prevalence, and host range of a new coral disease affecting Indo-Pacific reefs. Diseases of Aquatic Organisms 56: 95-100.

Savini D, Harding,JM, Mann R. 2002; Rapa whelk *Rapana venosa* (Valenciennes, 1846) predation rates on hard clams *Mercenaria mercenaria* (Linneaus, 1758). Journal of Shellfish Research 21: 777-779.

Shotts EB, Kleckner AL, Gratzek JB, Blue JL. 1976 Bacterial flora of aquarium fishes and their shipping waters imported from Southeast Asia. Journal of the Fisheries Research Board of Canada 33: 732-735.

Sindermann CL. 1993; Disease risks associated with importation of nonindigenous marine animals Fisheries review 54(3) 1-10.

Skóra K. 1999 *Gobius melanostomus*. In: Gollasch S, Minchin D, Rosenthal H & Voigt M (eds) Exotics Across the Ocean. Case Histories on Introduced Species, 69-73. Logos Verlag, Berlin.

Taylor A, Rigby G, Gollasch S, Voigt M, Hallegraeff G, McCollin T, Jelmert A. 2002; Preventative treatment and control techniques for ballast water. In: Gollasch, S., Minchin, D., Rosenthal H. & Voigt, M. (eds.): Exotics Across the Ocean. Case histories on introduced species, 484-510. Logos Verlag, Berlin.

Thresher RE. 1999; Diversity, impacts and options for managing invasive marine species in Australian waters. Journal of Environmental Management 6: 137-148.

UPM 2005; www.vet.upm.edu.my/english/research3.html (03 2005).

Ventilla RK. 1982; The scallop industry in Japan. Advances in Marine Biology 20: 309-382.

Wallentinus I. 2002 Introduced marine algae and vascular plants in European aquatic environments. In: Leppäkoski E., Olenin S & Gollasch S (eds): Invasive Aquatic Species of Europe: Distributions, Impacts and Management, 27-54. Kluwer, Dordrecht, the Netherlands.

Warren-Rhodes K, Sadovy Y, Cesar H. 2003; Marine ecosystem appropriation in the Indo-Pacific: a case study of the live reef fish food trade. Ambio 32(7): 481-488.

Williams EH, Sindermann CJ. 1992; Effects of disease interactions with exotic organisms on the health of the marine environment In: De Voe RM (ed) Proceedings of the Conference and Workshop on: Introductions and transfers of marine species: achieving a balance between economic development and resource protection. pp. 71-77. Sea Grant USA.

Williamson AT, Bax NJ, Gonzalez E, Geeves W (eds) 2002; Development of a regional risk management framework for APEC economies for use in the control and prevention of introduced pests. APEC Marine Resource Conservation Working Group. Asia-Pacific Economic Cooperation Secretariat.

WNT 2003; World navies today www.hazegray.org/worldnav/usa/decom.htm (accessed 04-02-2005).

Wolff WJ, Reise K. 2002; Oyster imports as a vector for the introduction of alien species into northern and western European coastal waters. In: Leppäkoski E, Olenin S & Gollasch S. (eds) Invasive Aquatic Species of Europe: Distributions, Impacts and Management, 27-54. Kluwer, Dordrecht, the Netherlands.

Würtz J, Taraschewski H. 2000; Histopathological changes in the swimbladder wall of the European eel *Anguilla anguilla* due to infections with *Anguillicola crassus*. Diseases of Aquatic Organisms 39: 121-134.

Zaitev Y, Ozturk B. 2001; Exotic species in the Aegean, Marmora, Black, Azov and Caspian seas. Turkish Marine Research Foundation, Istanbul, Turkey. Publication No 8 265pp.

In this chapter, we examine hull fouling on small craft as a transportation vector for NIS in widely dispersed regions including temperate and tropical environments. We provide summaries of NIS incursions associated with small craft movements, outline the factors that make small craft susceptible to fouling, document a recent general increase in the abundance of small craft and associated industries, discuss the likely dispersal routes of NIS by small craft and make recommendations for managing the risks of small craft fouling and NIS transportation.

2. Incursions of non-indigenous species associated with small craft movements

Over the past three decades, private and commercial small craft have been implicated in the spread of a number of aquatic NIS that includes both animals and plants (Table 1). Spread of these species has occurred in marine, brackish and freshwater environments, with transportation of organisms by hull fouling, entanglement on projecting equipment or from fragmentation or drop-off of individuals.

Fouling of hulls is the principal mode of transmission for many marine and freshwater animal and plant NIS, and this includes the movements of small craft with examples from most world regions (Table 1). Founder populations have become established within commercial shipping ports and marinas. These species may subsequently spread along coastlines or to remote embayments as a result of natural dispersal, aquaculture stock movements, shipping and small craft. Occasionally as in the case of the *Mytilopsis sallei* invasion to Darwin, Australia, the introduction will have been as a result of a primary inoculation, in this case by overseas yachts in March 1999 (Thresher 1999).

Small craft are also likely to have also contributed to the spread of NIS having become entangled in fishing gear or on anchor chains. The mutant form of the green alga (*Caulerpa taxifolia*) was first observed in the Mediterranean Sea near a public aquarium in 1984. By 2003, the species was found on the coasts of six Mediterranean countries: spreading from Monaco to France, Italy, Spain and Croatia and Tunisia (Ribera Siguan 2003). *C. taxifolia* is often retrieved with anchors and so could have been carried in this way from the Ligurian Sea to the Balearic Islands, a known leisure craft route (Cornell 2002). Even releases of small algal fragments, tolerant of being held within anchor lockers, can result in new populations (Sant *et al.* 1996). Fragments entangled in commercial fishing nets, often cleaned while in harbour, could also create new populations (Ceccherelli & Cinelli 1999; Relini *et al.* 2000; Meinesz *et al.* 2001).

It is indeed possible that diseases may be spread with small craft. At present the evidence for this is inconclusive but requires further investigation and this aspect should be included in risk assessments in the management of living resources. The viral disease, infectious salmon anaemia (ISA), can be transmitted without direct contact by wild and reared salmonids (Jones & Groman 2001) and well-boats (boats with slatted openings to cargo holds allowing a continuous water exchange) used to carry salmon smolts may be responsible for this virus transmission between fish-farms (Stagg *et al.* 2001). A further example is from barge movements that may have transmitted a disease of oysters to neighbouring bays on the southwest coast of Britain (Howard 1994). The spread of the white spot syndrome virus (WSSV) in the Indo-Pacific and Central America may also be spread to neighbouring areas by crustaceans acting as carriers fouling boat hulls.

Table 1. Documented cases of NIS introductions where small craft have been recorded or implicated in their spread.

Species	Native range	Introduced to	Spreading pattern	Impacts	References
Caulerpa taxifolia	Strain evolved from aquarium industry	Mediterranean Sea	Secondary spread with small craft and fishing vessels.	Forms dominant stands and excludes other biota.	Ribera Siguan 2003; Sant et al. 1996; Meinesz et al. 2001.
Sargassum muticum	Japan, Korea, China, SE Russia	NE Atlantic, NE Pacific,	Secondary spread with oyster movements and small craft.	Forms dominant stands, exclusion and hydrography changes.	Wallentinus 1999.
Undaria pinnatifida	Japan Sea, Korea, China, SE Russia	New Zealand, Australia, Northern Europe, Argentina	Port, marina and natural environments. Natural, boat and shipping mediated dispersal. Local spread with small craft and aquaculture stock movements.	Possible displacement of native algal species.	Hay & Luckens 1987; Fletcher & Farrell 1998; Curiel et al. 1999; Talman et al. 1999; Forrest et al. 2000, Wallentinus 1999.
Mytilopsis sallei	Central America and Caribbean	Australia, India	Invasive within enclosed environments and estuaries.	Nuisance fouling, displacement of native species.	Ganapati et al. 1971; Rao et al. 1989; Thresher 1999; NIMPIS 2002a.
Perna viridis	Indo-Pacific	Australia, Gulf of Mexico and Carribbean	Alongshore dispersal and with craft.	Nuisance fouling, displacement of native species.	NIMPIS 2002b; Neil et al. 2002.
Dreissena polymorpha	Ponto-Caspian river basins	Northern Europe and North America	Secondary spread via waterways or overland transport of small craft, downstream dispersal.	Nuisance fouling, displacement of native species.	Kinzelbach 1992; Minchin et al. 2003; Pollux et al. 2003.
Caprella scaura	Australia	Northern Adriatic Sea	Local spread by small craft fouling.	Unknown.	Hale 1929; Sconfietti & Danesi 1996; Sconfietti et al. (in press)
Tricellaria inopinata	Probably Pacific origin	Northern Europe	Secondary spread by fouling of naval and recreational vessels.	Nuisance fouling, displacement of native bryozoans.	d'Hondt & Occhipinti Ambrogi 1985; Occhipinti Ambrogi 1990; Dyrynda et al. 2000; De Blauwe & Faasse 2001; Fernández Pulpeiro et al. 2001.
Watersipora subtorquata	Largely unknown	Australasia	Fouling of commercial and recreational vessels. Provides a non-toxic substrate for other species.	Increased drag, enhanced fouling.	Allen 1953; Banta 1969; Gordon & Mawatari 1992; Hewitt et al. 1999; Floerl 2002; Wisely 1962; Ng & Keough 2003.
Elminius modestus	Australasia	Northern Europe	Secondary spread to harbours and remote bays by commercial and recreational vessels.	Increased drag.	Beard 1957; O'Riordan 1996.
Ficopomatus enigmaticus	Indo-Pacific	Europe, New Zealand, Australia	Secondary spread to estuaries and lagoons by ships and small craft.	Nuisance growths on hulls and navigational structures.	Kilty & Guiry 1973; Read & Gordon 1992.

3. Hull fouling on small craft: influencing factors and prevalence in locations worldwide

3.1 SMALL CRAFT MARINAS – SOURCES OF FOULING

Generally, private and commercial small craft are moored within designated small craft harbours ('marinas') or on pile or buoy moorings within or near commercial ports (Richardson & Ridge 1999; Floerl 2002). Most modern marinas consist of floating pontoons held in place by arrays of vertical piles or chains. Marinas are often surrounded by protective breakwaters or other wave-dampening devices (Floerl & Inglis 2003). These structures of concrete, wood and metal provide horizontal, vertical, subtidal and intertidal habitats for sessile organisms, under varying conditions of shelter. Most marinas give berthing for 100-1,000 vessels that supply an extensive surface area of artificial hard-substrate habitat; in some localities this may exceed the area for attachment granted naturally. A marina with 250 berths can have ~3,600 m^2 of pontoon and piling surfaces. The largest marina in the southern hemisphere, New Zealand's Westhaven Marina, has 1832 berths and supplies ~31,000m^2 of settlement space (Richardson & Ridge 1999). Pontoon and piling surfaces in most marinas are usually fully occupied by fouling assemblages (Floerl 2002).

3.2 HUMAN FACTORS INFLUENCING HULL FOULING

Any vessel used in the sea requires some means of reducing the growth of fouling or of damage to the hull by boring organisms. A frequent method for many small craft is to remove them from the water after each short period of use. However, for many craft this approach is not practical. Therefore, virtually all vessels that are permanently kept in seawater use toxic antifouling paints that are applied to submerged parts of the hull and that temporarily prevent the accumulation of fouling assemblages. Depending on the type of antifouling used, these paints can usually prevent hull fouling for periods of 9-18 months given regular use of the vessel (Floerl 2002). Data on antifouling paint ages on yachts sampled in New Zealand and Australia indicate that the potential for hull fouling is considerable. Of 920 yachts surveyed, 51% had an antifouling paint aged >9 months, and on 25% of these yachts the paint was older than 18 months.

Once fouling assemblages have developed on vessel hulls, they are normally removed via a combination of manual scraping or scrubbing with wire brushes, and using high-pressure water jets. Following fouling removal, new antifouling paint is usually applied to the hulls. However, this process is costly and generally amounts to US$~1,000 per boat annually. This is because vessels must be removed from the water and because good quality paints are expensive. The cleaning of hulls alone by using snorkellers/divers or by careening (tidal dry-out) is cheaper. However, there are only short-term benefits if the hull is not also painted. This is because the hull is seldom entirely cleaned using these methods. Commonly, traces of barnacle and oyster cement, byssal threads and soft tissues of sponges, hydroids or ascidians may remain and act as positive settlement cues and non-toxic platforms for a range of larvae (Burke 1986; Pawlik 1992). In an experiment conducted in Queensland, Australia, yacht hull surfaces that had been cleaned by scraping and brushing attracted six times more fouling organisms (recruits/cm^2) over a 14-day period than surfaces that had been cleaned using the same method but subsequently sterilised (Floerl et al. in press a). Small craft that periodically have their hulls cleaned manually in the water thus have an enhanced potential to transport fouling species between places. In Australia and New Zealand, 53% of 920 yachts sampled had their hull manually cleaned by divers in the past (Floerl 2002).

3.3 ENVIRONMENTAL FACTORS INFLUENCING HULL FOULING

The protection of many marina basins by continuous breakwalls can cause a local entrainment of water and propagules during periods of rising tides, resulting in a localised concentration of larvae and enhanced rates of recruitment (Floerl & Inglis 2003). Craft residing in such marinas tend to develop fouling assemblages that reflect the surrounding populations on pontoon and piling surfaces. Floerl and Inglis (2005) found a positive relationship between (i) the age of the antifouling paint on the vessels' hulls and the time they have resided in a marina, and (ii) the number of sessile species occurring on marina surfaces that had colonised the hulls of these vessels. However, this relationship may be altered by severe disturbance events. Severe freshwater flooding, such as that associated with the monsoonal season in the tropics, can kill off a substantial proportion of the fouling assemblages on marina surfaces and resident vessel hulls, overriding the influence of antifouling paint age on fouling abundance.

3.4 PREVALENCE OF HULL-FOULING ORGANISMS ON SMALL CRAFT IN LOCATIONS WORLDWIDE

Since it is difficult and expensive to prevent the colonisation of small craft hulls by sessile marine animals and plants, it is not surprising that the prevalence of hull fouling on these vessels is quite high (Table 2).

Table 2. Sessile marine taxa encountered on small craft hulls in Italy (five marinas), Ireland (four marinas), Spain (two marinas), Hawaii (two marinas) New Zealand (five marinas) and Australia (three marinas).

	Italy[1]	Ireland[2]	Spain[2]	Australia[4]	Hawaii[2,5]	New Zealand[4]
Number of craft sampled	15	132	96	70	139 & 12	783
Number of species	n/a	n/a	n/a	45	n/a	n/a
Macroalgae	+	+			+	+
Sponges	+	+	+	+	+	+
Hydroids		+	+	+	+	+
Anthozoans		+				
Serpulids	+	+	+	+	+	+
Other polychaetes				+	+	+
Barnacles	+	+	+	+	+	+
Amphipods	+	+		+		+
Isopods	+					
Tunicates	+	+	+	+	+	+
Bryozoans	+	+	+	+	+	+
Bivalves	+	+	+	+	+	+
Fish eggs	+					

[1] Savini and Occhipinti-Ambrogi, unpubl. Data, [2] Minchin, unpubl. Data, [3] Floerl et al. 2005; Floerl unpubl. Data, [4] Floerl 2002, [5] Godwin et al. 2004.

In Italy, 40% of 623 small craft sampled in marinas on the North Adriatic coast had fouling assemblages on their hulls (Savini & Occhipinti Ambrogi unpubl. data) (Figure 1). In Ireland, yachts used in competition were found devoid of noticeable fouling whereas those that were not used for racing could carry extensive fouling assemblages (Dan Minchin pers. obs.). All of the seventy domestic and international yachts surveyed by Floerl (2002) in Queensland

(Australia) were fouled. During 2002 and 2003, a total of 783 international yachts were surveyed upon their arrival in New Zealand; 14.5% of these vessels were found to have fouling organisms on their hulls (Floerl *et al.* 2005). In all six international locations, representative species of most sessile marine taxa were sampled from small craft hulls, and the area occupied by fouling assemblages ranged from 1% up to 97% of submerged hull surfaces (Floerl 2002; Table 2).

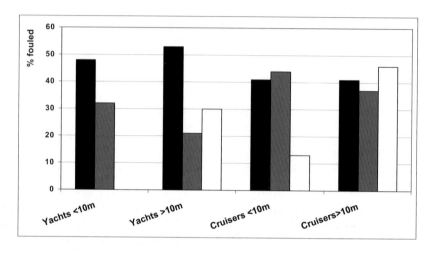

Figure 1. Percentage of craft (to 30 m) found fouled at three marinas in the Northern Adriatic Sea. Total boats examined: Aprilia (black) – 111, Cesenatico (grey) – 292 and Fano (white) – 220.

3.4.1. *Fouling of hulls*
Yachts within marinas in Ireland, Spain and Hawaii were examined under calm and good light conditions from pontoons. Levels of fouling were arbitrarily defined on a four-point scale from no noticeable fouling to heavy incrustation (Plate 1). These surveys were followed up with boatyard studies to score the principal fouling taxa.

Plate 1. Hull fouling of small craft showing relative fouling levels used in surveys. Left-top: no noticeable fouling, some diatomaceaous films some occasional algal filaments. Left-middle: some fouling – invertebrates and/or algae clearly visible with <33% cover. Left-bottom: moderate fouling with protrusion from the surface with <66% cover. Right-top: heavy fouling – extensive protrusion from the hull surface normally composed of several species and with up to 100% cover. Right-middle: fibre-glass hull demonstrating damaged gel-coat and osmosis blistering and basal plaques of barnacles which if untreated would provide a suitable surface for the development of heavy fouling over a short time-period. Right-bottom: Angling boat with zebra mussel attachment after two summers and a winter in water.

Greatest fouling occurred in the warmest water. There was an accretion of fouling on heavily fouled craft in Hawaii sufficient for pufferfish to feed on (Figure 2). Crusts of barnacles, molluscs and tube-worms were common and of red algae to 60 cm in length. Whereas in the Mediterranean Sea yachts had less fouling protruding from the hull but had algal films and fine filamentous green algae on which mullet appeared to feed. In Irish waters at Crosshaven, an estuary, the levels of fouling were less using the observation method used (Figure 2). However, on occasion, algae and tunicate fouling can be extensive on idle craft at sheltered marine sites. Thus, it would appear that the level of fouling on yachts in different world regions varies, with less in cooler regions. Reasons for this variation could be due to the influence of temperature on the generation time of the principal biota and paints becoming less effective sooner in warmer seas. There are some other contributory factors. Fibre-glass hulls should contain <0.5% water under the outer gel coat. This level can increase with prolonged immersion and can be prevented with periodic removal from the water. Where this is not done blisters can form (osmosis) and break away to leave roughened pits on the hull surface that can enhance opportunities for fouling biota attachment (Plate 1)

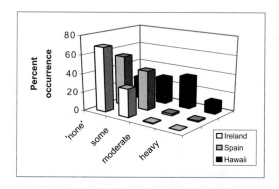

Figure 2. Relative levels of fouling at marinas in Crosshaven, Ireland (132 craft); Mallorca, Spain (139) and Honolulu, Hawaii (36) (Minchin unpub.).

3.4.2. Seasonal movements
Compilation of data from yachts visiting Hawaii during a 2-year period show that, from all source regions, there are trends by season. The main feature for craft is the avoidance of periods with tropical storms.

In Mallorca, berthing charges are high due to local congestion. In advance of the winter period many small craft overwinter at marinas in the south of Spain or in Turkey.

There are specially constructed ships for the transport of dozens of small craft at one time. These behave like floating dry-docks and, following changes in the water ballasting, enable small craft to 'swim-on' and 'swim-off'. Craft may be transported between north and south Europe, to either side of the Atlantic and between the Atlantic and Pacific Oceans in this way. These services are provided throughout the year and are popular for cruisers that otherwise would require large amounts of fuel for such journeys and in the smaller cruisers fuel space would compromise the living space aboard. It is not known what potential for NIS transmission such craft movements have. Movements from the Mediterranean to the Caribbean are favoured following the hurricane season.

4. Long distance routes and global patterns of small craft movements

4.1 VOYAGES

The majority of small craft long distance voyages are undertaken by sailing yachts. Yachts travel at comparatively low speeds compared with large engine powered commercial ships, and fouling assemblages acquired in coastal environments are generally not dislodged by drag when the vessels are under way. There are some specific patterns and situations that determine the passage of yachts across oceans:

- Yachts follow specific routes to take advantage of wind speed and direction and ocean current patterns to reduce the number of travelling days. The most practical overall route on world cruises is to sail westwards. Sailing craft generally follow the trading routes of former sailing ships that are likely to have regularly carried particular suites of fouling species (Cornell 2002).

- Tropical storms such as cyclones of hurricanes develop in tropical latitudes during parts of the year (the monsoonal period generally ranges from December to March). The tracks of these storms tend to be from east to west and veering northwards in the Northern Hemisphere and southwards in the Southern Hemisphere. Cruising in such areas at these times is unwise and normally avoided.

- A world trip will take two or more years. This requires extensive planning.

Given the popularity of established routes for intercontinental sailing trips, and the need for having to avoid certain regions on account of tropical storms, yachtsmen plan their routes to reduce the risks that can arise in poor sea states (Figure 3). For example, in New Zealand, visiting yachts move south from abroad to arrive between October and December to avoid the austral tropical cyclone season. The number of international yacht arrivals to New Zealand has increased considerably from <10 to over 450-550 over the past 30 years (Grant &Hyde 1991). Most (~ 95%) international yachts arrive at Opua, in the Bay of Islands. The previous destinations of 783 international arrivals sampled in 2002-2004 included 31 tropical South Pacific regions, notably Fiji (34.5% of all arrivals), Tonga (32%), a range of tropical Pacific island nations (20%), Australia (8.2%) and Vanuatu (2.8%). Prior to sailing to New Zealand, these yachts spent from 1 day to 6 years in these locations (median: 21 days; Floerl unpubl. data), and 14.3% arrived with visible fouling on their hulls (Floerl et al. 2005). Many of the fouling organisms on these yachts, acquired in tropical environments, are unlikely to survive in the colder New Zealand waters. However, some (~ 7% of total) yachts arrived from the similar temperate and sub-tropical Australian ports of Hobart, Sydney and Melbourne. These locations contain known invasive species that include the European crab (*Carcinus maenas*), the Japanese seastar (*Asterias amurensis*), the Mediterranean fanworm (*Sabella spallanzanii*) and the alga (*Caulerpa taxifolia*) (Hewitt et al. 1999). If imported to New Zealand on vessel hulls, these species may be able to survive in coastal waters.

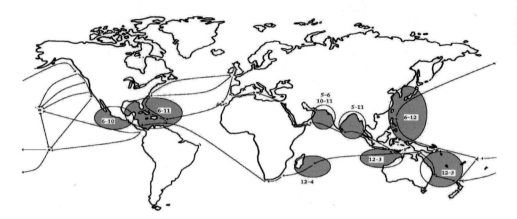

Figure 3. Principal world yachting routes (arrows) and storm regions (shaded areas). Months of storms arranged numerically. (Data from Cornell 2002)

Small craft arrive at the Big Island or Oahu in the Hawaiian Islands at all times of year. A study of a sample of yachts over a two-year period showed the majority arrived during the months from May to July (Godwin *et al.* 2004) (Figure 4). Arrivals at this time avoid the tropical storm periods on either side of the Pacific Ocean. The smaller numbers arriving during August to October may be due to planning to avoid the storm periods and will have arrived earlier. Interviews with five boat skippers living aboard in Waikiki Marina confirmed this view. The storm periods in Australasian waters covers a period from December to March. The majority of vessels (50%) arrive from the east (the main around-the-world route) and 33% from the south (Figure 5) (Godwin *et al.* 2004).

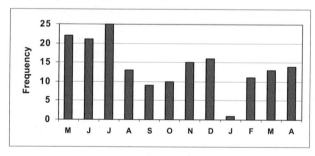

Figure 4. Arrivals of yachts by month to the Hawaiian Islands, May 2001 to May 2003 (data from Godwin *et al.* 2004). Tropical storm periods: East Pacific – June to October; Pacific – June to December; Australasian region – December to March

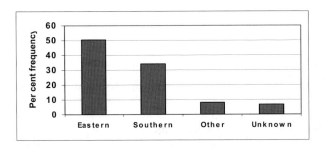

Figure 5. Frequency of likely routes used by yachtsmen to arrive in Hawaii (data from Godwin *et al.* 2004)

Ireland tends to receive international yachts predominantly from the Caribbean, the east-coast of America and from other northern European destinations. The majority arrive in Cork Harbour, Kilmore Quay or Waterford Harbour (Cornell 2002). In a sample of fourteen marinas on all Irish coasts there was a 100% prevalence of visiting boats from the British Isles, Northern Europe and Scandinavia, an 86% prevalence from North America, 64% from the Mediterranean, 28% from the Baltic Sea and 14% from South America. Six marinas had members who in the last three years had embarked on world tours. Vessels arriving from northern Europe tend to sail to the southwest coast of Ireland. Few vessels range to the northwest and north coasts as these areas have large expanses of unprotected coastline. Within the Irish Sea there are general movements to either side that often include the Isle of Man.

Italy receives world-trip traffic *en route* through the Mediterranean Sea (Cornell 2002). There are two principal gateways to the Mediterranean: Gibraltar, for craft arriving from Northern Europe and the Americas, and Port Said on the Suez Canal for vessels arriving from the Indo-Pacific. Yachts on world cruises will most usually enter the Mediterranean Sea *via* the Suez Canal where they remain for at least a season, usually the spring and/or summer. Yachts entering *via* the Strait of Gibraltar planning to reach the eastern Mediterranean Sea do so from late spring. The preferred routes include the Balearics, French Riviera, Sicily, Malta, Greece, Cyprus, Israel and then Port Said. Yachts sailing westwards from the Suez Canal also prefer the same destinations. Yachts entering the eastern Mediterranean from the Red Sea usually arrive in the early spring, almost 2-3 months before those arriving from the western Mediterranean.

4.2. DOMESTIC CRAFT

In many coastal locations around the world, domestic small craft are frequently moored within or in the vicinity of centres of international shipping activity. They are thus able to become colonised by NIS established in these locations or present on the hulls of international craft moored in their vicinity (Apte *et al.* 2000). However, there is strong variation in the potential of domestic craft to assist in the spread of marine NIS. This is principally because a large proportion of domestic small craft are seldom used and/or tend to remain locally rather than travel for long distances. In New Zealand, for example, 692 (=75%) of 923 domestic yachts sampled remained within 100 km of their homeports during a 12-month period, and 125 of these remained in their homeport without moving at all. Only 10% of the 923 yachts had undertaken 50 or more trips away from their homeport within 12 months. However, these boats in total visited 72 coastal locations around New Zealand, and they remained in these places for periods ranging from 1 day to 11 months. Generally, in New Zealand domestic yachts can be divided into three groups that differ in their potential to act as carriers of NIS: A first group (approximately 20% of 923 New Zealand yachts sampled) consists of yachts that are exclusively used in competition and are regularly cleaned to reduce drag and maximise speed and manoeuvrability (Yachting New Zealand pers. comm.; Floerl unpubl. data) and so are unlikely to pose a significant risk, especially since the great majority race only locally. A second group consists of yachts that are occasionally used for primarily short (<100km from homeport) day-trips but spend most of the time inactively moored in coastal marinas. Such craft are often likely to carry fouling assemblages on their hulls, and these may contain NIS occurring around their homeport, but these boats are unlikely to significantly spread them along the coastline. Amongst New Zealand yachts, this second group comprises approximately 75%. A third, small group of yachts are primarily used to travel between widely separated destinations (>100km to nationwide). During these trips, the yachts may spend days to months in foreign marinas. Maintenance of some of these craft is irregular and such boats could act as efficient dispersal vectors for NIS. In New Zealand, 5% of 923 yachts sampled fell into this category.

5. The increase in the private boating industry

In recent years, especially the 1980s, there has been a worldwide increase in the number of small craft and their associated marinas and moorings. In Queensland, Australia, for example, the number of coastal marinas has increased eleven-fold, from five in 1960 to 54 in 2000 (Floerl 2002). In Italy, small craft marinas and ports have increased in number by 73% between 1985 (403 marinas) and 2002 (716 marinas), and currently provide mooring space for approximately 116,000 small craft (Pagine Azzurre 1986, 2003). In Ireland, no marinas existed in the mid-1970s and coastal moorings were used for small craft. Today, there are 29 marinas along the coast and four further marinas planned. This increase in marina development was concomitant with an increase in the numbers of small craft. In Queensland (Australia), for example, the number of registered seaworthy yachts (>8m length) doubled between 1985 and 2000, from ~8,000 to ~16,000 boats (Floerl 2002).

Generally, small craft moorings are not uniformly distributed around the coastline but are instead characterised by regions of higher and lower densities. This can be described using the Index of Recreational Port Capability (RPCI). This index refers to the number of moorings/marina berths available within a given region of coastline in kilometres (Occhipinti-Ambrogi 2002). Concentrations of marinas and berthing space (high RPCI) are likely to be associated with high frequencies of small craft movements. Since the probability for establishment of NIS is a function of propagule supply (Ruiz *et al.* 2000), regions with a high RPCI may have a higher chance of harbouring or aiding in the spread of NIS compared with those regions with lower RPCI index. The average RPCI for Italy is 16, with the highest concentrations of berths occurring on the coasts of Veneto and Friuli (northwestern Adriatic) with a RPCI of 90 or higher. Occhipinti-Ambrogi and Savini (2003) identified the Northern Adriatic as one of the Mediterranean regions prone to invasion. RPCI values in this region have recently increased by 10-20%, which may lead to an increase in inoculation by NIS or their propagules. In this region the number of marina ports has almost doubled since 1985 (Figure 6).

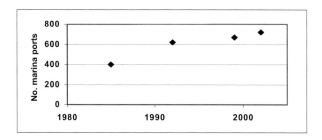

Figure 6. The number of marina ports on Italian coasts 1985-2003.

6. Risk assessment of leisure boat fouling

The Failure Modes and Effects Analysis (FMEA) is a system employed by industry to reveal risk areas that could lead to health and safety problems. Each identified component in the analysis develops a level of risk which accumulates a progressively higher risk level each time the same component is identified by an experienced operator. This approach has been used to find faults in equipment design, processing methods and where human behaviour may elevate risk. The first approach in using FMEA analysis for managing aquatic species transmissions was developed by Hayes (2002). He examined fishing vessels as vectors. The analysis consisted of in-depth discussion of each component of a boat that could act as a potential risk of carrying an organism, and included any carried water. These discussions were repeated in different ports and all opinions were reasoned into a series of levels of risk and mathematically arranged.

Here we have used the FMEA approach by examining the fouling of a single species, the zebra mussel (*Dreissena polymorpha*) on different leisure craft in Ireland. Each part of a boat was examined and scored according to the presence and relative number of zebra mussels present at a time when boats were lifted from the water for winter servicing. Study of bilge and toilet water was confined to a small number of craft because of difficulty of access. Areas examined included features found only on some boats (i.e. bow thrusters, rudder spaces) and spaces that developed over time with boat use (e.g. spaces that form between the keel-band and keel, especially of wooden lake-boats) (Table 3).

A further analysis was based on boat behaviour; different craft have different usage patterns according to their design, size and environment. For example, yachts are normally removed from the water each year and cleaned, generally remain faithful to lakes and seldom pass beneath bridges, so do not normally travel far and are unlikely to be significant vectors (Table 4). These two sets of gathered information were combined in the risk assessment to evaluate which craft and where the greatest relative risk of transporting zebra mussels took place. The analysis clearly shows that small open boat hull fouling poses the greatest risk of transmission (Plate 1). This analysis agrees well with the risks perceived as bearing the highest risk of transmissions by management. Hayes (2002) discussed the movement of zebra mussels in North America. There boat behaviour is different and different modes of transmission are perceived to be acting as vectors. In North America boats are regularly moved by trailer, so hull fouling does not create the greatest risk. This is because many boats are kept at home, or trailer parks, and do not remain in the water over long periods. The main transmission method is probably by plants with attached zebra mussels collecting on outboard motors and snagging on trailers when boats are removed from the water, later being transferred to a different water body. Johnson and Padilla (1996) claim that zebra mussels attached to a hull that is removed from the water will not survive more than five days' aerial exposure. In Ireland, however, under cooler and damper conditions, survival can occur for up to 18 days (Pollox *et al.* 2003). There was a further risk that larvae carried in live bait wells were being transmitted between water bodies (Johnson *et al.* 2001). In Ireland bait wells or buckets are not in general use because fishing using live fish as bait is illegal. Further factors that may be considered in risk assessments include distance of dispersal of trailered craft (Buchan & Padilla 1999) and lake size, landscape and facilities (Reed-Anderson *et al.* 2000).

Table 3. Relative hazard of transmissions based on field observations by examining sub-components of vessels as vectors, an approach modified from Hayes (2002). It is assumed that the inoculation capability is low where numbers transmitted are low. Normal font = samples from 1997/98, bold = 2004. 2004 data include 5 vessels imported from marine areas. Hazard levels: + – present unlikely to form an inoculum, ++ – inoculum size may be sufficient for an inoculum, +++ – inoculum likely to be effective. (See also text).

Vector	Boats examined Infested/ boat number	Life stage present	Principal risk	Potential numbers transmitted	Comment	Relative overall hazard
Hull surface	3/56 **7/79**	Juveniles, adults	Waterways international overland	10,000's	Frequent movements	+++
Beneath keel-bands	2/56 **3/79**	Juveniles, adults	Waterways international overland	1,000's	Movements noted, such fouling on old craft	+++
Rudder spaces and pinions	**2/79**	Juveniles, adults	Waterways international	1,000's	Movements noted	+++
Mooring buoy and chain	14/14	Juveniles adults	Waterways	10,000's	Movements unknown	++
Dinghy, tender	3/32	Juveniles adults	Waterways overland	1,000's	Some moved over winter	++
Bow-thruster tunnel	**5/14**	Juveniles, adults	Waterways, international	1,000's	Occurring on cleaned vessels	++
Outside of engine casing	**5/79**	Juveniles adults	Waterways overland	1,000's	On idle craft	++
Water intake ports	**4/79**	Juveniles, adults	Waterways international	100's	Vessels in water 1+ years	++
Propeller shaft	**1/79**	Juveniles adults	Waterways overland	1,000's	On idle craft	++
Propeller	**1/79**	Juveniles adults	Waterways overland	1,000's	On idle craft	++
Projecting instruments, anodes	**1/79**	Juveniles adults	Waterways international	100's	Normally on hull fouled vessels	++
Tyre fenders	4/5	Juveniles adults	Waterways overland	100's	Seldom moved	++
In submerged engine housing	**3/79**	Juveniles adults	waterways	100's	On idle craft	+
Anchors and chain lockers	1/1	adults	waterways	100's	Attached to debris, plants	+
Toilet water	2/3	larvae	Waterways international	10's	<5 litres capacity	+
Bilge water	0/1	larvae	Waterways overland	10's?	Bilge water quality varies	+
Bait bucket	Not sampled	larvae	Waterways overland	10's?	Bait buckets <10 litres	+

Table 4. Nature of craft in inland waters and potential for zebra mussel transmissions. In 1997/98 distinction between small and large cruisers was not made. Normal font = 1997-1998 data, bold = 2004 data. Hazard levels: + – present unlikely to form an inoculum, ++ – inoculum size may be sufficient for an inoculum, +++ – inoculum likely to be effective. (See also text).

Craft type	Boat no.	Maxi-mum fouling burden	Normal range	Comment	Usage pattern	Berthage pattern	Relative hazard
open craft <6m length	12/73	10,000+	Waterways, overland, Imports for special events	Used for angling and wildfowl hunting	Seasonal, Ranging brief to long immersions	Widely spread	+++
Small cruisers <10m length	(8/77) **2/15**	10,000+	Waterways, overland, regular imports	Sometimes trailered	Seasonal, removed in winter	Faithful to marinas, winter berths	+++
Large cruisers >10m length	(8/77) **6/15**	500,000+	Waterways, regular imports	Confined to waterways	Seasonal removed in winter	Faithful to marinas, winter berths	+++
Barges to 30m length	**23/33**	*ca* 20 million	Waterways, some imported from Britain, Netherlands, Germany	Remain in water for years accumulating large fouling burdens	seasonal	Not normally at marinas	+++
Narrow boats to 15m length	3/5	10,000+	Waterways, imported from Britain	Normally in canals generally lower fouling burdens	seasonal	Canal harbours, not usually marinas	+++
Yachts to15m length	**4/20**	100,000+	Waterways, overland, some imports	Confined to lakes on account of bridges	Seasonal, removed in winter	Adjacent to yacht clubs, marinas	++
Powerboats <6m	**2/28**	100's	Overland, waterways	Usually moved by trailer	Highly seasonal	variable	++
Sailing dinghies	34	None noted	Overland	Sailing events	Seasonal, brief immersions	Yacht clubs, public slipways	+
Hire cruisers	4/21	1,000+	Waterways	Well maintained	Seasonal, annual servicing	Special marinas	++

7. Discussion

Small craft worldwide are involved in the spread of a wide range of non-indigenous species of aquatic animals and plants. As we have shown, transport of marine species occurs predominantly via hull fouling on yachts or cruisers undergoing coastal or oceanic voyages. Hull fouling can also facilitate the spread of freshwater species, such as has occurred in the case of the zebra mussel (*Dreissena polymorpha*), in Ireland (Minchin *et al.* 2002). At the level of countries or biogeographic regions, small craft can facilitate the spread of NIS in two different ways: (1) by introducing new, previously absent species, and (2) by facilitating the secondary spread of already established NIS. New introductions can only be facilitated by craft arriving from overseas or from locations within the species introduced range. Secondary dispersal of established NIS can be accomplished by both domestic and foreign vessels (Inglis & Floerl 2002). In Italy and Ireland, small craft are likely to contribute primarily to the secondary spread of established NIS. It is likely that centres of shipping and boating activity, such as commercial ports, popular yachting destinations and international points of entry are at highest risk of containing or becoming inoculated with NIS or their propagules (Hewitt *et al.* 1999; Ruiz *et al.* 2000). During transfers a large proportion of hull fouling assemblages may partially or fully perish en-route following changes in water temperature or salinity (Minchin & Gollasch 2003). However, frequently a proportion of organisms survive coastal or inter-oceanic voyages of long duration and between different physical environments (Bertelsen & Ussing 1936; Allen 1953; Crisp 1958; Foster & Willian 1979; Gollasch & Riemann-Zuerneck 1996; Apte *et al.* 2000).

A global trend of increase in tourism and trade has been matched by a concomitant increase in private and commercial small craft and associated berthing infrastructure. The resulting increase in coastal artificial hard-substrate habitat and frequencies of vessel movements between coastal locations provides increased opportunities for the transfer and establishment of marine NIS (Carlton 1996). In some locations this has already been observed. For example, the sedimentary coastlines of the Northern Adriatic Sea are today protected by an almost continuous belt of stone breakwaters; many of these are associated with small craft marinas. These breakwaters have a total length of approximately 80.6km, offering >800,000m^2 of attachment surface to sessile marine biota. The alien predatory snail (*Rapana venosa*) and the alga *Codium bursa* have increased in abundance while progressively expanding their range along the breakwater series (Airoldi *et al.* 2004; Savini *et al.* 2004). *R. venosa* is dependent on hard surfaces on which to lay its eggs (Savini & Occhipinti Ambrogi 2004), and the provision of such habitat through breakwater systems has facilitated the regional spread of the species, which today occurs at a biomass of up to 7 kg/100 m^2 on breakwaters (Occhipinti Ambrogi & Savini, unpublished data). The shell of *R. venosa* provides a biogenic habitat for native epibiota, dominated by bryozoans, serpulids and barnacles (Savini & Occhipinti Ambrogi 2004).

There are several approaches to limiting the introduction and spread of NIS aquatic species by hull fouling on small craft. One would require an increase in hull maintenance by vessel owners and operators. However, this is difficult to achieve because more frequent dry-docking and antifouling paint renewal is likely to be unaffordable or impracticable to a large proportion of the boating population. A second approach would be the introduction of quarantine protocols for hull fouling on international vessel arrivals. With the exception of the port and marinas around Darwin, Australia, there are no existing quarantine procedures for arriving yachts to prevent the importation of non-indigenous fouling organisms into coastal countries worldwide. In both New Zealand and Australia, quarantine officials check arriving vessels for insects, plant seeds and pets (and their diseases) to limit quarantine risks to human health, agriculture and terrestrial environments (Grant & Hyde 1991; New Zealand and Australian Customs Services, pers. comm. 2001-2003). However, upon entry into Darwin's four marinas, internationally travelled vessels are required to go through another inspection by the Northern Territory Fisheries Department. Any vessel that has not renewed its antifouling paint since its arrival in (or return to) Australia is slipped and the hull inspected. If either the stage of the tide or the boat structure does not allow this method of inspection, then the hull is inspected by divers. In addition, the internal plumbing systems of internationally travelled vessels are subjected to a 5% detergent treatment held in by seacocks for a 14-hour period. In early 2003, approximately 30 potential pest introductions had been intercepted by the inspection of over 700 yachts (Marshall, pers. comm.). In all other Australian and New Zealand ports of entry, submerged hull surfaces of yachts are occasionally checked for illegal objects or attachments, but never for potential problem species they may carry.

Screening systems and risk assessment models have been developed for ballast water (e.g. the Australian Ballast Water Decision Support System) and imports of terrestrial plants and animals (Ruesink *et al.* 1995; Daehler & Carino 2000). Similar tools are required for managing the risk of introducing and spreading NIS on the hulls of ocean-going vessels. The development of predictive tools that allow quarantine officials to efficiently discriminate low-risk vessels from those that may pose a risk (i.e. those that are likely to carry NIS) would help reduce the number of propagules that reach native environments and, therefore, the number of NIS that may establish and (become) spread. Such tools are currently in development (Floerl *et al.* 2004) but have not been fully developed and verified.

In the absence of border quarantine protocols and/or resources to increase hull maintenance on small craft, outreach and education programmes for small craft users, port and marina operators and local and regional management authorities are required to increase their knowledge and awareness of invasive species and their potential impacts on ecosystems and industries, and to provide them with an incentive for reducing the frequency of transport of such organisms on the hulls of small craft worldwide.

Acknowledgments

The Irish Marine Institute supported annual studies on zebra mussels in Ireland assisted by Frances Lucy and Monica Sullivan. We thank Keith Hayes and Chad Hewitt for useful discussion on risk assessment. Part of this account by D. Minchin was supported by the European Union Framework 6th Programme, ALARM contract GOCE-CT-2003-506675. D. Minchin is also grateful for support from the Higher Education Agency, Ireland for funding through the National Development Plan PRTL3 grant 'Ecology and Human Transport' to attend a conference.

References

Airoldi L, Bulleri F, Abbiati M. 2004; Coastal-defence structures as vehicles of transfer of exotic species. Advisory Process\ACME\WorkingGroups\Sgbosv\Wgbosv04\Wgbosv04.Doc 07/05/04: 47.

Allen FE. 1953; Distribution of marine invertebrates by ships. Australian Journal of Marine and Freshwater Research 4: 307-316.

Apte S, Holland BS, Godwin LS, Gardner JPA. 2000; Jumping ship: a stepping stone event mediating transfer of a non-indigenous species via a potentially unsuitable environment. Biological Invasions 2: 75-79.

Banta WC. 1969; The recent introduction of *Watersipora arcuata* Banta (Bryozoa, Cheilostomata) as a fouling pest in southern California. Bulletin of the Southern California Academy of Sciences 68: 248-251.

Beard DM. 1957; Occurrence of *Elminius modestus* Darwin in Ireland. Nature 180: 1145.

Bertelsen E, Ussing H. 1936; Marine tropical animals carried to Copenhagen Sydhavn on a ship from the Bermudas. Videnskabelige Meddelelser fra Dansk naturhistorisk Forening i Kobenhavn 100: 237-245.

Buchan AJ, Padilla DK. 1999; Estimating the probability of long-distance overland dispersal of invading aquatic species. Ecological Applications 9: 254-265.

Burke RD. 1986; Pheromones and the gregarious settlement of marine invertebrate larvae. Bulletin of Marine Science 39: 323-331.

Carlton JT. 1996; Pattern, process, and prediction in marine invasion ecology. Biological Conservation 78: 97-106.

Carlton JT, Hodder J. 1995; Biogeography and dispersal of coastal marine organisms: experimental studies on a replica of a 16th-century sailing vessel. Marine Biology 121: 721-730.

Ceccherelli G, Cinelli F. 1999; The role of vegetative fragmentation in dispersal of the invasive alga *Caulerpa taxifolia* in the Mediterranean. Marine Ecology Progress Series 182: 299-303.

Champ MA, Lowenstein FL. 1987; TBT: The dilemma of antifouling paints. Oceanus 30: 69-77.

Cohen AN, Carlton JT. 1998; Accelerating invasion rate in a highly invaded estuary. Science 279: 555-558.

Coles SL, DeFelice RC, Eldredge LG, Carlton JT. 1999; Historical and recent introductions of non-indigenous marine species into Pearl Harbour, Oahu, Hawaiian Islands. Marine Biology 135: 147-158.

Cornell J. 2002; World cruising routes. McGraw Hill, London 624pp.

Coutts ADM. 1999; Hull fouling as a modern vector for marine biological invasions: investigation of merchant vessels visiting northern Tasmania. Dissertation, Australian Maritime College, Tasmania. 283pp.

Cranfield HJ, Gordon DP, Willan RC, Marshall BA, Battershill CN, Francis MP, Nelson WA, Glasby CJ, Read GB. 1998; Adventive marine species in New Zealand. NIWA Technical Report 34. National Institute of Water and Atmospheric Research, Wellington, 42pp.

Crisp DJ. 1958; The spread of *Elminius modestus* Darwin in North-West Europe. Journal of the Marine Biological Association of the United Kingdom 37: 483-520.

Curiel D, Bellemo G, Marzocchi M, Scattolin M, Parisi G. 1999; Distribution of introduced Japanese macroalgae *Undaria pinnatifida, Sargassum muticum* (Phaeophyta) and *Antithamnion pectinatum* (Rhodophyta) in the Lagoon of Venice. Hydrobiologia 385: 17-22.

Daehler CC, Carino DA. 2000; Predicting invasive plants: prospects for a general screening system based on current regional models. Biological Invasions 2: 93-102.

De Blauwe H, Faasse M. 2001; Extension of the range of the bryozoans *Tricellaria inopinata* and *Bugula simplex* in the North-East Atlantic Ocean (Bryozoa: Cheilostomatida). Neterlandse Faunistische Medelingen 14: 103-112.

Dyrynda PEJ, Fairall VR, Occhipinti Ambrogi A, d'Hondt J-L. 2000; The distribution, origins and taxonomy of *Tricellaria inopinata* d'Hondt and Occhipinti Ambrogi, 1985, an invasive bryozoan new to the Atlantic. Journal of Natural History 34: 1993-2006.

Fernández Pulpeiro E, Cesar-Aldariz J, Reverter Gil O. 2001; Sobre la presencia de *Tricellaria inopinata* d'Hondt and Occhipinti Ambrogi, 1985 (Bryozoa, Cheilostomatida) en el litoral gallego (N.O. España). Nova Acta Cientifica Compostelana. Bioloxia 11: 207-213.

Fletcher RL, Farrell P. 1998; Introduced brown algae in the North East Atlantic, with particular respect to *Undaria pinnatifida* (Harvey) Suringar. Helgolander Meeresuntersuchungen 52: 259-275.

Floerl O. 2002; Intracoastal spread of fouling organisms by recreational vessels. PhD Dissertation, James Cook University, Townsville. 293pp.

Floerl O, Inglis GJ. 2003; Boat harbour design can exacerbate hull fouling. Austral Ecology 28: 116-127.

Floerl O, Inglis GJ. 2005; Starting the invasion pathway: the interaction between source populations and human transport vectors. Biological Invasions 7(4):589-606.

Floerl O, Inglis GJ, Hayden BJ. 2005; A risk-based predictive tool to prevent accidental introductions of non-indigenous marine species. Environmental Management 35(6): 765-778.

Floerl O, Pool TK, Inglis GJ. 2004; Positive interactions between non-indigenous species facilitate transport by human vectors. Ecological Applications 14(6): 1724-1736.

Forrest BM, Brown SN, Taylor MD, Hurd CL, Hay CH. 2000; The role of natural dispersal mechanisms in the spread of *Undaria pinnatifida* (Laminariales, Phaeophyceae). Phycologia 39: 547-553.

Foster BA, Willian RC. 1979; Foreign barnacles transported to New Zealand on an oil platform. New Zealand Journal of Marine and Freshwater Research 13: 143-149.

Ganapati PN, Lakshmana Rao MV, Varghese AG. 1971; On *Congerie sallei* Recluz, a fouling bivalve mollusc in the Vishakhapatnam harbour. Current Science 40: 409-410.

Godwin LS. 2003; Hull fouling of maritime vessels as a pathway for marine species invasions to the Hawaiian Islands. Biofouling 19 Suppl. 123-131.

Godwin LS, Eldredge LG, Grant K. 2004; The assessment of hull fouling as a mechanism for the introduction and dispersal of marine alien species in the main Hawaiian Islands. Bernice Pauahi Bishop Museum, Hawai'i Biological Survey, Bishop Museum Technical Report No 28, Honolulu, Hawaii, 113pp.

Gollasch S. 2002; The importance of ship hull fouling as a vector of species introductions in the North Sea. Biofouling 18(2): 105-121.

Gollasch S, Riemann-Zuerneck K. 1996; Transoceanic dispersal of benthic macrofauna: *Haliplanella luciae* (Verrill, 1898) (Anthozoa, Actinaria) found on a ship's hull in a shipyard dock in Hamburg Harbour, Germany. Helgoländer Meeresuntersuchungen 50: 253-258.

Gordon DP, Mawatari SF. 1992;Atlas of marine fouling bryozoa of New Zealand Ports and Harbours. Miscellaneous Publications of the New Zealand Oceanographic Institute 107: 1-52.

Grant GE, Hyde NH. 1991; Quarantine risks associated with overseas yachts arriving at New Zealand's ports. Ministry of Agriculture and Fisheries, Wellington.

Hale HM. 1929; The crustaceans of South Australia. Part II. Adelaide, Government Printer pp. 201-380

Hay CH, Luckens PA. 1987; The Asian kelp *Undaria pinnatifida* (Phaeophyta: Laminariales) found in a New Zealand harbour. New Zealand Journal of Botany 25: 329-332.

Hayes KR. 2002; Identifying hazards in complex ecological systems. Part 2: infection modes and effects analysis for biological invasions. Biological Invasions 4: 251-261.

Hewitt CL, Campbell ML, Thresher RE, Martin RB. 1999; Marine Biological Invasions of Port Phillip Bay, Victoria. Centre for research on Introduced marine pests. *Technical Report No 20* CSIRO Marine Research, Hobart, 344pp.

d'Hondt J-L, Occhipinti Ambrogi A. 1985; *Tricellaria inopinata*, n. sp., un nouveau bryozoaire cheilostome de la faune méditerranéenne. P.S.Z.N.I: Marine Ecology 6(1): 35-46.

Howard AE. 1994; The possibility of long distance transmission of *Bonamia* by fouling on boat hulls. Bulletin of the European Association of Fish Pathologists 14(6): 211-212.

Inglis GJ, Floerl O. 2002; Risks to marine biosecurity associated with recreational boats. NIWA Client Report CHC02/23. National Institute of Water and Atmospheric Research, Christchurch, 47pp.

James P, Hayden BJ. 2000; The potential for the introduction of exotic species by vessel hull fouling: a preliminary study. NIWA Client Report No WLG 00/51. National Institute of Water and Atmospheric Research, Wellington. 61pp.

Johnson LE, Padilla DK. 1996; Geographic spread of exotic species: ecological lessons and opportunities from the invasion of the zebra mussel Dreissena polymorpha. Biological Conservation 78: 23-33.

Johnson LE, Riccardi A, Carlton JT. 2001; Overland dispersal of aquatic invasive species: a risk assessment of transient recreational boating. Ecological Applications 11(6): 1789-1799.

Johnstone IM, Coffey BT, Howard-Williams C. 1985; The role of recreational boat traffic in interlake dispersal of macrophytes: a New Zealand case study. Journal of Environmental Management 20: 263-279.

Jones SRM, Groman DB. 2001; Cohabitation transmission of infectious salmon anaemia virus among freshwater-reared Atlantic salmon. Journal of Aquatic Animal Health 13: 340-346.

Kilty GM, Guiry GM. 1973; Mercierella enigmatica Fauvel (Polychaeta, Serpulidae) from Cork Harbour. Irish Naturalists' Journal 11: 379-381.

Kinzelbach R. 1992; The main features of the phylogeny and dispersal of the zebra mussel Dreissena polymorpha. In: Neuman D & Jenner HA (eds) The zebra mussel Dreissena polymorpha, Ecology, Biological Monitoring and First Applications in Water Quality management, pp 5-17. Gustav Fisher, Stuttgart.

Meinesz A, Belsher T, Thibaut T, Antolic B, Mustapha K B, Boudouresque CF, Chiaverini D, Cinelli F, Cottalorda J-M, Djellouli A, El Abed A, Orestano C, Grau AM, Ivesa L, Jaklin A, Langar H, Massuti-Pascua lE, Peirano A, Tunesi L, de Vaugelas J, Zavodnik N, Zuljevic A. 2001; The introduced green alga Caulerpa taxifolia continues to spread in the Mediterranean. Biological Invasions 3: 201-210.

Minchin D, Gollasch S. 2003 Fouling and ships' hulls: how changing circumstances and spawning events may result in the spread of exotic species. Biofouling 19: 111-122.

Minchin D, Lucy F, Sullivan M. 2002; Zebra mussel: impacts and spread. In: Leppäkoski E, Gollasch S & Olenin S (eds) Invasive Aquatic Species of Europe: Distribution, Impact and Management. (eds) Kluwer Press, 135-146.

Minchin D, Maguire C, Rosell R. 2003; The zebra mussel (Dreissena polymorpha Pallas) invades Ireland: human mediated vectors and the potential for rapid intranational dispersal. Biology and Environment: Proceedings of the Royal Irish Academy 103B(1): 23-30.

Neil KM. 2002; Asian green mussel and Caribbean tubeworm survey within proposed dredge areas. Report prepared for the Cairns Port Authority by the Queensland Department of Primary Industries Northern Fisheries Centre through the CRC Reef Research Centre. 10 pp.

Ng TYT, Keough MJ. 2003 Delayed effects of larval exposure to Cu in the bryozoan Watersipora subtorquata. Marine Ecology Progress Series 257: 85.

NIMPIS. 2002a; Mytilopsis sallei species summary. National Introduced Marine Pest Information System (eds: Hewitt CL, Martin RB, Sliwa C, McEnnulty FR, Murphy NE, Jones T, Cooper S.). Web publication http://crimp. marine.csiro.au/nimpis.

NIMPIS. 2002b; Perna viridis species summary. National Introduced Marine Pest Information System (eds: Hewitt CL, Martin RB, Sliwa C, McEnnulty FR, Murphy NE, Jones T, Cooper S.). Web publication http://crimp. marine.csiro.au/nimpis.

Occhipinti Ambrogi A. 1990; The spread of Tricellaria inopinata into the Lagoon of Venice: an ecological hypothesis. In: F.P. Bigey (ed.) Bryozoaires actuels et fossiles: Bryozoa living and fossil. Bull. Soc. Sci. Nat. Ouest Fr. Mem. H.S l: 299-308.

Occhipinti-Ambrogi A. 2002; Susceptibility to invasion: assessing scale and impact of alien biota in the Northern Adriatic. In: Alien marine organisms introduced by ships in the Mediterranean and Black seas. CIESM Workshop Monographs, Monaco 20: 69-73.

Occhipinti Ambrogi A, Savini D. 2003; Bioinvasions as a component of global change in a stressed marine ecosystem. Marine Pollution Bulletin 46: 542-551.

O'Riordan RM 1996; The current status and distribution of the Australian barnacle Elminius modestus Darwin in Ireland. In: Keegan BF & O'Connor B (eds) Proceedings of the Irish Marine Science Symposium 1995. Galway University Press, Galway, Ireland, 207-218.

Pagine Azzurre. 1986; Il Portolano dei mari d'Italia, II, Copyright Pagine Azzurre s.r.l., 464pp.

Pagine Azzurre. 2003; Il Portolano dei mari d'Italia, XIX, Copyright Pagine Azzurre s.r.l., 830pp.

Pawlik JR. 1992; Chemical ecology of the settlement of benthic marine invertebrates. Oceanography and Marine Biology Annual Reviews 30: 273-335.

Pollux B, Minchin D, Van der Velde G, Van Allen T, Moon-Van der Staay SY, Hackstein J. 2003; Zebra mussels (Dreissena polymorpha) in Ireland, AFLP- fingerprinting and boat traffic both suggest an origin from Britain. Freshwater Biology 48: 1127-1139.

Rainer SF. 1995; Potential for the introduction and translocation of exotic species by hull fouling: a preliminary assessment. Centre for Research on Introduced Marine Pests, Technical Report No 1. CSIRO Division of Fisheries, Hobart, Tasmania, Australia.

Rao KS, Srinivasa VV, Balaji M. 1989; Success and spread of the exotic fouling bivalve *Mytilopsis sallei* (Recluez) in Indian waters. In: Exotic species in India, pp. 125-127. Asian Fisheries Society, Indian Branch, Mangalore.

Read GB, Gordon DP. 1992; Adventive occurrence of the fouling serpulid *Ficopomatus enigmaticus* (Polychaeta) in New Zealand. New Zealand Journal of Marine and Freshwater Research 25:269-273.

Reed-Anderson T, Bennett EM, Jorgensen BS, Lauster G, Lewis DB, Nowacek D, Riera JL, Sanderson BL, Stedman R. 2000; Distribution of recreational boating across lakes: do landscape variables affect recreational use? Freshwater Biology 43: 439-448.

Relini G, Relini M, Torchia G. 2000; The role of fishing gear in the spread of allochthonous species: the case of *Caulerpa taxifolia*. ICES Journal of Marine Science 57(5):1421-1427.

Ribera Siguan MA. 2003; Review of non-native marine plants in the Mediterranean Sea. In: Leppäkoski E, Gollasch S & Olenin S (eds) Invasive Aquatic Species of Europe: Distribution, Impact and Management, 291-310. Kluwer Press.

Richardson L, Ridge T. 1999; Marinas in New Zealand. A guide for visiting yachts-people, local cruisers, prospective berth owners, and marina operators. Capt. Teach Press, Auckland.

Ruesink JL, Parker IM, Groom MJ, Kareiva PM. 1995; Reducing the risk of nonindigenous species introductions. BioScience 45, 465-477.

Ruiz GM, Carlton JT, Grosholz ED, Hines AH. 1997; Global invasions of marine and estuarine habitats by non-indigenous species: mechanisms, extent, and consequences. American Zoologist 37: 621-632.

Ruiz GM, Fofonoff PW, Carlton JT, Wonham MJ, Hines AH. 2000; Invasion of coastal marine communities in North America: apparent patterns, processes, and biases. Annual Reviews in Ecology and Systematics 31: 481-531.

Sant N, Delgado O, Rodriguez-Prieto C, Ballesteros E. 1996; The spreading of the introduced seaweed *Caulerpa taxifolia* (Vahl) C. Agardh in the Mediterranean Sea: testing the boat transportation hypothesis. Botanica Marina 39: 427-430.

Savini D, Castellazzi M, Favruzzo M, Occhipinti-Ambrogi A. 2004; *Rapana venosa* (Valenciennes, 1846) in the Northern Adriatic Sea: population structure and shell morphology. Journal of Chemistry and Ecology 20: 411-424.

Savini D, Occhipinti-Ambrogi A. 2004; Spreading potential of an invader: *Rapana venosa* in the Northern Adriatic Sea Rapp. Comm int. Mer Mèdit. 37: 548.

Sconfietti R, Danesi P. 1996; Variazioni strutturali in comunità di Peracaridi agli estremi opposti del bacino di Malamocco (Laguna di Venezia). S.It.E. Atti 17: 407-410.

Sconfietti R, Mangili F, Savini D, Occhipinti Ambrogi A. 2005; Diffusion of the alien species *Caprella scaura* Templeton, 1836 (Amphipoda:Caprellidae) in the Northern Adriatic Sea. Biologia Marina Mediterranea 12(1): 335-337.

Skerman TM. 1960; Ship-fouling in New Zealand waters: a survey of marine fouling organisms from vessels of the coastal and overseas trade. New Zealand Journal of Science 3: 620-648.

Stagg RM, Bruno DW Cunningham CO, Raynard RS, Munro PD, Murray AG, Allan CET, Smail DA, McVicar AH, Hasdtings TS. 2001; Epizootiological investigations into an outbreak of infectious salmon anaemia (ISA) in Scotland. FRS Marine Laboratory Report No 13/01.

Talman S, Bite JS, Campbell SJ, Holloway M, McArthur M, Ross DJ, Storey M. 1999; Impacts of some introduced marine species found in Port Phillip Bay. In: Hewitt CL, Campbell ML, Thresher RE & Martin RB (eds) Marine Biological Invasions of Port Phillip Bay, Victoria. Centre for Research on Introduced Marine Pests. Technical Report No.20 pp. 261-274 CSIRO Marine Research, Hobart.

Thresher RE. 1999; Diversity, impacts and options for managing invasive marine species in Australian waters. Journal of Environmental Management 6: 137-148.

Visscher JP. 1927; Nature and extent of fouling of ships' bottoms. Bulletin of the Bureau of Fisheries 43: 193-252.

Wallentinus I. 1999. *Sargassum muticum. Undaria pinnatifida.* In: S. Gollasch, D Minchin, H Rosenthal, M. Voight (eds) Exotics across the ocean: case histories on introduced species. Logos Vertlag Berlin,74pp.

Wisely B. 1962; Effect of an antifouling paint on a bryozoan larva. Nature 193: 543-544.

CHAPTER 7: THE ENVIRONMENTAL IMPACTS OF PRIVATE CAR TRANSPORT ON THE SUSTAINABILITY OF IRISH SETTLEMENTS

RICHARD MOLES, WALTER FOLEY & BERNADETTE O'REGAN
Chemical and Environmental Sciences Department, University of Limerick, Limerick, Ireland

1. Introduction

"Missing in relation to transport is information on the relationship of settlement patterns and traffic, emissions and environmental services. Settlement patterns...are important areas where the balance between dispersion and concentration needs to be carefully struck. Planning laws... need regular assessment in terms of their effects: are more and longer journeys being promoted when all current policies are taken together?" (Scott 2004)

This chapter reports on a study on the impact of private transportation on the sustainability levels in a cross section of Irish towns and a city. It explains the scope and purpose of the study, the 'Sustainability and Future Settlement Patterns in Ireland' (SFSPI), and why transport related issues are important to the sustainable development of Ireland. Research methods adopted are described briefly, and an analysis is provided of some of the results of this study, on the effects on sustainability of private transport choices in Irish settlements. The chapter ends with a brief discussion of the current status of private transport, and possible means of reducing future pollution resulting from private car use. Where comparisons are made with settlements in other countries, these are generally restricted to the developed world. The research was carried out in the Centre for Environmental Research (CER), University of Limerick, and was funded by the Environmental Protection Agency, under the National Development Plan.

The chapter aims to derive a general framework for planning for more sustainable transport:

- Explain and evaluate methods developed to analyse the extent to which transport is sustainable in settlements classified by size, location, functionality and place within the national settlement hierarchy

- Examine the extent to which transport sustainability is related to the size, location, functionality and place within the national settlement hierarchy

- Examine the competing idea that transport sustainability increases with settlement population size (Goodbody 2000)

- Examine the idea that accessibility plays a crucial role in the development of more sustainable settlements. Accessibility can be defined as a function of the mobility of the individual, and of spatial location of activity opportunities relative to the residence of an individual (Odoki *et al.* 2001)

- Examine the effects of urban development on sustainable transport and the relationship between rate of population growth and sustainable transport

Within a pilot project study undertaken in 2001, entitled 'Methodologies for the Estimation of Sustainable Settlement Size' (EPA 2003), of 11 settlements, also undertaken by the CER, tentative conclusions were drawn concerning the relationships between settlement size, as measured by population, and sustainable development, as quantified through indicators. Settlements were classified on the basis of population size alone. Conclusions included

John Davenport and Julia L. Davenport, (eds.), The Ecology of Transportation: Managing Mobility for the Environment, 119–164,
© 2006 *Springer. Printed in the Netherlands*

(O'Regan *et al.* 2002):

- There appears to be a lower threshold of about 20,000-30,000 population at which internal bus services become economically viable

- In relation to Ireland's National Spatial Strategy (NSS), strategic towns may benefit from being increased to a population size which supports additional social, economic and environmental infrastructure

- Traffic congestion increases in larger settlements

- Sustainability may be enhanced as density of population within settlements increases

Starting in 2002, the SFSPI project included 80 settlements, and was undertaken to build on the methods developed in the pilot project. An inventory of individual settlements and settlement networks was created through collation of published and otherwise available data, but large data gaps were found, and these were filled where possible through a detailed questionnaire survey of more than 8,000 individuals resident in these settlements, with a response rate of about 40%. The 80 settlements (listed in Table 1) are located in three clusters in the Limerick, Sligo and Midland regions. The selection of cluster boundaries was based on National Spatial Strategy (NSS) data illustrating travel hinterlands around urban centres (http://www.irishspatialstrategy.com). This facilitated the achievement of substantial national coverage, and the investigation of contrasting settlement types within polycentric and monocentric spatial hierarchies. The locations of these settlements are provided in Figure 1. Note that Dublin city and region was excluded from the study. This decision was taken in order that the study could focus on under-examined settlement patterns outside the Dublin metropolitan region. However, the primacy of Dublin was such that its influence was evident in the travel patterns in some Midland Cluster settlements.

Table 1. Settlements in the three clusters

Limerick cluster	Limerick cluster	Midlands cluster
Limerick and Suburbs	Caherconlish	Athlone & Environs
Ennis & Environs	Annacotty	Mullingar & Environs
Shannon & Environs	Borrisoleigh	Tullamore & Environs
Nenagh & Environs	Foynes	Longford & Environs
Tipperary & Environs	Cratloe	Ballinasloe & Environs
Newcastle & Environs	Glin	Birr & Environs
Kilrush	Cappamore	Roscommon & Environs
Templemore & Environs	Ardnacrusha (Castlebank)	Edenderry & Environs
Rathkeale		Portarlington
Newmarket-on-Fergus	*Sligo cluster*	Mountmellick & Environs
Abbeyfeale	Sligo & Environs	Moate
Castleconnell	Ballyshannon & Environs	Banagher
Kilmallock	Boyle & Environs	Mountrath
Kilkee	Carrick-on-Shannon	Ferbane
Sixmilebridge	Bundoran & Environs	Granard
Adare	Swinford	Portumna
Patrickswell	Ballaghaderreen	Lanesborough-Ballyleague
Croom	Tubbercurry	Kilcormac (Frankford)
Killaloe	Manorhamilton	Castlepollard
Ennistymon	Ballymote	Ballymahon
Newport	Ballinamore	Meathas Truim (Edgeworthstown)
Askeaton	Strandhill	Rochfortbridge
Borrisokane	Inniscrone	Daingean
Kilfinane	Charlestown-Bellaghy	Kilbeggan
Hospital	Drumshanbo	Strokestown
Bruff	Ballisodare	Ballygar
	Collooney	Kinnegad

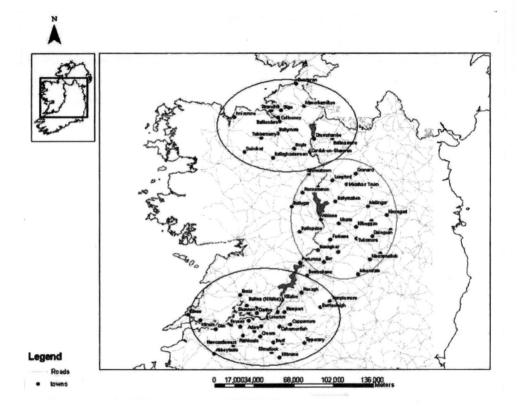

Figure 1. Locations of Settlements. Reproduced here with the permission of DoEHLG

2. The development of settlements from the Industrial Revolution to the Age of Sustainable Development

Today, the key components of transport pollution are carbon dioxide (CO_2) and other gases produced by the combustion of hydrocarbon fuels. Expanded road networks and increasing private car use facilitate the mass movement of people between locations of work, leisure, shopping and residence, encouraging rapid urbanisation, allowing decentralised development and the emergence of new trends in personal mobility. Environmental impacts associated with urbanisation grew in importance during the Industrial Revolution (Gulick 1989). At first the Industrial Revolution had no striking effect on urban growth. Typical industrial landscapes consisted of a sprawl of small industries across an area that was essentially still rural (Hall 1993). The use of coal as a fuel changed this situation and concentrated industries in locations close to coal resources or transport networks (Hall 1993).

Then, settlements grew rapidly in size and in terms of their share of national populations. Daily per capita energy consumption was approximately 50,000kcals (Haughton & Hunter 1994). In more recent times global interdependence developed as the global economy became more economically and environmentally interconnected. Settlements acted as nodal points for unprecedented flows of resources, wastes, traded products and services, finance capital and labour. Daily per capita energy consumption rose to approximately 300,000kcals, and has tended to increase up to the present as settlements expand in size and car ownership rates increase (Haughton & Hunter 1994). Mass transportation is a product of increasing industrialisation and urbanisation. The two are inextricably linked, as urbanisation has increased almost in direct proportion to industrialisation (Mumford 1995). Three major factors have facilitated an unparalleled physical expansion of modern settlements and transportation flows (Haughton & Hunter 1992):

- The increased use of motorised road transport allowed industry and residences to locate on the peripheries of settlements

- The growth in size and influence of businesses and government agencies has been important in creating new large-scale factories and industrial estates, and office blocks, each generating large flows of commuters and goods

- The use of functional zoning to plan for the separation of industry and commerce from housing

Rising urban populations must be provided with adequate housing, transport and other facilities (Yokohari et al. 2000), so that urban expansion is necessary. Moreover, urbanisation and developments in transport technology have contributed to the bringing together of resources, components and expertise, which have contributed to a growth in wealth (Douglas et al. 2002). The World Commission on the Environment and Development, which sponsored The Brundtland Report, argued for "more rapid economic growth in both industrial and developing countries", with the expectation that "a 5- to 10-fold increase in world output can be anticipated by the time the world population stabilizes some time in the next century" (Button 2002). There is no reason to suppose that Ireland will not follow this trend of economic growth, and that increasing urbanisation will go hand in hand with it. Urban generated environmental impacts will be exacerbated unless measures are taken to enhance the sustainability of the growing urban areas (Haughton & Hunter 1994).
Most of the energy needed to power transportation is generated through burning fossil fuels. Burning fossil fuels has resulted in increased emissions to the atmosphere of greenhouse gases, which most scientists believe is linked to adverse climate change (Botkin & Keller 1995). Therefore, CO_2 emissions are a serious cause for concern, and in this chapter are adopted as the unit of measurement of pollution caused by transport, and as an indicator for the overall impact of transport on the environment.

3. The importance of accessibility, population density and mixed land uses to a sustainable transport system

The SFSPI project explores possible links between the availability of services, accessibility to these services, and land-use within a settlement and the extent to which transport is sustainable. Recent population growth is a focus of the study.

"Urban sprawl may negate technical advances [in transport]." (Scott 2004)

Given that Irish settlements are likely to increase in size, questions may be posed. First: "What are the specific effects of recent growth on sustainability of transport?" Second: "Does the proliferation of housing estates in smaller settlements and the creation of new residential settlements result in increased transport flows in contemporary Ireland?" The growth centre debate is not restrained to a simple concentration versus dispersion issue. The appropriate policy for one region or settlement may not necessarily be suitable for another region or settlement (O'Farrell 1979).

"Between these end points [compact and dispersed] of the concentration-dispersal continuum lie an almost infinite variety of alternative strategies" (O'Farrell 1979).

One factor, which is recognised as central to the debate on settlement planning and in relation to movement between settlements, is accessibility (Acutt & Dodgson 1997; Moles *et al.* 2002a). Lau and Chiu (2004) suggest that transfers during a journey have the greatest impact on the accessibility of a place of work. A transfer occurs when a traveller changes from one mode of transport to another such as park and ride, or simply walks to a bus stop. Therefore, accessibility to job locations is different in a compact city such as Hong Kong, with a hierarchical transport network, from that in cities in Europe and the US, which are generallymore sprawling. The continuing increase in private car use in Ireland stresses the potential importance of accessibility. Furthermore, access to public transport increases accessibility to services, perhaps especially for those unable to afford private transport (Sanchez 2002 cited in Lau & Chiu 2004). Public transport is usually taken to be more sustainable than private transport, as less fuel is combusted per capita. Factors which have been identified as playing a positive role in promoting the use of public transport are:

- Higher population densities (Cervero & Kockelman 1997)

- Mixed land uses (Jacobs 1961)

- Compact settlements (Goodbody 2000)

- The concentration of jobs and services along transport corridors (Goodbody 2000)

Cervaro and Kockelman (1997) examined the reasons for people making journeys, and ways in which the number and distance of journeys by motorised transport in the San Francisco Bay Area were reduced. They concluded: "Creating more compact, pedestrian-orientated, and diverse neighbourhoods in combination can meaningfully influence how [people] travel". That is, mixed land uses in a compact urban structure encourage reduced car journeys and the use of public and non-motorised transport modes, and especially for non-work journeys, increased the proportion of people choosing to walk or cycle. On the other hand, they found that large residential areas without job locations and services created additional private transport journeys. Other authors have pointed to the negative impacts on sustainability of not controlling traffic expansion (McLaren 1992), and have identified the compact settlement (that is, with high population density throughout) as the optimal urban structure to facilitate this.

"The district, and indeed as many of its internal parts as possible, must serve more than one primary function; preferably more than two. These must ensure the presence of people who can go outdoors on different schedules and are in place for different purposes, but who are able to use many facilities in common" (Jacobs 1961).

A British study concluded that, in the reduction of number and distance of journeys, the key land-use variables are density, degree of centralisation and size of settlement (McLaren 1992). A recommendation of this study was that new development should be concentrated in urban areas, that urban centres should be maintained and revitalised, and that constraints should be placed on the development or extension of small settlements. However, there is by no means consensus on this view. It has been argued that increasing urban density without consideration of urban form makes little sense (Haughton & Hunter 1994). Schumacher (1973) argued for a return to small communities and producers:

"Although even small communities are sometimes guilty of causing serious erosion, generally as a result of ignorance, this is trifling in comparison with the devastations caused by gigantic groups motivated by greed, envy and the lust for power. It is moreover obvious men organised in small units will take better care of their bit of land or other natural resources than anonymous companies or megalomaniac governments which pretend to themselves that the whole universe is their legitimate quarry." (Schumacher 1973)

This view still finds support today, for example the Irish Rural Dwellers Association encourages positive planning for the open countryside. However, there is no disagreement that mobility and transport mode provision are crucial in achieving high quality urban environments (Jenks *et al.* 1996). There is more general support for the need for planning to ensure that neighbourhoods are economically successful, convenient and attractive, and to support and improve the public transport which serves them (Jenks *et al.* 1996). At the same time it is often recognised that some services will generate car journeys, whatever the location:

"For certain high trip attracting uses a more appropriate location may be a suburban centre…bypass and orbital locations tend to be associated with longer journey distances, travel times and levels of fuel consumption" (McLaren 1992).

In the early 1990s, within UK settlements with populations over 250,000 people, average weekly travel was 141km, of which 14% was by public transport. In smaller settlements (of less than 3,000 people) the equivalent value was 211km (of which 8% was by public transport). In somewhat larger settlements (3,000-25,000 people) the proportion of journeys by public transport was equally small (McLaren 1992). A large proportion of Irish settlements fall into this latter category of population size (CSO 1996; CSO 2002a). Between 1991 and 1996 all Irish settlements showed an increase in the length of the journey to work/education. In 1991, there was some evidence that daily journeys were shorter in smaller settlements than in larger settlements, but by 1996 this difference had largely disappeared. This suggests that all Irish settlements are not moving towards more sustainable modes of transport. However, there is some evidence that bus transport provision is more likely to be economically viable in larger settlements, and a minimum settlement size requirement of 15,000-30,000 population has been suggested (Moles *et al.* 2002b). SFSPI investigates the contemporary Irish situation with regard to settlements of different sizes and functions.

It has been argued that decentralised sub-urbanisation has been a dominant and successful mechanism for reducing inner settlement traffic congestion, and that as industry and services move to the suburbs so does the labour force, allowing workers to make shorter daily journeys (Gordon & Richardson 1997). In addition, Gordon and Richardson argue that that sub-urbanised areas are the preferred lifestyle locations of the majority of Americans and recent urbanisation patterns suggest this preference may extend to contemporary Ireland. The extent to which people's wishes may be thwarted within a democracy is an issue raised by some authors. Furthermore, "the economic and resource efficiency of compact development has never been adequately demonstrated" (Gordon & Richardson 1997).

Gordon and Richardson (1997) discount the advantages of compactness, arguing that many previous research findings are not convincing in relation to negative aspects of sub-urbanisation. However, there is firm evidence that in some settlements in which employment locations follow residential areas out of larger settlements to dormitory (or satellite) settlements, there may be associated marked increases in the length of journeys both for work and other purposes (McLaren 1992). This is because residential areas may be situated far from employment locations, or people may choose not to live near their place of work (Jenks *et al.* 1996; Lau & Chiu 2004). There is no clear method for tackling this problem: many settlements have become too large to function adequately with just one centre. To counteract this problem McLaren suggests the concept of decentralised concentration, which focuses development in accessible centres within the urban transport network (McLaren 1992). This is suggested because larger single centre settlements often suffer from severe traffic congestion in central areas (McLaren 1992). However, while congestion diminishes from centre to suburbs, and vehicle fuel consumption per kilometre therefore improves, per capita fuel use by residents increases markedly (McLaren 1992). Congestion may decrease fuel efficiency but it also may make public transport more attractive (McLaren 1992).

> "Congestion may result in any city where land-use policies are employed without traffic reduction policies, but if transport costs for the private car are high, and modal shift easy because of high densities and good public transport, then congestion will be less intense and less frequent" (McLaren 1992).

O'Regan and Moles (1997) identified a well-known knock-on effect of building new roads, namely, latent traffic. New roads encourage additional vehicle usage and in turn add to congestion. This additional congestion is normally addressed through building yet more roads rather than addressing the fundamental issue of creating policies to reduce car usage (O'Regan & Moles 1997).

Frustratingly, in Ireland during the 20th century another associated localised trend emerged. What was initially a marginal withering of rail services developed into a rapid decline (McCutcheon 1970). This change saw road replace rail as the preferred mode of travel especially for freight traffic. The reasons for the rail closures were economic: the stagnation of urban growth centres, spiralling operating costs and competition from the roads (Baker 1972). The most obvious solution to the financial problems besetting the railways was closure.

During the mid 20th century in the North of Ireland a substantial amount of money was made available and invested in the road network. This investment in roads coincided with a decline in the rail network.

> "Motor transport was...better suited to many of the tasks which railways once performed...it was difficult [however] to rid the impression that the Government was hell bent on wiping out the entire rail network" (Baker 1972).

It can be rationally argued that the improvements in Irish roads from the second half of the 20th century have until recently been accomplished to the neglect of the rail network.

There appears to be no clear consensus emerging with regard to finding optimal forms of urbanisation, associated patterns of transportation, and the impact on sustainability. Many argue that higher population densities provide scope for policies designed to restrict and reduce personal transport usage. There may be environmental, health and social benefits associated with such a modal change. Gordon and Richardson (1997) argue that high-rise buildings (that is, high density urbanisation) exist where the high costs of erecting and maintaining them are viable economically and where accessibility, communications and interaction are considered to be worth the high initial investment. Camagni *et al.* (1997) claim concentration of activities and proximity to residences is not only a precondition for efficient social interaction and economic activity, but also, to some degree, it is a source of increasing returns in the use of scarce, non-renewable resources. They suggest that it is paradoxical that people are tempted to move out of the centre to peripheries. Whilst their individual well-being may rise temporarily, damage to the environment will increase due in part to fewer economies of scale and increased transportation. Land use change to increase densities is relatively slow but on the timescale of sustainable development, even a rate of 1% yearly renewal or development may be acceptable (McLaren 1992).

On the other hand, increased suburbanisation as a result of population decentralisation may result in reduced welfare, for example through higher crime rates (Burton 2000). Even decentralised concentration is travel and energy intensive under prevailing conditions of mobility (Owens & Rickaby 1992). Regarding the environmental, health and social benefits of higher densities, Gordon and Richardson (1997) found that 30% of urban households living in suburban areas (in the US) had annual incomes below $25,000. This proportion is lower than for the city centre but shows that suburban households are not necessarily benefiting financially. This they argue negates the belief that sub-urbanisation is the result of the rich escaping the problems of the central city, as was suggested by Lau and Chiu (2004), a process which some argue results in segregation of the population (Gordon & Richardson 1997). Some argue that sub-urbanisation is much more positive and efficient than is implied by concepts such as 'edge city', 'satellite settlements' and 'polycentricity'. The negative effects of more compact areas (air, noise and water pollution, destruction of green areas, traffic congestion and intensive energy usage) are highly visible and damaging to welfare (Camagni et al. 1997). If the same mass of economic activity were to take place over a wider area, then the spatial concentration of emissions would be lower, allowing dissipation, which might counterbalance the higher consumption of natural resources (Camagni et al. 1997).

4. Transport and sustainability in Ireland

Historically, in Ireland a strong anti-urban bias and a commitment to rural tradition had been associated with the Irish nationalist movement, which had close links to the "land question" (Bannon 1984).

> "In a free Ireland…the population will expand to twenty millions…towns will be spacious and beautiful…but since the country will chiefly rely for its wealth on agriculture and rural industry there will be no Glasgows or Pittsburghs" (Pearse 1916).

However, Ireland has not developed as envisaged by Pearse. Contemporary growth in transport and energy consumption are two primary concerns of Irish national environmental policy development (Moles et al. 2002a). In 1990, emissions to atmosphere from Irish road transport combustion of petrol and diesel were 2,760 and 1,901 Gg-CO_2 respectively. By the year 2000, Gg-CO_2 totals rose to 4,925 for diesel and 4,610 Gg-CO_2 for petrol (EPA 2002). Sustainable development is now an important aim of land use and transportation policy in Ireland (EPA 2000b; EPA 2000c), but growth in transport makes achievement of sustainability targets difficult. Ireland's commitment to reduce greenhouse gas emissions under the Kyoto Protocol allows a 13% increase on 1990 levels by 2008-2012. It is now evident that this level of increase had in fact already accrued by the time the Kyoto Protocol was signed in 1997, and the annual growth rate in emissions in the early years of the 21st Century was over 4%. Greenhouse gas emissions for the period 1990-2001 in Ireland increased by 124%, compared with an EU average of 21% (EEA 2003a). By 2001, Irish greenhouse gas emissions were about 31% above the 1990 level: transport was a major contributor to the total, and the sector that produced the greatest increase (see Table 2). This rate of increase in emissions has major implications for Ireland's ability to comply with the Kyoto Protocol primary objective. Even if Irish greenhouse gas emissions are curbed to comply with the Kyoto commitment, by the end of the first commitment period, they will still be among the highest per-capita of all OECD countries (EPA 2002).

Table 2. Greenhouse Gas (GHG) Emissions by Sector (Data from ICF & DoEHLG 2004). Base year is 1990

Sector	Kyoto Reference (Base) Year [Mt]	Sector % Base Year	2001 [Mt]	Sector % 2001
Energy	11.778	22.05	18.059	25.79
Residential	7.020	13.14	6.724	9.60
Industrial	9.601	17.97	12.235	17.47
Agriculture	18.656	34.93	20.126	28.74
Transport	5.143	9.63	11.531	16.47
Waste	1.218	2.28	1.342	1.92
GHG Emissions excl. net CO_2 from LUCF	53.353		70.018	

Car ownership in Ireland is increasing rapidly in line with global trends (Mackett & Edwards 1998). Many factors contribute to this increase in ownership. A very important factor in Ireland is an increase in disposable income over the past decade (54% in the period 1995-2000 (CSO 2002b)). Car ownership also provides accessibility to a wider choice of employment, shopping and leisure facilities (Mackett & Edwards 1998). Many new shopping centres are located at the periphery of settlements and are poorly served by public transport. Investment in public transport may have been insufficient to provide attractive alternatives to travel by car, as modern cars provide a level of comfort and safety that is difficult to match. The Sustainable Energy Ireland (SEI 2003) Report 'Analysis of New Car Registrations in Year 2000' focuses on methods to improve fuel consumption and reduce CO_2 emissions in private cars. The report is confined to the 1990-2001 time period and to energy related CO_2 emissions from the national fleet of private cars (SEI 2003). Between the period 1990 and 2001, the total number of vehicles on Irish roads rose by 68% to 1.77 million. During the same period the private car fleet total increased by 74% to almost 1.4 million (SEI 2003). This equates to a doubling of CO_2 emissions from transport sources, which are largely accounted for by road traffic, due to growth in vehicle numbers and total distance traveled. In the latter years fuel-tourism is important: a significant proportion of the automotive fuels sold in Ireland are used by vehicles in the UK and other countries. The proportion was estimated to be approximately 6% for petrol in 2000 but it may have been as high as 20% for diesel. This reverses the situation in 1990, when there was significant movement of automotive fuels into Ireland from the UK (EPA 2002).

The density of private cars for the year 2000 was 348 cars per thousand of population, and this increased to 361 in 2001 (SEI 2003). This compares with 469 cars per thousand of population in 2000 for the EU-15 average (SEI 2003). Between 1995 and 2002 the growth rate in passenger car use in Ireland was 4.6%, compared with an EU average of 1.6%. Moreover, comparing the time periods January-April 2003 with January-April 2004, registrations in Ireland increased by over 8% compared with an EU average of 3% (ACEA 2004). The number of new vehicles registered fell from 225,269 in 2000 to 142,992 by 2003 (CSO 2005), but initial indications suggest a steep rise again in 2005. While it is the case that technology has improved the energy and pollution efficiencies of modern cars, the rapid growth in car ownership is likely to continue to offset resultant environmental gains.

In 2001, transport accounted for approximately 30% of Ireland's primary energy consumption (4.3 million tonnes of oil equivalent (Mtoe)). This represents an 8% rise above 1991 levels. Road transport accounted for 82% of this total, half of which was consumed by the private car fleet (SEI 2003). Thus, it is estimated that private cars are responsible for 12% of Ireland's gross energy consumption. Studies have indicated that demand by the transport sector is expected to increase by 67.2% between 1990 and 2015 (EPA 2000b). In Western Europe the transport sector was responsible for approximately 30% of total energy use in 2001 (EEA 2003b).

Private cars account for approximately 40% of all transport energy demand and 50% of road transport energy (SEI 2003). This figure is based on the assumption that 90% of cars are petrol fuelled, and 10% diesel fuelled. The average fuel efficiency of new cars bought in 2000, calculated on the basis of proportion of total sales in each engine size class, was 7.16L/100km (7.23 for petrol cars, 6.47 for diesel cars). The average CO_2 efficiency was 166g/km for both petrol and diesel vehicles. The most efficient new petrol cars registered in 2000, in relation to CO_2 emissions, were in the 900-1200cc range of engine size, which formed 23% of the total number of cars bought in 2000 (SEI 2003). Between the years 1990-2000, a change in the structure of the Irish car fleet resulted in an additional 169kt CO_2 being emitted. (SEI 2003). In 1990 the most common new car engine size was in the 900-1200cc range, accounting for 39% of the total car fleet. By 2003, this proportion had dropped to 20% (DoEHLG 2003). In 2002, the most common engine size (40% of the total fleet) was in the 1200-1500cc range (SEI 2003). Between 1990 and 2003, the proportion of cars in the 1701-1900cc engine size range increased from 4 to 11%.

The National Climate Change Strategy (NCCS) predicted that emissions from the Irish transport sector will reach 14Mt (14,000Gg) CO_2 by 2010, an increase of 179% on 1990 levels (see Figure 2). The NCCS identified abatement measures to deliver a reduction of 0.35Mt CO_2 per annum, which include modal shift and demand management measures. A further 0.5Mt CO_2 abatement is anticipated from increases in private car registration charges and increases in annual motor tax. An additional 0.1Mt CO_2 reduction is projected resulting from fuel economy and CO_2 emission labelling as required under EU directive 1999/94/EC. On the basis of an annual average driving distance of 20,000km for all cars (NRA 2003), the SEI (2003) estimated a national emissions total of 4.4Mt CO_2 for 2000, assuming all cars in that year were modern models. The effect of an annual per driver 2,000km reduction in distance travelled by new cars registered in 2000 was estimated to be a reduction in CO_2 emissions of 0.7Mt. For all private cars, the equivalent abatement value was estimated as 0.4Mt CO_2 per annum.

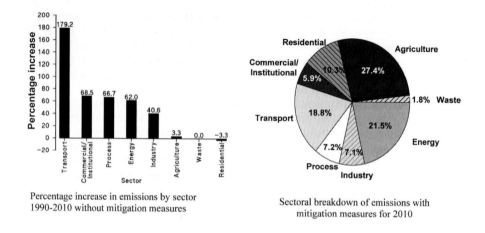

Percentage increase in emissions by sector
1990-2010 without mitigation measures

Sectoral breakdown of emissions with
mitigation measures for 2010

Figure 2. (Data from DoEHLG 2000)

In order to meet Kyoto Protocol requirements, the EU adopted a strategy to reduce average CO_2 emissions by new cars by about 33% (EEA 2002). Improved fuel efficiency gained through advances in technology-reduced emissions markedly up to 1998, but more slowly thereafter: such 'technological savings' were estimated as 12Mtoe in 1998 and 13Mtoe in 2001. Between 1995 and 2001, fuel used per distance traveled reduced from 7.5 to 6.8L/100 km (7.9 to 7.3L/100km for petrol cars and 6.9 to 5.9L/100km for diesel cars). This reduction was in part achieved as a result of an agreement on carbon emissions negotiated with three car manufacturer associations (European-ACEA, Japanese-JAMA, Korean-KAMA), with an EU target of 140g/km for new cars by 2008. Reaching this target will require additional reduction of 20-25% in the fuel consumption (to reduce CO_2 emissions by a further 25%) of new cars of between 20% and 25% (EEA 2003b; ICF & DoEHLG 2004; SAVE-Odyssee 2003).

It is clear that major additional savings in energy consumption by the transport sector will be made with difficulty if reliance is placed on technological advances alone. The interlinking of land use and transport policy is one area where considerable advancements are possible. Since 1998 reductions in the distance travelled have contributed to a reduced energy demand within the EU. These "behavioural savings" are estimated to have reduced the energy requirement by 11Mtoe in 2001. As Irish greenhouse gas emissions for the period 1990-2001 increased by 124%, compared with an EU-15 average of 21% (EEA 2003a) the need for such behavioural changes within Ireland is clear.

5. The SFSPI study of the sustainability of transport in Ireland

The purpose of the SFSPI study was to assess the extent to which Irish settlements are moving towards or away from sustainability, and to identify possible differences between types of settlement by comparing modes of transport used and distances travelled. Of special interest was the identification of those settlements that tended to encourage relatively long journeys. The focus of this chapter is on analyses of private car use: other aspects of transport will be published elsewhere. In order to address the important issues, the study undertook characterisation of the types of cars used and the total distance travelled in settlements. Between settlements differences in car travel were analysed at two scales of aggregation. The first analysis was based on aggregated settlement data within geographical distance buffers around large settlement centres within monocentric clusters around Limerick and Sligo, which does not have a single dominant settlement. The second analysis considered settlements classified into classes on the basis of a set of attributes, including population, distance to nearest Gateway settlement, spatial extent, population change over time, and functionality. In relation to many attributes, Limerick city was exceptional, so that in the classification the city is allocated to its own settlement class. The seven settlement classes are referred to in this chapter as 'Settlement Classes'. Data were required on distances travelled to work and shops by individuals per week, and total weekly distances travelled. Data available from the Central Statistics Office (CSO), Small Area Population Statistics (SAPS), and data from the National Road Authority (NRA) were utilised where available. Major gaps in available information were found to exist in relation to transport at settlement scale, and these were filled through questionnaire surveys undertaken in each of the 80 settlements.

Here results are presented for a set of analyses. First, settlements were aggregated in terms of location within the spatial hierarchy, with the primary purpose of relating travel habits to the spatial distribution of settlements. For this analysis, settlements were aggregated by the creation of buffers based on distance from Limerick and Sligo within the monocentric clusters. These buffers were created using Geographical Information System (GIS) software. Each buffer width is 5km, but where necessary buffers were aggregated to ensure an adequate sample of settlements. Limerick and Sligo clusters were compared: both are monocentric, but the central settlements differ in size, and may be expected to exert differing influences on surrounding smaller settlements. Second, settlements were classified according to attributes to permit analysis of the relationship between travel behaviour and functionality, especially the availability of tertiary level services. Third, a functionality index was calculated to analyse transport patterns in relation to the functions offered by settlements. Fourth, an analysis was undertaken of the relationship between population growth rate within settlements and travel behaviour. Fifth, to calculate transport CO_2 emissions for settlements, data for settlements aggregated into Settlement Classes. For the calculation of CO_2 emissions, a range of data sources was available: Sustainable Energy Ireland (SEI), CSO, Department of the Environment, Heritage and Local Government (DoEHLG), and SFSPI survey results.

5.1 SETTLEMENT CLASSES

Settlements were classified to facilitate analysis and interpretation of results (Table 3). This aggregation of settlements increased sample size for data analysis without compromising the integrity of the data, allowing for more robust comparisons and the identification of trends over time. Presentation of results was made clearer, as comparisons among 80 settlements tended to be cumbersome and sometimes difficult to interpret. Data selected on which the classification was based shared the following characteristics: they were available for a wide range of the settlements, they reflected the differences and similarities of settlements across a spectrum of socio-economic and environmental issues, they were reliably accurate, and they related to attributes which showed variations amongst settlements. Attributes on which the classification was based are listed in Table 4. Table 3 describes the 7 different classes, which were derived from MVSP analyses. The classes are firstly described under their title, that is, Classes 1-6 and the Limerick Class. This simple numbering system is sequentially based on a dendrogram branching pattern, which resulted from the analyses and separated the settlements according to the similarities between applicable settlement variables. Secondly, the classes are described and differentiated according to size and functionality averages. In the final column, transport and general characteristics are shown for each class.

Table 3. Class Type and Description *(w/s/c: work/school/college)

Class	Size and functionality	Growth/ Decline in population	Transport details and general characteristics
1	This class includes the NSS Gateway settlements of Tullamore, Mullingar, Athlone and Sligo and the Hub settlement of Ennis. Ennis is the second largest SFSPI settlement. The average population size for these settlements is 16888. The service level index is 65. These settlements are generally important regional and county administrative and employment centres.	Class 1 settlements in general had stable population growth for the 1996-2002 period at a mean growth rate of 14%. The median rate of growth was 11%.	Settlements suffer congestion problems with 69% of individuals experiencing frequent or continual congestion problems. A strong negative correlation between percentages using foot and car suggest public transport is either not widely available or not widely used.
2	The Class 2 settlements are smaller in size and service index level score than Class 1. The largest settlement in this class is the Shannon which has a population size of 8561. The average population size is 5329. The service index score is 36. These settlements are important county administrative and employment centres. For example, due to decentralisation Ballinasloe is a transport administration centre. Nenagh is a local government county council centre.	Population growth rate in Class 2 settlements was similar to Class 1 at a mean rate of 11% and median rate of growth of 9%. These rates indicate an influx of individuals to both Class 1 and Class 2 settlements, which is most likely due employment opportunities. All settlements in Class 1 and Class 2 have positive growth rates.	49% of individuals experience frequent or continual traffic congestion. There is an inverse relationship between percentage of individuals using foot for w/s/c* and population growth.
3	Class 3 settlements are smaller in size than Class 2 with an average population size of 1290 and a functionality score of 17. This class has a wide variety of settlement sizes and service index scores.	The 38% growth rate of Class 3 settlements is not a true reflection of Class 3 growth rates. There is large growth rate differential between settlements. This is due to the presence of Annacotty and Rochfortbridge, which experience population explosions relatively, speaking for the period 1996-2002. The median population growth rate was 4%.	There is an inverse relationship between percentage using a car to w/s/c and distance to Gateway. This indicates that the closer a Class 3 settlement is to a Gateway settlement the more common car use becomes. These Gateway settlements provide the services which many larger yet functionally underdeveloped class 3 settlements lack.
4	Class 4 represents the smallest sized settlements with an average population of 934 people. It also has the lowest service index score of 15.	Class 4, which at first glance has similar characteristics to Class 5 has a mean growth rate of just 1% and a median population decline of -1.5%.	There is a very strong relationship in Class 4 settlements between population growth and percentage of individuals using a car to travel to w/s/c. There is also a strong inverse relationship between population growth and percentage of individuals unemployed.

Class	Size and functionality	Growth/ Decline in population	Transport details and general characteristics
5	Class 5 has similar size and functionality characteristics to Class 4 with a population of 957 people and a service index score of 17. A number of satellite settlements are contained in this class.	Class 5 has a median growth rate of 6% suggesting that unlike Class 4 it contains settlements of a more expansive nature.	There is a strong relationship in Class 5 settlements between "education ceased after 21" and percentage of individuals using a car to travel to w/s/c. Where car use is high distance to Gateway is low, possibly due to the presence of satellite settlements. As the service index decreases car use predictably increases, as shown for Class 5 settlements.
6	Class 6 has a service index score of 29 and an average population size of 2159 people. The settlements in this class are similar in size to some Class 3 settlements but have on average a higher functionality score. Results indicate that these settlements have stand-alone potential as employment bases and service centres. These settlements are also most isolated on average with respect to average distance to Gateway.	Class 2 has a median growth rate of just under than 2%. These settlements are therefore generally static in growth terms. This may be a reflection of their isolation but also of their ability to provide a substantial and stable employment base thus resistance to population decline.	There are very few significant relationships between selected variables in Class 6 settlements. There is a positive relationship between the service index and percentages of individuals in professional employment. Surprisingly, the percentage of individuals using a car to travel to w/s/c correlates positively with service index score, further highlighting the abnormality of this class.
Limerick	Limerick was classified separately from other SFSPI settlements. It has a population of 86998 (including environs in County Clare).	Limerick had a growth rate of 10% for the 1996-2002 period. This was mainly due to growth in the suburbs, which was above 20%, as within the city limits the growth rate is essentially static.	Limerick differs from all other settlements due to its size. Its functionality level is not considerably bigger than Class 1 settlements, however it has a much lower recreation and shop per capita CO_2 emission indicating that the difference in functionality is nonetheless important. Limerick has 59% individuals experiencing traffic congestion.

Table 4. Attributes Adopted for the Settlement Classification

Settlement attributes				
Distance to Settlement of Equal or Greater Size, and Average Distance Travelled to Work	Distance to Capital City (Dublin)	Population based on 2002 Census	Percentage Population Change 1996-2002	Average Monthly Income
Percentage Unemployed	Percentage Education Ceased pre age 15	Percentage Education Ceased after age 21	Percentage Employed in Services	Percentage Employed in Agriculture
Percentage Employed in Construction	Percentage Employed in Manufacturing	Percentage Employed in Commerce	Percentage Employed in Transport	Percentage Employed in Public Admin.
Percentage Workforce Professional	Population Density (Persons per sq. Km)	Percentage Homes Built after 1996	Distance to nearest NSS "Gateway" Town	Service index Score

5.1.1 Class Correlations

Correlations were run within and between SFSPI generated data and small area population statistics (SAPS) from the CSO 'Census 2002'. There were found to be significant correlations between a number of variables.

For example, there was a predictable positive correlation between "% employed in professional work" and "% of individuals whose education ceased after age 21." There is a negative correlation of $r = -0.444$ at $p \leq 0.01$ between "% of individuals whose education ceased after age 21" and "% using foot to w/s/c".

Furthermore, another trend indicated by Table 5 for individuals who use their private car for work, school or college journeys or have an automobile available for these specific journeys is that these individuals are much less likely to travel by foot.

In addition, Table 5 shows that, where population growth is higher, the percentage of individuals using a car as the travel mode also increases.

Table 5. For all classes relationship between percentage using car for work, school or college from CSO 2002 and selected other variables

	% education ceased after age 21	% using foot to w/s/c	Population change	Distance to gateway settlement
% Using car for w/s/c	$r = 0.603$ at $p \leq 0.01$	$r = -0.801$ at $p \leq 0.01$	$r = 0.549$ at $p \leq 0.01$	$r = -0.453$ at $p \leq 0.01$

One interesting statistic is that, as distance to Gateway settlement increases, the percentage of individuals using a car as their travel mode decreases and those using foot as means of travel increases. This possibly suggests another reason why Class 6 settlements perform well with regard to road transport CO_2 emissions.

The most significant correlations are shown for each class below. Though some of these correlations are significant, in reality they may convey very little useful information. The majority however are readily interpretable.

5.1.1.1 *Class 1*. Correlations at $p \leq 0.05$ are as follows. There is a direct negative correlation between percentage using foot to w/s/c and percentage using car for w/s/c of $r = -0.930$ at $p \leq 0.05$. This may seem a predictable correlation at first glance. However, further examination of the data and related correlations shows that there is no significant relationship between car usage and public transport usage. This would indicate that travel decisions are made on a binary basis, foot or car and that public transport is not considered as an option.

5.1.1.2 *Class 2*. Correlations at $p \leq 0.01$ are as follows. A number of unexpected correlations were discovered for these analyses. Firstly, there is a significant negative correlation between percentage using foot to w/s/c and percentage using public transport of $r = -0.767$. Secondly, there is a similar negative correlation of $r = -0.754$ between percentage using foot for w/s/c and population growth. These correlations indicate that within settlements contained in Class 2 there is a common relationship whereby if foot usage is high for w/s/c trips then public transport usage is correspondingly lowered. Moreover in settlements with high population growth the percentage using foot to w/s/c is relatively low.

5.1.1.3 *Class 3*. Correlations were taken from the correlation analyses at both the $p \leq 0.01$ and $p \leq 0.05$ significance levels. There are significant negative relationships between percentage car use and distance of settlement to Gateway settlement, percentage using foot to w/s/c, and the service index of r = -0.903, -0.878 and -0.884 respectively at the $p \leq 0.01$ level. This suggests that the closer a settlement is to a Gateway settlement then the higher the car use and that this increase in car use directly and negatively effects the proportion of individuals using foot to w/s/c. The negative correlation suggests that for 'index of services' Class 3 settlements' car use is high due in part to a lack of services. The relationship between increasing distance to Gateway settlement and increasing service index is significant, r = 0.886 at $p \leq 0.01$.

5.1.1.4 *Class 4*. Correlations were taken from the correlation analyses at both the $p \leq 0.01$ and $p \leq 0.05$ significance levels. There is a significant correlation at $p \leq 0.01$ of 0.994 between "% education ceased at 15" and "% using public transport." Population growth correlates negatively with percentage individuals unemployed at $p \leq 0.01$, with a score of r = -0.994. At $p \leq 0.05$ significance level there is a correlation of 0.985 between population growth and percentage using car to w/s/c. Unlike between Class 5 settlements the service index score does not correlate with percentage using car to w/s/c.

5.1.1.5 *Class 5*. Showed a number of significant correlations. Some of the relationships highlighted were more expected than others. Percentage individuals unemployed correlates negatively with percentage ceasing their education after age 21, r = -0.699 at $p \leq 0.01$ significance level. For individuals whose education ceased after age 21 and percentage individuals using a car to w/s/c there is a significant relationship at $p \leq 0.01$ level of r = 0.870. There is a negative correlation of r = -0.698 at $p \leq 0.01$ between percentage individuals using a car to w/s/c and distance to nearest Gateway settlement. Where car usage is high, distance to Gateway is low indicating the possible presence of satellite settlements. At the $p \leq 0.05$ significance level the percentage of individuals using a car for w/s/c negatively correlates with "index of services" scores. This indicates that for settlements in this class, as service index score increases the percentage of individuals using a car decreases. There is a correlation of r = 0.751 at the $p \leq 0.01$ significance level between distance travelled to work of "15-29 miles" (24-46km) and population growth.

5.1.1.6 *Class 6*. Is unusual in that there are very few relationships of significance between the variables analysed. At the $p \leq 0.05$ level of significance the percentage of individuals using a car to w/s/c correlates positively with service index score. This contrasts with Class 5 and may suggest that settlements of different size or position in the spatial hierarchy may react differently to changes in functionality or size. The percentage of individuals in professional employment correlates with the service index score at the $p \leq 0.05$ level (r = 0.895)

5.1.1.7 *Limerick Class* is unique due to its size and functionality. It is not paired with any other settlements

The level of tertiary services available in a settlement was quantified through calculation of an index of services, described in the next section. Physical attributes were measured through population density mapping, data on new homes built and percentage growth. Geographical location was evaluated through use of distance to Dublin, nearest Gateway Settlement and nearest greater size neighbour. Socio-economic dimensions of sustainability were quantified by employment, income and job description data. Adoption of a broad spectrum of variables provided a robust classification matrix, which facilitated inter-settlement comparison. Data were input to a multivariate statistical package (MVSP 2003), which achieves classification through iterative stages of correlation coefficient calculations, grouping settlements with large numbers of significant correlations, and providing output in the form of a dendrogram in which the closeness of position reflects the similarity of the settlements. The Unweighted Pair Group Method with Arithmetic Mean was adopted so that the need to normalise data was eliminated. Classes were identified from the dendrogram. This was in part a subjective process, as it was necessary to select a manageable number of Classes. Too many would make interpretation of results difficult and involve comparisons amongst settlements which differed little. Too few would hide differences between settlements. On this pragmatic basis, seven classes were selected. Settlements forming classes are identified in Table 7. Population size played a dominant role in classifying settlements. Limerick is substantially larger than any other settlement included in the study, with approximately 65,000 more people than the next largest settlement, Ennis.

However, other attributes contributed to the separation of the classes (Table 6). For example, "Distance to nearest Gateway" was important in separating out Class 6 settlements. Gateways are settlements that have a strategic location nationally and relative to their surrounding areas (NSS 2003). A Hub is defined as a settlement that can act as both a support to Gateway settlements on a national and international scale, and a stimulant for development locally and in nearby smaller settlements (NSS 2003). Class 6 settlements are distinctive in that they maintain themselves in relatively isolated spatial locations. It should be noted that Ennis is not a Gateway but a Hub: if it had been a Gateway, then "Average distance to nearest Gateway" for Class 1 settlements would have been zero km. The results of the classification are provided in Table 6.

Table 6. Average settlement size and distance to Gateway for settlement classes

Settlement class	Average population	Average distance to nearest Gateway (km)
Limerick	86,998	0
Class 1	16,888	10
Class 2	5,329	45
Class 3	1,291	39
Class 4	935	40
Class 5	943	42
Class 6	2,185	67

5.2 THE SIGNIFICANCE OF SERVICES

Quantification of the level of services available within a settlement was important for two reasons. First, the index value was an input into the classification of settlements (Table 3) and second, the relationship between index values and transport behaviour was analysed for individual settlements and Classes. The method for calculating the index was developed in part on the basis of previously published work, especially that of Grove and Huszar (1964). Service Index values for settlements are provided in Table 7.

Settlements offer services not available in their hinterlands, and may therefore be thought of as "central places" (Holton 1986; O'Farrell 1968). Christaller's study on central places assumed central places are regularly distributed over an isotropic resource, transportation and income surface, serving evenly distributed and rational consumers with identical consumption functions (Glasscock & Stephens 1970). Reality is of course much more complex, but the theory is of continuing relevance.

Other studies on centrality, based in Ireland, attempted to explain the distribution of settlements by measuring the relative importance of possible causal factors F (Glasscock & Stephens 1970). A measure of centrality of settlements in County Tipperary, based on the number of functions provided, was calculated and the number of central functions of settlements which provided more than one function was established through direct observation (O'Farrell 1968). Analysis revealed that the distance between two settlements of similar population size was governed by the population of the two settlements and that as the number of functions provided by settlements increased, so did the distance between these settlements. Thus, the greater the population size or centrality, the greater the distance between the two towns (O'Farrell 1968). These studies provided an important basis for analyses reported in this chapter.

Table 7. Settlements forming classes, and populations (Data from CSO 2002a)

Class	Settlements	Population in 2002	Service index
Limerick	Limerick	86998	79
Class 1	Athlone	15936	70
	Ennis	22051	63
	Mullingar	15621	62
	Sligo	19735	76
	Tullamore	11098	54
Class 2	Ballinasloe	6219	46
	Birr	4436	40
	Edenderry	4559	27
	Longford	7557	46
	Mountmellick	3361	17
	Nenagh	6454	43
	NewcastleWest	4017	35
	Portarlington	4001	21
	Roscommon	4489	55
	Shannon	8561	41
	Tipperary	4964	30
Class 3	Annacotty	1342	4
	Ballinamore	687	21
	Croom	1056	18
	Drumshanbo	623	17
	Kilrush	2699	34
	Kinnegad	1296	11

Table 8. Settlements in buffer ranges defined by straight-line distance from Limerick City

Buffer ranges				
5-10km	*11-20km*	*21-30km*	*31-40km*	*Over 40km*
Cratloe	Caherconlish	Newmarket-on-Fergus	Kilfinane	Kilkee
Ardnacrusha	Newport	Rathkeale	Tipperary	Kilrush
Annacotty	Adare	Askeaton	Nenagh	Glin
Castleconnell	Croom	Bruff	Ennis	Abbeyfeale
Patrickswell	Pallaskenry	Cappamore	Foynes	Borrisokane
	Shannon	Kilmallock	Newcastle West	Templemore
	Ballina/Killaloe	Hospital		Borrisoleigh
	Sixmilebridge			

For settlements within these buffer ranges, data on residence to workplace journeys were generated through the SFSPI survey. Average (both mean and median) journey lengths were calculated for settlements within each buffer range: the degree of skewness in the data suggested that the median was a more informative measure of average (Table 9). Limerick City, with largest population and highest services index, had the lowest median journey length, at 4.8km.

Table 9. Median and mean residence to workplace values for settlements in each buffer range. *Value excludes Shannon Town

	Buffer Ranges					
	Limerick	*5-10km*	*11-20km*	*21-30km*	*31-40km*	*Over 40km*
Median distance (km)	5.0	12.0	16.0* 16.0	13.0	6.5	6.5
Mean distance (km)	10.1	13.2	16.41* 15.7	16.3	12.6	13.0
Ratio of skewness to standard error of skewness	10.7 (>2)	7.7 (>2)	6.6 (>2)	3.1 (>2)	9.3 (>2)	7.6 (>2)

The results suggest that many Limerick residents also work in the city. The median distance of 12.0km travelled by individuals in the 5-10km buffer range suggest that many individuals resident in this buffer range also work in Limerick, though they may also work in other settlements. For the settlements in the 11-20km buffer range median distance travelled increased to 16.0km. This indicates that for residents of this buffer range, Limerick continues to have a strong influence on choice of workplace. However, within the 21-30km and 31-40km buffer ranges, the median distance travelled is lower. For the 21-30km buffer range, the median travel distance of 13.0km indicates that most individuals living in these settlements do not travel to work in Limerick. The presence of Shannon (which offers many industrial jobs) in the 11-20km range may be an important factor in reducing the travel to work distance. For residents within the 31-40km buffer range, the travel to work distance falls to a median of 6.5km. The location of larger settlements, such as Ennis, Nenagh, Newcastle West and Tipperary Town, in this range may be a contributory factor in reducing travel distance. The reason for reduced travel to work distance in the over 41km buffer range is less apparent. The presence of coastal towns offering specialised coastal- and tourism-related jobs, such as Kilkee and Kilrush, may provide local jobs.

An analysis of the proportion of individuals in each buffer range who indicated in questionnaire responses that they worked in Limerick City provided additional insight into the role of Limerick City as a provider of jobs for the region (Table 10).

Table 10. Proportion of respondents in buffer ranges who work in Limerick city

Range	Percentage of respondents
5-10km	75%
11-20km	63%
21-30km	49%
31-40km	30%
Over 40km	14%

The proportion of people working in Limerick but resident within the 5-10km buffer range is perhaps surprisingly small, but can be explained in part by the presence of Shannon, a clearly atypical settlement with an international airport and large industrial estates. Of respondents resident in this buffer range, 10% work in Shannon. When responses for residents in buffer ranges 11-20km and 21-30km were combined, it was found that 30% of respondents work in Shannon but reside in another settlement. Of respondents resident in Limerick but working in another settlement, 43% identified Shannon as their employment location.

5.3.2. The Sligo cluster

Sligo and Limerick clusters differ in important ways. While Sligo town has a population much larger than any other settlements in its cluster, it has a markedly smaller population than Limerick (Table 1 and Table 7). Of Sligo County's population, 34% reside in Sligo town (CSO 2002a), while 60% of Limerick city and County population resides in Limerick city and suburbs. The buffer ranges for the Sligo cluster differed from those used in the Limerick cluster analysis: this was necessary to maintain sample size given that the Sligo cluster was formed of a lower number of smaller settlements (Table 11).

Table 11. Settlements in Sligo Cluster Buffer Ranges. Straight-line distance from Sligo town used

5-15 km	16-25km	Buffer Ranges 26-35km	36-50km
Strandhill	Manorhamilton	Tubbercurry	Charlestown-Bellaghy
Collooney	Ballymote	Bundoran	Carrick-on-Shannon
Ballisodare		Ballyshannon	Inniscrone
		Boyle	Ballaghderreen
			Ballinamore
			Swinford

Table 12 shows the proportions of respondents living and working in the same settlement within each buffer range. Thirty five percent of respondents living in the 5-15km buffer work and reside the same settlement. This figure may appear relatively high, given the proximity to Sligo Town. There are two possible explanations for this. First, as was found in the case of Annacotty in the Limerick cluster, respondents in Sligo satellite settlements may in error record their place of residence as Sligo Town, as they perceive these satellite settlements now to be suburbs. Second, Strandhill in many ways acts as Shannon in the Limerick cluster. It has a coastal location, is a tourism centre and is close to Sligo regional airport, and has a growing population, again increasing the proportion of respondents who live and work in the same settlement.

Table 12. Sligo Cluster: For Each Buffer Range the Proportion of Respondents Working and Residing in the Same Settlement

Buffer Range	Work Location Same as Residence	Work Location Different from Residence	Respondents With More Than One job Location
Sligo town	94%	5%	1%
5-15 km	35%	65%	0%
16-25 km	52%	48%	0%
26-35 km	67%	32%	1%
36-50 km	60%	39%	1%

Evidence that Sligo Town is the primary source of employment for respondents not working and residing in the same settlement is provided in Table 13. As in the case of the Limerick cluster, with increasing distance from Sligo Town there is a decrease in the proportion of respondents resident in other settlements, but working in Sligo Town. A marked reduction occurs in the 26-35km buffer range, which contains the medium sized settlements of Bundoran, Ballyshannon, Tubbercurry and Boyle. A comparable decline was noted for the

Limerick cluster within the 31-40km buffer range (Table 10), also associated with the presence of medium sized settlements.

Table 13. Proportion of respondents not working in their settlement of residence who work in Sligo Town

Buffer Range	Percentage of Respondents
5-15km	85%
16-25km	57%
26-35km	39%
36-50km	10%

An examination of the average home to work distances (Table 14) for the Sligo cluster suggests that the majority of individuals resident in Sligo Town find work there and the small median distance to work for respondents in the 5-15km buffer range suggests that many of them also find work in Sligo Town. In the 16-25km buffer range, the proportion residing and working in different settlements was comparable with the value for the 21-30km buffer in the Limerick cluster. For respondents in the 26-35km buffer range, 13% only work in Sligo town and 67% reside and work in the same settlement, with a median travel distance of 4km: this underlines the importance of the medium sized settlements as providers of jobs. Of respondents resident in the 36-50km buffer range, 60% also work in this range. Only 4% of individuals in this buffer range work in Sligo. Carrick-on-Shannon is an administrative centre in Co. Leitrim. When data from residents of this settlement are excluded from the analysis, the median travel to work distance increases to 13km, confirming its importance as an employment provider. Questionnaire results also indicate that 12% of respondents in this buffer range who do not reside and work in the same settlement travel to Castlebar in Co. Mayo.

Table 14. Median distance to work for residents in Sligo cluster buffers (*With Carrick-on-Shannon excluded)

| | Sligo Town | Buffer ranges | | | |
		5-15km	16-25km	26-31km	30-50km
Median distance (km)	3	8	16	4	6 (13*)

5.4 RESULTS FOR ANALYSES OF SETTLEMENT CLASSES

Results of analyses for all 80 settlements classified into 7 Classes are described in this section. First analysed was the proportion of questionnaire respondents who work and reside in the same or different settlements.

Table 15. Percentage of individuals working and residing in same or different settlement for each class

	Number of respondents resident in class	Proportion who work and reside in same settlement	Proportion who work and reside in different settlements	Proportion with more than one job	Mean service index for class
Limerick	152	77%	23%	0%	79
Class 1	273	80%	19%	1%	65
Class 2	486	70%	29%	1%	36
Class 3	250	43%	55%	2%	17
Class 4	260	50%	48%	2%	15
Class 5	676	43%	54%	3%	17
Class 6	178	70%	29%	1%	29

Limerick and settlements in Class 1 are important centres of employment (Table 15). The proportion of respondents who work and reside in the same settlement is slightly lower for Limerick than for Class 1, perhaps because of sample size differences, or the job opportunities available in Shannon. However, as a rule the greater the services index value for a settlement Class, the greater the proportion of respondents who both reside and work in that settlement.

Table 16. Median distance to work for all settlement classes

Settlement Class	Distance to workplace	
	Median (km)	Mean (km)
Limerick	3	8.3
Class 1	3	8.8
Class 2	4	12.5
Class 3	7.5	12.6
Class 4	8	12.9
Class 5	8	13.8
Class 6	2.5	8.8

An inverse relationship was found between the service level (functionality level) of a settlement (as measured by the Service index) and the median distance travelled to work by respondents in that settlement (Figure 3). Essentially as the Service index score increases the median distance to work deceases. In Figure 3 the values for median distance to work were inversed and increased by a factor of 100 in order to clearly demonstrate the relationship. Figure 3 is thus a visual representation of the aforementioned trend showing the interconnectedness of service provision and reduced work distances. Median distances to work are provided in Table 16. Class 6 represents an exception to the general pattern, as work distances were shorter than anticipated. The average population in Class 6 settlements is 2,185. However the average distance to the nearest Gateway settlement is 66.5km, which is markedly longer than for other classes (Table 6). It may be that these settlements have developed so as to be more independent of others, removing the need to travel, and that larger settlements are too far away for daily commuting.

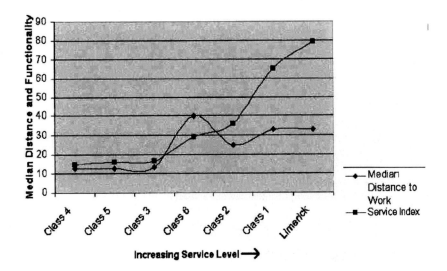

Figure 3. Relationship between service level and distance travelled to work. (Values for median distance to work were inversed and increased by a factor of 100)

Limerick represents a further exception to the general pattern. The reasons for this are unclear, but as suggested earlier may be due to the extent of Limerick, and within it the segregation of services, industry, commercial activities and residential areas. Limerick is the only settlement studied that has an urban public bus service. This provides a greater degree of mobility and thus accessibility to workplaces. Due to this segregation and the availability of public transport, a greater number of residents may travel longer distances to work. The place of work is likely to be within Limerick as 77% of respondents in Limerick both reside and work in the city.

Allowing for the exceptional case of Class 6, it is firstly apparent that Class 1 settlements showed a markedly greater mean service index value than Class 2 or Class 3 settlements. Yet, the median distance to work is similar for Class 2 and Class 1 settlements. There is a distinct drop in distance travelled to work from Class 3 to Class 2 and/or Class 1: this manifests as a change in slope of the trend line in Figures 3 and 4. This suggests that where a settlement reaches a higher level of functionality, shorter journeys to work are the rule, that is, at Class 2 level. This analysis does not take into account other journey destinations, such as to shops, schools, recreational facilities and other locations.

Figure 4 shows mean values for distance of journey to work in settlement Classes, and compares this with the mean service index within these Classes. For functionality and mean distance to work, Classes in general are ranked very similarly, though Class 6 settlements do not follow the general trend in relationships. Analyses were repeated, this time for a sample restricted to respondents residing within 1.6km of settlement centres (see Table 17).

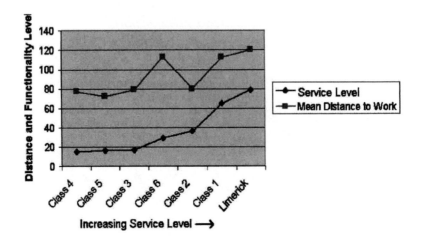

Figure 4. Relationship between settlement service level and mean distance travelled to work. Functionality is measured by the service index. Values for mean distance to work were inversed and increased by a factor of 100

Table 17. Mean and median distance to work values for settlement classes for residents living within 1.6km of settlement centres

Settlement class	Mean distance to work (km)	Median distance to work (km)
Limerick	7.0	3.0
Class 1	13.5	3.0
Class 2	17.5	3.0
Class 3	18.0	11.0
Class 4	18.5	15.5
Class 5	20.0	15.5
Class 6	13.5	3.0

Median distance values for Limerick and Class 1, 2 and 6 settlements suggest that many of these respondents reside close to their place of work. This suggests that settlements within these classes have centres with the capacity to provide both employment and residences. In relation to mean distances, Limerick clearly has the lowest value, and a journey of 7km may often be within the urban area in which public transport is available.

5.5 THE EFFECTS OF RECENT POPULATION CHANGE AND PLACE IN THE SPATIAL HIERARCHY ON TRAVEL MODE CHOICE

Table 18 shows the proportion of journeys to w/s/c made by car in 47 of the settlements, for which Small Area Population Statistics (SAPS) data were available (CSO 2002a).

Table 18. Settlements growth rates (1996-2002 and % car use (Driver only) (CSO 2002a)

Settlement	Code in Figure 5	% car use compared with other modes of transport	% population growth 1996-2002
Granard	1	24.8	-13.6
Rathkeale	2	26.6	-11.9
Ferbane	3	37.5	-5.7
Kilkee	4	31.1	-5.3
Castleconnell	5	45.2	-5.0
Newmarket-on-Fergus	6	39.0	-3.0
Ballyshannon	7	31.2	-2.2
Boyle	8	33.8	-0.8
Templemore	9	31.3	1.2
Tipperary	10	28.7	2.3
Athlone	11	33.5	2.5
Mountrath	12	29.3	2.5
Bundoran	13	31.5	2.6
Kilrush	14	29.3	4.0
Croom	15	35.6	4.7
Moate	16	37.5	4.7
Adare	17	41.5	5.8
Birr	18	33.5	5.8
Sligo	19	31.9	6.6
Tubbercurry	20	33	7.5
Shannon	21	37.3	7.8
Swinford	22	30.2	8.0
Longford	23	33	8.2
Ballinasloe	24	34.3	8.7
Nenagh	25	36.4	9.1
Banagher	26	31.1	9.8
Limerick	27	34.6	9.9
Tullamore	28	35.1	10.5

Settlement	Code in Figure 5	% car use compared with other modes of transport	% population growth 1996-2002
Kilmallock	29	31.8	10.6
Newcastle	30	37	11.0
Abbeyfeale	31	30	13.3
Ballaghaderreen	32	26.5	13.5
Roscommon	33	38.6	14.7
Mountmellick	34	34	15.4
Sixmilebridge	35	42.8	16.0
Edenderry	36	35	19.2
Carrick-on-Shannon	37	40.3	20.3
Portarlington	38	32.9	20.5
Killaloe	39	34.9	20.8
Ennis	40	43.2	24.4
Mullingar	41	37.3	25.0
Portumna	42	37.2	25.5
Strandhill	43	52.4	31.2
Rochfortbridge	44	43.6	90.9
Ballina*	45	47.4	98.2
Annacotty	46	46.5	129.0
Kinnegad	47	40.6	150.7

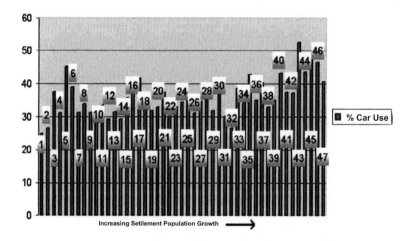

Figure 5. Rate of population growth in settlements and proportion of individuals selecting cars for journey to w/s/c (CSO 2002a) (Numbers correspond to those shown in Table 18)

To aid explanation, data in Table 18 are plotted in a bar chart in Figure 5. While the relationship between population growth and travel mode choice is complex, in general values rise from left to right. However, there are some peaks in values which depart from this trend. The codes 5 and 17 represent Castleconnell and Adare, both satellite settlements for Limerick City, and both with high car usage. The code 35 represents Sixmilebridge which functions as a satellite town for both Shannon and Limerick, code 43 represents Strandhill, which functions as a satellite settlement for Sligo and code 6 represents Newmarket-on-Fergus, which functions as a satellite settlement for both Ennis and Shannon. Some of these satellite settlements increased in population between 1996-2002, some remained at similar population, and some declined in population. They appear to share some other attributes which encourage relatively high car usage rates. However, respondents in all settlements with population growth during 1996-2002 greater than 30% use cars for more than 40% of journeys to w/s/c. In Strandhill, respondents indicated that cars are used in 52% of these journeys. The population growth in this settlement is striking when compared with that of Sligo County, which experienced 4% growth (Sligo Co. Co. 2002).

Abbeyfeale (code 31), Ballaghaderreen (code 32) and Swinford (code 22) are also exceptions to the general trend shown in Figure 5, but in these cases car use is lower than might be expected on the basis of population size and growth. These mid-sized settlements share a number of similar attributes (see Table 19). They are relatively isolated from larger settlements, and the nearest larger settlements are not markedly larger than they are (Ballina, population 8,006 in 2002, Newcastle West, population 4,017 in 2002, and Boyle, population of 2,222 in 2002). They are located at considerable distances from the nearest Gateway settlements. Relatively high services indices suggests that these settlements have the capacity to support a range of functions despite their relative isolation.

Table 19. Selected Attributes of Swinford, Abbeyfeale and Ballaghaderreen

Settlement	Population in 2002	Distance to nearest larger settlement (km)	Distance to nearest Gateway (km)	Services index	% car use (Driver)	% population growth 1996-2002
Swinford	1497	40	77	26	30	8
Abbeyfeale	1683	28	89	21	30	13
Ballaghaderreen	1416	32	67	18	27	13

6. Calculation of carbon dioxide emissions from transport

Transport emissions were calculated for settlement Classes so as to increase sample size and facilitate comparisons among settlements. Analysis of SFSPI questionnaire returns showed that, as a rule, there was little regional variation in the engine fuel type or capacity. Analysis of CSO data for new cars registered in 2002 (CSO 2003) corroborated this result. 'Mile' was the unit of distance adopted in the SFSPI questionnaire, as it was considered that respondents might not estimate journey length accurately in kilometres.

Transport emissions were calculated for settlement Classes so as to increase sample size and facilitate comparisons among settlements. The following method was adopted in calculating CO_2 emissions for journeys to work, shop, school and recreation. Individual questionnaire responses were divided on the basis of the settlement Class in which respondents lived. For each settlement Class, responses were selected in which car was the travel mode for journeys to work mode, either driver alone or with passengers. Cars registered in 2000 or subsequently were selected, as Irish data are more complete for this time period. Car fuel type, petrol or diesel, was selected. Car engine capacity was selected for petrol engines.

- Engine capacities were aggregated into capacity classes using fuel consumption as the primary control.

- For each settlement Class, data on the number of cars in each engine class were linked with data on distance, duration and frequency of journeys made. This allowed calculation of CO_2 emissions for journeys of home to work, home to shops, home to school and home to recreation. Journeys made by each respondent were classed as either urban or extra urban, according to speed given in miles per hour (mph). If the speed was more than 25.5mph (41km/h) then the journey was classified as extra urban, and if below this speed, the journey was classified as urban.

- CO_2 emissions by petrol cars for urban journeys were thus calculated for each respondent as: total kilometres x urban fuel consumption x 2320g. 2320g CO_2 represents the amount of CO_2 produced by combustion of a litre of petrol. Thus: (2.32kg/litre) x total km x L/km = Total kgCO_2. CO_2 emissions for extra urban journeys were calculated similarly.

- Calculation of CO_2 emissions by diesel cars was carried out for the same journey destinations as for petrol cars. Because many fewer respondents owned diesel engine cars (18% is the national proportion) (DoEHLG 2003), the calculation for diesel emissions was based on average fuel consumption values for all vehicles classes (engine sizes) combined. Other than that, steps followed were identical to those described for petrol cars. However, CO_2 emissions for urban journeys were calculated for each respondent as: total kilometres x fuel consumption x 2680g. Combustion of a litre of diesel results in emission of 2680gCO_2. Thus: (2.68 kgCO_2/litre) x total km x L/km = Total kgCO_2. CO_2 emissions for extra urban journeys were calculated similarly.

Analysis to this stage provided CO_2 emissions for each respondent. The next stage was to determine monthly CO_2 emissions for the private car fleet within each settlement, using the following method. For each SFSPI settlement, the value for per capita CO_2 emissions per month was multiplied by the number of people who use their car to travel to work, college or school. Data on number of people using a car are provided in SAPS for the majority of SFSPI settlements. Calculation of total kilometres travelled (all journeys combined) and corresponding CO_2 emissions were also calculated using SFSPI questionnaire data. The method of calculation was different from that used to quantify home to work, shop, school and recreation journeys. This is because speed of travel could not be estimated and therefore urban and extra urban were not differentiated. Steps in the calculation method were as follows:

- Data for respondents resident in each settlement Class were aggregated.
- Cars owned were classified on the basis of year of registration: before 1991, 1991-1995, 1996-1999 and 2000-2004. Corresponding fuel consumption values for these years were applied.
- The CO_2 emission in g/km was then calculated by multiplying fuel consumption for a given time period by the CO_2 emissions factor.

- The average annual distance of car mode travel for respondents resident in each settlement Class was known, as was the number of these cars falling into each of the four age classes. Using these data together with average emissions, annual CO_2 emissions per car traveller were calculated.

- The proportion of residents of settlements using cars for journeys was provided in SAPS data for a majority of settlements. (For SFSPI settlements for which SAPS data were unavailable, estimates were made based on comparisons with similar settlements for which data were available.) For each settlement, average per capita CO_2 emissions were multiplied by the number of residents using cars for journeys to work, college, school or recreation. (Values for each settlement were also then used to calculate settlement Class average).

6.1 CO_2 EMISSIONS FOR SETTLEMENT CLASSES

Per capita CO_2 emissions varied amongst settlement Classes (Table 20).

Table 20. Per capita private car emissions for settlement classes

Settlement Class	Per capita weekly emissions, KgCO$_2$	Per capita annual emissions, KgCO$_2$
Limerick	43	2,234
Class 1	43	2,220
Class 2	58	3,004
Class 3	59	3,093
Class 4	58	3,036
Class 5	59	3,098
Class 6	64	3,353

CO_2 emissions are lower for Limerick and Class 1 settlements. Limerick and Class 1 settlements have substantially larger populations than those in other Classes (see Table 7). Note however that annual emissions per capita are slightly higher in Limerick than in Class 1 settlements: it may be that as populations increase in size above those of Class 1 settlements, emissions rise because of longer urban journey distances. Of course, within Limerick more people use bus transport, and the effects of this on per capita emissions have not been analysed to date. Very small differences were found amongst per capita CO_2 emissions for settlement Classes 2 to 5. These are formed of settlements with average populations of 5,329 people for a Class 2 to 944 people for Class 5 (from Table 7 data). Class 6 settlements, with an average population of 2,159, appear to be exceptional. This suggests that the location of these settlements within the spatial hierarchy, relatively isolated from larger and Gateway settlements, is of importance. Further analysis on individual journey types was undertaken (Table 21). While analyses of annual distances travelled provides a general assessment of CO_2 emissions of settlement Classes, they do not discriminate amongst journey destinations. This was undertaken in an attempt to identify the contribution to emissions made by journeys to differing destinations.

Table 21. Per capita annual emissions by private cars for selected journey destinations

Settlement Class	To recreation (kgCO$_2$)	To school (kgCO$_2$)	To shop (kgCO$_2$)	To work (kgCO$_2$)	Total (kgCO$_2$)	Return journey (kgCO$_2$)
Limerick	98	287	84	634	1,103	2,206
1	289	303	109	598	1,299	2,598
2	115	292	187	789	1,383	2,766
3	148	260	206	839	1,453	2,906
4	185	242	232	813	1,472	2,944
5	170	380	195	887	1,632	3,264
6	127	401	207	639	1,374	2,748

Journeys may have several purposes and involve more than one destination. This is very difficult to account for in data collection, and here it is assumed that each journey was to one location only. The assumption was also made that for each outward journey a return journey was made, so that values derived for single journeys were doubled to quantify return journey distances. These values show that per capita CO$_2$ emissions were lowest for Limerick City, and increased up to Class 6, which had a value comparable with Class 2 settlements. Journey destinations were chosen in order to be compatible with those for which data were published in Census 2002 (CSO 2002a) but for journeys to shops and recreation, data were generated through SFSPI questionnaires alone. Table 21 shows the relative importance of these destinations to total CO$_2$ emissions. For example, in Class 1 settlements, journeys to recreation and school account for 22% and 8% of total CO$_2$ emissions, respectively. For Class 4 settlements the comparable values are 13% and 16%. Comparing Tables 20 and 21, CO$_2$ emissions for Limerick and Class 1 settlements once again are lowest, but differences amongst Classes are not so marked with respect to Table 21 data. Of note, Class 6 settlements, for which CO$_2$ emissions were reported as highest in Table 20, appears to have lower emissions when CO$_2$ are calculated for the journey destinations included in Table 21. A reason for this disparity may be that journeys are made with greater frequency in Class 6 settlements, or that journeys are not so often made to multiple destinations. While Class 6 settlements have low per capita CO$_2$ emissions for journeys to work and recreation, they have the highest emissions for journeys to school and second highest for journeys to shops. The services index value of Class 6 settlements is lower than those for Limerick, Class 1 or Class 2 settlements. This may result in longer or more frequent journeys to larger settlements that provide a larger range of services. A comparison of the proportion of journey destinations which are (a) urban and (b) extra urban for settlement Classes 1, 2 and 6 was carried out (Table 22). This was a complex calculation and time constraints prevented inclusion of all settlement Classes.

Table 22. Proportions of journeys to selected destinations which are (a) Urban and (b) Extra Urban for Classes 1, 2 and 6

Urban travel Class	Journey destination			
	Recreation	School	Shops	Work
1	77%	74%	85%	77%
2	66%	75%	73%	52%
6	75%	77%	65%	45%

Extra urban travel Class	Journey destination			
	Recreation	School	Shops	Work
1	23%	26%	15%	23%
2	34%	25%	27%	48%
6	25%	23%	35%	55%

For respondents in Class 6 settlements, a relatively high percentage of journeys to shops are extra urban, which suggests that these journeys are longer for this Class, thus explaining high per capita CO_2 emissions for journeys to shops. In Class 1 settlements, a relatively high proportion of journeys to recreational destinations are in urban settings. However, this fails to explain the relatively high CO_2 emissions for Class 1 journeys to recreation. To ascertain the reasons for this would require much fuller analysis of the lifestyles of residents. It may be that the availability of a wider range of recreation services in these higher order settlements encourages more travel to recreation. Perhaps time saved in making the relatively short journeys to work, school or shops is then available for recreation.

Table 23. Per capita CO_2 journey to work emissions and class service indices

Settlement Class	Service index	Per capita CO_2 journeys to work
Limerick	79	634
Class 1	65	598
Class 2	36	789
Class 3	17	839
Class 4	15	813
Class 5	17	887
Class 6	29	639

The service index value for Class 1 settlements is markedly greater than for all other Classes except Limerick (Table 23). Per capita CO_2 emissions for work journeys appear to be related inversely to the level of services available within settlements. A similar conclusion was drawn in relation to length of journeys (see Figure 3). This is to be expected as both analyses used distance to work data. However, for per capita CO_2 emissions, a range of other variables was also considered in calculations, such as frequency of journey and fuel consumption. The similarity between results in Table 23 and Figure 3 suggest that distance to work is especially important in explaining inter-Class differences in CO_2 emissions. While for Class 1 settlement residents, a high percentage of respondents work and live in the same settlement, for respondents resident in Classes 3, 4 and 5 settlements, a higher proportion work and live in different settlements, which results in longer home to work distances and thus higher CO_2 emissions. Class 6 settlements, which have lower journey to work CO_2 emissions than those forming Classes 3, 4 and 5, are once again exceptional. In Class 6 settlements, 66% of respondents work and live in the same settlement, compared with 67% for Class 2 respondents, yet per capita CO_2 emissions are much lower in Class 6 settlements. The difference may be explained in part by the high number of journeys in an extra urban setting made by Class 6 residents.

For Limerick City, per capita emissions of CO_2 for journeys to recreation and shop destinations are lower than for any other Class, and its service index value is highest (Table 20 and Table 23). Per capita CO_2 emissions for journeys to work are higher than for Class 1 settlements, which tends to confirm previous results on travel to work distances (Table 15). Again, the very tentative conclusion may be drawn that if settlements grow larger than those in Class 1, distance to work will not fall further and CO_2 emissions may rise. For Limerick, the lower CO_2 emission values for journeys to recreation and shops suggest that these facilities are more centralised whereas employment locations are more scattered. Recent expansion of the Limerick urban area has reduced extra urban distances to Ardnacrusha and Annacotty. These settlements, which have respective population growth rates for the period 1996-2002 of 42% and 129%, are satellite settlements of Limerick City (Table 8 and Table 9). Population growth and income level for all SFSPI settlements were not found to be statistically correlated, but Ardnacrusha and Annacotty respondents have annual incomes in excess of €30,000, substantially above the SFSPI respondent average of €18,800. This suggests the movement of more affluent individuals to satellite settlements, many of whom retain jobs in Limerick city.

Class 1 settlements support relatively large populations and have higher service index values. Of respondents in this Class, 77% reside and work in the same settlement and use private cars as their mode of travel. This contributes to higher traffic volumes so that even though journeys are shorter, they are fuel intensive. For settlements in Class 2 (average population 5,329 people) the proportion of urban journeys is lower (52%), but average per capita CO_2 emissions are greater. What this suggests is that in larger settlements such as those forming Class 1, transport is potentially more sustainable if there were changes to less polluting travel modes.

Traffic congestion is a major problem confronting many commuters and shoppers in contemporary Ireland. Perception of the importance of traffic congestion is subjective, and may be influenced by local factors, such as the presence of a narrow street or busy thoroughfare. Respondents in Limerick and Class 1 settlements are most likely to suffer traffic congestion frequently or on a daily basis, and in general respondents in smaller settlements with lower Services Indices experience least congestion (Table 24). It is also interesting to note that the reported experience of congestion in Class 6 settlements is similar to that for Class 2 settlements. Settlements in Class 2 and Class 6 have similar Services Indices (Table 23), are on average furthest from Gateway settlements (Table 6) and have similar CO_2 emissions for journeys to work, school, shop and recreational combined (Table 21). This suggests that Class 2 and 6 settlements share characteristics despite being different in average population size (Table 6). A combination of functionality and position within the spatial hierarchy may result in a concentration of activity and resultant traffic congestion.

Table 24. Proportions of respondents in classes experiencing traffic congestion

Class	Proportion of respondents reporting experience of congestion				
	Never	Occasionally (1-10 times yearly)	Sometimes (1-10 times monthly)	Frequently (1-10 times weekly)	Always (Daily)
Limerick	9%	15%	14%	25%	37%
Class 1	10%	11%	10%	32%	37%
Class 2	14%	25%	12%	28%	21%
Class 3	29%	23%	14%	21%	13%
Class 4	25%	26%	15%	19%	15%
Class 5	30%	23%	12%	19%	16%
Class 6	12%	24%	18%	27%	19%

6.2 CO_2 EMISSIONS FOR INDIVIDUAL SETTLEMENTS

To calculate emissions per settlement, SAPS data on proportions of individuals using private car as their mode of transport were combined with SFSPI questionnaire data on private car use in journeys to work, school and college. This allowed calculations to include the number of people in each settlement who use private cars for journeys. Previous analyses assumed uniform car usage across all classes and settlements. Emissions for journeys to shops and recreation are substantial and differ among settlement classes, but could not be included in this analysis. Taking an average value for all cars (in relation to both engine size and year of registration) and making the assumption that the private car fleet accounts for 50% of road transport, an estimate for annual road transport emission was derived for each settlement (Table 25).

Table 25. Settlement annual CO_2 emissions for journeys to w/s/c only

Class	Settlement	Population in 2002	Road transport CO_2 emissions in tonnes	Per capita road transport CO_2 emissions in tonnes
Limerick	Limerick	86,998	91,390	1.05
Class 1	Athlone	15,936	16,941	1.06
	Sligo	19,735	22,693	1.15
	Tullamore	11,098	13,316	1.20
	Mullingar	15,621	19,877	1.27
	Ennis	22,051	33,047	1.50
Average				1.21

Class	Settlement	Population in 2002	Road transport CO_2 emissions in tonnes	Per capita road transport CO_2 emissions in tonnes
Class 2	Tipperary	4,964	4,592	0.93
	Birr	4,436	4,125	0.93
	Ballinasloe	6,219	6,701	1.08
	Longford	7,557	8,209	1.09
	Portarlington	4,001	4,644	1.16
	Mountmellick	3,361	3,922	1.17
	Edenderry	4,559	5,788	1.27
	Nenagh	6,454	8,544	1.32
	NewcastleWest	4,017	5,343	1.33
	Roscommon	4,489	6,106	1.36
	Shannon	8,561	12,865	1.50
Average				1.19
Class 3	Kilrush	2,699	2,494	0.92
	Croom	1,056	1,378	1.30
	Patrickswell	998	1,412	1.41
	Ballinamore	687	972	1.41
	Drumshanbo	623	882	1.42
	Annacotty	1,342	1,948	1.45
	Newmarket-on-Fergus	1,496	2,331	1.56
	Kinnegad	1,296	2,100	1.62
	Rochfortbridge	1,382	2,367	1.71
	Sixmilebridge	1,327	2,379	1.79
Class 4	Rathkeale	1,362	1,223	0.90
	Granard	1,013	922	0.91
	Caherconlish	616	656	1.06
	Pallaskenry	550	586	1.07
	Kilfinane	779	830	1.07
	Manorhamilton	927	988	1.07
	Ardnacrusha	926	987	1.07
	Kilcormac	879	937	1.07
	Hospital	621	662	1.07
	Borrisokane	832	887	1.07
	Edgeworthstown	726	774	1.07
	Banagher	1,553	1,844	1.19

Class	Settlement	Population in 2002	Road transport CO_2 emissions in tonnes	Per capita road transport CO_2 emissions in tonnes
	Kilmallock	1,362	1,648	1.21
Class 5	Ballaghaderreen	1,416	1,438	1.02
	Swinford	1,497	1,615	1.08
	Mountrath	1,331	1,465	1.10
	Kilkee	1,260	1,397	1.11
	Tubbercurry	1,171	1,499	1.28
	Killaloe	1,174	1,506	1.28
	Ferbane	1,198	1,704	1.42
	Portumna	1,235	1,861	1.51
	Collooney	619	934	1.51
	Glin	560	845	1.51
	Inniscrone	668	1,008	1.51
	Lanesborough-Ballyleague	943	1,423	1.51
	Ballymahon	827	1,248	1.51
	Cratloe	656	990	1.51
	Foynes	491	741	1.51
	Kilbeggan	652	984	1.51
	Askeaton	921	1,390	1.51
	Ballygar	642	969	1.51
	Bruff	695	1,049	1.51
	Castlepollard	895	1,351	1.51
	Newport	887	1,339	1.51
	Daingean	777	1,173	1.51
	Ballymote	981	1,481	1.51
	Charlestown-Bellaghy	753	1,137	1.51
	Ballisodare	853	1,288	1.51
	Borrisoleigh	598	903	1.51
	Cappamore	684	1,033	1.51
	Strokestown	631	953	1.51
	Moate	1,520	2,406	1.58
	Adare	1,102	2,188	1.99
	Castleconnell	1,343	2,801	2.09
	Ballina	1,185	2,535	2.14

Class	Settlement	Population in 2002	Road transport CO_2 emissions in tonnes	Per capita road transport CO_2 emissions in tonnes
	Strandhill	1,002	2,392	2.39
Class 6	Bundoran	1,842	1,867	1.01
	Ballyshannon	2,715	2,866	1.06
	Boyle	2,205	2,367	1.07
	Abbeyfeale	1,683	1,810	1.08
	Templemore	2,270	2,568	1.13
	Carrick-on-Shannon	2,237	3,286	1.47

Comparison of settlement populations in Census '96 and Census 2002 data allowed calculation of percentage increase in population per settlement between 1996 and 2002. Settlements with high population growth generally produce greatest transport related per capita CO_2 emissions, though there were exceptions to this rule. For settlements with per capita CO_2 emissions of between 0.9 and 1.5 tonnes CO_2, mean population growth between 1996 and 2002 was 6%. For settlements with per capita emissions between 1.5 tonnes CO_2 and 2.4 tonnes CO_2, mean population growth was 21%. Settlements with the 10 highest per capita CO_2 emissions in general shared important attributes (Table 26). There is also a general trend evident in Figure 6 that as settlement population growth increases CO_2 emissions correspondingly increase.

Table 26. Population growth in settlements with high CO_2 emissions

Settlement	Per capita tonnes-CO_2	% population growth 1996-2002	Satellite to:
Annacotty	1.55	129	Limerick
Newmarket-on-Fergus	1.56	-3	Shannon/ Limerick
Moate	1.58	5	Tullamore/Athlone
Kinnegad	1.62	151	Dublin/Mullingar
Rochfortbridge	1.71	91	Dublin
Sixmilebridge	1.79	16	Shannon/Limerick
Adare	1.99	6	Limerick
Castleconnell	2.09	-5	Limerick
Ballina	2.14	98	Limerick
Strandhill	2.39	31	Sligo

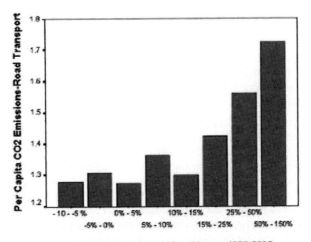

Figure 6. Increasing annual emissions with increasing population growth

All of these settlements are located within 10-15km of the Gateway or Hub settlement for which they function as a satellite. Rochfortbridge and Kinnegad are both located on national primary roads, which serve Dublin. All but two of these small size settlements experienced high population growth during 1996-2002. Both SAPS and SFSPI data show no clear relationship between settlement population size and the proportion of residents who use public transport. People are becoming increasingly dependent on cars for transport, and as satellite settlements provide few services, journeys are tending to become longer and more frequent. This is evident even within Limerick City and environs. A small scale SFSPI study in 2003/2004 sought information from 50 respondents within an area of 3 hectares, 0.5km from the city centre and 60 respondents within an area of 11 hectares, 5-6km from the centre to identify possible differences in travel habits. Mean income for suburban respondents was estimated to be €32,000, and €26,000 for city centre respondents and age of respondents was similar in both neighbourhoods. The median distance travelled to work 8.0km for suburban residents and 3.2km for city centre residents. The mean time taken for the journeys was 22 minutes for suburban residents and 13 minutes for the city centre residents. The mean distance travelled to shops for suburban residents was 1.2km, and 0.8km for city centre residents. Mean time for the journey was 7 minutes in the suburb and 9 minutes in the centre: however, 97% of suburban respondents use a private car as their mode of travel to shops, but 33% of city centre respondents walked to shops.

7. Conclusions

- The classification of settlements into Classes proved to be a useful tool in analysis of the degree of sustainability achieved by contemporary settlements. The classification allowed analysis of the distinctiveness of Limerick city, and the differences between Classes 1 and 6, for example in relation to CO_2

emissions, provided insights into the roles of population size, population growth rate, functioning and place in the spatial hierarchy in influencing the sustainability of settlements. In relation to per capita CO_2 emissions, Limerick city and Class 1 settlements created less pollution from transport than did smaller settlements. Differences which emerged between Classes 2, 3, 4 and 5 provided additional insights, especially in relation to analysis of journey destinations. For these settlement Classes, the distance of the journey to work increased with population size.

- Literature review had suggested that in general larger settlements were to be expected to show more sustainable attributes. In relation to Limerick City, by far the largest settlement in the study, results of analyses did not always corroborate this view. The distance of journey to work differed little between Limerick City and Class 1 settlements. Within the larger area of Limerick city, people living in one neighbourhood but working in another often travelled relatively long journeys. Limerick has a local bus service, but this appeared to make little impact on car use. Distinction was drawn between near centre neighbourhoods, in which walking (to shops) was a frequent travel mode, and suburbs, in which almost all journeys are by car. However, journeys to both shops and recreation in Limerick resulted in the lowest CO_2 emissions: this suggests that these services are widely available in the city, requiring shorter journeys from residential areas. In this regard, the primacy of Limerick in the region did appear to enhance sustainability. Traffic congestion was perceived to be equally serious in Limerick and Class 1 settlements, it may be that city residents have a different view on what constitutes a notable hold-up in traffic. As a rule, in smaller settlements, fewer people regard traffic congestion as a problem.

- Both Limerick City and Sligo Town act as central places within their regions. Both had high Services Index scores. Both settlements supported populations clearly larger than in surrounding counties, and relatively large proportions of these populations were both resident and working in them. They both provided employment for many people living in other settlements: many travelled up to 15km to work in Sligo, and up to 20km for Limerick.

- In important ways, Class 6 settlements emerged as distinctive. With populations generally in the range 2,000-2,500, they were comparable in size with Class 3 settlements, but behaved markedly differently. They varied in relation to attributes such as population density and average income, but all are relatively isolated from Gateway settlements (and thus have no obvious satellite function), and have higher services index scores than might be expected on the basis of population size. Traffic congestion was similar to that for Class 2 settlements, even though population size was smaller. This may be related to the structure of these settlements, typically with one main street, also functioning as a busy through route, with many functions and activities concentrated in it. With regard to the journey to work and shops, per capita CO_2 emissions were low, the distance was relatively short, and walking was a more popular travel mode choice. With regard to journeys to recreation and schools, a different picture emerged. For these journeys, per capita emissions were relatively high. This suggests that these settlements do not provide centrally all the functions required by residents. Additional evidence came from analysis of the split between urban and extra urban travel: for journeys to both work and shops, there was a relatively high proportion of extra urban car travel. These journeys avoided congestion and were undertaken at more efficient speeds, so it may be that emissions were lower even though distances were longer. Questionnaire responses for Class 6 settlements produced a problem: total distance travelled per week was asked for in one question, and distances to specific destinations (work, school, shops, college, recreation) in other questions. Responses did not tally, in that the total distance travelled was rather greater than the combined distances for specific destinations. It may be that there are other destinations of importance, not recognised in this analysis. These findings for Class 6 settlements underline the importance of location in the spatial hierarchy as a vital consideration in the analysis of the sustainability of transport.

- Of all settlements studied, those with highest population growth between 1996 and 2002 (that is, over 30%) were satellites, providing dormitory functions for larger settlements nearby. Some had recent rapid population growth, but others did not. All showed relatively high dependence on private car transport (over 40% of journeys to work, school and college were by car), little use of public transport, and few journeys on foot. As a result, per capita CO_2 emissions related to transport were high. Of the ten settlements with the highest per capita CO_2 emissions, all were satellites within 5-15km of a larger settlement, with Kinnegad and Rochfortbridge being close to a main road serving Dublin. Of respondents residing in a satellite but working in a regional centre, more than 70% travelled 5-10km to Sligo and 5-15km to Limerick. Very few both resided and worked within the satellites: they are not sited around industrial or other sources of employment. Satellite settlements had low Service Index scores so that longer journeys were not only required to work destinations, but to all other destinations. Some satellite settlements had relatively higher population densities, but no correlation was found between density and distance or transport mode of journey to work. Some of the satellite settlements had frequent bus services, but provision of public transport did not alter transport related per capita CO_2 emissions. Apart from Limerick City, in all settlements no relationship was found between population size or density and use of public transport. There is no evidence that public transport use will increase as these satellite settlements grow in population, until they reach a size able to sustain a local bus service. Average income of respondents in some satellite settlements was relatively high, and respondents in all such settlements in general were relatively affluent. This may indicate that individuals and families with higher incomes are moving to satellite settlements. There is no reason to suppose that recent population growth in these settlements will not continue into the future. As a rule, for all settlements studied, those with strong population growth had higher per capita CO_2 emissions related to transport: further population growth in all settlements, including satellites, may be expected to result in higher per capita CO_2 emissions.

- Information sources for these analyses were both published data and data derived from questionnaires. Over 8,000 questionnaires were distributed with a return rate of just under 40%. Collection of these data employed three researchers for 20 months. While this has resulted in the creation of the largest database in Europe on attributes of settlements and their residents, sample size in each settlement was necessarily limited. This limitation was overcome to an extent by aggregating results for settlements forming Classes, but additional research is required to test further the conclusions drawn.

- Given this caveat, the overall conclusion is that, in contemporary Ireland, transport within and between settlements is an important source for carbon dioxide, a greenhouse gas adopted in this study as an indicator for sustainability. Recent patterns of change in the distribution of people, likely to continue into the future, are increasing the extent to which personal transport is unsustainable by enhancing dependency on private cars and increasing distance between residences and journey destinations. While satellite settlements are perhaps the most obvious examples of such change, developments in others contribute to the problem. The number of private cars in Ireland increased by 25% between 1996 and 2000 (NDP 2002): increasing numbers of cars and distances travelled, and the occurrence of traffic congestion, negate improvements in car fuel efficiency resulting from technological advances in engine design. Because settlements differ in attributes in important ways, successful strategies designed to reverse the trend away from sustainability in transport will be specific to each settlement or settlement Class. With regard to settlement pattern parameters of size, location, place within the national settlement hierarchy and functionality, results indicate the importance of satellite settlements, functionality (the provision of services) and size. The relationship between size and per capita CO_2 emissions however is not entirely linear. Settlements of similar size have been shown to have dissimilar per capita emissions. The distinguishing factor is population growth. With these points in mind, the results and trends shown, therefore, provide direction for future policy decisions.

References

ACEA 2004; New Passenger Car Registrations. (Brussels) European Automobile Manufacturers Association.
Acutt M, Dodgson J. 1997; Controlling the Environmental Impacts of Transport: Matching Instruments to Objectives. Transportation Research Part D: Transport and Environment 2(1) 17-33.
Baker M.1972; Irish railways since 1916. Shepperton, Ian Allan, UK.
Bannon, MJ. 1984; Promise and Performance - Irish environmental policies analysed. Blackwell J & Convery FJ. (eds) The resource and environmental policy centre (Dublin).
Botkin D, Keller E. 1995; Environmental Science - Earth as a Living Planet. John Wiley & Sons, Ontario, Canada.
Burton E. 2000; The Compact City. Urban Studies 37 (11): 1969-2001.
Button K. 2002; City Management and Urban Environmental Indicators. Ecological Economics 40 (2): 217-233.
Camagni R, Capello R, Nijkamp P. 1997; The Co-Evolutionary City. International Journal of Urban Sciences 1 (1): 32-46.
Cervero R, Kockelman K. 1997; Travel Demand and the 3Ds: density, diversity, and design. Transportation Research Part D: Transport and Environment 2 (3): 199-219.
CSO 1996; Census 96. Central Statistics Office, Dublin.
CSO 2002a; Census 2002. Central Statistics Office, Dublin.
CSO 2002b; Household Budget Survey 1999-2000 Final Results. Central Statistics Office, Dublin 2.
CSO 2003; Vehicles Licensed For The First Time 2002. Central Statistics Office, Dublin.
CSO 2005; Vehicles Licensed For The First Time 2004. Central Statistics Office, Dublin.
DoEHLG 2000; Executive Strategy National Climate Change Strategy. Department of Environment, Heritage and Local Government, Dublin.
DoEHLG 2003; Irish Bulletin of Vehicle and Driver Statistics 2003. Department of Environment, Heritage and Local Government, Shannon.
Douglas I, Hodgson R, Lawson N. 2002; Industry, Environment and Health Through 200 years in Manchester. Ecological Economics 41(2): 235-255.
EEA 2002; TERM 2002 27 EU - Overall energy efficiency and specific CO_2 emissions for passenger and freight

transport (per passenger km and per tonne-km and by mode. European Environment Agency, Copenhagen.
EEA 2003a; TERM 2003 02 EEA-31 - Transport emissions of greenhouse gases by mode. European Environmental Agency, Copenhagen.
EEA 2003b; TERM 2003 27 EEA 31 - Overall energy efficiency and specific CO_2 emissions for passenger and freight transport. European Environmental Agency, Copenhagen.
EPA 2000a; Corine Land Cover 2000. Environmental Protection Agency, Wexford.
EPA 2000b; Indicators for Transport and the Environment in Ireland. Environmental Protection Agency, Dublin.
EPA 2000c; Ireland's Environment - A Millennium Report. Environmental Protection Agency, Dublin.
EPA 2002; National Inventory Report on Greenhouse Gas Emissions 1990-2000. Environmental Protection Agency, Wexford.
EPA 2003; Methodologies for the estimation of sustainable settlement size. Environmental Protection Agency, Dublin.
Glasscock R, Stephens N. 1970; Irish Geographical Studies in honour of E. Estyn Evans. Queens University of Belfast, Belfast.
Goodbody 2000; Sustainable Travel Demand - Executive Report. Goodbody Economic Consultants, Dublin.
Gordon P, Richardson HW. 1997; Are Compact Cities a Desirable Planning Goal? American Planning Association 63(1): 95-106.
Grove D, Huszar L. 1964; The Towns of Ghana - the role of service centres in regional planning, Ghana University Press, Accra, Ghana.
Gulick J. 1989; The Humanity of Cities. Bergin & Garvey Publishers, New York.
Hall P. 1993; Urban and Regional Planning 3rd Edition. Routledge, London.
Haughton G, Hunter C. 1994; Sustainable Cities. Jessica Kingsley, London.
Holton RJ. 1986; Cities, Capitalism and Civilisation. Allen & Unwin Ltd., London.
ICF, DoEHLG 2004; Determining the Share of National Greenhouse Gas Emissions. Department of Environment, Heritage and Local Government, Dublin.
Jacobs J. 1961; The Death and Life of Great American Cities. Vintage Books, USA.
Jenks M, Burton E, Williams K. 1996; The Compact City: A Sustainable Urban Form. E & FN Spon, London.
Lau JCY, Chiu, CCH. 2004; Accessibility of Workers in a Compact City: the Case of Hong Kong. Habitat International, 28 (1): 89-102.
Mackett R, Edwards M. 1998; The impact of new urban public transport systems: will the expectations be met? Transportation Research Part A: Policy and Practice 32(4): 231-245.
McCutcheon A. 1970; Ireland: Railway History in Pictures. David & Charles Limited, Newton Abbott, UK.
McLaren D. 1992; Compact or Dispersed: Dilution is no Solution. Built Environment 18(4): 268-282.
Moles R, Kelly R, O'Regan B, et al. 2002a; Methodologies for the Estimation of Sustainable Settlement Size in Ireland- Synthesis Report. EPA, Wexford.
Moles R, Kelly R, O'Regan B, et al. 2002b; Methodologies for the Estimation of Sutainable Settlement Size in Ireland - Final Report. EPA, Wexford.
Mumford L. 1995; The City in History: Its origins, its transformations, and its prospects. MJF Books, New York.
NRA 2003; Future Traffic Forecasts 2002-2040. National Roads Authority, Dublin.
NSS 2003; National Spatial Strategy, 2002-2020 People, Places & Potential. Government Stationery Office, Dublin 2.
Odoki JB. Kerali HR, Santorini F. 2001; An integrated model for quantifying accessibility-benefits in developing countries. Transportation Research Part A 35(7): 601-623.
O'Farrell PN. 1968; A Multivariate Analysis of the Spacing of Central Places in County Tipperary. Irish Geography 5: 428-439.
O'Farrell PN. 1979; National Economic and Social Council-Urbanisation and Regional Development Ireland - Urbanisation and Regional Development in Ireland. Government Stationary Office, Dublin.
O'Regan B, Moles R. 1997; Applying a Systems Perspective to Environmental Policy. Journal of Environmental Planning and Management 40(4): 535-538.
O'Regan B, Moles R, Kelly R. 2002; Developing indicators for the estimation of sustainable settlement size in Ireland. Environmental Management and Health 13(5): 450-466.
Owens SE, Rickaby PA, 1992; Settlements and Energy Revisited. Built Environment 18(4): 247-252.
Pearse 1916; The Spark 3: 62.
SAVE-Odyssee 2003; Energy Efficiency in the European Union. ENERDATA, Gières, France.
Schumacher E. 1973; Small is Beautiful: a study of economics as if people mattered. Blond & Briggs Ltd., London.
Scott S. 2004; Research Needs of Sustainable Development (Working Paper 162). ESRI, Dublin.
SEI 2003; Analysis of New Car Registrations in Year 2000. Sustainable Energy Ireland, Dublin.
Sligo_Co.Co., 2002; Strandhill Local Area Plan. Development Planning Unit, Sligo.
Yokohari M, Takeuchi K, Watanabe T, et al. 2000; Beyond Greenbelts and Zoning: a new planning concept for the environment of Asian mega-cities. Landscape and Urban Planning 47(3-4): 159-171.

Statistical packages used:
MVSP Version 3.13G, Kovach computing services, 2003.
SPSS for Windows 10.0.5, SPSS Inc, 2005.

Some relevant websites (accessed 04-05-2005):
http://www.vcacarfueldata.org.uk/information/consumption.asp.
http://www.dpiwe.tas.gov.au/inter.nsf/WebPages/MCLE-5WV8R7?open.
http://www.carbonneutral.com.au/calculate%20your%20emissions.htm.
http://www.dft.gov.uk/stellent/groups/dft_susttravel/documents/pdf/dft_susttravel_pdf_024796.pdf.
http://themes.eea.eu.int/Sectors_and_activities/transport/indicators/technology/TERM27,2003.09/index_html.
http://www.cso.ie/principalstats/princstats.html#transport.
http://www.cso.ie/publications/transport/vehlica.pdf.

CHAPTER 8: MORTALITY IN WILDLIFE DUE TO TRANSPORTATION

ANDREAS SEILER & J-O HELLDIN

Swedish University of Agricultural Sciences, Dept. of Conservation Biology, Grimsö Wildlife Research Station, SE-73091 Riddarhyttan, Sweden

1. Introduction

No other ecological impact of transportation is as perceptible as its deadly toll on wildlife. A Swedish moose smashed into the windscreen of a car; an elephant dissected by a passenger train in northern India; a fin whale pierced on the bow of a cruise ship in Portugal; or a flock of geese sucked into the turbines of a Boeing – incidents such as these are probably among the most eye-catching examples of the ecological impact of modern transportation. In fact, billions of vertebrates are mutilated or destroyed every year in collisions with cars, trains, boats and airplanes, and the numbers are steadily increasing as infrastructure networks expand, vehicle speed increases and the fleet of vehicles swells larger from year to year.

The badger is one of the species that are vulnerable to road traffic in Europe. Especially during spring and early summer, badger carcasses are a common sight on Swedish roads. Photograph by J-O Helldin

Ever since the advent of the automobile during the early years in the 20th century, wildlife casualties on roads have received particular public as well as scientific attention (e.g. Stoner 1925; Dickerson 1939; Haugen 1944; McClure 1951; Ueckermann 1964; Way 1970; Hansen 1982; Caletrio *et al.* 1996; Seiler *et al.* 2004). Today dead animals alongside roads are a common sight and are probably seen by many more people than their living conspecifics in zoos or native habitats. Humouristic cookbooks (Petersen & McLean 1996) and guides to the flattened fauna (e.g. Knutson 1987; Hostetler 1997) are just some expressions of this public interest. Collisions between trains and wildlife are often less spectacular and less dangerous to passengers, but are even more common than animal-vehicle collisions on roads (Van der Grift 1999; Trocmé *et al.* 2003). Less obvious are accidents between ships and large marine mammals; such incidents are becoming alarmingly frequent as a result of the increasing number and speed of large ships (e.g. Laist *et al.* 2001; Nowacek *et al.* 2004). Bird strikes by

John Davenport and Julia L. Davenport, (eds.), The Ecology of Transportation: Managing Mobility for the Environment, 165–189,
© 2006 *Springer. Printed in the Netherlands*

airplanes are another, recently acknowledged problem (Kelly 2001; Chapter 1). Modern transportation has become one of the deadliest activities on earth; even for humans.

There is evidence of a substantial effect of traffic mortality on the persistence of certain wildlife populations including amphibians (Fahrig et al. 1995), turtles and other reptiles (Haxton 2000; Steen & Gibbs 2004), large whales (Laist et al. 2001) and some large carnivores (e.g. McLellan & Schackleton 1988; Forman et al. 2003). Accidents with most other large vertebrates, especially ungulates, are often of lesser relevance to conservation than to game management, traffic safety and thus economy (Groot-Bruinderink & Hazebroek 1996). From an animal welfare point of view, of course, any incident is troublesome as it causes unnecessary and partially avoidable suffering and damage to the animal involved, and thus challenges international laws on animal protection (e.g. Sainsbury et al. 1995). Wildlife mortality in traffic clearly conflicts with our endeavour to achieve an environmentally sound and sustainable development.

What are the reasons for this tragic development? What are the driving factors that cause collisions in the first place? And what can be done to counteract or remedy the impact? How significant is the problem, and where and when are preventive measures necessary? In this chapter, the extent of transport-related mortality in wildlife, mainly regarding animal-vehicle collisions on roads is reviewed. Patterns and factors that help to explain where, when and why collisions occur, and how they can be mitigated are identified. Special emphasis is put on how collisions and thus mortality in wildlife can be evaluated, because this aspect has so far received only little attention. Decisions on whether counteraction is needed or not is partly political but should rely on best knowledge of critical threshold levels in this instance. To establish such levels is an outspoken challenge to applied science.

2. The extent of animal-vehicle collisions

Animal carcasses alongside roads have become a common sight in almost any country in the world. The 'flattened' fauna includes a broad variety of terrestrial animal species, regardless of whether they are rare or common, have a backbone, wings or four legs. To some extent, road-kill statistics actually give a crude but useful index of wildlife occurrences in a region (e.g. Göransson & Karlsson 1979; Seiler et al. 2004).

The number of casualties appears to be steadily growing as traffic increases and infrastructure networks expand. In their review on ecological effects of roads, Forman and Alexander (1998) concluded that "sometime during the last three decades, roads with vehicles probably overtook hunting as the leading direct human cause of vertebrate mortality on land". In fast developing and mobilising countries like China or India, one may expect road traffic to soon become a significant threat to biodiversity.

National road-kill estimates range from some hundred thousand to some hundred million casualties each year (Table 1) and the number of railroad-kills may be almost as large (Havlin 1987; Child et al. 1991; Modafferi 1991; Van der Grift 1999). Estimates made by the Human Society of the USA during the 1960s suggested a minimum of one million animal road-kills *per day* (Lalo 1987). In Spain, extensive surveys revealed a minimum of 10^7 vertebrate traffic kills per year (about 31 incidents per km road; Caletrio et al. 1996). In The Netherlands, Van den Tempel (1993) suggested as many as 2 million bird kills per year. Belgium estimates point to more than 4 million vertebrate road-kills each year (Rodts et al. 1998). A rough estimate on avian road-kills in Sweden suggested as many as 8.5 million bird victims in 1995 (Svensson 1998).

Unfortunately, most national road-kill estimates are not related to the density of the infrastructure network, to traffic work (driven mileage), or animal density. In addition, most estimates are extrapolations of rather limited data (but see Caletrio *et al.* 1996; Rodts *et al.* 1998) obtained by field inventories, drivers' interviews, or expert estimates. Due to differences in quality and uncertainties in these numbers, a quantitative comparison among countries is not feasible.

In many countries, statistics on vehicle collisions with large ungulates provide the most detailed and extensive road-kill estimates, because accidents with these species often involve material damage and considerable risk of human injury. In the United States, it was estimated that more than half a million automobile collisions with deer (*Odocoileus* spp.) occurred in 1991 (Romin & Bissonette 1996). Within Europe (excluding Russia), approximately 500,000 automobile-ungulate collisions are recorded each year, with Sweden contributing the greatest number, followed by Austria and Germany (Groot-Bruinderink & Hazebroek 1996).

Table 1. Estimates of annual nationwide road-kills in wildlife, as obtained from field inventories or drivers enquiries

Species	Road kills *	Country	Year/Period	Source
vertebrates	365	USA	1960's	Humane Society 1960, in Lalo 1987
	32	BRD	1987-1988	Fuellhaas *et al.* 1989
	100	ES	1990's	Caletrio *et al.* 1996
	6.5	FI	2002	Manneri 2002
birds	8.5	SE	1998	Svensson 1998
	5.0	BL	1983	Nankinov & Todorov 1983
	4.0	GB	1966	Hodson 1966
	4.0	BE	1994	Rodts *et al.* 1998
	3.7	DK	1981	Hansen 1982
	2.5	GB	1965	Hodson & Snow 1965
	2.0	NL	1993	Van den Tempel 1993
	1.0	SE	1970's	Göransson *et al.* 1978
	0.6	NL	1977	Jonkers & De Vries 1977
large & medium sized mammals	1.5	DK	1980	Hansen 1982
	0.5	SE	1970's	Göransson *et al.* 1979
	0.2	SE	1992	Seiler *et al.* 2004
	0.2	NL	1977	Jonkers & De Vries 1977
amphibians	5.0	AUS	1983	Ehmann & Cogger 1983, in Bennett 1991
	3.0	DK	1982	Hansen 1982

* in millions per year, nationwide

In Sweden, about 4,500 moose and 25,000 roe deer accidents were registered annually by the police during the 1990s, which was over 60% of all police reported road accidents and between 5-10% of the annual game bag in these species (Seiler 2004). However, the true number of collisions was more likely to be twice as large because not all collisions are detected by the driver, reported to the police, or registered by the road and railroad administrations (Almkvist *et al.* 1980; Seiler *et al.* 2004). The smaller the species, the greater the risk that accident numbers are severely underestimated. Collisions with small mammals and songbirds are rarely noticed by drivers, and carcasses of these species are hard to find in field counts. In the few studies reporting the proportion of smaller species, they often constitute more than 60% of the total kill of both birds and mammals (Figure 1a and b). A wide variety of birds are found in European road-kill studies. In general, species distributions among road-kills reflects the bird fauna in the surrounding landscape, and accordingly, species occurring in roadside habitat are overrepresented (e.g. Adams & Geis 1973). For example, house sparrow (*Passer domesticus*) and blackbird (*Turdus merula*) almost invariably top the lists of road victims in Europe (Bergmann 1974; Hansen 1982; Nankinov & Todorov 1983; Rodts *et al.* 1998).

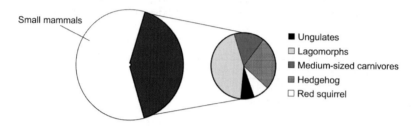

Figure 1a. Distribution of mammal taxa in road-kill counts from northern Europe. Average data based on i) left circle: Hodson (1960), Bengtsson (1962), Fehlberg (1994), Holsbeek *et al.* (1999), and ii) right circle: Göransson *et al.* (1978), Fehlberg (1994), Jensen (1996), Rodts *et al.* (1998), Seiler *et al.* (2004)

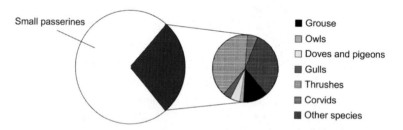

Figure 1b. Distribution of bird taxa in road-kill counts from Europe. Average data based on Hodson (1960), Hodson & Snow (1965), Bengtsson (1962), Bergmann (1974), Bolund (1974), Göransson *et al.* (1978), Hansen (1982), Nankinov & Todorov (1983)

Estimates of animal-train collisions are few, and include in the main incidents with large ungulates (Child *et al.* 1991; Modafferi 1991; Van der Grift 1999; Gundersen & Andreassen 1998; Johansson & Larsson 1999, but see Havlin 1987). However, these studies indicate that railroads exert a similar level of mortality as roads on a regional or national scale.

3. Evaluating animal-vehicle collisions

The number of animal-vehicle collisions is immense, but is this impact on animal life and traffic safety really significant and always worthy of the costs of mitigation? How great a number of collisions are we willing to accept? How great a loss can a species cope with? There are no simple answers to these questions, no simple rules or thresholds defined thus far that could guide decision-making. In many cases, it is probably a political issue rather than a biological problem, to determine whether the extent of animal-vehicle collisions is critical and whether counteraction is necessary.

Evaluating the importance of animal-vehicle collisions is a complex task and must involve ecological, economical, social and technical perspectives, and consider both broad and small scales. The results will vary between perspectives and between species (Figure 2). For example, vehicle collisions with large ungulates, such as moose, may be insignificant for the conservation of these species, but may very well conflict with harvest-management goals and be unacceptable for traffic safety and thus economic reasons. For other, non-harvestable but common species, neither harvest nor traffic safety reasons may lead to a demand for a reduction of collision numbers. Still, there may be a strong public interest in keeping incidents rare. People certainly would not want road and rail traffic to be among the most important sources of human induced mortality in wildlife. Traffic losses in certain small and rare species may be unnoticed and disregarded by the public, but may be of ecological significance. For red-listed species, there are strong legislative incitements that require counteraction.

Finding a sound balance between the various interests and perspectives in animal-vehicle collisions and establishing appropriate thresholds for the maximum tolerable impact is a clear challenge to mitigation planning (Spellerberg 1998).

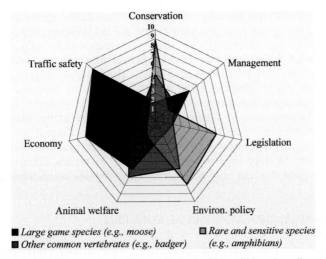

Figure 2. Illustration of the differences in significance of animal-vehicle collisions depending on which species is considered and which perspective applied. Ranking significances according to a common scale (here deliberately chosen from 1-10) helps to compare and weight relative differences among perspectives and species and may help to decide upon whether mitigation is needed.

3.1 ETHICAL, POLITICAL AND LEGAL ASPECTS

Most people consider a collision with an animal on the road or elsewhere in traffic as a very unpleasant experience, not only because of safety risks. People are concerned especially as the damage to and suffering of the animal is unnecessary, unintended and partly avoidable. Many feel pity and dismay when seeing carcasses of mammals and birds alongside roads or an approaching train sprinkled with blood from the animal it hit. These are not scenes road authorities or railway companies want to be remembered; and it certainly does not fit their promotion of a modern, safe, and environmentally sound transportation. In fact, the European public seems generally to be more concerned about killing and injuring wildlife than about habitat destruction or air pollution (Kirby 1997). Simply from an aesthetical and ethical point of view, there should thus be interest enough to justify preventive measures against animal-vehicle collisions. This interest can influence environmental policy making and be used to establish new quality objectives for the environmentally sustainable development of the transport sector (e.g. Seiler & Eriksson 1997; Nilsson & Sjölund 2003).

Also, the deadly impact of traffic on wildlife conflicts with the principle of animal protection and welfare (e.g. Sainsbury *et al.* 1995). Yet, animal protection laws at a European level make it an offence to cause unnecessary suffering only in captive or domestic but not in free-living animals (Animal Welfare Act 1988). Only Norway and Finland provide legal protection for the welfare of wild animals: (Norwegian Animal Welfare Act (Code: 750.000, 16.06.95); Finnish Act on Animal Protection (247/1996)). In general, there is better legal protection for the physical environment than for the welfare of the organisms living within it. Sainsbury *et al.* (1993) illustrated this inconsistency in a large number of cases across the European Community, where the welfare of wildlife had been compromised due to human activities, including collisions with vehicles and hunting activities. The authors developed a methodology for the assessment of wildlife welfare (Kirkwood *et al.* 1994) and suggested it to be integrated in environmental impact assessment (EIA). However, resistance to include welfare of wildlife under legal protection is strong, as it may have considerable consequences to e.g. hunting and fishing practices, recreational land use and transport planning.

In European policies and directives, the issue of animal-vehicle collisions is typically considered as part of the ecological problem of habitat fragmentation caused by transport infrastructure (e.g. OECD 1994; Trocmé *et al.* 2003). The European Strategic Environmental Impact Assessment (SEA) Directive (2001/42/EC) enforces the integration of ecological aspects in future planning and programming of infrastructure. Recently, a new 'Code of Practice for the Incorporation of Landscape and Biodiversity in the Planning of Linear Transport Infrastructure' has been developed by the European Council for endorsement in 2003 that includes recommendations for integration of the Pan-European Biological and Landscape Diversity Strategy in environmental impact assessment (Damard *et al.* 2003). These efforts are good, but can only be the first steps towards an ecologically sustainable development.

3.2 TRAFFIC SAFETY AND ECONOMICAL CONCERN

In many European countries, animal-vehicle collisions make up only a small proportion of the total number of automobile traffic accidents. In The Netherlands, for example, only 0.3% (29 out of 11,124) reported accidents involving personal injury or death were due to animals (Borer & Fry 2003). This is in contrast with the situation in Sweden where about 8% of all road accidents that involved human injuries or deaths and more than 60% of all police reported road accidents were due to collisions with moose (*Alces alces*) and roe deer (*Capreolus capreolus*) (Seiler & Folkeson 2003). Although the socio-economic costs of animal-vehicle collisions are likely to be underestimated, it is traffic safety and economic concern that drives mitigation efforts against these accidents (e.g. Romin & Bissonette 1996; Putman 1997).

Of course, there is a direct relationship between the seriousness of collisions with animals and the size of the animals involved. The larger the species, the greater the risk for material damage and human injury. Large ungulates such as the moose or the red deer (*Cervus elaphus*) are among the most dangerous animals in this respect. Their tall shoulder height and heavy body weight can result in collisions where the whole body mass of the animal strikes directly against the windshield pillars and the front roof of the vehicle. Such accidents can cause severe head and neck injuries to front seat passengers (Björnstig *et al.* 1986). In Sweden, for example, about 2.5% (about 600 per year) of all police-reported accidents with moose and the smaller roe deer involve human injuries; about 10-15 people are killed each year in result of these collisions. Collisions with moose account for 82% of these human fatalities and for 75% of all human injuries in reported wildlife-vehicle collisions. Altogether, the annual cost of these 4,500 moose and 23,000 roe deer collisions in Sweden approximates €100 million (Seiler & Folkeson 2003). In Germany, Anonymous (1997) estimated 200,000 wildlife accidents on roads annually, producing a socio-economic cost of 600 million DM (about €315 million). At the European level (excluding Russia), automobile collisions with ungulates have been estimated to exceed half a million incidents each year, including 300 human fatalities and 30,000 injuries, producing a total cost of more than €1 billion (Groot-Bruinderink & Hazebroek 1996). Also, in the US, deer-vehicle collisions generate significant economic losses, estimated at more than $1 billion each year (Conover 1997).

Such economic estimations usually include costs for material damage and human injuries or fatalities. However, they rarely account for 'external' costs such as loss of meat or hunting opportunities, call-out costs for veterinarians, gamekeepers and police to deal with injured or dead animals, costs for ambulances and any subsequent human medical costs, or societal costs of traffic delays (e.g. Borer & Fry 2003). In Norway, for example, the annual cost of moose-vehicle collisions (about 1,200) has been estimated to between €11 and 17 million, depending on whether only the material damage to the vehicle is assessed or if the cost to moose management is included (Stikbakke & Gaasemyr 1997). Schwabe and Schuhmann (2002) demonstrated how different approaches in assessing the value of deer to hunters result in estimates of the annual cost of deer-vehicle collisions in the US that range from $23 million to nearly $1 billion. They concluded that large economic gains could be made by reducing the number of animal-vehicle collisions. Understanding the external costs of animal-vehicle collisions and internalising these costs in infrastructure management will help to develop more just and adequate evaluations of the environmental impact, and, in turn, the funds that should be made available for further research or mitigation (Johnson 1995).

3.3 POPULATION MANAGEMENT AND SPECIES CONSERVATION

Despite the great public interest in road-kills, and the high annual costs of collisions with larger mammals, traffic has probably only had a small effect on the survival of common wildlife populations. Rodents, lagomorphs, medium sized carnivores and many songbirds, often dominate road-kill statistics from Europe, but also ungulates, especially roe deer, are frequently victims of road and rail traffic. Yet, traffic usually contributes less than 5-10% to the overall (direct) mortality in these species (e.g. Bennett 1991; Seiler 2003a). The frequent carnage of these animals is appalling enough, but may simply indicate that they are very abundant and widespread, and can probably cope with losses due to traffic. In most ungulates, road mortality is thus a management issue rather than a conservation problem (Groot-Bruinderink & Hazebroek 1996). In game species, traffic losses should be considered in harvest goals in order to balance the management of these populations against environmental, social and economic constraints. Norwegian deer management strategies, for example, aim at a ratio of traffic losses to harvest numbers of 4% at national and 10% at municipal level, although eventually, in certain hunting districts and during certain years, more moose may be killed by vehicles than shot by hunters (Stikbakke & Gaasemyr 1997). Year to year changes in collision numbers in these species are often strongly correlated with changes in population sizes (Seiler 2004; Mysterud 2004).

However, there is evidence for a significant effect of traffic mortality on certain wildlife populations including some rare species (e.g. Forman *et al.* 2003). For instance, traffic is especially dangerous to herpetofauna (Blaustein & Wake 1990). Road density has a proven negative effect on survival and recruitment of amphibian populations and the risk for local extinctions increases with proximity of breeding ponds to well-travelled roads (Fahrig *et al.* 1995; Vos & Chardon 1998). Traffic mortality is by far the most significant source of mortality in the endangered Florida panther (*Felis concolor*), accounting for more than 50% of all known deaths (Harris & Gallagher 1989; Harris & Scheck 1991). The Iberian lynx (*Felis pardina*) suffers 6-10% mortality due to road traffic, which is considered as the second most important mortality factor (Ferreras *et al.* 1992). Traffic stood for 7-25% of the known annual mortality in wolf (*Canis lupus*), and for almost all known mortality in bear (*Ursus arctos*) in Italy between 1974 and 1984 (Boscali 1987). Road traffic is a major source of mortality in badgers (*Meles meles*) across Europe (Griffiths & Thomas 1993; Neal & Cheeseman 1996). In The Netherlands, road traffic was held responsible for a nationwide decline of badgers during the 1980s (Van der Zee *et al.* 1992). Reported badger road casualties in The Netherlands during the early 1990s accounted for 10-16% of the summer population, but the total loss, including unreported accidents and death of juveniles that lost their mother on a road, probably exceeded 25% (Bekker & Canters 1997), which is more than 50% of the annual reproduction (Lankester *et al.* 1991). Traffic mortality is also considered significant to population recruitment in the Florida Key deer (*Odocoileus virginianus clavium*) (Calvo & Silvy 1996). Other well-known examples of species that are heavily affected by road traffic include hedgehog (*Erinaceus europaeus*) (Huijser 2000), otter (*Lutra lutra*) (Madsen 1996), moorhen (*Gallinula chloropus*) (Rodts *et al.* 1998), barn owl (*Tyto alba*) (Newton *et al.* 1997) and several other birds of prey (Van der Zande *et al.* 1980; Van den Tempel 1993). Some of these species are often found among road-kills and traffic is indeed one of the key mortality factors. In others, road casualties are rarer, but nonetheless important to population recruitment or dispersal.

The significance of occasional traffic mortality to population development may easily be underestimated if traffic causes a disproportionate loss in especially sensitive or important cohorts of individuals, such as female turtles on nesting migrations (Steen & Gibbs 2004), territorial male otters (Madsen 1996) or lactating and pregnant female badgers during spring (Seiler *et al.* 2003). Road-kill statistics seldom provide a representative sample of the demography and structure of living animal populations, nor do they give a correct picture of the existing fauna. Certain cohorts of individuals and certain species are more likely to be exposed to traffic than others (see also Jahn 1959; Hodson & Snow 1965; Dixon *et al.* 1996). What distinguishes species that are vulnerable to road traffic from those that are not? When does traffic mortality become a threat to the survival or management of a species? Finally, how great a loss of individuals should be tolerated politically or ecologically?

In theory, wide-ranging species such as lynx or wolf are more exposed to traffic, since they are more likely to encounter roads or railroads than more stationary species with smaller home ranges such as roe deer or moose. Species that cope with or are attracted to disturbed habitats such as human settlements and transportation corridors may be more exposed than species that avoid the vicinity to humans. Species that do not fear vehicles are more exposed than those that do. If these traits are combined with small population sizes, high adult survival, low fecundity and late reproductive age, then the species in question is prone to suffer due to the effects of human transportation.

To evaluate the significance of traffic mortality (like any other mortality factor) from an ecological standpoint, it should be studied in the context of population demography, considering sex- and age-specific mortality and fecundity rates. As a rule of thumb, the larger the percentage of road-kills on all deaths, the more likely traffic is a 'key factor', unless traffic mortality is compensated for by increased survival or fecundity of the remaining individuals, or mainly affects the already 'doomed surplus' (Southwood & Henderson 2000). However, most mortality factors, including traffic, are neither completely compensatory nor completely additive. Therefore, the percentage of road-kills may eventually produce a misleading picture. Similar to the assessment of a maximum sustainable harvest in game or fish (Robinson & Bodmer 1999), estimation of an ecologically 'sustainable' level of traffic mortality should relate to population growth rather than to the size of the population or the proportional kill. A 'sustainable' loss takes the interest in population growth and does not affect the population capital, i.e., the population size. The higher the growth rate, the larger a loss can be sustained without changes in population density. However, if population growth rate is already close to stationary, even a small, uncompensated increase in mortality can be significant and provoke a decline of the population.

Thus, species that are sensitive to traffic mortality are typically slow reproducing (low growth rate) and long-lived, while species that are most exposed to road traffic are wide-ranging or migratory animals, or those attracted to the infrastructure corridor (Verkaar & Bekker 1991). These behaviours and life history traits are indeed typical for many medium to large wildlife species, and therefore careful attention must be paid to the effects of infrastructure mortality on these species.

4. Factors and patterns in animal-vehicle collisions

So why does the chicken cross the road? Well, movement is one inherent property of animal life, and after some time, any movement leads almost inevitably to a road or railroad in the landscape. The question is thus not why but rather where and when the chicken crosses the road – and why it is eventually run over. If the amount of animal-vehicle collisions is considered significant from an ecological or human perspective, specific knowledge about the why, where and when is needed to develop adequate counteractive measures.

Factors responsible for the occurrence of animal-vehicle collisions can be summarized under three major categories: a) the animal, its abundance, ecology and behaviour, b) the traffic, its density and velocity, and c) the environment including both road and surrounding landscape (Table 2). The interplay of these factors creates a complex, species specific pattern in the spatial and temporal distribution of animal-vehicle collisions that must be understood before effective counteraction can be designed and employed (e.g. Putman 1997; Forman *et al.* 2003).

Table 2. Factors potentially responsible for the occurrence of animal-vehicle collisions

Animal factors	Traffic factors	Environmental factors
Individual behavior	*Vehicle/Driver*	*Road corridor*
- sex, age, status	- vehicle speed	- corridor width
- dispersal, mating, foraging movements	- road surface	- road side habitat
	- visibility	- fences, gullies
- explorative, defensive, aggressive behaviour	- reaction time	- bridges, tunnels
		- road lighting
Species ecology	*Traffic*	*Landscape*
- abundance	- density	- topography
- solitary / group-living	- continuous / clumped	- linear features
- habitat utilisation	- velocity	- adjacent habitat
- areal needs	- diurnal / seasonal pattern	- landscape composition
- migratory movements		- microclimate
- nocturnal / diurnal		

Spatial patterns in collisions relate to local variations in animal abundance and activity, habitat distribution, landscape topography, road design and alignment in relation to landscape structure, and traffic characteristics (e.g. Madsen *et al.* 1998; Hubbard *et al.* 2000; Clevenger *et al.* 2003; Nielsen *et al.* 2003; Malo *et al.* 2004). Temporal patterns reflect seasonal and diurnal variations in traffic volume, weather and light conditions, and the different biological periods that influence a species' activity, such as the daily rhythm of foraging and resting, seasons for mating and breeding, dispersal of the young-of-the-year, or seasonal migration between winter and summer habitats (e.g. Gundersen *et al.* 1998; Inbar & Mayer 1999; Haikonen & Summala 2001; Joyce & Mahoney 2001; Mysterud 2004).

4.1 ANIMAL ABUNDANCE AND ACTIVITY

In theory, the number of animal-vehicle collisions should be a function of the density and activity of animals and vehicles. Various studies have confirmed that broad-scaled trends in the frequency of animal-vehicle collisions reflect patterns in animal abundance and, to a lesser degree, traffic volume (e.g. Groot-Bruinderink & Hazebroek 1996).

In Norway, the increase in vehicle collisions with red deer from 1971 to 2001 strongly correlated with the increase in population size (Mysterud 2004). McCaffery (1973) observed significant correlations between the number of collisions involving white-tailed deer (*Odocoileus virginianus*) and the number of antlered buck harvested in 28 of 29 management areas in Wisconsin, USA. Similarly, Seiler (2004) found strong correlations between the densities of vehicle collisions with moose and roe deer and the average annual harvests of these species among 22 Swedish counties as well as among moose hunting districts within counties. The significant change in ungulate-vehicle collisions that occurred in Sweden over the past 3 decades could also be attributed to increasing ungulate densities. Harvest statistics, as an index of ungulate abundance, were the primary correlate with collision numbers at national level, whereas the steadily increasing traffic explained most of the residual variation and kept collision numbers at a high level when population sizes declined again (Seiler 2004).

The strong, broad-scaled relationship between trends in animal-vehicle collisions and animal densities suggests that road-kill data could be used as index to monitor wildlife populations (Jahn 1959; Göransson & Karlsson 1979; Hicks 1993; Loughry & McDonough 1996). According to Baker *et al.* (2004) road-kill statistics can be a useful index for certain species, but it needs further research into the validity and precision of this technique before it can be used in national monitoring programmes because the exposure to traffic is species as well as age and gender specific.

Measures of animal density usually refer to comparatively large areas, encompassing many individuals' home ranges. At a local scale though, at the scale of an individual's use of space, other parameters will be needed to identify and foresee collision risks. Animal-vehicle collisions are spatially and temporally aggregated (e.g. Finder *et al.* 1999; Hubbard *et al.* 2000; Madsen *et al.* 2002). This implies that patterns observed at a broad scale may not necessarily apply at a finer scale where emerging local factors and increased variance and error obstruct the picture (Seiler 2004). Consequently, different criteria need to be studied to understand the pattern at different scales. For local risk assessment, measures of animal movement and activity will be more appropriate than any density measure. In part, these can be derived from knowledge about the species' habitat utilisation and distribution of preferred habitat in the landscape. Indeed, several studies have demonstrated how data on landscape composition, road and traffic features can be used to predict the risk for collisions with deer and other wildlife (e.g. Clevenger & Wierzchowski 2002; Malo *et al.* 2004; Seiler 2005).

4.2. TRAFFIC INTENSITY AND VEHICLE SPEED

The second category of factors influencing numbers and likelihood of animal-vehicle collisions is related to traffic density and vehicle speed. Increasing traffic has been held responsible for the growing number of animal-vehicle collisions worldwide (e.g. Forman & Alexander 1998). Trend analyses and comparisons of field inventories made during different decades seem to support this idea (e.g. Jonkers & De Vries 1977; Hansen 1982; Van den Tempel 1993; Newton *et al.* 1997; Seiler 2004), but the significance of traffic is usually inferior to the effect of changes in animal abundances (see above).

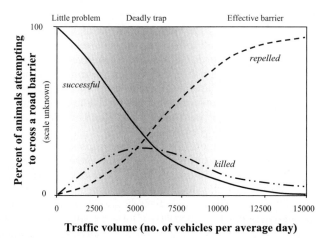

Figure 3. Conceptual model on the effect of traffic volume on the percentage of animals that successfully cross a road, are repelled by traffic noise and vehicle movement, or get killed as they attempt to cross. The model is partly based on empirical data on moose-vehicle collisions in Sweden (changed after Seiler 2003b)

At broad scale, collision numbers increase linearly with traffic volume. Locally, however, the effect is not necessarily linear and can be confounded by population dynamics, animal behaviour, spatial and temporal factors (e.g. Groot-Bruinderink & Hazebroek 1996). For example, studies on amphibians (Van Gelder 1973; Kuhn 1987), small mammals and birds (Oxley *et al.* 1974; Wilkins & Schmidly 1980), carnivores (Clarke *et al.* 1998; Rosell Pagès & Velasco Rivas 1999), and ungulates (Skölving 1987; Berthoud 1987; Seiler 2005) showed a higher density of collisions on intermediate roads than on major highways or on local access roads. This may be due to an increasing repellence of intensive traffic that reduces accident risks (Figure 3). At low traffic volumes, animals may not waver to enter a roadway and only few individuals may collide with vehicles. With increased traffic, more animals will be killed while trying to cross a road. On very busy roads, however, approaching animals will more likely be repelled by traffic noise or vehicle movement, which leaves fewer to be run over but also reduces the number of successful crossings. This interaction between mortality risk and repellence produces a barrier effect that increases exponentially with traffic volume. Busy highways with traffic volumes above 10,000 vehicles per day and multiple lane railroads with over 15 trains per hour are therefore considered as an insurmountable barrier to most terrestrial animals (e.g. Müller & Berthoud 1997; Righetti & Malli 2004; compare Chapter 10).

Once an animal has taken a step onto the road, the risk of colliding with vehicles increases with traffic volume, vehicle speed, road width and the presence of obstacles such as gullies or central wire railings that slow down the animal or prevent its exit from the road. Simulation models suggest that traffic volume and the velocity of the animal are the most important factors determining the immediate risk of collisions (van Langevelde & Jaarsma 2004). Similarly, Hels and Buchwald (2001) calculated collision risks for amphibians and small mammals as the product of animal speed, vehicle width, traffic density and the angle at which the animal crosses the road (i.e. a measure of road width). Their model suggested that traffic volumes of less than 4,000 vehicles per average day might kill most amphibians trying to cross a road, while faster animals such as hares (*Lepus lepus*) would not be significantly affected. Clearly, mobile animals may have a greater chance to slip through the time gap between following cars than slow moving species. Mader (1981) suggested that the chance of survival increases logarithmically with the velocity of the animal. These data clearly point out that temporal and local reduction of traffic volume and vehicle speed can be effective mitigation options against animal-vehicle collisions (van Langevelde & Jaarsma 2004).

The above models presume that neither the car driver nor the animal tries to avoid a collision, and that the animal is not aware of fast moving vehicles or even recognises the danger they imply. If animals react to vehicles, however, they may be able to learn, finding safe time gaps between cars. On the other hand, if an animal has grown afraid of vehicles, it may need a long time to recover before attempting a new crossing.

Vehicle speed has little effect on the distribution of time gaps in traffic, but it may have an influence on how animals respond to traffic. The faster the vehicle, the more difficult it may be for the animal to assess distance and speed and the lesser chance there is for the driver to avoid colliding with animals. Indeed, vehicle speed is assumed to be a major reason for road mortality in e.g. raccoons (*Procyon lotor*) in Indiana (Rolley & Lehman 1992), and Tasmanian devils (*Sarcophilus harrisii*) and Eastern quolls (*Dasyurus viverrinus*) in Tasmania (Jones 2000). The combination of high speed and low traffic volumes, i.e. long time gaps between vehicles, may also partly explain the comparably high mortality rate in wildlife along railways that have been observed in Sweden (Seiler 2003b). A fast train that approaches silently, and often after a long interval of quietness, may therefore be more unexpected and also more dangerous to wildlife than a busy highway. There are indications from accident reports of the Swedish Rail Authority suggesting that per kilometre railway more wildlife is killed than per kilometre average road in Sweden.

4.3 ENVIRONMENTAL FACTORS

The distribution of animal-vehicle collisions is non-random, and spatially and temporally aggregated. Beside changes in animal density and traffic volume, that can create annual and inter-annual trends in collision numbers, there are various environmental factors that produce a local pattern of hot-spots (e.g. Caletrio *et al.* 1996; Madsen *et al.* 1998). Once the influence of spatial and temporal factors is understood, it should be possible to delineate dangerous road sections and periods, and then employ local and temporary mitigation measures such as speed control, road closure or traffic rerouting (e.g. Malo *et al.* 2004; Seiler 2005).

Design and placement of infrastructure corridors and the contrast between the open verges and the adjacent vegetation are crucial factors affecting the behaviour of animals. These factors are also more easily adapted to reduce traffic mortality than traffic flow or animal abundance. Naturally, animal-vehicle collisions are most likely to occur where infrastructure traverses habitats rich in wildlife, dissects dispersal corridors or natural linkages, or provides attractive resources to wildlife that are missing in the surrounding landscape.

A Spanish study on vertebrate mortality on roads suggested that environmental measures and habitat quality had a much higher effect than road features on the frequency of fauna casualties (Gonzalez-Prieto *et al.* 1993). The highest collision frequencies occurred in undisturbed areas remote from human habitations. In Sweden, Göransson *et al.* (1978) observed increased frequencies of road-killed mammals and birds in suburban and forest habitats compared with open, agricultural areas; a pattern reflecting differences in animal abundances between these habitats. Forest habitat is also an important prerequisite for deer-vehicle accidents (e.g. Almkvist *et al.* 1980; Berthoud 1987; Finder *et al.* 1999; Hubbard *et al.* 2000). Clevenger *et al.* (2003) found that small terrestrial road-kills were more likely to occur close to vegetative cover and far from wildlife passages or culverts. Clearly, the effect of habitat on collision risks depends on the composition of the wider landscape and the juxtaposition of the road relative to important landscape elements. Where the preferred habitat is extensive and common, animal-vehicle accident sites tend to be more randomly distributed (e.g. Allen & McCullough 1976; Bashore *et al.* 1985). Accident sites are more aggregated and collision risk locally increased, where favourable habitat is patchy and coincides with infrastructure corridors; and where linear landscape features such as riparian corridors, fence rows or other transport infrastructure funnels animals alongside or across infrastructure (e.g. Feldhamer *et al.* 1986; Lehnert *et al.* 1996; Lodé 2000; Clevenger *et al.* 2003; Nielsen *et al.* 2003). The probability of deer-vehicle collisions in Illinois, USA, is significantly increased where public recreational land is near roads and the presence of adjacent gullies and riparian travel corridors are intersected by roads (Finder *et al.* 1999). Hubbard *et al.* (2000) observed that the likelihood of accidents with white-tailed deer was increased where highways bridged over (riparian) travel corridors for deer. Also, where exclusion fences terminate or are interrupted by interchanges and connecting infrastructure, collision risk will be increased (e.g. Ward 1982; Feldhamer *et al.* 1986; Clevenger *et al.* 2001). Traffic casualties in otters (*Lutra lutra*), for instance, were most likely to occur where roads cross over watercourses along which otters move (e.g. Philcox *et al.* 1999). In The Netherlands road-kills of hedgehogs were more likely to occur where railway corridors, along which hedgehogs foraged, intersected with trafficked roads (Huijser 2000). In Denmark, more road-killed foxes and roe deer were found near interchanges than elsewhere along the studied highways (Madsen *et al.* 1998). This was explained by the design of the interchanges, including the extent of forestation and the fencing between the roads that together first attract but then trap animals in the road junction. In Sweden, the risk of moose-vehicle collisions was slightly increased where private roads connected to highways (Seiler 2004). This may be due to moose using minor roads as travel corridors. However, where private roads connected to bridges or tunnels providing a safe passage across highways, collisions occurred less frequently.

Many species may indeed take advantage of roads and railroads as transportation corridors (Pienaar 1968; Kolb 1984; Modafferi 1991, Thurber *et al.* 1994). Others are attracted to the infrastructure as it provides food resources, such as carrion for scavengers, spilled grain, herbal forage, or de-icing salt sought after by herbivores or nesting sites for birds and other small vertebrates (e.g. Way 1977; Bennett 1991; Auestad *et al.* 1999; Meunier *et al.* 2000; compare Chapter 11). Clearing roadsides of vegetation that either obscures the visibility of approaching wildlife or provides attractive forage for deer may reduce the risk for animal-vehicle collisions considerably (Nilsson 1987; Jaren *et al.* 1991; Rea 2003). Several studies have even revealed higher numbers of some open area birds along roadsides than compared with the surrounding landscape (Laursen 1981; Havlin 1987; Helldin & Seiler 2003). However, density can be a misleading index for habitat quality, and many of these 'roadside populations' will probably suffer from decreased fitness and survival due to traffic disturbances. Under such circumstances, infrastructure corridors act as sink habitats or ecological traps (e.g. Reijnen *et al.* 1997; Battin 2004; compare Chapters 11, 12).

In addition, other environmental factors such as weather and light conditions (affecting visibility and detectability), temperature and precipitation (attractiveness of road surface for e.g. amphibians or snakes) can create local and temporary patterns in the distribution of collision risks. Almkvist *et al.* (1980), for example, suggested that during daylight the risk of moose-vehicle collisions in forested areas was four times greater than in open habitats, whereas the risk during night was about the same in both habitat types. Increasing darkness reduces the drivers' ability to detect approaching animals (Haikonen & Summala 2001). Studies on moose-train collisions in Norway demonstrated strong interactions between lunar phase, snow cover, temperature, and time of day in determining collision sites (e.g. Andersen *et al.* 1991; Gundersen & Andreassen 1998; Gundersen *et al.* 1998).

To conclude, numerous factors affect the distribution of animal-vehicle collisions in both space and time. The most important factors are the abundance of animals, traffic volume and probably speed, and road and landscape characteristics that vary in importance from species to species. Hot spots or periods in collision pattern typically occur through a combination of several independent factors. These factors operate at different scales and must be considered accordingly. If spatial and temporal factors are understood, it may be possible to identify and mitigate the most risky spots during especially risky times. Understanding these patterns is the first step towards effective mitigation, but it must be followed by an appropriate choice of counteractive measures.

5. Mitigation against animal-vehicle collisions

The transport sector must provide a safe infrastructure for traffic. Various measures to counteract animal-vehicle collisions have been tested throughout the years, yet only few have proven effective (Reed & Ward 1987; Romin & Bissonette 1996; Putman 1997). Most counteractive measures implemented today seek to keep animals off the transportation corridor in the first place (Figure 4). This is done by means of technical measures such as fences, gullies, and olfactory, optical and acoustical repellents or by adaptations of the corridor design e.g. through suppressing attractive vegetation that could provide food, shelter or nesting sites (e.g. Putman 1997; Keller *et al.* 2003; Rea 2003). Other measures aim at increasing the awareness of the driver or the animal to the danger and enhance their ability to detect each other in time (by means of warning signs, roadside clearance, optical reflectors or olfactory signals, road lightening combined with infra-red detectors, and public education) (e.g. Ujvari *et al.* 1998; Huijser & McGowen 2003). Rarely, mitigation aims at altering traffic patterns through speed limits or temporary road closing (e.g. Romin & Bissonette 1996; Forman *et al.* 2003). Animal population control in the vicinity of roads has also been tested, but the success has been limited and the effect may eventually be opposite to what has been intended if animals increase their movements when population density is reduced (e.g. Almkvist *et al.* 1980).

However, even if risky road sections can be foreseen and mitigated, it is hardly possible to completely erase the risk of collisions with wildlife – or human pedestrians. Simply because vehicles share their world with humans and animals, there will always be the need for both to cross each other's paths. Yet, in contrast to humans, most animals will never perceive cars or trains – or boats for that matter (e.g. Laist *et al.* 2001) – as a threat to their lives. Unless they avoid the open infrastructure corridor, they are likely to suffer from traffic mortality. On the other hand, species that fear vehicles and avoid transport corridors for this or other reasons (such as risk of predation, pollution, human disturbance), may not be exposed to traffic at all, but instead experience infrastructure as a strong barrier to their movements. Depending on the species' life history, population size, and behaviour, mitigation measures should focus on either preventing mortality or reducing the barrier effect (e.g. Keller *et al.* 2003). Under certain conditions, erecting movement barriers and fences to prevent mortality may be of less importance to population persistence than mitigating against isolation by allowing a certain loss to occur (Jaeger & Fahrig 2004). Barrier and mortality effects are closely linked, as an increase in barrier effect from infrastructure often leads to reduced exposure to traffic. Traffic mortality also contributes to the overall barrier as it prevents animals from reaching the other side of the road.

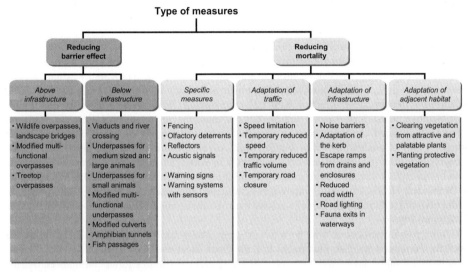

Figure 4. Mitigation options against barrier and mortality effect of linear infrastructure and traffic. Information from Keller *et al.* (2003)

5.1. EXCLUSION FENCES AND FAUNA PASSAGES

Typically, mitigation measures against animal-automobile collisions aim at increasing human traffic safety rather than reducing traffic mortality in wildlife. Exclusion fencing has so far proven to be by far the most cost- effective and practical measure against ungulate-vehicle collisions (e.g. Putman 1997; Clevenger *et al.* 2001). Experiments and field inventories in Sweden suggest that road fencing may locally reduce the rate of accidents with moose by more than 80%, with reindeer (*Rangifer tarandus*) by more than 50% and with roe deer by up to 55% (Almkvist *et al.* 1980; Skölving 1985). Fencing of the Trans Canada Highway at Banff National Park reduced ungulate vehicle collisions by 94-97% (Wood 1990). Of course, conventional exclusion fences are designed for large mammals only, because the risk for human injuries is greatest with these species. If fences are to be effective for smaller mammals as well, fences must be completed with smaller mesh sizes near the bottom or combined with concrete protection barriers (compare Iuell *et al.* 2003).

Fencing, if effective, implies a local dispersal barrier. Extensive fencing without provisions for animals to pass necessarily leads to isolation of habitat and animal accumulations which may entail secondary effects such as increased browsing pressure on forest regeneration (Ball & Dahlgren 2002; Seiler *et al.* 2003), or increased predation (Wood 1990). In addition, extensive fencing may be less efficient as animals that are determined to cross the road and that are physically able to break through a fence will probably do so if they cannot find a safer passage within reasonable distance. Once the fence has been opened, however, animals may be trapped inside the fenced road corridor and run high risk of colliding with passing vehicles. This problem is especially apparent where fenced roads cut through seasonal migration routes of large mammals such as moose and the animals are highly motivated to continue their movements (e.g. Nilsson 1987; Kastdalen 1999; Seiler et al. 2003). Too short fences, on the other hand, may not actually reduce the risk for accidents but only shift the problem towards the end of the fences (e.g. Ward 1982; Foster & Humphrey 1995; Clevenger et al. 2001). Economic drawbacks and poor planning and design may also result in suboptimal fences that are only partially effective and therefore give a false impression of traffic safety.

Based on various negative experiences with exclusion fences, the Swedish National Road Administration, among others, now promotes a new perspective on road fencing (Anonymous 2004). Rather than attempting a wildlife-proof barrier and focusing on keeping animals off the road corridor, fencing shall now aim at directing animals towards locations where a safer passage is possible. The effect of fences on traffic safety and wildlife mortality can indeed be enhanced if fences are combined with culverts, tunnels, or bridges that allow continuous animal movements (Keller et al. 2003). For instance, fencing in combination with culverts effectively reduced road mortality in Florida Desert tortoises (93% reduction; Boarman & Sazaki 1996). Construction of a barrier wall-culvert system in the Paynes Prairie State Preserve, Florida, resulted in a 93.5% reduction in vertebrate mortality (except in treefrogs; Dodd et al. 2004). Inventories of road-kills and the use of culverts under the Pacific Highway at Brunswick Heads, in New South Wales, confirmed that exclusion fencing in combination with culverts facilitated safe passage across the road for a range of wildlife species (Taylor & Goldingay 2003).

Wildlife crossing structures will be most efficient where traffic mortality is frequent or the need for continued wildlife movement is high. Such locations exist especially where fences terminate and linear landscape elements and topographical features funnel animals across the roadway (e.g. Madsen et al. 1998; Finder et al. 1999; Hubbard et al. 2000; Cain et al. 2003). Appropriate localisation of the crossing structures and an adapted design appeared to be crucial factors in determining their utilisation by wildlife (e.g. Bekker & Canters 1997; Keller et al. 2003; Clevenger & Waltho 2005), however, the requirements differ from species to species. On the other hand, various observations suggest that even conventional road bridges, tunnels, and drainage culverts are used by wildlife to some degree and thus may contribute to reduce the number of animals crossing roads at-grate (e.g. Olbrich 1984; Rodriguez et al. 1996; Clevenger & Waltho 2000; Seiler 2004; compare Chapter 10).

A perforated barrier system, i.e. fences and passages combined, is probably the only large-scale cost-efficient solution to the deadly conflict between transportation infrastructure and wildlife on land. Although other measures that address the behaviour or perception of animals and drivers can have some effect (e.g. Huijser & McGowen 2003), it may need a physical separation to prevent accidents in the long run. Vehicles and animals simply do not mix. The foremost task of the 'ecological engineer' is thus to determine how safe and how perforated the infrastructure corridor needs to be; or how isolating or unsafe it may be in order to keep mortality, traffic safety and secondary barrier effects within acceptable limits.

5.2 GUIDELINES FOR EVALUATION AND MITIGATION

In the course of the infrastructure planning process, the first question to ask in this matter is whether the impact on animal life and traffic safety is significant enough to justify expensive counteraction. If so, one needs to know what mitigation option would prove most effective in reducing the impact below an acceptable limit. Various examples of such options are given in guidelines and handbooks, such as the European Handbook on Traffic and Wildlife (Iuell *et al.* 2003, see Figure 4). Like any other mitigation, efforts to reduce animal-vehicle collisions should be evaluated with respect to their overall benefits and costs; and this includes tangible as well as intangible, pecuniary as well as non-monetary values. Depending on the viewpoint from which the impact is evaluated, mitigation costs and benefits will carry different weight (compare Figure 2).

A simple chain of reasoning is proposed that considers different perspectives in an ecologically-biased order (Figure 5). One question to begin with is whether the species involved are red-listed or otherwise threatened. If so, any additional traffic mortality should be seen as a substantial problem even if it is a rare event. Typically, as soon as red-listed species are concerned, there is a legal incitement for counteraction. Mitigation options of choice here are primarily the avoidance of any conflict with these species, and this includes imposing dispersal barriers to prevent mortality. Occurrence of red-listed species in an area can indeed dislocate new infrastructure corridors or justify construction of expensive ecoducts (e.g. Iuell *et al.* 2003). Without the involvement of endangered species, motives for mitigation efforts must be found in environmental policy, economics, traffic safety or ethical considerations.

Also in other than red-listed species, traffic may still be a threat to their local survival, especially if it hits disproportionately many individuals from cohorts that are valuable to population recruitment or dispersal (such as pregnant females). If the proportion of traffic-kills is large in relation to other sources of mortality, such as hunting, or if traffic mortality is likely to be additive and the realised growth rate of the population is close to stationary – then traffic has the potential to be a regulating or even limiting factor for populations of these species. These precarious levels of population are, practically speaking, beyond the effects of human interference and could easily suffer a fast and unexpected decline. Under such conditions, mitigation should focus on reducing mortality below the critical limit or on compensating for traffic mortality by integrating these losses in e.g. harvest quota. Also, in game species, a greater loss due to traffic could be accepted if size, age structure and sex ratio of these populations are well monitored.

If traffic mortality is neither a conservation nor a population management issue, there may be traffic safety or economic aspects that limit the acceptable and most cost-efficient level of collision numbers. Mitigation efforts may provide socio-economic benefits in terms of reduced human injury or damage to private property, improved traffic flow, and greater reliability of the transport mode. In fact, economical and safety concerns may often put more conservative limits to the number of collisions with wildlife than would be needed from an ecological point of view. Counteractive measures in such cases could include fencing, warning systems, repellents and population control.

If there is no legal or economic reason for counteraction, then environmental policy, public opinion, and ethical considerations may still provide sufficient motives for implying counter-active measures. In these respects it is much less a scientific task to establish critical limit for the maximum tolerable impact, than it is a question of what we want to achieve, what impact on wildlife and what frequency of accidents with wildlife we are willing to accept in our modern society. These are important issues which must be resolved if tomorrow's transport infrastructure is to be an integrated and sustainable part of our natural and social environment.

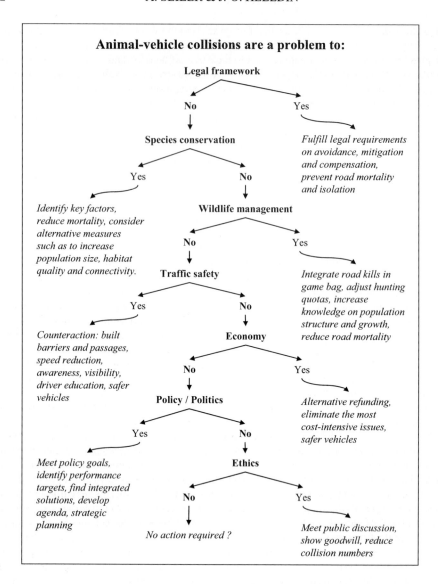

Figure 5. Flow chart illustrating possible steps in the evaluation of animal-vehicle collisions. If, under a given situation, animal-vehicle collisions are not a problem to the conservation or management of a species, there may very well be traffic safety or economic reasons that encourage counteraction. Also environmental policies, including protection of biological diversity, or animal welfare considerations, may set a limit to the maximum acceptable number of collisions

6. Conclusions

Collisions between vehicles and wildlife clearly conflict with the effort to achieve an environmentally sound and sustainable transport system. In many respects, collision numbers are close to the limit that we can tolerate in a modern society. Locally, the impact can be tremendous and the decision about mitigating the problem may be easy. At broader scale, however, and for broader purposes such as the making of environmental policies for the transport sector, it needs a clearer definition of where and when and why mitigation is required. Clearly, transportation is most hazardous to rare wildlife with large area requirements, pronounced migratory movements, small population sizes, and slow reproduction. If such species are attracted to vehicles (cars and ships) or infrastructure facilities (verges, road salt), transportation is acting as an ecological trap and is responsible for a significant part of the total mortality in this species.

The conflict is ubiquitous, as vehicles and animals practically co-occur in all types of environments. It may need a complete physical separation of traffic and wildlife in terms of a 'perforated barrier system' to overcome the risk of collisions for good. Such a system has been proposed for major roads and railroads, but it is hardly applicable to minor infrastructure and fully inappropriate for transportation in air or on water. On the other hand, animal-vehicle collisions are usually aggregated in time and space. That implies that a major part of collisions in these types of transport could eventually be prevented if mitigation is focused on these high-risk sites, areas or periods. A small number of collisions could probably be accepted. To accomplish a significant reduction, it needs a deeper understanding of the factors behind collisions, i.e. the animals' ecology and behaviour, the traffic flow and the vehicle and driver, and the surrounding environment that funnels or attracts wildlife to infrastructure. Mitigation measures mainly address the presence of animals near and on the road, or the awareness and behaviour of animal and driver.

Clearly, effective counteraction against animal-vehicle collisions calls for an integrated approach in infrastructure design, traffic regulation, wildlife population management, and land use (e.g. Schwabe *et al.* 2002; Sullivan & Messmer 2003). There is no single simple solution; measures must be adjusted to spatial, temporal and species-specific constraints and will thus vary from site to site.

Hence, the responsibility for counteracting animal-vehicle collisions is in part with the transport authorities (who decide about road localisation, design and traffic velocity), in part with the driver (who determines the actual travel speed, detects and react to animals and can avoid collisions), and in part with the landowner or wildlife manager (who influence wildlife population densities and the distribution of wildlife habitat relative to a road and thereby affects animal abundances and movements). If knowledge about these influences is combined wisely, transport infrastructure may not only be safer, but also provide new and valuable resources to wildlife and become a more integrated part in the landscape.

References

Adams LW, Geis AD. 1973; Effects of roads on small mammals. Journal of Applied Ecology 20: 403-415.
Allen R, McCullough A. 1976; Deer-car accidents in southern Michigan. Journal of Wildlife Management 40: 317-325.
Almkvist B, André T, Ekblom S, Rempler SA. 1980; Slutrapport Viltolycksprojekt. Swedish National Road Administration, TU146:1980-05, Borlänge, Sweden.
Andersen R, Wiseth B, Pedersen P, Jaren V. 1991; Moose-train collisions: effects of environmental conditions. Alces, 27: 79-84.
Anonymous 1997; Gefahr aus dem Strassengraben. ADAC 9: 36-37.

Anonymous 2004 Vägars och gators utformning (VGU 2004). Vägverket Publikation 2004:80, Borlänge.

Auestad I, Norderhaug A, Austad I. 1999; Road verges - species-rich habitats. Aspects of Applied Biology, 54: 269-274.

Baker PJ, Harris S, Robertson CPJ, Saunders G, White PCL 2004; Is it possible to monitor mammal population changes from counts of road traffic casualties? An analysis using Bristol's red foxes Vulpes vulpes as an example. Mammal Review, 34: 115-130.

Ball JP, Dahlgren J. 2002; Browsing damage on pine (Pinus sylvestris and P-contorta) by a migrating moose (Alces alces) population in winter: Relation to habit at composition and road barriers. Scandinavian Journal of Forest Research, 17: 427-435.

Bashore TL, Tzilkowski WM, Bellis ED. 1985; Analysis of deer-vehicle collision sites in Pennsylvania. Journal of Wildlife Management 49: 769-774.

Battin J. 2004; When good animals love bad habitats: Ecological traps and the conservation of animal populations. Conservation Biology 18: 1482-1491.

Bekker GJ, Canters KJ. 1997; The continuing story of badgers and their tunnels. In: Canters K, Piepers A & Hendriks-Heersma A. (eds) Proceedings of the international conference on "Habitat fragmentation, infrastructure and the role of ecological engineering" Maastricht & DenHague 1995, 344-353. Ministry of Transport, Public Works and Water Management, Road and Hydraulic Engineering division, Delft, The Netherlands.

Bengtsson SA. 1962; En fallviltundersökning. Fältbiologen 15: 9-15.

Bennett AF. 1991; Roads, roadsides and wildlife conservation: a review. In: Saunders DA & Hobbs RJ (eds) Nature conservation 2: The role of corridors, 99-118. Surrey Beatty & Sons, Chipping Norton.

Bergmann HH. 1974; Zur Phänologie und Ökologie des Strassentods der Vögel. Vogelwelt, 95: 1-21.

Berthoud G. 1987; Impact d'une route sur une population de chevreuils. In: Bernard J-M, Lansiart M, Kempf C & Tille M (eds)Actes du colloques "Route et fauna sauvage". Strasbourg 1985, 167-170. Ministère de l'Équipement, du Longement, de l'Aménagement du Territoire et des Transports, Colmar, France.

Björnstig U, Eriksson A, Thorson, J, Bylund PO. 1986; Collisions with passenger cars and moose in Sweden. American Journal of Public Health, 76: 460-462.

Blaustein AR, Wake DB. 1990; Declining amphibian populations: a global phenomenon? Trends in Ecology and Evolution 5: 203-204.

Boarman WI, Sazaki M. 1996; Highway mortality in desert tortoises and small vertebrates: success of barrier fences and culverts. In: Evink G, Ziegler D, Garrett P & Berry J. (eds) Transportation and wildlife: reducing wildlife mortality and improving wildlife passageways across transportation corridors. Proceedings of the Florida Department of Transportation/Federal Highway Administration transportation-related wildlife mortality seminar; 1996 April 30-May 2; Orlando, FL., 169-173. U.S. Department of Transportation, Federal Highway Administration, Washington, DC.

Bolund L.1974; Trafikdöd. Djurskyddet, 82: 92-95.

Borer F, Fry GL. 2003; Safety and Economic Considerations. In: Trocmé M, Cahill S, De Vries JG, Farall H Folkeson L, Fry GL, Hicks C & Peymen J. (eds) COST 341 - Habitat Fragmentation due to transportation infrastructure: The European Review, 175-182. Office for Official Publications of the European Communities, Luxembourg.

Boscali G. 1987; Wolves, bears and highways in Italy. In: Bernard J-M, Lansiart M, Kempf C & Tille M (eds)Actes du colloques "Route et fauna sauvage". Strasbourg 1985, 237-239.Ministère de l'Équipement, du Longement, de l'Aménagement du Territoire et des Transports, Colmar, France.

Cain AT, Tuovila VR, Hewitt DG, Tewes ME. 2003; Effects of a highway and mitigation projects on bobcats in Southern Texas. Biological Conservation 114: 189-197.

Caletrio J, Fernandez JM, Lopez J, Roviralta F. 1996: Spanish national inventory on road mortality of vertebrates. Global Biodiversity 5: 15-18.

Calvo RN, Silvy NJ. 1996; Key deer mortality, U.S.1 in the Florida Keys. In: Evink GL, Garrett P, Ziegler D & Berry J (eds) Highways and movement of wildlife: Improving habitat conditions and wildlife passageways across highway corridors, 287-296. Proceedings of the Florida Department of Transportation/Federal Highway Administration transportation-related wildlife mortality seminar. Orlando April 30-May 2, 1996, Florida Department of Transportation, FHWA-PD-96-041, Tallahassee, Florida.

Child KN, Barry SP, Aitken DA. 1991; Moose mortality on highways and railways in British Columbia. Alces 27: 41-49.

Clarke GB, White PCL, Harris S. 1998; Effects of roads on badger Meles meles populations in south-west England. Biological Conservation 86: 117-124.

Clevenger AP, Chruszcz B, Gunson KE. 2001; Highway mitigation fencing reduces wildlife-vehicle collisions. Wildlife Society Bulletin 29: 646-653.

Clevenger AP, Chruszczc B, Gunson KE. 2003; Spatial patterns and factors influencing small vertebrate fauna road-kill aggregations. Biological Conservation, 109: 15-26.

Clevenger AP, Waltho N. 2000; Factors influencing the effectiveness of wildlife underpasses in Banff National Park, Alberta, Canada. Conservation Biology 14: 47-56.

Clevenger AP, Waltho N. 2005; Performance indices to identify attributes of highway crossing structures facilitating movement of large mammals. Biological Conservation 121: 453-464.

Clevenger AP, Wierzchowski J. 2002; GIS-based modeling approaches to identify mitigation placement along roads. In: Proceedings of the International Conference on Ecology and Transportation, Keystone, CO, September 24-28, 2001. Raleigh, NC, 134-148. Center for Transportation and the Environment, North Carolina State University (March 2002),

Conover MR. 1997; Monetary and intangible valuation of deer in the United States. Wildlife Society Bulletin 25: 298-305.

Damard T, Bouwma I, VanStraaten D. 2003; Policy Development and Future Trends. In: Trocmé M, Cahill S, De Vries JG, Farall H, Folkeson L, Fry GL, Hicks C & Peymen J. (eds) COST 341 - Habitat Fragmentation due to transportation infrastructure: The European Review, 183-187. Office for Official Publications of the European Communities, Luxembourg.

Dickerson LM. 1939; The problem of wildlife destruction by automobile traffic. Journal of Wildlife Management 3: 104-116.

Dixon N, Shawyer C, Sperring C. 1996; The impact of road mortality on barn owl populations: its significance to road development schemes. Raptor 23: 37-40.

Dodd CK, Barichivich WJ, Smith LL. 2004; Effectiveness of a barrier wall and culverts in reducing wildlife mortality on a heavily traveled highway in Florida. Biological Conservation 118: 619-631.

Fahrig L, Pedlar JH, Pope SE, Taylor PD, Wegner JF. 1995; Effect of road traffic on amphibian density. Biological Conservation 73: 177-182.

Fehlberg U. 1994; Ecological barrier effects of motorways on mammalian wildlife - an animal protection problem. Deutsche Tierärztliche Wochenschrift 101: 125-129.

Feldhamer GA, Gates JE, Harman DM, Loranger AJ, Dixon KR. 1986; Effects of interstate highway fencing on white-tailed deer activity. Journal of Wildlife Management 50: 497-503.

Ferreras P, Aldama JJ, Beltran,JF, Delibes M. 1992; Rates and causes of mortality in a fragmented population of Iberian lynx (Felis-pardina Temminck, 1824). Biological Conservation 61: 197-202.

Finder RA, Roseberry JL, Woolf A. 1999; Site and landscape conditions at white-tailed deer/vehicle collision locations in Illinois. Landscape and Urban Planning 44: 77-85.

Forman RT, Alexander LE. 1998; Roads and their major ecological effects. Annual Review of Ecology and Systematics 29: 207-231.

Forman RT, Sperling D, Bissonette JA, Clevenger AP, Cutshall CD, Dale VH, Fahrig L, France R, Goldman CR, Heanue K, Jones JA, Swanson FJ, Turrentine T, Winter TC. 2003; Road ecology - Science and Solutions. Island Press, Washington.

Foster ML, Humphrey SR. 1995; The use of highway underpasses by Florida panthers and other wildife. Wildlife Society Bulletin 23: 95-100.

Fuellhaas U, Klemp C, Kordes A, Ottersberg H, Pirman M, Thiessen A, Tschoetschel C, Zucchi H. 1989; Untersuchungen zum Strassentod von Vögeln, Säugetieren, Amphibien und Reptilien. Beitr.Naturk.Niedersachsens 42: 129-147.

Gonzalez-Prieto S, Villarino A, Freán MM. 1993; Mortalidad de vertebrados por atropello en una carretera nacional del no de españa. Ecología 7: 375-389.

Göransson G, Karlsson, J. 1979; Changes in population densities as monitored by animals killed on roads. In: Hytteborn H (ed) The use of ecological variables in environmental monitoring, 120-125. Proceedings from the first Nordic Oikos conference, 2-4 October 1978, Uppsala, Sweden. National Swedish Environmental Protection Board, PM 1151,

Göransson G, Karlsson J, Lindgren A. 1978; Vägars inverkan på omgivande natur. II Fauna. Swedish Environmental Protection Agency, SNV PM 1069, Stockholm.

Griffiths HI, Thomas DH. 1993; The status of the Badger (Meles meles) in Europe. Mammal Review 23: 17-58.

Groot-Bruinderink GWTA, Hazebroek E. 1996; Ungulate traffic collisions in Europe. Conservation Biology 10: 1059-1067.

Gundersen H, Andreassen HP. 1998; The risk of moose Alces alces collision: A predictive logistic model for moose-train accidents. Wildlife Biology 4: 103 -110.

Gundersen H, Andreassen HP, Storaas T. 1998; Spatial and temporal correlates to norwegian moose - train collisions. Alces 34: 385-394.

Haikonen H, Summala H. 2001; Deer-vehicle crashes - Extensive peak at 1 hour after sunset. American Journal of Preventive Medicine 21: 209-213.

Hansen L. 1982; Trafikdræbte dyr i Danmark. Dansk Ornitologisk Forenings Tidskrift 76: 97-110.

Hansen L. 1982; Trafikdræbte dyr i Danmark. (In Danish with English summary: Road-kills in Denmark.). Dansk Ornitologisk Forenings Tidskrift 76: 97-110.

Harris LD, Gallagher PB. 1989; New initiatives for wildlife conservation: the need for movement corridors. In: Mackintosh G. (ed) Preserving communities and corridors, 11-34. Defenders of Wildlife, Washington, DC.

Harris LD, Scheck J. 1991; From implications to applications: the dispersal corridor principle applied to the conservation of biological diversity. In: Saunders DA & Hobbs RJ (eds) Nature conservation 2: The role of corridors, 189-220. Surrey Beatty & Sons, Chipping Norton.

Haugen AO. 1944; Highway mortality in southern Michigan. Journal of Mammalogy 25: 177-184.

Havlin J. 1987; On the importance of railway lines for the avifauna in agrocoenoses. Folia Zoologica 36: 345-358.

Haxton T. 2000; Road mortality of Snapping Turtles, Chelydra serpentina, in central Ontario during their nesting period. Canadian.Field Naturalist 114: 106-110.

Helldin JO, Seiler A. 2003 Effects of roads on the abundance of birds in Swedish forest and farmland. - submitted. Landscape and Urban Planning,

Hels T, Buchwald E. 2001; The effect of road-kills on amphibian populations. Biological Conservation 99: 331-340.

Hicks AC. 1993; Using road-kills as an index to moose population change. Alces 29: 243-247.

Hodson NL. 1960; A survey of vertebrate road mortality. Bird Study 7: 224-231.

Hodson NL. 1966; A survey of road mortality in mammals (and including data for the grass snake and common frog). Journal of Zoology, 148: 576-579.

Hodson NL, Snow DW. 1965; The road deaths enquiry, 1960-1961. Bird Study 12: 168-172.

Holsbeek L, Rodts J, Muyldermans S. 1999; Hedgehog and other animal traffic victims in Belgium: results of a countrywide survey. Lutra 42: 111-119.

Hostetler M. 1997; That gunk on your car: a unique guide to insects of North America. Ten Speed Press, Berkeley, California.

Hubbard MW, Danielson BJ, Schmitz RA. 2000 Factors influencing the location of deer-vehicle accidents in Iowa. Journal of Wildlife Management 64: 707-713.

Huijser MP. 2000; Life on the edge. Hedgehog traffic victims and mitigation strategies in an anthropogenic landscape. PhD thesis, Wageningen University, Wageningen, NL.

Huijser MP, McGowen PT. 2003; Overview of animal detection and animal warning systems in North America and Europe. In: Irwin CL, Garrett P & McDermott KC (eds) 2003 Proceedings of the International Conference on Ecology & Transportation (ICOET), 368-382. Center for Transportation and the Environment, North Carolina State University.

Inbar M, Mayer RT. 1999; Spatio-temporal trends in armadillo diurnal activity and road- kills in central Florida. Wildlife Society Bulletin, 27: 865-872.

Iuell B, Bekker H, Cuperus R, Dufek J, Fry GL, Hicks C, Hlavac V, Keller J, Le Marie Wandall B, Rosell Pagès C, Sangwine T, Torslov N. (eds). 2003; Wildlife and Traffic - A European Handbook for Identifying Conflicts and Designing Solutions. Prepared by COST 341 - Habitat Fragmentation due to Transportation Infrastructure. Ministry of Transport, Public Works and Water Management, Road and Hydraulic Engineering division, Delft, The Netherlands.

Jaeger JAG, Fahrig L. 2004; Effects of road fencing on population persistence. Conservation Biology 18: 1651-1657.

Jahn IR. 1959; Highway mortality as an index of deer population change. Journal of Wildlife Management 23: 187-179.

Jaren V, Andersen R, Ulleberg M, Pedersen P-H, Wiseth B. 1991; Moose-train collisions: the effects of vegetation removal with a cost-benefit analysis. Alces 27: 93-99.

Johansson A, Larsson PO. 1999; Djurpåkörningar. Banverket Norra Regionen, Luleå.

Johnson CJ. 1995; Method for estimating the dollar values of lost wildlife diversity and abundance resulting from wildlife-road vehicle collisions. Ministry of Transportation and Highways, Planning Services Branch, Victoria, British Columbia.

Jones ME. 2000; Road upgrade, road mortality and remedial measures: impacts on a population of eastern quolls and Tasmanian devils. Wildlife Research 27: 289-296.

Jonkers DA, De Vries GW. 1977; Verkeersschlachtoffers onder der fauna. Nederlandsa Vereniging tot bescherming van vogels, Zeist, NL.

Joyce TL, Mahoney SP. 2001; Spatial and temporal distributions of moose-vehicle collisions in Newfoundland. Wildlife Society Bulletin 29: 281-291.

Kastdalen L. 1999; Gardermoen Airport development - Evaluation of mitigation measures for moose, Rapport 26:1999 (in Norwegian) edn. Hogskolen i Hedmark, Norway, Oslo.

Keller V, Bekker H, Cuperus R, Folkeson L, Rosell Pagès C, Trocmé, M. (2003) Avoidance, Mitigation and Compensatory Measures and their Maintenance. In: Trocmé M, Cahill S, De Vries JG, Farall H Folkeson L, Fry GL, Hicks C & Peymen J. (eds) COST 341 - Habitat Fragmentation due to transportation infrastructure: The European Review, 129-174. Office for Official Publications of the European Communities, Luxembourg.

Kelly TA. 2001; Key data from bird strikes. Aviation Week & Space Technology 154: 1-20.

Kirby KJ. 1997; Habitat fragmentation and infrastructure: Problems and research. In: Canters K, Piepers A & Hendriks-Heersma A. (eds) Proceedings of the international conference on "Habitat fragmentation, infrastructure and the role of ecological engineering" Maastricht & DenHague 1995, 32-39. Ministry of Transport, Public Works and Water Management, Road and Hydraulic Engineering division, Delft, The Netherlands.

Kirkwood JK, Sainsbury AW, Bennett PM. 1994; The welfare of free-living wild animals: methods of assessment. Animal Welfare 3: 257-273.

Knutson R. 1987; A field guide to common animals of roads, streets, and highways. Ten Speed Press, Berkely, USA.

Kolb H-H. 1984; Factors affecting the movement of dog foxes in Edinburgh. Journal of Applied Ecology 21: 161-173.

Kuhn J. 1987; Strassentod der Erdkröte (Bufo bufo): Verlustquoten und Verkehrsaufkommen, Verhalten auf der Strasse. Beih.Veröff.Naturschutz & Landschaftspflege in Baden-Württemberg, 41: 175-186.

Laist DW, Knowlton AR, Mead JG, Collet AS, Podesta M. 2001; Collisions between ships and whales. Marine Mammal Science 17: 35-75.

Lalo J. 1987; The problem of road-kill. American Forests 72: 50-52.

Lankester K, Van Apeldoorn RC, Meelis E, Verboom J. 1991; Management perspectives for populations of the Eurasian badger (Meles meles) in a fragmented landscape. Journal of Applied Ecology 28: 561-573.

Laursen K. 1981: Birds on roadside verges and the effect of mowing on frequency and distribution. Biological Conservation 20: 59-68.

Lehnert ME, Romin L, Bissonette JA. 1996; Mule deer and highway mortality in northeastern Utah: causes, patterns, and a new mitigative technique. Transportation and wildlife: reducing wildlife mortality and improving wildlife passageways across transportation corridors. In: Evink G, Ziegler D, Garrett P & Berry J. (eds) Transportation and wildlife: reducing wildlife mortality and improving wildlife passageways across transportation corridors. Proceedings of the Florida Department of Transportation/Federal Highway Administration transportation-related wildlife mortality seminar; 1996 April 30-May 2; Orlando, FL., 101-107. U.S. Department of Transportation, Federal Highway Administration, Washington, DC.

Lodé T. 2000; Effect of a motorway on mortality and isolation of wildlife populations. Ambio 29: 163-166.

Loughry WJ, McDonough CM. 1996; Are road-kills valid indicators of armadillo population structure? American Midland Naturalist 135: 53-59.

Mader HJ. 1981; Der Konflikt Strasse-Tierwelt aus ökologischer Sicht. Schriftenreihe für Landschaftspflege und Naturschutz, Heft 22, Bundesamt für Naturschutz, Berlin.

Madsen AB. 1996; Otter (Lutra lutra) mortality in relation to traffic, and experience with newly established fauna passages at existing road bridges. Lutra 39: 76-88.

Madsen AB, Fyhn HW, Prang A. 1998; Trafikdræbte dyr i landskapsøkologisk planlægning og forskning. (In Danish with English summary: Traffic killed animals in landscape ecological planning and research.). Rapport 228, National Environmental Research Institute, DMU, Denmark, Kalö, Denmark.

Madsen AB, Strandgaard H, Prang A. 2002; Factors causing traffic killings of roe deer Capreolus capreolus in Denmark. Wildlife Biology 8: 55-61.

Malo JE, Suarez F, Diez A. 2004; Can we mitigate animal-vehicle accidents using predictive models? Journal of Applied Ecology 41: 701-710.

Manneri-Martin A. 2002; Pienten ja keskikokoisten selkärankasiten liikennekuolleisus Suomessa. Finnish Road Administration, Tiehallinto, TIEH 3200758, Helsinki, Finland.

McCaffery KR. 1973; Road-kills show trends in Wisconsin deer populations. Journal of Wildlife Management 37: 212-216.

McClure HE. 1951; An analysis of animal victims on Nebraska's highways. Journal of Wildlife Management 15: 410-420.

McLellan BN, Schackleton DM. 1988; Grizzly bears and resource extraction industries: effects of roads on behaviour, habitat use, and demography. Journal of Applied Ecology 25: 451-460.

Meunier FD, Verheyden C, Jouventin P. 2000; Use of roadsides by diurnal raptors in agricultural landscapes. Biological Conservation 92: 291-298.

Modafferi RD. 1991; Train moose kill in Alaska: Characteristics and relationships with snowpack depth and moose distribution in Lower Susitna Valley. Alces 27: 193-207.

Müller S, Berthoud G. 1997; Fauna/Traffic Safety - Manual for civil engineers. LAVOC, Lausanne, Switzerland.

Mysterud A. 2004; Temporal variation in the number of car-killed red deer Cervus elaphus in Norway. Wildlife Biology 10: 203-211.

Nankinov DN, Todorov NM. 1983; Bird casualties on highways. The Soviet Journal of Ecology 14: 288-293.

Neal E, Cheeseman CL. 1996; Badgers. T & AD Poyser, Natural History, London.

Newton I, Wyllie I, Dale L. 1997; Mortality causes in British Barn owls (Tyto alba), based on 1101 carcasses examined during 1963-1996. In: Duncan JR, Johnson DH & Nicholls TH (eds) Biology and Conservation of Owls of the Northern Hemisphere, Second International Owl Symposium, February 5-9, 1997, 299-306. USDA Forest Service, General Technical Report NC-190, Winnipeg, Manitoba, Canada.

Nielsen CK, Anderson RG, Grund MD. 2003; Landscape influences on deer-vehicle accident areas in an urban environment. Journal of Wildlife Management 67: 46-51.

Nilsson J. 1987; Effekter av viltstängsel. Viltolyckor pp. 65-69. Nordiska trafiksäkerhetsrådet, Rapport 45, Linköping.

Nilsson L, Sjölund A. 2003; Targets and measures for consideration of natural and cultural heritage assets in the transport system. In: Irwin CL, Garrett P, McDermott KC & Rappole JH (eds) Proceedings of the International Conference on Ecology and Transportation, 626-641. Center for Transportation and the Environment, North Carolina State University,

Nowacek DP, Johnson MP, Tyack PL. 2004; North Atlantic right whales (Eubalaena glacialis) ignore ships but respond to alerting stimuli. Proceedings of the Royal Society of London Series B-Biological Sciences 271: 227-231.

OECD (1994) Environmental impact of roads. Road Transport research, Organisation for Economic Co-operation and development, OECD Scientific Expert Group, Paris, France.

Olbrich P. 1984; Untersuchung der Wirksamkeit von Wildwarnreflektoren under der Eignung von Wilddurchlässen. Zeitschrift für Jagdwissenschaft 30: 101-116.

Oxley DJ, Fenton MB, Carmody GR. 1974; The effects of roads on populations of small mammals. Journal of Applied Ecology 11: 51-59.

Petersen B, McLean JA. 1996; The totalled roadkill cookbook. Celestial Arts, Berkeley, California.

Philcox CK, Grogan AL, Macdonald DW. 1999; Patterns of otter Lutra lutra road mortality in Britain. Journal of Applied Ecology 36: 748-762.

Pienaar Ud. 1968; The ecological significance of roads in a nature park. Koedoe 11: 169-174.

Putman RJ. 1997; Deer and road traffic accidents: Options for management. Journal of Environmental Management 51: 43-57.

Rea RV. 2003; Modifying roadside vegetation management practices to reduce vehicular collisions with moose Alces alces. Wildlife Biology 9: 81-91.

Reed DF, Ward AL. 1987; Efficacy of methods advocated to reduce deer-vehicle accidents: research and rationale in the USA. In: Bernard J-M, Lansiart M, Kempf C & Tille M (eds)Actes du colloques "Route et fauna sauvage". Strasbourg 1985, 167-170. Ministère de l'Équipement, du Longement, de l'Aménagement du Territoire et des Transports, Colmar, France.

Reijnen R, Foppen R, Veenbaas G. 1997; Disturbance by traffic of breeding birds: Evaluation of the effect and considerations in planning and managing road corridors. Biodiversity and Conservation 6: 567-581.

Righetti A, Malli H. 2004; Einfluss von ungezäunten (Hochleistungs-) Zugstrecken auf Wildtierpopulationen. COST-341 Synthesebericht - PiU GmbH, Bern, CH.

Robinson JG, Bodmer RE. 1999; Towards wildlife management in tropical forests. Journal of Wildlife Management 63: 1-13.

Rodriguez A, Crema G, Delibes M. 1996; Use of non-wildlife passages across a high speed railway by terrestrial vertebrates. Journal of Applied Ecology 33: 1527-1540.

Rodts J, Holsbeek L, Muyldermons S. 1998; Dieren onder onze wielen. Koninklijk Belgisch Verbond voor de Bescherming van de Vogels, Vubpress, Brussel.

Rolley RE, Lehman LE. 1992; Relationships among raccoon road-kill surveys, harvests, and traffic. Wildlife Society Bulletin 20: 313-318.

Romin LA, Bissonette JA. 1996; Deer-vehicle collisions: Status of state monitoring activities and mitigation efforts. Wildlife Society Bulletin 24: 276-283.

Rosell Pagès C, Velasco Rivas JM. 1999; Manual on preventing and remedying the impact of roads on fauna. Generalitat de Catalunya, Departement de Medi Ambient, Secretaria General, Barcelona.

Sainsbury AW, Bennett PM, Kirkwood JK. 1993; Wildlife Welfare in Europe. European Commission, Contract No. B4-3060(92)12596, Brussels.

Sainsbury AW, Bennett PM, Kirkwood JK. 1995; The welfare of free-living wild animals in Europe: harm caused by human activities. Animal Welfare 4: 183-206.

Schwabe KA, Schuhmann PW. 2002; Deer-vehicle collisions and deer value: an analysis of competing literatures. Wildlife Society Bulletin 30: 609-615.

Schwabe KA, Schuhmann PW, Tonkovich M. 2002; A dynamic exercise in reducing deer-vehicle collisions: Management through vehicle mitigation techniques and hunting. Journal of Agricultural and Resource Economics 27: 261-280.

Seiler A. 2003a; Effects of infrastructure on nature. In: Trocmé M, Cahill S, De Vries JG, Farall H Folkeson L, Fry GL, Hicks C & Peymen J. (eds) COST 341 - Habitat Fragmentation due to transportation infrastructure: The European Review, 129-174. Office for Official Publications of the European Communities, Luxembourg.

Seiler A. 2003b; The toll of the automobile: Wildlife and roads in Sweden. Ph.D. thesis at the Department for Conservation Biology, SLU, Uppsala.

Seiler A. 2004; Trends and spatial pattern in ungulate-vehicle collisions in Sweden. Wildlife Biology 10: 301-313.

Seiler A. 2005; Predicting locations of moose-vehicle collisions in Sweden. Journal of Applied Ecology, in press.

Seiler A, Cederlund G, Jernelid H, Grängstedt P, Ringaby E. 2003; The barrier effect of highway E4 on migratory moose (Alces alces) in the High Coast area, Sweden. In: Turcott E (ed) Proceedings of the IENE conference on "Habitat fragmentation due to transport infrastructure" 13-14 November 2003 (http://www.iene.info), 1-18. Brussels.

Seiler A, Eriksson I-M. 1997; New approaches for ecological consideration in Swedish road planning. In: Canters K, Piepers A & Hendriks-Heersma A. (eds) Proceedings of the international conference on "Habitat fragmentation, infrastructure and the role of ecological engineering" Maastricht & DenHague 1995, 253-264. Ministry of Transport, Public Works and Water Management, Road and Hydraulic Engineering division, Delft, The Netherlands.

Seiler A, Folkeson L. (eds) 2003; COST 341 - Habitat Fragmentation due to transportation infrastructure: The Swedish National Review. Office for Official Publications of the European Communities, Luxembourg.

Seiler A, Helldin JO, Eckersten T. 2003; Road mortality in Swedish Badgers (Meles meles). Effect on population. The toll of the automobile: Wildlife and roads in Sweden. Ph.D. thesis at the Department for Conservation Biology, SLU, Uppsala.

Seiler A, Helldin JO, Seiler Ch. 2004; Road mortality in Swedish mammals - Results of a drivers' questionnaire. Wildlife Biology 10: 225-233.

Skölving H. 1985; Viltstängsel - olika typers effekt och kostnad. Swedish National Road Administration, Report TU 1985: 2, Borlänge, Sweden.

Skölving H. 1987; Traffic accidents with moose and roe-deer in Sweden Report of research, development and measures. In: Bernard J-M, Lansiart M, Kempf C & Tille M (eds)Actes du colloques "Route et fauna sauvage". Strasbourg 1985, 317-327. Ministè de l'Équipement,   du Longement, de l'Aménagement du Territoire et des Transports, Colmar, France.

Southwood TR, Henderson PA. 2000; Ecological Methods, Third edition, Blackwell Science Ltd., Oxford.

Steen DA, Gibbs JP. 2004; Effects of roads on the structure of freshwater turtle populations. Conservation Biology 18: 1143-1148.

Stikbakke H, Gaasemyr I. 1997; Innsatsplan mot viltpåkörsler. Rapport TTS 16/1997, Statens vegvesen - Vegdirektoratet, Oslo.

Stoner D. 1925; The toll of the automobile. Science 61: 56-58.

Sullivan TL, Messmer TA. 2003; Perceptions of deer-vehicle collision management by state wildlife agency and department of transportation administrators. Wildlife Society Bulletin 31: 163-173.

Svensson S. 1998; Birds kills on roads: is this mortality factor seriously underestimated? Ornis Svecica 8: 183-187.

Taylor BD, Goldingay RL. 2003; Cutting the carnage: wildlife usage of road culverts in north-eastern New South Wales. Wildlife Research 30: 529-537.

Thurber JM, Peterson RO, Drummer TD, Thomasma SA. 1994; Gray Wolf Response to Refuge Boundaries and Roads in Alaska. Wildlife Society Bulletin 22: 61-68.

Trocmé M, Cahill S, De Vries JG, Farall H Folkeson L, Fry GL, Hicks C & Peymen J. (eds) COST 341 - Habitat Fragmentation due to transportation infrastructure: The European Review, 129-174. Office for Official Publications of the European Communities, Luxembourg.

Ueckermann E. 1964; Wildverluste durch den Strassenverkehr und Verkehrsunfälle durch Wild. Zeitschrift für Jagdwissenschaft 10: 142-168.

Ujvari M, Baagoe HJ, Madsen AB. 1998; Effectiveness of wildlife warning reflectors in reducing deer- vehicle collisions: A behavioral study. Journal of Wildlife Management 62: 1094-1099.

Van den Tempel R. 1993; Vogelslachtoffers in het wegverkeer. (in Dutch with English summary)Vogelbescherming Nederland, Ministerie van Verkeer en Waterstaat, Directoraat-Generaal Rijkswaterstaat., Delft, The Netherlands.

Van der Grift EA. 1999; Mammals and railroads: Impacts and management implications. Lutra 42: 77-98.

Van der Zande AN, ter Keurs WJ, Van der Weijden WJ. 1980; The impact of roads on the densities of four bird spesies in an open field habitat - evidence of a long-distance effect. Biological Conservation 18: 299-321.

Van der Zee FF, Wiertz J, Terbraak CJ, Van Apeldoorn RC. 1992; Landscape change as a possible cause of the badger (Meles meles L.) decline in the Netherlands. Biological Conservation 61: 17-22.

Van Gelder JJ. 1973; A quantitative approach to the mortality resulting from traffic in a population of Bufo bufo. Oecologia 13: 93-95.

van Langevelde F, Jaarsma CF. 2004 Using traffic flow theory to model traffic mortality in mammals. Landscape Ecology 19: 895-907.

Verkaar HJ, Bekker GJ. 1991; The significance of migration to the ecological quality of civil engineering works and their surroundings. In: Van Bohemen HD, Buizer DAG & Littel D (eds) Nature engineering and civil engineering works, 44-62. Pudoc, Wageningen, The Netherlands.

Vos C, Chardon JP. 1998; Effects of habitat fragmentation and road density on the distribution pattern of the moor frog Rana arvalis. Journal of Applied Ecology 35: 44-56.

Ward AL. 1982; Mule deer behaviour in relation to fencing and underpasses on Interstate 80 in Wyoming. Transportation Research Records 859: 8-13.

Way JM. 1970; Wildlife on the motorway. New Scientist 47: 536-537.

Way JM. 1977; Roadside verges and conservation in Britain: a review. Biological Conservation 12: 65-74.

Wilkins KT, Schmidly DJ. 1980; Highway mortality of vertebrates in southeastern Texas. Texas Journal of Science 4: 343-350.

Wood JG. 1990; Effectiveness of fences and underpasses on the Trans-Canada Highway and their impact on ungulate populations. Environment Canada Parks Service, Natural History Research Division, Alberta, CA.

CHAPTER 9 : HABITAT FRAGMENTATION DUE TO TRANSPORT INFRASTRUCTURE: PRACTICAL CONSIDERATIONS

EUGENE O'BRIEN

University College Dublin, Department of Civil Engineering, Belfield, Dublin 4

1. What is habitat fragmentation?

In simple terms, roads and railways divide land into a number of discrete parts. A busy transport route can constitute an impenetrable barrier that animals of a species, particularly shy ones, cannot cross. Over time, the population becomes divided into a number of sub-populations separated by roads or railways. Some of the sub-populations will be too small to be viable – it is well known that, below a certain critical number, populations are far more susceptible to extinction. This can happen for example due to a virus. In a large population, a virus may wipe out a species in a given locality but it will be recolonised over time by individuals from neighbouring areas. However, if the locality is isolated by a network of impenetrable roads or railways, the species can become locally extinct.

In addition to the risk of local extinction, fragmentation of habitats can lead to a reduction in genetic diversity (inbreeding) within sub-populations. This can make the whole population susceptible to extinction. If there is to be a mixing of genes and the possibility of recolonising areas that are locally extinct, then animals have to cross transport corridors. The 'permeability' of roads and railways is a measure of the extent to which animals can cross them. A permeable road will constitute a disincentive but may allow some crossings. The ranges of mule deer (*Odocoileus hemionus*) have been shown to be bounded by highways. The highway, through noise and fencing, is a disincentive to travel and forms a natural boundary to the animals' range. Two types of animal movement are important, daily or occasional foraging trips and annual movements made during the mating season and when juveniles leave the family group in search of new habitat.

Figure 1. Red Squirrel, Stockholm, Sweden

John Davenport and Julia L. Davenport, (eds.), The Ecology of Transportation: Managing Mobility for the Environment, 191–204,
© 2006 *Springer. Printed in the Netherlands*

Animals on heat or those who have been driven out of a family group, will overcome natural fears and will attempt movements that they would not consider on a daily basis. Transport infrastructure provides a barrier to movement at a number of levels: noise, artificial lighting and the presence of vehicles can induce fear in animals but there is also the physical danger of being killed or injured and there are sometimes physical barriers such as fences. It is generally accepted that rail infrastructure is a considerably more permeable barrier than road. Canals also constitute a barrier, particularly to some animals. The volume of traffic on a road or railway is important and the barrier becomes less permeable as the volume increases. For example, there are many animals that will overcome their fear to cross a main road when there are few vehicles present. However, when the traffic is always busy, then it may simply not be possible to cross safely.

Hence, there are two problems with transport infrastructure, traffic volume and number of routes. The volume of traffic on a route makes it less permeable to animals. On the other hand, if there are a greater number of roads and railways, habitats become more fragmented with ever smaller fragments which leave animals more prone to local extinction. This is not an easy problem – the European Commission White Paper on a common transport policy, published in September 2001, states that heavy goods vehicle traffic on European roads could increase by nearly 50% by 2010 relative to its 1998 level. The paper outlines a number of measures to reduce the growth of road haulage to 38%. The situation with passenger cars and trains is similar and is likely to grow with European Gross Domestic Product. Trains are a more effective environmental solution than roads, both for passenger and freight transport. However, even with a significant growth in train traffic, growth in road traffic seems inevitable.

The phenomenon of habitat fragmentation has come to the attention of highway authorities in many developed countries in recent years. In Europe, the Infra Eco Network Europe (IENE – http://www.iene.info/) group was established in 1996 to bring together experts involved in habitat fragmentation caused by linear transport infrastructure (motorways, railways and canals). IENE aims to promote an efficient, sustainable and safe pan-European transport network which maintains biodiversity and reduces traffic collisions with fauna. It has representatives from 21 countries and includes experts from a range of disciplines.

In 1998, IENE secured funding from the European Commission for a COST (CoOperation in Science and Technology) action. COST is a framework that coordinates the research efforts of member countries on a wide range of approved topics. The new action, known as 'COST 341 – Habitat Fragmentation due to Transportation Infrastructure', ran for 5 years and produced:

- A review of current state-of-the-art Knowledge of habitat fragmentation in Europe

- A best practice handbook on habitat fragmentation.

The recommendations of the COST 341 handbook (Iuell *et al.* 2003) are summarised in this chapter.

2. Impact of roads/railways on wildlife

The impact of roads and railways on the environment is not just the barrier effect. Roads in particular can take up a significant area of land, ranging from as little as 0.3% in Norway to as much as 5% in The Netherlands. Hence, there can be a significant local reduction in habitat. Furthermore, the habitat immediately adjacent to a major road is often of reduced quality due to disturbance and air pollution effects.

Roadside verges and excess land can have important ecological benefits and some authorities have programmes to enhance them. For example, it is possible to plant roadsides with species that will be attractive to butterflies – 50% of all species of butterfly in The Netherlands occur in roadside verges. However, there is a conflict here as there is a risk of attracting animals to an area where they may be killed or cause accidents. For example, raptors are attracted to roadsides by the abundance of small mammals in roadside verges and are often killed as they fly low in search of their prey.

2.1 THE BARRIER EFFECT

The creation of a barrier is undoubtedly the most significant effect of transport infrastructure on wildlife. Its importance is greatly influenced by the type of animal being considered. Many mammal species (Figure 2) will attempt to cross roads and are generally only blocked by wildlife fencing. For small invertebrates on the other hand, the road itself can be a major barrier. Animal behaviour also features. Some bats fly low and are significantly affected by a transport corridor while others fly high and are much less affected. Some animals are clearly more shy than others. Wild reindeer (*Rangifer tarandus*) in Norway stay away from good grazing that is within 5km of a road.

Figure 2. Irish Hare, Dublin, Ireland: (a) Family group; (b) Regular fencing with gaps underneath will not restrict movement

The most vulnerable animals to a barrier are:

- Rare species with small local populations and large individual ranges such as large carnivores.

- Species with daily or seasonal migrations that involve the crossing of a barrier. Some amphibians need to cross roads to reach their breeding ponds and some deer species e.g. fallow deer (*Dama dama*), migrate daily from the shelter of woodlands to grazing lands nearby. In such cases, a road alongside a wood can be a problem.

- Species that undertake long-distance seasonal migrations such as moose (*Alces alces*) and reindeer.

The permeability of a barrier is significantly affected by the volume of traffic passing along it (Table 1). As the volume of traffic increases, the degree to which animals can cross diminishes. While a quiet road with less than 1,000 vehicles per day can be crossed by most species, a busy motorway can be virtually impenetrable to all.

Table 1. Influence of traffic density on permeability (from COST 341 Handbook)

Traffic volume	Permeability
Roads with less than 1,000 vehicles per day	Permeable to most species
Roads with 1,000 to 4,000 vehicles per day	Permeable to some species but avoided by more sensitive species
Roads with 4,000 to 10,000 vehicles per day	Strong barrier: Noise and movement will repel many individuals. Significant road kill.
Motorways with more than 10,000 vehicles per day	Impermeable to most species

There is some debate about the advantages and disadvantages of having parallel roads and railways close to each other. This arrangement reduces the number of barriers and reduces the area of poor habitat that typically occurs between the road and the railway. However, the barrier effect is amplified by the combination and, without mitigating measures, can become a problem.

2.2 ROAD KILL

Road kill (Figure 3) is often a very visible result of the conflict between roads and wildlife. However, where species are common and are not deterred by shyness, there can be significant animal casualties without any threat to the population. For example, only 1-4% of the mortality of common rodents is thought to be from road kill. On the other hand, 40% of the badger population is killed annually on the roads of Flanders. Clearly, road kill can be a significant threat to a population in such circumstances.

Figure 3. Pine Martin, Kyparissia, Greece

For larger mammals, road kill is also a serious safety issue. Hence, even if a road is permeable to wildlife, there can be a safety problem. Roads authorities have responded to this by attempting to deter animals from approaching or crossing roads. For example, light reflectors are used to reflect the lights from vehicles onto the roadside and hence warn the animals. However, the only method known to work consistently is fencing which exacerbates the barrier problem unless accompanied by wildlife crossings. For invertebrates, fencing is clearly not the solution and it is necessary to design traps and passages to allow the animals to safely cross the road.

2.3 OTHER EFFECTS

There are other impacts from roads and railways on the environment. Noise and vibration can be locally important, particularly for shy species. In The Netherlands, noise in excess of 50 dB has been shown to reduce the density of bird populations. Some species of bird occur in high densities in disturbed areas but have less breeding success. Light can also be a problem for nocturnal species. For example, lights can attract insects which in turn can attract bats which can be killed by traffic (Figure 4). Lights can be screened from the roadside verges to reduce the disturbance to animals there. Alternatively, a natural screen of vegetation can be very effective in shielding the surrounding land from noise and light and can result in a good quality of habitat quite close to a busy road.

Figure 4. Bat attracted to swimming pool, Kyparissia, Greece

Noise can be alleviated quite successfully by lowering the road level below the surrounding countryside and providing raised embankments with the excavated soil. A barrier of 2m is enough to shield cars. However, embankments are quite demanding of space as vegetation will only succeed if they are at reasonably shallow slopes. A maximum slope of 1:2 is recommended to ensure that natural vegetation will grow and it makes for a more natural landscape if the slopes are varied.

As roads must be drained, lowering the road level has the effect of lowering the groundwater locally. This can potentially lead to the draining of local wetlands. For example, a section of the M7 motorway in Ireland was built at a reduced elevation 4.5km from an alkaline wetland. It was calculated that the road drainage would reduce the water table at the wetland by as much as 30cm, which could threaten three rare species of whorl snail. After considerable delay and much debate, an extensive section of the motorway was lined with a waterproof material to prevent any significant reduction in the water table. This was expensive but provided a long-term solution nonetheless to a difficult combination of conflicting requirements.

Chemical pollution due to vehicle exhaust emissions and de-icing salts can also be detrimental to the local ecology. Salt can get into local waterways and local vegetation and animals are subject to a range of air-borne chemicals. There are also indirect impacts of transport infrastructure which relate to the planning framework. Transport provides access which increases the prospects of human settlement and disturbance due to recreational use of land. This can be reduced if transport routes are chosen through areas where disturbance is already high, hence preserving those regions that have poor access and which are likely to have higher value for wildlife.

3. Route selection framework

Habitat fragmentation is only one of a wide range of issues to be considered during the design of a transportation route. It is common practice today to consider a number of alternative routes and to assess each one using a range of measures. While cost is an important issue, the problem generally comes down to one of selecting the route which minimises adverse impacts on communities or individuals and avoids sensitive areas. Areas can be considered sensitive due to the presence of an important habitat, an archaeological site or a scenic landscape. Other areas may be less sensitive but more expensive to travel through or may not provide a feasible route option because of impacts on homes, communities and farms. Topology is a significant cost issue as it can involve the need to cut through hills and fill or bridge valleys. There is generally a need to roughly balance cut and fill to avoid a soil disposal problem. Furthermore, there are costs associated with the movement of fill material so it is best to keep the cuts near to the fills. There are other direct and indirect costs associated with roads/railways. The quality of the soil is important: poor soil may need to be strengthened or even excavated and replaced, both of which are expensive. There is a tendency to avoid human habitation to avoid the need to demolish houses and to avoid problems due to noise and the fragmentation of communities.

The end result is that engineers will try to identify routes that avoid as many sensitive sites as possible. It is clearly important to have ecological conservation areas identified as being sensitive from an early stage. However, it is also important to highlight the need to connect such sensitive areas to other supporting habitat fragments. At a regional level, there may be pockets of habitat that link together to support animal populations. Individual animals may travel between them, particularly during mating or when juveniles are in search of new territories. Hence, the maps which identify conservation areas should include the linkages which are important to them. Such maps can form part of an ecological plan for a region. This can be an extremely useful tool in conservation and can provide a means to incorporate wildlife planning into the greater land use plan for future development.

Most of the problems of routing roads/railways can be overcome through additional expenditure. For example, soft ground can be spanned with bridges and noise problems for housing can be alleviated with noise barriers. This is also true for problems of habitat fragmentation. A range of measures are described in the next section which can be taken to make transport barriers more 'permeable' to wildlife and hence to connect habitats that have otherwise been severed. Of course, such measures are not as effective as a rerouting to avoid the cutting of important habitat linkages in the first place.

3.1 DECISION MAKING PROCESS

The decision making structures in transportation planning are important to ensure that habitat impacts such as fragmentation are taken into account in the route selection process. Engineers involved in the selection process work with maps. If better alternatives are not available, they will work with existing lists such as the Special Areas of Conservation (SAC's) which member states of the European Union (EU) are required to provide. The importance of connectivity will only be clear if maps of the important areas include the important connections between them. The focus at present is in avoiding SAC's; it needs to be developed to also avoid severing the connections.

It is important that decision makers understand the key ecological issues and the relative impact of alternative transportation solutions. Many decisions in the past have been made in a highly confrontational environment and have tended to focus on rare species. This can have damaging public relations consequences, particularly if the rare species are seen by the public as obscure (such as a small insect or invertebrate). The emergence of the lists of important habitats provides a welcome change of focus from species to habitats. However, there remains a lack of appreciation of the important issues in many regions and confrontations have led to an emphasis on blocking development rather than seeking best solutions to complex problems. The situation has been improved in many cases by the development of multi-disciplinary teams including ecologists to assess the impact of alternative routes from the earliest stages.

Some countries have produced practical guidelines for the minimisation of ecological impact. An example is the issue of providing for wildlife corridors alongside rivers and streams – it is well known that many animals use such natural boundaries to travel between habitat fragments. Clearly it is desirable to provide extra space under bridges for animals to pass alongside watercourses. This is not a major cost issue for a long bridge where the increase in span is relatively small. However, it can add significant cost to a short bridge – for example, a 5m wide culvert would need to be increased to 11m if it were to incorporate 3m pathways on each side. This would approximately double the cost. Short-span bridges and culverts generally span small watercourses and it could be argued that these often constitute less important corridors. Hence, an engineering solution would be a guideline that allows lesser passageways alongside streams (e.g. only one 3m passageway for a 5m bridge) and proportionally wider passageways alongside bigger rivers. Of course, this is very simplistic and ignores species-specific issues but it is the kind of guideline that, if implemented at the preliminary stages, can avoid what appear to be expensive changes to the design later. Similarly, it is useful to have simple guidelines for the numbers of mammal tunnels per kilometre so that these costs can be incorporated in a routine way in the preliminary design stages. Taking this a stage further, links between habitats could be ranked in importance and the degree of permeability of any severance specified as a function of the importance of the link. This would range from putting the whole road/railway on a viaduct (at great cost) to some minor permeability measures that may have minimal cost.

3.2 COMPENSATION

It is now an established principle that an area protected by the EU Birds and Habitats Directive cannot be destroyed without the provision of a habitat of equivalent value. It is generally accepted among ecologists that habitat compensation is much less desirable than avoidance of the original habitat or mitigation of the damage and should be a last resort. Nevertheless, it is a useful mechanism for expressing in simple terms the cost of habitat destruction. Creating a habitat is an expensive process which will encourage Engineers to seek less expensive alternatives. Figure 5 illustrates a wetland in Ireland which was artificially constructed in compensation for impacts on a SAC nearby.

(a)   (b)

Figure 5. Translocation of marsh soils at Kilmacanogue, Co. Wicklow as part of the N11 Road scheme: (a) Before; (photograph by Richard Nairn, NATURA) (b) After (from Cullinane *et al.* 2003)

Unfortunately, the principle of habitat compensation does not take account of Habitat Fragmentation issues. It is legally possible to create a fragment of compensating habitat that is not connected to any other fragment and which will therefore be of limited ecological value. If this situation is to be addressed, the principle should be established that an important habitat can only be severed if there is compensation for any isolated parts. Clearly the compensating habitat will have to be well linked to the other part. Furthermore, it will be necessary to draw up maps illustrating the key existing linkages between habitats throughout Europe and to provide some measure – in cost terms – of the damage that severance causes.

3.3 OTHER ECONOMIC AND PUBLIC RELATIONS ISSUES

As road and rail design decisions are strongly linked with cost considerations, the cost-related issues are important when making the case for conservation. Measures to improve the permeability of transportation corridors may be expensive but they are considerably less expensive to incorporate at the construction stage than to construct later on in a live traffic situation. Of course there can be legislative requirements, codes of practice and guidelines which impose minimum conditions for compliance. However, beyond such ecological constraints and in many situations where there are conflicting requirements, economic issues will feature. In such cases, public relations become important. There are many benefits, both direct and indirect, to conservation. Clearly, if 22 young people play in a wood over the course of an hour or two, then it has a similar recreational value to a football field and can be judged to be at least as important.
Eco-tourism is another economic argument for conservation and the prevention of habitat fragmentation. Worldwide there are many examples where a specialist tourist industry has been established as a result of the local wildlife. Clearly there are potential conflicts and a balance has to be struck between leisure/educational activities and conservation. Eco-tourism has significant potential benefit for local business and employment. The development of local ownership and pride in a special habitat can transform attitudes to the benefits of conservation. Local interest can be developed through careful consultation and education. In the context of transport planning, this may involve public meetings between the designers and community groups.

It is important to monitor the effectiveness of permeability measures, long term. There are many examples of measures, such as mammal tunnels, incorporated at the construction stage, that do not work in practice. It is important for public relations, as well as ecological benefit, to monitor permeability measures long term and to report the success stories in the local media.

4. Minimising habitat fragmentation – permeability measures

Where habitats are severed by roads/railways, as they inevitably will be to a greater or lesser extent, measures can be taken to ensure that the barriers are permeable, that is, to ensure that animals can pass safely through them. Permeability measures can generally be divided into two groups, depending on whether the animals pass over or under the road/railway. There is no hard and fast rule about which is better. It depends on many factors such as local topography and the behavioural preferences of the target species. Some mammals such as badgers are comfortable with tunnels and will happily use a 500mm diameter pipe to cross under a road. Other animals such as deer, are shy about crossing under low bridges and are much more likely to use a crossing over a road or railway. The local vegetation plays an important role and the approaches to a crossing. Clearly fencing is also important – many animals such as moose, on migration, will only use a safe crossing if they are prevented from using unsafe, but more convenient, alternatives.

4.1 GENERAL PRINCIPLES

Regardless of the type of crossing to be provided or the target species, some general principles apply. Many animals have a natural fear of the noise and light associated with transport routes and will not use a wildlife crossing just because it is there. It is therefore important to integrate the crossing into the local habitat. To encourage animals to approach the crossing, it is useful to provide linking fragments of habitats so that they do not have to cross large open spaces in the approaches. Vegetation that is attractive to the species in question can be used to attract the animals and some can be trained to use a crossing by providing food treats or scents. It seems obvious that the approaches to animal crossings should not be broken by service roads or walking tracks which may discourage animals from approaching.

When animals are near a crossing point, steps need to be taken to prevent them from walking onto the road or railway. This means fencing – alternatives have been tried but are generally considered to be ineffective. Guidelines on the recommended mesh sizes and heights of fences are given in the COST 341 Handbook (Iuell et al. 2003). At the entrance point to the crossing, it is helpful to provide a recess to discourage the animal from passing by as they travel parallel to the fence. The fence should be used to guide the animals towards the entrance. Some animals are shy and are encouraged by having a source of cover such as a hedgerow nearby. Again, the hedge should be used to guide the animals towards the crossing point. Other natural guides are also possible such as existing streams or valley bottoms.

For overpasses, i.e. where animals pass over the road, shielding against the noise and light of traffic is an issue. For underpasses, i.e. where animals pass under the road, drainage becomes important. It is essential that culverts or pipes under a road/railway do not get flooded after heavy rains and that they do not support standing water.

Animals may be discouraged by the presence of steep slopes in the approaches to crossings. In general, the Handbook recommends gentle slopes. It reports successful examples with 16% in flat areas and 25% in mountainous areas. Clearly, where possible, slopes are better avoided so that animals move naturally from the surrounding area into the crossing. The details at the entrance are important. There are examples of wildlife crossings which end in a step that some animals cannot negotiate. Similarly, it is important that noise/light screens at the sides of overpasses are connected thoughtfully to fencing at the side of the road/railway.

Maintenance is an important issue and one that should be considered at the design stage. There is clearly no benefit from providing expensive permeability measures if they become dysfunctional due to a lack of maintenance or abuse by local landowners – there are examples where wildlife crossings have been fenced off or used for storage. Normal maintenance consists of the removal of litter or water that is hindering the functioning of the crossing, the repair of damaged fencing, the maintenance of vegetation in accordance with the original plan, etc.. There is also a need for access and facilities for wildlife structures to be inspected on a regular basis. Whether maintenance is carried out by transport authority personnel, by land owners or conservation volunteers, there needs to be an agreed programme and those carrying out the work must have a good understanding of the objectives of the crossing.

4.2 OVERPASSES

Overpasses are a relatively expensive solution to the problem of habitat fragmentation but there are circumstances, particularly when the road/railway is low-lying, where no viable alternative exists. The ideal overpass is very large and connects habitats at the ecosystem level. These are sometimes referred to as landscape bridges and are in excess of 80m wide. At this scale, the wildlife bridge over the road/railway becomes quite expensive and a cut-and-cover tunnel through the habitat may be a better solution. Of course, cut-and-cover tunnelling causes major disturbance and a significant barrier during construction. In the longer term however, it can be an excellent solution with good quality linking habitat re-established over the road/railway. Bored tunnels, which are bored without excavating the soil above, are of course an ideal ecological solution. However, they are very expensive and would normally be seen only as a last resort.

In the majority of roads and railways there are no landscape bridges. In such cases, less effective solutions may be employed to reduce the barrier effect. The most common wildlife overpass is considerably less wide than a landscape bridge. Unfortunately, larger animals tend not to use narrow overpasses. The Handbook recommends a minimum width of 40m to 50m but acknowledges that widths as low as 20m have been used to good effect in certain circumstances. Bridge cost is approximately proportional to its area so width is a big cost issue. A funnel shaped crossing has been proposed as an economical solution where the width at the narrowest point can be as little as 20m. The length to width ratio is also an important factor for animals – longer crossings need to be wider – and it is recommended that the width should not be less than 80% of the length. Width is not an important issue for small invertebrates for whom the key issue is continuity of habitat (vegetation) across the bridge.

Vegetation is an important issue for wildlife crossings. The key principle is to reflect the habitat of the adjacent areas, hence providing continuity across the bridge. The roots of large trees can be a problem for a structure. Hence, if they are really necessary, they should be enclosed in large pots above ground. Bushes and shrubs are generally a better solution although they require a greater depth of topsoil than grass/herbs (60cm as opposed to 30cm). In general, the top surfaces of bridges are substantially flat which suggests a constant depth of topsoil. However, where the road/railway is culverted – see Figure 6 – variable depth topsoil is a natural result of the form of construction and helps to promote a varied vegetation on top. In general, sowing the soil is not necessary and good results have been reported from allowing a natural re-vegetation of top soil on the overpass. However, planting of food species that attract the target animals can clearly be beneficial.

Figure 6. Culverted road, United Kingdom (photograph E.O'Brien courtesy of ABM Construction)

Screens are generally necessary to shelter animals from the noise and light of vehicles/trains, particularly in quiet areas that are otherwise undisturbed. Hedgerows are best, preferably on small mounds, and can also serve to guide animals over a crossing. These should be placed at the edges to provide the maximum width between them. However, hedges are quite demanding of space and where the crossing width is limited, artificial screens may be a better solution. Screens need to be about 2m high to be effective. However, in narrow crossings, excessively high screens can create a tunnel effect which may discourage animals from using them.

4.2.1 *Multiple-use overpasses*
A crossing used for several purposes is not ideal for wildlife due to disturbance effects, etc. However, it is a cheap solution and can provide useful supplementary permeability to a road or railway where the key habitats have already been connected. Some crossings are for farm access purposes and will have a very low density of traffic. In such cases, small changes can make them attractive as wildlife crossings. The simplest enhancement is to add a strip of natural vegetation for invertebrates, continuous across the full length and connecting to vegetation on both ends. Screening from traffic is clearly an advantage and it is also useful if the vegetated strip can be located at an edge and separated from the other users with an earth wall. If a wildlife crossing is also to be used by walkers, it is helpful to provide a path to concentrate the disturbance in one part of the bridge.

4.3 UNDERPASSES FOR MEDIUM/LARGE ANIMALS

When the road/railway is elevated, an animal passage under it – an underpass – is a more natural solution than an overpass. When the landscape is hilly, a viaduct over a wide valley is sometimes necessary. This is good news for habitat linkage as there are often good quality open spaces under the viaduct that allow the free movement of animals. These should have their value recognised and should be protected against future development.

Many large animals such as deer tend to follow valley bottoms or rivers that pass under a viaduct. (Furthermore, small animals that would not use unvegetated tunnels or pipes, will have a continuity of vegetation through which to move under the road/railway). Ideally, at least 10m of bank vegetation should be provided on either side of a river passing under a viaduct to promote animal movement.

Common sense measures such as removing any blockages under the viaduct and screening off local sources of disturbance will promote the movement of animals. There are circumstances where viaducts are constructed purely for wildlife purposes, such as when a road/railway cuts through a wetland. In such cases, it is recommended that the viaduct spans the wetland completely.

Modern highway bridges are generally 'twinned', that is, separate parallel bridges are used for each carriageway. This allows for traffic to be diverted temporarily onto one of the bridges if the other one needs to be repaired or replaced. Twinned viaducts with a narrow gap between them can be disturbing to animals due to the sudden bursts of traffic noise from the gap. It is therefore recommended to use wide gaps. This also allows light through and reduces the continuous width of habitat that will be in shadow. If viaducts are low over the ground, light can be a problem and vegetation may not survive. Nevertheless, it is recommended to use simple earth cover in preference to the less natural alternatives such as concrete or gravel.

Engineers often culvert rivers under viaducts to prevent piers from being undermined during floods. However, this is not good for wildlife and vertically walled culverts in particular can become a death trap for animals attempting to cross the river. A far better solution is to provide an island around the pier to protect it and keep the river banks natural. If culverts have to be used, sloped sides are much better than vertical walls.

In flat terrain, viaducts are less likely and, if the road/railway is elevated, animal underpasses will be needed to link habitats. Large and medium sized animals are often wary of passing through narrow passages. The Handbook defines an 'openness index' as the cross-sectional area of an underpass divided by its length in metres and recommends a minimum openness index value of 1.5. In other words, if it is longer then it needs to be wider and/or higher. It also recommends a minimum width of 15m and a minimum height of 3-4m. Vegetation should be encouraged, particularly at the entrance, but may not be possible due to the lack of light. As for land under viaducts, an earth base is recommended when vegetation is not possible. Logs or dead vegetation can be a useful means of providing cover.

Drainage of underpasses is important – standing water or seasonal flooding can be a serious obstacle to movement. Thus, underpasses should be designed above the water table. If there is a watercourse, then a dry shelf should be provided alongside it to allow animals to cross in all seasons.

As for overpasses, hedgerows perpendicular to the road/railway are good for providing cover and for guiding animals towards the underpass. Fencing is also necessary in the area of the underpass to prevent animals for taking an apparently easier route across the road/railway. Carefully designed fencing with, for example, a recess at the entrance to the underpass, can encourage use. As for overpasses, access to the entrance should be as level as possible and there should be no obstacles to animal movement.

4.3.1 *Multiple-use underpasses*
There are circumstances where animals will use underpasses designed for other purposes. For example, an unsealed road may pass under the road/railway which, if not heavily used by humans, may be used by animals. Shared use underpasses are generally only recommended if the passage is wide and not too long. However, there are often circumstances where existing underpasses can be improved for animal use provided they are not too long (less than 25-30m). As for overpasses, an earth strip is the simplest form of enhancement but other improvements may be possible such as a source of shelter and connections at the ends.

4.4 UNDERPASSES FOR SMALL ANIMALS

Small animals such as badgers and otters, are easier to cater for than larger animals such as deer. Small underpasses are an obvious solution when the road/railway is raised on an embankment but are also possible when it is on level ground. They can be round or rectangular and in the range of 0.4-2m in diameter/width. A diameter of 1.5m or a width of 1-1.5m is good for many species. Burrowing animals such as badgers will use pipes as little as 0.3-0.5m but such small underpasses are difficult to maintain and the number of species that benefit is limited.

Round pipes are generally easier to construct, particularly if they are to be jacked under an existing road/railway. However, rectangular tunnels are better, particularly for amphibians who use the vertical side as a guide. Rabbits and some carnivores will avoid metal and plastic is not favoured by some species. Concrete is good although the bottom surface needs to be covered with soil. Where concrete is precast, particular care needs to be taken to ensure smooth joints where the lengths meet. A small slope (1% or more) is important for drainage purposes but passages should not slope at more than 1:2.

Inspection and maintenance are important and must be considered at the design stage. The Handbook recommends inspections from 2 to 10 times per annum depending on importance. Maintenance is also necessary to remove litter/blockages and to control vegetation and water. As for all crossings, fencing is necessary in the vicinity and the details are all-important. The fence should ideally be recessed in the vicinity of the entrance and should extend for a considerable length on either side. For badgers, fencing is required for a minimum of 10m on either side of the crossing. It should be dug into the ground and have a small rectangular mesh (details of fencing for a range of species are given in the Handbook). For otters, fencing is required for 25-50m on either side. The ends of fences are important to prevent animals from entering and being trapped on the road side of the fence.

5. Conclusions

The threat of transport infrastructure to wildlife is considerable because of the implications for habitat fragmentation. However, solutions are possible which, in many cases, are relatively inexpensive. These must be carefully designed and thought through and require inspection and maintenance on an ongoing basis. However, they can work and there are an increasing number of examples of successful permeability measures which improve the connectedness of animal habitats. Lessons have been learned – the details, particularly where the crossing meets the surrounding habitat – are all important.

There are opportunities to get it right, particularly in countries where transport infrastructure is being expanded. Some countries are currently retrofitting permeability measures in existing roads/railways. This is much more difficult and expensive than incorporating measures in new construction. Where new infrastructure is being built, routes can be selected which minimise the fragmentation problem. Low maintenance and easily inspectable permeability measures can be integrated into the design from the outset. Land use planning measures can be taken to protect key crossing points from damage due to future development. With good design, the negative impact of transport on wildlife can be considerably reduced and a sensible compromise reached between the need for transportation and the need for a healthy natural environment.

References

Iuell, B., Bekker, G.J., Cuperus, R., Dufek, J., Fry, G., Hicks, C., Hlaváč, V., Keller, V., Rosell, C., Sangwine, T., Tørsløv, N., Wandall, B.M., *Wildlife and Traffic: A European Handbook for Identifying Conflicts and Designing Solutions*, KNNV Natural History Publishers, Utrecht, The Netherlands (www.knnvuitgeverij.nl), ISBN: 90-5011-186-6, 2003.

Keller, V., Bekker, G.J., Cuperus, R., Folkeson, L., Rosell, C., Trocmé, M. 2003; Avoidance, mitigation and compensatory measures and their maintenance. In: Trocmé, M., Cahill, S., de Vries, J.G., Farrall, H., Folkeson, L., Fry, G., Hicks, C. & Peymen, J. (eds) COST Action 341 - Habitat fragmentation due to transportation infrastructure: The European review, 129-173. Office for Official Publications of the European Communities, Luxembourg.

CHAPTER 10: RESTORING HABITAT CONNECTIVITY ACROSS TRANSPORT CORRIDORS: IDENTIFYING HIGH-PRIORITY LOCATIONS FOR DE-FRAGMENTATION WITH THE USE OF AN EXPERT-BASED MODEL

EDGAR VAN DER GRIFT & ROGIER POUWELS

Alterra, Wageningen University and Research Center, P.O. Box 47, NL-6700 AA Wageningen, The Netherlands.

1. Introduction

Roads, railroads and canalized waterways are, as well as causing changes in land use, major sources of habitat fragmentation. These transport corridors, divide landscapes into discrete sections, thus decreasing connectivity between habitat patches. The extent to which habitat connectivity is affected (i.e. the cohesion between patches) is species-specific and dependent on the characteristics of both the transport corridor itself and its traffic. For most animal species a transport corridor is a partial barrier, with the exchange of animals between habitat patches on opposite sites of the barrier being a small fraction of the exchange in the pre-fragmented situation. It is not uncommon, however, for transport corridors to create an absolute barrier, cutting off all animal movements across the transport corridor or prohibiting any successful crossings. If all necessary habitat types (e.g. breeding, foraging, resting habitat) for a species are still present within a section, unnatural mortality at the transport corridors will be low or may be fully prevented. If the isolated section is still large enough to facilitate the existence of a viable population, it is not likely that the species will become locally extinct. However, if the transport infrastructure network is dense enough to result in isolated habitat patches that are too small to support viable populations, and traffic-induced mortality is considerable, the risk of local extinction is high resulting in a loss of biodiversity (see Chapter 9).

To counteract the impact of habitat fragmentation by transport corridors, a variety of mitigation measures have been developed (see e.g. Iuell *et al.* 2003). The barrier effect of roads and railroads, as well as road-kills due to collisions with traffic, are usually mitigated by the construction of wildlife-crossing structures, such as wildlife overpasses or underpasses in combination with the construction of wildlife fences that guide the animals towards the passages and keep them off the road or railway (Figure 1; Chapter 9). Although many such so-called mitigation measures have been constructed, and at many of these sites use of wildlife-crossing structures by wildlife have been assessed, only a few studies have addressed the effectiveness of these measures (Foster & Humphrey 1995; Pfister *et al.* 1997; Dodd *et al.* 2004). There are especially few studies in which the attributes of crossing structures, the infrastructural barrier as well as the landscape, are correlated with the acceptance of crossing structures and wildlife-crossing rates (but see Yanes *et al.* 1995; Clevenger & Waltho 2005). Even more surprising is the low number of studies that address suitable methods for determining the placement of wildlife-crossing structures along transport corridors. The dearth of such studies is disturbing since the few studies that have been conducted indicate that the placing of such structures is often more important than other factors, such as the design or dimensions of the passageways themselves (Foster & Humphrey 1995; Land & Lotz 1996; Rodriguez *et al.* 1996).

John Davenport and Julia L. Davenport, (eds.), The Ecology of Transportation: Managing Mobility for the Environment, 205–231,

Figure 1. Wildlife-crossing structures that provide animals a safe passage across transport corridors: A wildlife underpass (*left*) that links Florida panther habitat across State Road 29 in Big Cypress National Preserve, US, and wildlife overpass 'Woeste Hoeve' (*right*), constructed to help ungulates across highway A50 in The Netherlands.

The impact of habitat fragmentation by transport infrastructure is now widely recognised and in many countries locations for de-fragmentation, i.e. locations where wildlife-crossing structures should be constructed, have been identified (Hicks & Peymen 2003; Van Bohemen *et al.* 2004). It is not unusual to find that, within one country or region, different stakeholders have come up with different lists of priority locations for de-fragmentation. The differences in identified bottleneck locations may have been the result of differences in ambitions, research method, or scale of detail at which de-fragmentation measures were identified (Van der Grift 2005). Furthermore, only a few studies have quantified the ecological benefits of mitigation measures for each bottleneck location separately. Consequently, a proper comparison of the different sets of bottleneck locations, and an objective prioritisation of actions to counteract habitat fragmentation, are usually difficult to perform.

In this chapter applied methods to identify de-fragmentation locations, i.e. approaches described in the literature to assess locations where wildlife-crossing structures can best be constructed, are first reviewed. Second, an expert-based model that may help spatial planners, transport infrastructure managers and wildlife conservationists to identify the best locations for de-fragmentation measures, both from an ecological and a cost-benefit point of view, is introduced. Third, the use of the introduced model is illustrated by a case study from The Netherlands. Fourth, a framework for a research approach is suggested in which our model and some of the earlier applied methods are combined. Lastly, some challenges concerning habitat fragmentation that are ahead of us, including determining what research is needed to help us deal with the problem of habitat fragmentation in a better and more efficient way are addressed.

2. Applied methods to identify de-fragmentation locations

Since the impact of habitat fragmentation by transport corridors became known, a variety of methods have been used to identify where the construction of wildlife-crossing structures is needed. The applied methods to assess the de-fragmentation locations differ greatly in complexity, theoretical basis, and ecological indicators used. Basically, the methods can be separated into two groups: (1) methods in which wildlife movements across existing or future infrastructural barriers are surveyed, (2) methods in which wildlife movements across existing or future infrastructural barriers are predicted.

2.1 SURVEYING CROSS-BARRIER WILDLIFE MOVEMENTS

The placement of wildlife-crossing structures can be effectively judged if detailed knowledge of wildlife movements prior to the construction of the transport corridors, i.e. daily or seasonal migration routes or dispersal paths are available (e.g. Singer & Doherty 1985; Scheick & Jones 1999). It is, after all, widely recognised that wildlife-crossing structures should be located where wildlife naturally migrate. The advantage of a pre-construction survey is that the assessment of wildlife migration paths is not biased by the presence of the infrastructural barrier itself, as is the case in a post-construction assessment. It is well known that the presence of a road, railway or canalized waterway may cause changes in habitat use or movement patterns. For example, roads – especially those with high traffic volume – have proven to be partial barriers to the movements of mule deer and elk (Rost & Bailey 1979), wolves (Paquet & Callaghan 1996), bobcats and lynx (Lovallo & Anderson 1996), bears (Servheen et al. 1998; Chruszcz et al. 2003; Kaczensky et al. 2003), as well as many small mammal species (Oxley et al. 1974), amphibians (Vos & Chardon 1998; Reh & Seitz 1990) and invertebrates (Baur & Baur 1990; Mader et al. 1990; Munguira & Thomas 1992). As a result, cross-barrier migration paths may have been abandoned in the post-construction situation, which will cause an underestimation of the barrier effect if wildlife movements are only surveyed after the construction of the transport corridor.

The approach for identifying de-fragmentation locations at existing transport corridors is somewhat different, although in some studies historical data of pre-construction wildlife migration routes have been used to assess the best locations for wildlife-crossing structures (e.g. Bekker et al. 1995). Post-construction assessments of places to construct crossing structures are usually based on surveys of locations where animals frequently try to cross the transport corridor successfully or unsuccessfully (Foster & Humphrey 1995; Becker 1996; Neal & Cheeseman 1996; Carey 2001; Gibeau et al. 2001; Van Manen et al. 2001). Such crossing sites are assessed by a variety of methods, such as the mapping of well-established wildlife paths (e.g. badger trails), animal track surveys along transport corridors e.g. with the use of sand beds or by snow-tracking during winter, through analyses of radio-telemetry or infrared camera data on wildlife movement patterns, surveys of wildlife mortality due to wildlife-vehicle collisions (roads/railroads) or drowning (canalized waterways), or by a combination of these methods.

The advantage of mortality data is that, in contrast with surveys that address spatial distribution of movement patterns of live animals, mortality sites are usually easy to assess. The disadvantage of the method is that the results are biased by the restricted focus on sites where, during the survey period, animal crossings were not successful. Mortality data, especially when collected over a limited time period, do not provide full insight into the spatial distribution of cross-barrier movements. Crossing rates may be higher at locations where no dead animals are found, but as a consequence of the survey-method such locations will not be recognised as suitable for de-fragmentation measures. Furthermore, the absence of dead animals at certain parts of transport corridors does not necessarily mean that at these locations, there is no fragmentation problem, i.e. that animals are able to cross here without any hindrances. The absence of dead animals may after all simply be the result of the absence of all cross-barrier movements due to the physical or behavioural barrier the transport corridor induced.

2.2 PREDICTING CROSS-BARRIER WILDLIFE MOVEMENTS

The relatively recent recognition of fragmentation problems that accompany the construction of transport corridors means that pre-construction data on wildlife movements are often lacking in the case of existing infrastructure. Furthermore, acquisition of pre-construction data on wildlife movements, as well as most of the described post-construction survey methods to assess cross-barrier movements, require a considerable effort, both in time and manpower, in order to be useful tools to plan mitigation measures. Consequently, not all transportation planners or wildlife managers are able to start initiatives to acquire such data on wildlife movements, crossing locations, or hot-spots for wildlife-vehicle collisions. Therefore, it is not surprising that methods have been developed to predict the sites where cross-barrier movements of animals will most likely occur. These methods, however, vary considerably in complexity and consequently in the accuracy of their predictions of suitable de-fragmentation locations.

A first group of predictive methods is to superimpose transport networks on maps with areas that are of ecological importance. One of the most simple applications of this approach, and one which is widely used, is to base the placement of wildlife-crossing structures on approved nature policy plans, such as spatial plans for the development of ecological networks. Some practisioners plan wildlife-crossing structures at any location where the human transport network intersects the core areas or corridors of proposed ecological networks (Carr et al. 1998; Hicks & Peymen 2003). Others use a similar approach, but instead of superimposing transport networks on proposed ecological networks, crossing-points are mapped between transport networks and (1) actual distribution patterns of animal species that have been proven susceptible for fragmentation by transport corridors, or (2) existing habitat and habitat linkage zones. In other studies, intersections between transport networks and nature areas with a national or international protection status, or other areas of conservation interest are identified (e.g. Van der Fluit et al. 1990; Ruediger et al. 1999). Such overlay techniques usually give a broad outline of possible bottleneck locations, but often lack sufficient detail to provide specific recommendations for the placement of wildlife-crossing structures. In addition, if the exact placement or aims of the proposed ecological networks have not yet been worked out in detail, only an indicative assessment of de-fragmentation locations is possible. Species distribution patterns may be a relic of former distribution patterns, thus underestimating the number of bottleneck locations. Furthermore, little information can be derived from such analyses about the importance, i.e. ecological benefit, of each identified de-fragmentation location.

A second group of predictive methods is the use of habitat models. With the help of such models, favourable locations for crossing structures are identified by analyses of size, quality and spatial configuration of suitable habitat and habitat linkages. In such models crossing locations are predicted using either empirical data or expert information about habitat suitability, landscape and topographical features, roadway features, and land use (Klein 1999; Kobler & Adamic 1999; Kohn et al. 1999; Tremblay 2001; Clevenger et al. 2002; Barnum 2003). Empirical data to construct such models may be derived either from general ecological studies that address habitat use and movement patterns of species, or from studies in which wildlife-vehicle collision sites or crossing sites are correlated with features of the surrounding landscape (e.g. Lehnert et al. 1996; Singleton & Lehmkuhl 1999; Smith 1999; Craighead et al. 2001; Malo et al. 2004). The common argument for the development of such models is that the placement of wildlife-crossing structures should not be based solely on road-kill and tracking data, but should fit in the overall patterns of the landscape and coincide with effective zones of landscape connectivity (Forman et al. 2003). Also the ability of animals to learn and adjust their movement patterns should be included. Thus, wildlife-crossing structures at locations with high road-kill rates may be less effective than ones that are placed at locations where high-quality habitat is bisected by the transport corridor.

A third group of predicting methods involves the use of movement models to simulate individual animal movements through the landscape (e.g. Schippers et al. 1996; Bakker et al.

1997; Vos *et al.* 1999). Such models are based on information about habitat use, habitat preference, movement patterns, and dispersal distances of species. A strong point in favour of the use of these models is that through running high numbers of simulations, the favourability of different crossing locations can be quantified and thus possible locations for de-fragmentation can be better compared and prioritized.

The disadvantage of both habitat linkage models and animal movement models is that the applicability of the models very much depends on a careful validation of the models. Most studies, however, lack such validation (but see Clevenger *et al.* 2002), which may be the consequence of the lack of empirical base-line information. Therefore, trustworthy models can be developed only for a limited number of well-studied species at the present time. Furthermore, the models cannot always easily be used in areas other than the one for which they were developed, due to differences in habitat use or movement patterns of the studied species.

3. Population viability analysis (PVA) as a tool to identify de-fragmentation locations

3.1 WHY INCLUDE PVA?

All the described methods that have been applied to identify de-fragmentation locations are based on identifying locations where animals (used to) cross the transport corridor or are likely to cross. The aim of this approach is usually to restore all, or as many as possible, of the pre-construction movement patterns. Besides the limitations of each method, as described above, such an approach only implicitly addresses the consequences of fragmentation or de-fragmentation for population viability or species survival, thus missing a clear link with the overall aim of such interventions: preserving sustainable habitat networks in which population viability for all species concerned is safeguarded and with it the conservation of biodiversity.

Resources may not be sufficient to restore all disturbed movement patterns or prevent all wildlife-vehicle collisions. The shortage of resources is a major factor, especially when attempts are taken to retrofit existing, often extensive, transport corridors in order to mitigate the mistakes of the past. Consequently, choices have to be made as to what, where and when mitigation measures should be put in place. To be able to make such choices carefully, knowledge is needed about the positive impact of de-fragmentation measures on the viability of fragmented populations. After all, efforts and expense are best spent on constructing de-fragmentation measures at locations where the resulting improvement of population viability is highest. However, even if the aim is to eliminate all bottleneck locations, i.e. restore all pre-construction movement patterns and prevent all wildlife mortality due to collisions or drowning, knowledge of the impact of each de-fragmentation measure on population viability is needed in order to assess the locations where mitigation initiatives should be started.

It therefore seems useful to include PVA in both bottleneck assessments and methods of prioritizing actions for de-fragmentation. The approach can be simple: compare analyses of population viability where the infrastructural barriers are present with those where the barrier effect is mitigated. Locations where mitigation measures are expected to result in a significant improvement of population viability can be identified as priority de-fragmentation locations. Furthermore, the extent to which population viability will change at different locations can be compared. Where a significant improvement is expected, the information can be used to set priorities. An important benefit of such an approach is the ease with which different scenarios for both transportation network and green network planning can be analysed and compared. PVA, for example, may be based on actual or proposed distribution of suitable habitat, or on actual or proposed habitat linkages, offering the opportunity to see how de-fragmentation of transport corridors relates to expected changes in land use or initiatives to restore ecological networks. Transport corridor plans that differ in routing of future infrastructure or the placement and number of wildlife-crossing structures can be easily compared and valued with PVA. This enables optimisation of plans.

3.2 HOW TO INCLUDE PVA?

We have developed a method utilising the comparative approach described above, using expected shifts in population viability due to mitigation as an indicator for identification of de-fragmentation locations at transport corridors. In our method seven steps can be distinguished:

1. Selection of species or species groups for which shifts in population viability will be analysed.

2. Habitat modelling.

3. Assessment of local populations.

4. Assessment of habitat networks.

5. Population viability analysis.

6. Identification of potential de-fragmentation locations.

7. Setting priorities, based on differences in ecological benefit between de-fragmentation locations.

With the exception of the selection of species or species groups, all steps are made operational in the spatially explicit expert-based GIS model LARCH (= Landscape ecological Analysis and Rules for the Configuration of Habitat) (Verboom *et al.* 2001; Opdam *et al.* 2003). This model allows an analysis of the configuration and persistence of habitat networks for a variable set of species. LARCH uses thresholds to determine whether these habitat networks can support viable populations or not (Verboom & Pouwels 2004). Although such a static model is not able to predict trends in population viability over time, and its use of thresholds results in discrete classifications of viability, the model runs on a more limited number of assumptions and parameters than dynamic population models and consequently better enables us to execute multi-species analyses. Furthermore, the model is more suitable for aggregating population viability analyses for different species or species groups due to the simpler model design (Van der Grift *et al.* 2003a). Because input in LARCH is partly based on expert-knowledge, output should be treated as such as well, and precise predictions of future population viability should not be expected. We would like to stress that models such as LARCH can work best in comparative studies, e.g. where different land use scenarios or transport corridor networks are compared (see also Verboom & Wamelink 2005). The LARCH model results can be used to prioritise scenarios, based on relative rather than absolute differences between locations or scenarios.

3.2.1 *Step 1: Selection of species or species groups*
The selection of animal species is one of the most essential steps in the analysis. The number and type of species determine, after all, the scope of the analysis and are decisive for the extent of the identification of de-fragmentation locations. If an analysis of de-fragmentation locations for a certain species is needed, e.g. as part of the development of a species protection plan, selection of that particular species is axiomatic. If the identification of de-fragmentation locations within an ecological network – with a variety of ecosystems and multitude of species – is the objective, then species selection demands more consideration.

We suggest three selection criteria: species sensitivity to transport corridor impact, species representative of ecosystems studied, species dispersal capacity and habitat area requirements. First, species should be selected that are most sensitive to transport corridor impacts. Each type of transport corridor may have a different impact on a species (see e.g. Jaeger *et al.* 2005), hence the pool of species that are known to be sensitive to fragmentation impacts by transport corridors can, if the focus of the study is limited to one type of transport corridor, be downsized to species that show susceptibility to that particular type of transport corridor. If different types of transport corridors are to be included in the analyses, species selections should include species that are affected by the infrastructural barrier type under study that causes the least impact. Second, selected species should represent the variety of ecosystems in the ecological network of the study. If only species of forest or heathland ecosystems are selected no de-fragmentation locations will be identified for species that depend on wetlands or grasslands. Third, the selected species should fit the chosen scale of the analyses. For example, species with limited dispersal capacity and limited spatial habitat requirements will be less differentiating in analyses that aim to assess the impact of highways on population viability at a national scale. Most habitat networks within the matrices of the highway network will be large enough to provide such species with viable populations. Similarly, species with habitat networks on a national or even international scale will not be very differentiating in bottleneck assessments at the local scale of a region or county.

Even when a selection of species has been made based on these criteria, an assessment of de-fragmentation locations for all selected species is not usually feasible. In order to further downsize the number of analyses one may decide to focus on a group of indicator species that are selected in such a way that they represent all initially selected species. An alternative approach can be to use so-called 'ecological profiles' that represent clusters of species differing in habitat preference and sensitivity to habitat fragmentation (Vos *et al.* 2001; Opdam *et al.* 2003). The use of ecological profiles differs from the use of indicator species in that the parameter values for ecological profiles are not based on the values for a single species but on the mean value of the species group it represents. This limits the potential to draw conclusions for individual species, but better allows analysis of the impact of de-fragmentation on species assemblages and ecosystems. Furthermore, it downsizes considerably the number of analyses necessary, and increases the possibility of aggregating multi-species study results. Each ecological profile can be described in terms of three parameters: (1) habitat preference, (2) individual habitat area requirements, (3) average dispersal distance (Vos *et al.* 2001; Figure 2). To optimise the robustness of bottleneck analyses or de-fragmentation location assessments, ecological profiles should be selected for all habitat types present in the study area, and should cover the whole spectrum of habitat area requirements and dispersal capacity (Simberloff 1998; Opdam *et al.* 2003).

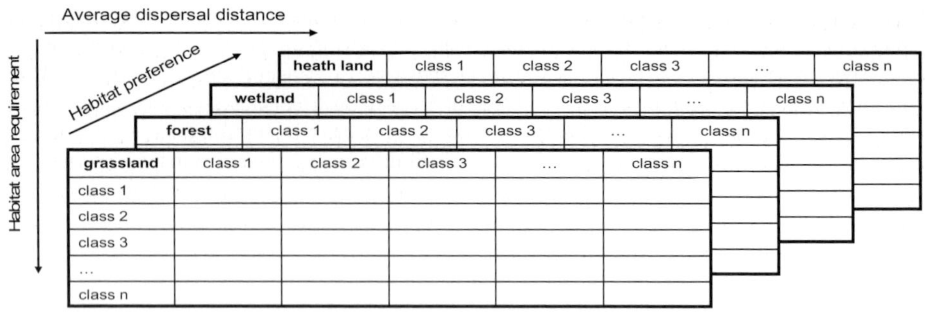

Figure 2. Species can be ordered into ecological profiles by their habitat preference, individual habitat area requirements, and dispersal capacity. In the given matrices each cell represents one ecological profile.

3.2.2 *Step 2: Habitat modelling*

The second step in the method is the construction of a habitat map for each selected species or ecological profile. Such habitat maps will form the basis for the analyses of size and spatial configuration of metapopulations and, consequently, the assessment of metapopulation viability. In this step all patches of nature in the study area are classified as unsuitable, marginal, suitable, suboptimal or optimal habitat, based on empirical knowledge of habitat requirements and habitat preferences of species, combined with a land cover or vegetation map. The more empirical knowledge available about habitat selection of species, and the more detail in maps with spatial configurations of biotopes, the better the assessment of habitat suitability (Verboom & Pouwels 2004). For each habitat patch, carrying capacity is calculated, based on size and habitat quality of the patch. Carrying capacity is not expressed in terms of the number of individuals, but the number of reproductive units (RU) per area unit. One RU is usually a breeding pair, but may consist of more than two individuals when a species forms social groups in which only a limited number of individuals take part in reproduction (e.g. badger, red deer). In unsuitable habitats carrying capacity is 0. In marginal, suitable and suboptimal habitats, the number of RUs at carrying capacity is, in this study, assumed to be 10%, 50% and 70% respectively of the number at carrying capacity in optimal habitats of the same size. The number of RUs at carrying capacity in optimal habitats is derived from empirical studies. Patches suitable as habitat, but too small to contain one RU, are not rated as habitat patches.

The habitat map can be based on the present situation, or may reflect future scenarios for nature development, e.g. if land cover maps with planned instead of existing ecological networks are used. The choice of habitat map depends on the research question that needs to be addressed: is the focus limited to solve short term fragmentation problems in relation to existing nature areas, or is an analysis needed for future schemes for spatial development or ecological networks? Note that habitat maps derived from land cover maps as described here, provide an overview of patches where a species may potentially occur. This is in accordance with metapopulation theory which emphasises that suitable habitat patches may not be inhabited at certain times due to local extinctions (Hanski & Gilpin 1991). Even in viable metapopulations, up to half of the habitat patches may be uninhabited at times (Vos *et al.* 2001). One may prefer, however, to focus on fragmentation problems in habitat networks in which a species is still present in at least one patch. Habitat networks in which the species have become extinct can, after all, by definition only be recolonised with the help of humans, i.e. by species reintroduction or by linking an 'empty' habitat network with a habitat network in which the species still occurs, after which recolonisation can take place. If such actions are not planned or part of the evaluation, leaving such 'empty' networks out will provide a more realistic view of bottleneck locations and therefore provide a better tool to prioritise de-fragmentation measures. Again, the research question will determine the approach.

3.2.3 *Step 3: Assessment of local populations*
The third step is to assess which habitat patches belong to the same local population. Some habitat patches will be located so close to each other that the inhabitants can be considered as a single interbreeding population in which random mating will occur. LARCH analyses size and location of such local populations with the use of a species-specific patch merging distance (PMD). This PMD is the distance below which about 90% of all home-range movements take place, based on empirical data (Verboom & Pouwels 2004). Transport corridors may prevent the merging of habitat patches into local populations. This depends on both the species and the type of barrier. Typically, roads with low traffic volumes are not a significant barrier to mobile species with large home-ranges. River otters or pine martens, for example, often cross local and even provincial roads during home-range movements, but tend to cross major highways only on dispersal. The classification of transport corridors as barriers for local populations should preferably be based on empirical studies or, if this is not possible, estimated by expert-judgement. The presence of existing wildlife-crossing structures can be included in the assessment of local populations if it is believed that these measures facilitate 'random mating' of individuals that inhabit habitat on opposite sides of the transport barrier. Usually, however, such mitigation measures are used by only a small proportion of the population. This is especially true for territorial species that may exclude all other individuals from using a crossing structure within their own home-range. In such cases it is preferable to consider the transport corridor as a barrier to the merging of patches into local populations, whether mitigation measures have been constructed or not.

3.2.4 *Step 4: Assessment of habitat networks*
A habitat network is a set of habitat patches that are situated close enough to each other that a reasonable level of inter-patch dispersal occurs. In LARCH, habitat networks are delineated with the use of a species-specific network merging distance (NMD). This NMD is based on empirical data on dispersal capacity of species, and reflects the distance below which about 90% of all dispersal movements take place (Verboom & Pouwels 2004). We ignore rare, long distance dispersal events, as these will contribute little to the equilibrium dynamics of a metapopulation (Opdam *et al.* 2003). The NMD is a Euclidian distance, i.e. the distance between two points in a straight line ('as the crow flies'). Transport barriers, urban areas or other forms of land use may, however, force a species to detour or decrease the permeability of the landscape for a species. Therefore, in LARCH, landscape permeability is mapped and the NMD is ecologically scaled with the use of cost-distance analysis (provided as a standard function in GIS packages), in which the Euclidian distance between habitat patches is

modified into an 'effective distance' (Adriaensen *et al.* 2003; Opdam *et al.* 2003). If a barrier such as a built-up area forces the animals to detour, effective distances increase and with them the chances that the habitat patches will be part of different habitat networks (Chardon *et al.* 2003; Verbeylen *et al.* 2003).

LARCH differentiates between absolute and partial barriers. Absolute barriers can only be passed by an outflanking movement. If such outflanking movements are not possible, e.g. when an absolute barrier has a linear shape (such as transport corridors with high-traffic volumes), habitat patches at opposite sides of the barrier will be part of different habitat networks. Partial barriers will only reduce the exchange of individuals to a certain extent. Partial barriers are modelled in LARCH by calculating expected dispersal events between habitat patches. Such dispersal events are dependent on patch quality, patch size, distance between patches and permeability of the landscape matrix. High resistance of the landscape matrix will reduce the dispersal between patches and hinder population dynamic processes in the metapopulation. Such high resistance may be caused by the presence of unsuitable habitat or land use types that are usually avoided by the species, as well as the presence of transport corridors that will inhibit a proportion of individuals from dispersal crossing. At a certain threshold level dispersal between two habitat patches will be too low to consider the two patches as part of one habitat network. Unfortunately, values for such a threshold have not been empirically assessed yet. In LARCH we thus choose a practical approach and used a conservative threshold, defined as the expected number of dispersal events between a key patch (KP) and one-tenth of a KP when the distance between the patches is equal to the NMD and the landscape matrix has no resistance. Here a KP is defined as a patch in the habitat network large enough for a key population, i.e. a relatively large local population which is viable under the condition of one immigrant per generation.

As mentioned before, the resistance of transport corridors is species-specific: what may be an impregnable obstacle for one species may be easy to pass for another. Furthermore, the resistance of transport corridors may be different for individuals that move within their home-range and individuals on dispersal. That is why a transport corridor may delineate the local population but not the habitat network or metapopulation. The presence of mitigation measures, such as wildlife-crossing structures, will affect the assessment of habitat networks. Because dispersal movements are assumed to be less affected by territorial behaviour of individuals that inhabit habitats around transport corridors, such measures are expected to reduce the barrier effect of transport corridors effectively, and cause the merging of patches on opposite sides of the transport barrier into one habitat network.

3.2.5 *Step 5: Population viability analysis*

After habitat networks have been delineated, LARCH calculates total carrying capacity of each network and compares these with thresholds for minimum viable metapopulations (MVMPs) in order to determine whether a habitat network is expected to be sustainable or not. An MVMP is here defined as a metapopulation with an extinction probability of exactly 5% in one hundred years. Thresholds for MVMPs (expressed in RUs) are species specific and depend on the configuration of the metapopulation: when the metapopulation includes a KP, the threshold for an MVMP can be considerably lower than in a case where no KP is present (Verboom *et al.* 2001). Three possible situations can be described in which a habitat network is expected to be sustainable: (1) if one or more patches in the network is large enough to support a minimum viable population (MVP), (2) if one or more patches in the network is large enough to be a KP and total carrying capacity of the network is high enough to provide sufficient immigrants in the KP, (3) if no patches large enough for an MVP or KP are present, but total carrying capacity in the network is great enough to compensate for the high degree of fragmentation (see also Opdam *et al.* 2003; Verboom & Pouwels 2004).

Standards for sustainability of habitat networks i.e. thresholds for MVMPs, MVPs and KPs have been derived from empirical data on population turnover and presence/absence data, as well as model simulations in order to determine how much habitat is needed for sustainable networks in different habitat configurations (for a detailed description, see Verboom *et al.* 2001). Small, medium-sized and large animals have different thresholds. LARCH simply compares the calculated total carrying capacity of each habitat network with these standards, including the specific configuration of each habitat network, e.g. whether a network includes a MVP or KP or not. If the habitat network carrying capacity is between 1 and 5 times the standard for a MVMP, the metapopulation is considered viable (extinction probability 1-5% in 100 years). If the carrying capacity of the network is five times or more the standard for a MVMP the metapopulation is considered highly viable (extinction probability <1% in 100 years). Non-viable metapopulations have an extinction probability >5% in 100 years.

3.2.6 *Step 6: Identification of potential de-fragmentation locations*

Transport corridor stretches where a significant shift in population viability is predicted solely as a result of the construction of wildlife-crossing structures are all identified as de-fragmentation locations. What is meant by 'a significant shift' is a matter of definition. We suggest that all locations should be classified as de-fragmentation locations where population viability would shift from non-viable to either viable or highly viable, and also where population viability would shift from viable to highly viable. After eliminating the barrier effect of transport corridors, the exchange of individuals between habitat networks will be improved. Habitat networks that were originally separated by the transport corridor will form larger and more sustainable habitat networks in the new, de-fragmented situation.

In some situations shifts in population viability can be achieved in different ways, i.e. by restoring habitat connectivity across different transport corridors. Figure 3a shows a habitat patch (A) of a certain species that is too small to support a viable population, as well as two habitat patches (B and C) that are large enough for a persistent population. Although the dispersal capacity of the species is great enough to connect patch A into a habitat network with patch B and/or C, the presence of transport corridors, here assumed to be absolute barriers, inhibit the exchange of animals, hence keeping the population in patch A isolated. The construction of a wildlife-crossing structure at the transport corridors will link patch A with patch B and patch C respectively (Figure 3b). After restoring linkage A-B as well as linkage A-C, population viability in patch A will shift from non-viable to viable. Constructing a wildlife-crossing structure at both locations, however, would be unnecessary because viability in patch A is already reached by restoring a connection with one of its neighbouring habitat patches. Hence a choice has to be made at what location a wildlife-crossing structure is likely to be most effective. We suggest such choices should be based on the differences in habitat connectivity, i.e. differences in expected exchange rate of animals between patches. Such habitat connectivity estimates are based on (1) dispersal capacity of the species, (2) the size of the source populations i.e. populations where individuals emigrate, (3) the distance between patches, and (4) the resistance of the intermediate landscape, including the presence of linear barriers such as transport corridors. Habitat connectivity is high if the dispersal capacity of the species is high, the source populations are large (i.e. the number of individuals on dispersal is high, under the assumption that this number is a fixed proportion of a population), the distance between patches is small, and the resistance of the intermediate landscape is low. In our example, dispersal capacity is not different because we compare the two wildlife-crossing locations for the same species. Also the resistance of the intermediate landscape is identical, assuming that the characteristics of both transport corridors and the landscape matrix are similar at both locations. However, the distance between the habitat patches as well as the size of the source populations differs. Patch B is closer to patch A than patch C. Furthermore, patch B is larger than patch C. Both aspects result in a higher connectivity, i.e. a higher number of dispersal events, between patch A and B than between patch A and C (Figure 3c). Linking A and B by constructing a wildlife passageway at the transport corridor that is between them will therefore result in the highest exchange rate of individuals between patches. Consequently, this location should be identified as a de-fragmentation location to improve population viability in patch A.

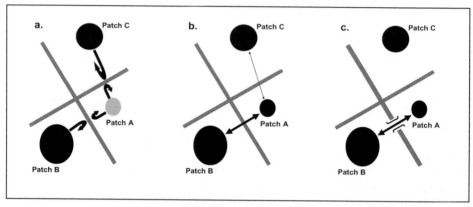

Figure 3. Differences in habitat connectivity can be used to identify the best location for de-fragmentation if population viability in a habitat patch can be improved in more than one way. a. Habitat patches (A, B, and C) in a fragmented landscape. The population in patch A is non-viable (*grey*). Populations in patches B and C are both highly viable (*black*). The transport corridors (*dark grey*) form an absolute barrier for migrating animals, represented by the *black arrows*. b. Connecting habitat patch A with either B or C into a habitat network will result in a shift from a non-viable towards a highly viable population in patch A. c. A wildlife-crossing structure will be most effective between habitat patch A and B because the expected degree of exchange (i.e. habitat connectivity) between these patches, represented by the size of the arrows in b, is higher than the expected dispersal events between A and C.

3.2.7 *Step 7: Setting priorities*

The next step is to prioritise actions for the identified de-fragmentation locations. We suggest two steps to set such priorities: (1) assess whether there is a direct or indirect change in population viability, (2) assess the ecological profit for each location where a shift in population viability will occur, i.e. the extent in which population viability will improve.

At some locations mitigation measures immediately result in a shift in population viability, regardless of mitigation measures elsewhere. At other locations such shifts can only be achieved if mitigation measures have been taken elsewhere first. This difference allows for a first prioritisation of locations, with locations that cause an immediate shift in population viability receiving the highest priority. Figure 4 provides an example. Three habitat networks of a certain species are separated from each other by two transport corridors. The threshold for a viable population is, in this example, assumed to be 100 RU. In the fragmented situation (Figure 4a) only one of the habitat networks is large enough to provide habitat for a viable population. The others, although considerably different in size (50 and 25 RU), are both too small to sustain independently a viable population. Mitigation measures at location 1 will immediately result in a shift in population viability in the central habitat network, because this habitat network will merge with the already viable habitat network on the left (Figure 4b). On the other hand, mitigation measures at location 2 will not result in such a shift in population viability, because, even after restoring the exchange between the central habitat network and the habitat network on the right, the threshold for a minimum viable population is not met: the newly established network provides habitat for only 75 RU (Figure 4c). If mitigation measures, however, are first constructed at location 1, resulting in a shift towards viability in the central habitat network, and subsequently measures are taken at location 2, population viability in the habitat network on the right can also be reached (Figure 4d).

The ecological profit can be expressed in the number of RUs (i.e. breeding pairs, social groups; see Step 2: Habitat modelling) that benefit from the shift in population viability due to de-fragmentation measures. The larger the number of RUs that benefit from de-fragmentation measures, the higher the priority of the location. A shift in population viability may occur in one of the habitat networks between which connectivity is restored or in both habitat networks. In the case of the latter the ecological profit is the summation of RUs of both networks (Figure 5). The number of priority classes one may distinguish and the method to define these classes is a matter of choice and primarily dependent on how detailed priorities are chosen to be set. Priority classes may be defined by numerical classifications such as spreading all de-fragmentation locations evenly over a certain number of classes (i.e. quartiles in case of four priority classes), or one may choose to define the priority classes by ecologically relevant thresholds. An example of the latter is when e.g. thresholds for KPs or MVMPs are used to divide priority classes.

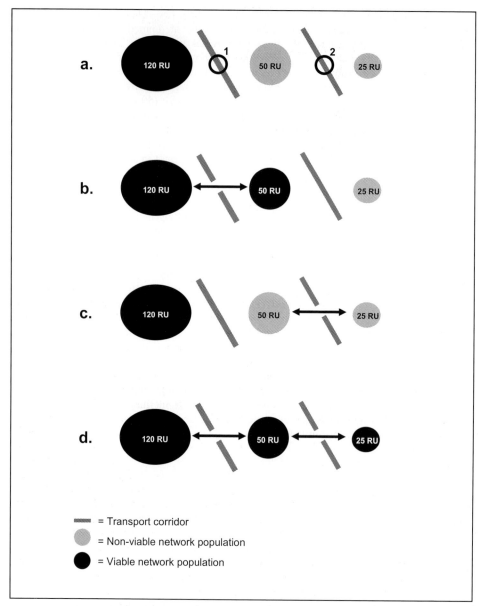

Figure 4. The first step in setting priorities for de-fragmentation locations: assessment whether a shift in population viability is reached independently of mitigation measures elsewhere or not. a. Fragmented situation; location 1 and 2 are identified de-fragmentation locations. Population size in each habitat network given in RUs. Threshold for viability: 100 RU. b. Population viability in case of de-fragmentation measures at location 1. c. Population viability in case of de-fragmentation measures at location 2. d. Population viability in case of de-fragmentation measures at both locations.

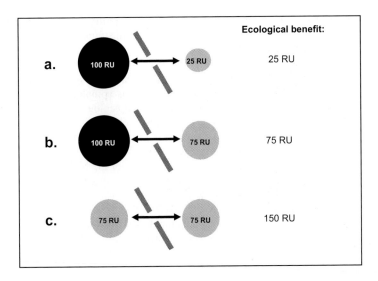

Figure 5. The second step in setting priorities for de-fragmentation locations: assessment of the ecological profit (in RU) for each location where a shift in population viability will occur. The threshold for viable network populations (*black*) is, in this example, assumed to be 100 RU. Non-viable network populations in *grey*. The ecological profit can be a result of a population shift in one of the habitat networks (a and b) or in both habitat networks (c). The ecological profit is highest in situation c, which should consequently receive the highest priority in de-fragmentation programmes.

4. Case study: De-fragmentation of transport corridors in The Netherlands

The method presented above was used to identify high-priority locations for de-fragmentation at existing transport corridors (roads, railroads and canalized waterways) in The Netherlands (Van der Grift *et al.* 2003b). The research was initiated by the national government in order to compose a long-term programme for de-fragmentation that would help identify and prioritise problem locations in a more standardized way. Several bottleneck analyses had already been published that addressed conflict points between ecological networks and human transport networks, but with widely varying approaches and scales of detail. This hindered a good comparison of the outcome of these studies and made an objective prioritisation of de-fragmentation locations impossible. Furthermore, in most studies the ecological benefit of mitigation measures was not quantified, making it hard for decision-makers in governmental agencies to decide where the available money could be best invested. Another reason for composing such a long-term programme for de-fragmentation was the understanding that de-fragmentation will be most successful if a regional planning approach is chosen in which different modes of human transport are included. In a highly fragmented country such as The Netherlands, de-fragmentation at one location will, after all, have only little ecological benefit if no measures are taken at neighbouring infrastructural barriers. By addressing the problem of habitat fragmentation simultaneously for main roads, railroads and canalized waterways, it was believed that de-fragmentation and biodiversity policy goals could be reached in a more effective way.

The objective in the Dutch study was not to assess de-fragmentation locations with the present configuration of habitat as starting point, but to focus on the future situation (expected in 2018) in which the proposed National Ecological Network (NEN) will have been completed. With this approach the long-term programme for de-fragmentation would address not only locations that can be considered bottlenecks at present, but also locations that are expected to affect population viability of target species when the NEN is completed. The assessment of de-fragmentation locations was done with a selection of ten ecological profiles, each named after one conspicuous species within the group of species it represents. The selected ecological profiles represent wetlands as well as terrestrial ecosystems that are known to be most affected by fragmentation due to transport corridors (Figure 6). With this selection of ecological profiles about 75% of the Dutch NEN is covered. Most parts of the NEN that are not covered consist of natural areas that provide habitat for species that are, in general, believed to be less susceptible to the barrier effect of transport corridors. The selected ecological profiles differ in habitat area requirements and dispersal capacity, and are thus tuned to both the regional and national scale of research. Figure 7 illustrates the output of the different research steps for the ecological profile 'slow worm'. Figure 8 shows two ways in which the results of the ecological profile analyses were aggregated, in order to prioritise de-fragmentation actions.

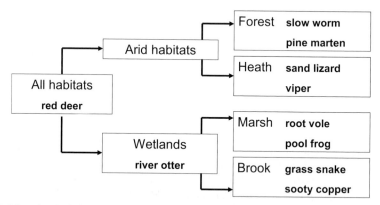

Figure 6. Selected ecological profiles and the habitat types they represent. Two ecological profiles represent more than one habitat type: the otter profile represents both marshland and brook habitat, while the red deer profile represents all four habitat types.

Figure 7. (next page). Example of de-fragmentation location analyses with LARCH for the ecological profile 'slow worm' in the central parts of The Netherlands. a. Habitat map. Differences in colour represent differences in habitat suitability, i.e. from moderately suitable habitat (*light green*) to optimal habitat (*dark green*). b. Spatial configuration of local populations: habitat too small for a local population (*pink*), small populations (*light green*), key populations (*green*), and minimum viable populations (*dark green*). c. Landscape permeability, based on resistance of land use patterns in between habitat patches and infrastructural barriers, ranging from low permeability (*light green*) to high permeability (*dark green*). d. Habitat networks: each colour represents a different network. e. Population viability in the situation with transport corridors as barriers: habitat too small for a local population (*pink*), non-viable populations (*red*), viable populations (*green*), highly viable populations (*dark green*). f. Population viability in the situation without transport corridors as barriers, i.e. the situation in which all fragmentation effects of transport corridors are mitigated: habitat too small for a local population (*pink*), non-viable populations (*red*), viable populations (*green*), highly viable populations (*dark green*). g. Identified de-fragmentation locations, i.e. locations where a shift in population viability from non-viable to viable or highly viable, or from viable to highly viable occurs due to mitigation measures. h. Priority to address each de-fragmentation location, based on differences in ecological benefit: low (*yellow*), moderate (*orange*), medium (*red*), above medium (*blue*), and high priority (*black*). In g and h the proposed NEN is in *green*.

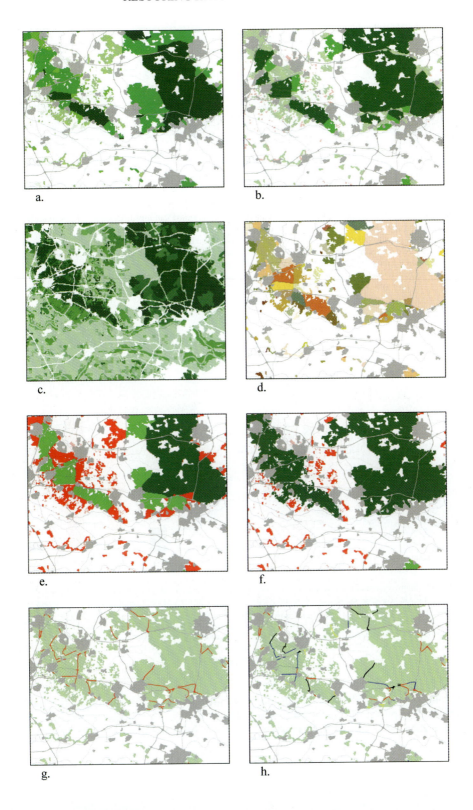

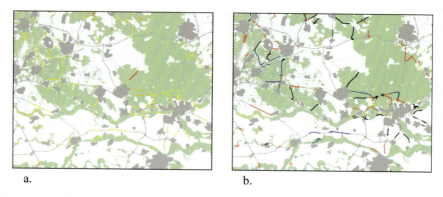

a.   b.

Figure 8. To prioritise de-fragmentation actions the results of population viability analyses of species or ecological profiles can be aggregated in several ways. The maps show the central parts of The Netherlands. a. De-fragmentation locations are classified by the number of ecological profiles (1 ecological profile = *yellow*, 2 ecological profiles = *orange*, 3 ecological profiles = *red*) for which the location is identified as de-fragmentation locations. b. De-fragmentation locations are classified by the ecological benefit of a location, based on the ecological profile with the highest ecological benefit at that location (small benefit = *yellow*, small to medium benefit = *orange*, medium benefit = *red*, high benefit = *blue*, very high benefit = *black*). The proposed NEN in *green*.

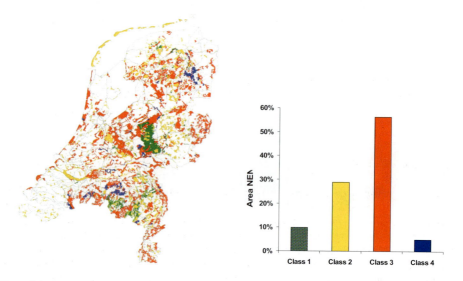

Figure 9. Improvement of network population viability for the selected ecological profiles if habitat linkages have been restored at all identified de-fragmentation locations. Class 1 (*green*): even without de-fragmentation measures populations are highly viable. Class 2 (*yellow*): despite de-fragmentation measures no improvement in population viability can be expected. Class 3 (*red*): improvement of population viability from non-viable to viable or highly viable, or from viable to highly viable for one or two ecological profiles. Class 4 (*blue*): improvement of population viability from non-viable to viable or highly viable, or from viable to highly viable, for three to five ecological profiles.

At many locations in The Netherlands, de-fragmentation of transport corridors will result in a significant improvement in wildlife population viability. Our analyses resulted in the identification of 1126 de-fragmentation locations. Most de-fragmentation locations are identified at roads, both in absolute numbers and per unit road length (Figure 10). At about 75% of these locations, mitigation measures will result in an immediate increase in population viability of one or more ecological profile. At the other locations similar results may be achieved, but only if other de-fragmentation locations are addressed first. In 23% of the cases, a de-fragmentation location is a bottleneck for two or more ecological profiles at the same time. The maximum number of ecological profiles for which one location is identified as bottleneck is five. Almost a quarter of all identified de-fragmentation locations are classified high-priority for the construction of wildlife passages and restoring habitat connectivity. If all de-fragmentation locations are successfully addressed, population viability in more than 60% of all wildlife habitat within the proposed NEN of The Netherlands will be improved for one or more ecological profiles (Figure 9).

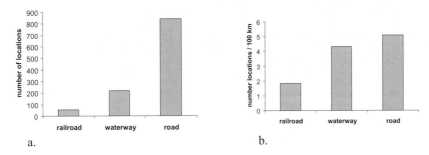

Figure 10. Number of identified de-fragmentation locations at existing roads, railroads and canalized waterways in The Netherlands. a. Total number of de-fragmentation locations. b. Average number of de-fragmentation locations per 100km of transport corridor.

5.   Combination of methods

The method of identifying high-priority de-fragmentation locations presented in this chapter could be usefully combined with some of the earlier applied methods to assess the best locations for wildlife-crossing structures. Moreover, we believe that it would be essential to use a combination of methods in order to be able to identify specific sites for the construction of mitigation measures. Our method results in de-fragmentation locations at the scale of road, railroad or canal *stretches*, which may differ in length from a few hundred metres to many kilometres. Such stretches are usually delineated by intersections with other transport corridors on either side, or by points where the characteristics of the infrastructural barrier change in such a way that it no longer significantly affects population viability. It will not usually not be possible to restore habitat connectivity over the full length of the identified de-fragmentation stretch, although some examples occur in which transport corridors are constructed in tunnels or on viaducts over large distances to prevent fragmentation of wildlife habitats (Keller *et al.* 2003). Before installing wildlife overpasses or underpasses, a further evaluation of de-fragmentation locations is needed in which knowledge of the local situation is used, e.g. knowledge of the distribution and movement patterns of species, the landscape, and the characteristics of the transport corridor.

We propose a framework encompassing five components:

1. Assessment of potential de-fragmentation locations in which significant shifts in population viability due to the removal of (existing or planned) transport barriers are used as an indicator, as presented in this chapter. The identified de-fragmentation locations consist of transport corridor stretches that may be up to many kilometres long.

2. Selection of de-fragmentation stretches where habitat connectivity i.e. the expected amount of dispersal between two habitat networks is highest. This step applies when significant shifts in population viability in a habitat network can be reached in alternative ways (see Section 4.6 and Figure 3).

3. Specification of mitigation sites within high-connectivity stretches with the use of detailed knowledge about pre-construction or cross-barrier movements, i.e. data on wildlife movements and successful or unsuccessful crossing sites. Where clearly defined animal trails exist, or locations can be identified where animals frequently cross the road or fall victim to vehicle-animal collisions, wildlife-crossing structures should be placed as close to them as possible.

4. Further specification of mitigation sites by landscape features. Differences in landscape patterns and/or land use may increase or decrease the permeability of the area for a species and therefore the suitability of a site for restoring a habitat linkage. In particular, the presence of suitable habitat and linear natural elements (e.g. stream, hedgerow, woodland) that are intersected by transport corridors may be used to specify the placement of mitigation measures at a local scale.

5. Fine tuning of favourable locations for crossing structures based on physical characteristics of the transport corridor itself. Technical characteristics of transport corridors (e.g. vertical alignment, width of verges, presence of overpasses or underpasses that can be retrofitted for co-use by wildlife) should be included in decision-making about mitigation measure placement in order to reduce costs and allow for optimisation of crossing structure design and dimensions.

Our framework is primarily applicable in situations where de-fragmentation locations need to be appointed and prioritised at existing transport corridors in highly fragmented landscapes. In landscapes where there is little fragmentation, the removal of infrastructural barriers will be less likely to result in a significant shift in population viability because most habitat networks are already large enough to provide habitat for viable populations (see also Jaeger & Holderegger 2005). De-fragmentation, however, may still be considered necessary, e.g. to preserve wildlife movement patterns and habitat use, prevent unnatural mortality due to wildlife-vehicle collisions, or simply because of a zero-tolerance policy for any fragmentation of pristine wildlife habitat. Caution is needed if the method is applied in evaluations of proposals for new transport corridors. Construction of new transport corridors in little fragmented landscapes is not likely to result in a significant change in population viability. Straightforward application of our method is then undesirable because habitat fragmentation will not be considered a problem, i.e. will not result in the identification of de-fragmentation locations, until the threshold for a minimum viable population is reached. Similarly, new transport corridors in highly fragmented landscapes may not change population viability because populations often do not meet the threshold for a minimum viable population to begin with, hence non-viable populations will still be non-viable after the construction of a new road. Consequently, a better approach when constructing new transport corridors is to choose a 'no-net-loss' principle: post-construction habitat connectivity should be similar to pre-construction connectivity levels. Both in cases of evaluations of existing transport corridors in little fragmented landscapes and evaluations of proposals for new transport corridors, steps 2 to 5 of our framework can be used to identify favourable locations for the construction of wildlife-crossing structures.

Theoretically one wildlife-crossing structure at an identified de-fragmentation location, with the assumption that it is within reach of the populations on either side of the transport corridor, is sufficient to restore dispersal events across the transport corridor for the species studied, thus reducing the chance of local extinction and increasing the probability of recolonisation if local extinction occurs. One crossing structure may also be sufficient to facilitate genetic exchange between populations on both sides of a transport corridor in order to prevent loss of genetic variation. Only a small number of individuals that exchange between populations is needed to provide gene flow and increase the levels of genetic variation (Mills & Allendorf 1996; Vucetich & Waite 2000). However, knowledge is still limited of the effect concerning wildlife-crossing structures on colonisation probabilities for most species. Furthermore, improving colonisation probabilities and ensuring genetic exchange between populations are not the only reasons for installing de-fragmentation measures. Transport corridors may block random mating within local populations and change habitat use patterns which may result in less frequently used areas and consequently in a change of ecosystems. Reduced habitat connectivity for large herbivores, for example, will affect grazing pressure and accordingly vegetation development. Therefore the number of habitat linkages, i.e. the number of wildlife-crossing structures, needed at one de-fragmentation stretch should be geared to the level of habitat cohesion or ecosystem restoration aimed for, based on knowledge about the ecology of species.

6. Future challenges in restoring habitat connectivity

6.1 INTEGRATION OF PVA IN TRANSPORT PLANNING

To avoid future conflicts between ecological networks and transport networks, population viability analysis should be an integrated part of all transport planning. Without an elaborate study of what the impact of (future) transport corridors will be on the level of wildlife populations, no reliable conclusions about fragmentation impacts can be drawn. After all, even if not a single habitat patch is touched by a proposed road or railroad, it does not necessarily mean that populations will not be affected. If the exchange between habitat patches is partly blocked by transport corridors, the balance between local extinctions and recolonisations may be disturbed and the chances that, in the long run, the metapopulation, or a significant part of it, become extinct will increase. De-fragmentation may not always improve population viability. Moreover, when a "population sink" is added to the habitat network, it may even cause a decline in viability. The routing of transport corridors and the placement of de-fragmentation measures should therefore never be solely based on static geographic data for nature areas, habitat patches or species distributions, but should be based on the ecological processes that connect these areas. Spatially explicit analyses of population viability can thus be viewed as indispensable for careful comparisons of routing alternatives for proposed transport corridors.

6.2 PLANNING RESEARCH IN EARLY STAGES

It is of utmost importance to plan de-fragmentation well ahead of any rail/road construction. As argued in this chapter, placement of de-fragmentation measures should preferably be based on pre-construction surveys of wildlife migration routes. Such surveys take time. A survey period of one year is the minimum because it includes possible seasonal variation in migration patterns. However, it is better to conduct surveys over several years because animal movements may vary over the years. Similar recommendations can be made in cases of the up-grading of existing transport corridors where pre-extension surveys of both successful and unsuccessful crossings can provide valuable information for the fine-tuning of crossing structure placement. Obviously, it demands an early planning of inventory research, something that is so far the exception rather than the rule in the planning of transport corridors. The advantages of such an approach are two-fold: it enables better identification of de-fragmentation locations at the transport corridor under study, and results in better and more effective assessments of de-fragmentation locations in future projects because it allows validation of GIS-models, such as LARCH, that predict the best locations for wildlife-crossing structures. Furthermore it is always cheaper to construct de-fragmentation measures simultaneously with the construction of the transport corridor itself.

6.3 DISSEMINATION OF KNOWLEDGE AND BEST-PRACTICES

One of the main challenges ahead of us is to avoid the mistakes we have made in the past. Most knowledge and technologies for counteracting the impacts of habitat fragmentation due to transport corridors have been developed in Western industrial countries in which the problem of fragmentation reached such proportions that biodiversity seriously started to decline and species became extinct. In such countries, money and public awareness were available to initiate improvements to landscape cohesion and restore habitat linkages across transport corridors. It is different in countries that are still less developed economically. More limited financial resources, less explicit public awareness and support for environmental issues, and the presence of large natural areas usually make the need for immediate action less obvious. Economic growth and improvement of social welfare are usually higher on the political agenda than environmental protection or nature conservation. In countries that can be considered 'runners-up' on the economic ladder, significant efforts are put into planning extensions of human transport networks, often aimed at attracting foreign financial support. An example is the development of Pan-European Transport Corridors as part of the Trans-European Transport Network, which will result in a significant extension of the road and railroad network in Eastern-Europe (European Communities 2002). If, simultaneously, no measures are planned to prevent habitat fragmentation, it will not be long before these countries have the same fragmentation problems as experienced in Western countries. Therefore, national policies for de-fragmentation should be developed by these countries and implemented in spatial and transport planning. Ecological consequences of transport corridor construction should be recognised and comprehensively described at early stages of the planning process. If fragmentation of wildlife habitats is expected, alternatives should be explored and evaluated. The emphasis should be on prevention of habitat fragmentation: mitigation measures, such as the construction of wildlife-crossing structures, should only be viewed as options if impacts cannot be avoided. The same applies to compensatory measures, such as the development of a habitat elsewhere that is similar in size and quality to the habitat to be lost due to the construction of a transport corridor. Within Europe, the flow of money from west to east should be accompanied by a flow of preconditions to safeguard biodiversity as well as a flow of knowledge, best-practices, tools and technologies that enable policy-makers and transport planners to prevent loss of habitat connectivity and better foster the development of a more environmental-friendly human transport network.

6.4 CONSTRUCTING MORE ROBUST MITIGATION MEASURES

Another challenge we face is to raise awareness, and underpin the need for more robust de-fragmentation measures. The design and dimensions of wildlife-crossing structures should preferably not be determined by the needs of one target species but by the objecti\ connecting habitats at an ecosystem level. Instead of basing the size of a crossing structur the minimum requirements of a species to be physically able to accept and use the passageway, one should aim for dimensions that result in a 'no-change-in-behaviour'- passageway for a multitude of species. Hence, instead of narrow wildlife overpasses up to a few dozen metres wide, on which space allows for the development of only one or two habitat types and the presence of the transport corridor and its traffic is always obvious, the construction of so-called landscape bridges or landscape connectors should be more seriously considered, which include transport corridors between 80 and several hundred metres wide (Kramer-Rowold & Rowold 2001; Forman *et al.* 2003; Iuell *et al.* 2003). Landscape bridges link habitats on such a scale that migrating animals no longer notice that they are crossing a transport corridor. This is of special importance for species with low dispersal capacity, i.e. species that may need several generations to cross the full width of a transport corridor and therefore depend on a continuous habitat corridor on the crossing structure. Landscape bridges also allow other horizontal ecological flows (e.g. of nutrients, pollen or seeds). Furthermore, landscape bridges can be viewed as valuable from a recreational point of view. Small wildlife- crossing structures shared by hikers, cyclists or horse riders are of debatable benefit because of the negative impacts of human use on the acceptance and use of the passageway by wildlife. More robust crossing structures allow human co-use and provide better opportunities to establish not only an ecological network across transport corridors, but also a recreational network.

6.5 ACQUIRING PUBLIC UNDERSTANDING AND SUPPORT

Finally, a serious challenge is to create public support for de-fragmentation initiatives. The issue of habitat fragmentation by transport corridors may not always receive much understanding or acceptance. Although in general people may support ideas and initiatives to preserve natural areas and to improve opportunities for nature-oriented leisure activities, specific plans may still be viewed as a threat to acquired 'rights' or personal interests. An example is the protest by a small group of citizens in The Netherlands who raised objections to the construction of a wildlife overpass that would connect two nature reserves across a provincial road, railroad, and railroad construction site. Their main fear was that the construction of the railway overpass, on which pet dogs would not be allowed, would reduce the area adjacent to the planned overpass in which dogs were allowed to run freely. It so happens that in most of these nature areas dogs are only allowed on a leash to prevent disturbance of deer populations and semi-domestic cattle that graze the area as a form of nature management. Although the fears were a little too hasty - rearrangement of the reserved area for free running dogs had always been planned - it well illustrates how relatively small personal interests may incite people to object a project that will clearly have a positive impact on the quality of life.

Another often-heard complaint is the cost of de-fragmentation projects, especially if more robust measures are taken, such as wildlife overpasses or landscape bridges. Usually, money needed to construct wildlife-crossing structures or other measures that prevent impacts on landscape and natural values is still a very small percentage of the total budget for construction of a transport corridor. Nevertheless, people's perception may be that money is spent carelessly to save 'a few rabbits'. Such perceptions often develop because the general public is not well informed about the ecological impacts related to transport corridors, and the necessity to offer solutions such as wildlife-crossing structures to maintain biodiversity.

To avoid these misconceptions, open communication and active public involvement, at early stages of a project, is of decisive importance, especially because, in highly urbanised and densely populated countries, the construction of de-fragmentation measures is likely to demand more and more changes in land use. Frequently, agricultural land, industrial sites or even residential areas cannot be avoided in an attempt to (re)connect wildlife habitats. In the heart of a nature conservation area in the central parts of The Netherlands, for example, a provincial road, together with the built-up areas on both sides of the road – private houses, small industries and health institutions – is blocking all wildlife movements between the northern and southern parts of the nature area. Research has shown that de-fragmentation measures are necessary to prevent wildlife populations from extinction in the long run, and a wildlife overpass has been proposed (Van der Grift 2004). However, such a passage for wildlife can only be created if existing land uses disappear, i.e. if houses or industrial buildings are re-located. Such rigorous changes in spatial planning can only be justified to the public if the action is of major public interest and no alternatives with less impact are available. It not only demands early communication of plans with the public, but also emphasises the importance of providing convincing arguments for the sense and necessity of a proposed de-fragmentation measure. More frequently, specific arguments will be required to justify why a wildlife-crossing structure is needed and why the proposed location was chosen. Providing insight into the effects on wildlife population viability of decisions to construct de-fragmentation measures, as argued and illustrated in this chapter, will then be an indispensable tool that may be of greatest importance to the completion of planned de-fragmentation programmes.

References

Adriaensen F, Chardon JP, De Blust G, Swinnen E, Villalba S, Gulinck H, Matthysen E. 2003; The application of 'least-cost' modelling as a functional landscape model. Landscape and Urban Planning 64: 233-247.
Bakker J, Knaapen JP, Schippers P. 1997; Fauna dispersal modelling: a spatial approach. In: Canters KJ, Piepers AAG & Hendriks-Heersma D (eds) Habitat fragmentation & infrastructure. Proceedings of the international conference on habitat fragmentation, infrastructure and the role of ecological engineering, 183-192. Road and Hydraulic Engineering Division, Delft, The Netherlands.
Barnum SA. 2003; Identifying the best locations to provide safe highway crossing opportunities for wildlife. In: Irwin CL, Garrett P & McDermott KP (eds) 2003 Proceedings of the International Conference on Ecology and Transportation, 246-252. North Carolina State University, Center for Transportation and the Environment, Raleigh, USA.
Baur A, Baur B. 1990; Are road barriers to dispersal in the land snail Arianta arbustorum? Canadian Journal of Zoology 68: 613-617.
Becker DM. 1996; Wildlife and wildlife habitat impact issues and mitigation options for reconstruction of US Highway 93 on the Flathead Indian Reservation. In: Evink GL, Garrett P, Zeigler D & Berry J (eds) Trends in addressing transportation related wildlife mortality. Proceedings of the Transportation Related Wildlife Mortality Seminar, 96-104. Florida Department of Transportation, Tallahassee, USA.
Bekker H, van den Hengel B, van Bohemen H, van der Sluijs H. 1995; Nature across motorways. Road and Hydraulic Engineering Division, Delft, The Netherlands.
Carey M. 2001; Addressing wildlife mortality on highways in Washington. In: Anonymous (eds) Proceedings of the International Conference on Ecology and Transportation, 605-610. North Carolina State University, Center for Transportation and the Environment, Raleigh, USA.
Carr MH, Zwick PD, Hoctor T, Harrell W, Goethals A, Benedict M. 1998; Using GIS for identifying the interface between ecological greenways and roadway systems at the State and sub-State scales. In: Evink GL, Garrett P, Zeigler D & Berry J (eds) Proceedings of the International Conference on Wildlife Ecology and Transportation, 68-77. Florida Department of Transportation, Tallahassee, USA.
Chardon JP, Adriaensen F, Matthysen E. 2003; Incorporating landscape elements into a connectivity measure: a case study for the speckled wood butterfly (Pararge aegeria L.). Landscape Ecology 18: 561-573.
Chruszcz B, Clevenger AP, Gunson KE, Gibeau ML. 2003; Relationships among grizzly bears, highways, and habitat in the Banff-Bow Valley, Alberta, Canada. Canadian Journal of Zoology 81: 1378-1391.
Clevenger AP, Waltho N. 2005; Performance indices to identify attributes of highway crossing structures facilitating movement of large mammals. Biological Conservation 121: 453-464.
Clevenger AP, Wierzchowski J, Chruszcz B, Gunson K. 2002; GIS-generated, expert-based models for identifying wildlife habitat linkages and planning mitigation passages. Conservation Biology 16 (2): 503-514.

Craighead AC, Craighead FL, Roberts EA. 2001; Bozeman Pass wildlife linkage and highway safety study. In: Anonymous (eds) Proceedings of the International Conference on Ecology and Transportation, 405-422. North Carolina State Univesity, Center for Transportation and the Environment, Raleigh, USA.

Dodd CK, Barichivich WJ, Smith LL. 2004; Effectiveness of a barrier wall and culverts in reducing wildlife mortality on a heavily travelled highway in Florida. Biological Conservation 118: 619-631.

European Communities. 2002; Trans-European Transport Network. TEN-T priority projects. Office for Official Publications of the European Communities, Luxembourg.

Forman RTT, Sperling D, Bissonette JA, Clevenger AP, Cutshall CD, Dale VH, Fahrig L, France R, Goldman CR, Haenue K, Jones JA, Swanson FJ, Turrentine T, Winter TC. 2003; Road ecology – Science and solutions. Island Press, Washington, USA.

Foster ML, Humphrey SR. 1995; Use of highway underpasses by Florida panthers and other wildlife. Wildlife Society Bulletin 23 (1): 95-100.

Gibeau ML, Clevenger AP, Herrero S, Wierzchowski J. 2001; Effects of highways on grizzly bear movement in the Bow River watershed, Alberta, Canada. In: Anonymous (eds) Proceedings of the International Conference on Ecology and Transportation, 458-472. North Carolina State Univesity, Center for Transportation and the Environment, Raleigh, USA.

Groot Bruinderink G, van der Sluis T, Lammertsma D, Opdam P, Pouwels R. 2003; Designing a coherent ecological network for large mammals in Northwestern Europe. Conservation Biology 17 (2): 549-557.

Hanski I, Gilpin M. 1991; Metapopulation dynamics: brief history and conceptual domain. Biological Journal of the Linnean Society 42: 3-16.

Hicks C, Peymen J. 2003; Habitat fragmentation due to existing transportation infrastructure. In: Trocmé M, Cahill S, de Vries JG, Farrall H, Folkeson L, Fry G, Hicks C & Peymen J (eds) COST Action 341 - Habitat fragmentation due to transportation infrastructure: The European review, 73-113. Office for Official Publications of the European Communities, Luxembourg.

Iuell B, Bekker GJ, Cuperus R, Dufek J, Fry G, Hicks C, Hlaváč V, Keller V, Rosell C, Sangwine T, Trøsløv N, le Maire Wandall B (eds). 2003; Wildlife and traffic: a European handbook for identifying conflicts and designing solutions. KNNV Publishers, Utrecht, The Netherlands.

Jaeger J, Holderegger R. 2005. Schwellenwerte der Landschaftszerschneidung. GAIA 14 (2): 113-118.

Jaeger JAG, Bowman J, Brennan J, Fahrig L, Bert D, Bouchard J, Charbonneau N, Frank K, Gruber B, Tluk von Toschanowitz K. 2005. Predicting when animal populations are at risk from roads: an interactive model of road avoidance behaviour. Ecological modelling 185: 329-348.

Kaczensky P, Knauer F, Krze B, Jonozovic M, Adamic M, Gossow H. 2003; The impact of high speed, high volume traffic axes on brown bears in Slovenia. Biological Conservation 111: 191-204.

Keller V, Bekker GJ, Cuperus R, Folkeson L, Rosell C, Trocmé M. 2003; Avoidance, mitigation and compensatory measures and their maintenance. In: Trocmé M, Cahill S, de Vries JG, Farrall H, Folkeson L, Fry G, Hicks C & Peymen J (eds) COST Action 341 - Habitat fragmentation due to transportation infrastructure: The European review, 129-173. Office for Official Publications of the European Communities, Luxembourg.

Klein L. 1999. Usage of GIS in wildlife passage planning in Estonia. In: Evink GL, Garrett P & Zeigler D (eds) Proceedings of the Third International Conference on Wildlife Ecology and Transportation, 179-184. Florida Department of Transportation, Tallahassee, USA.

Kobler A, Adamic M. 1999; Brown bears in Slovenia: identifying locations for construction of wildlife bridges across highways. In: Evink GL, Garrett P & Zeigler D (eds) Proceedings of the Third International Conference on Wildlife Ecology and Transportation, 29-38. Florida Department of Transportation, Tallahassee, USA.

Kohn B, Frair J, Unger D, Gehring T, Shelley D, Anderson E, Keenlance P. 1999; Impacts of a highway expansion project on wolves in northwestern Wisconsin. In: Evink GL, Garrett P & Zeigler D (eds) Proceedings of the Third International Conference on Wildlife Ecology and Transportation, 53-65. Florida Department of Transportation, Tallahassee, USA.

Kramer-Rowold EM, Rowold WA. 2001; About the effectiveness of wildlife passages at roads and railroads. Informationsdienst Naturschutz Niedersachsen 21 (1): 2-58. [in German]

Land D, Lotz M. 1996; Wildlife-crossing designs and use by Florida panthers and other wildlife in southwest Florida. In: Evink GL, Garrett P, Zeigler D & Berry J (eds) Trends in addressing transportation related wildlife mortality. Proceedings of the Transportation Related Wildlife Mortality Seminar, 323-328. Florida Department of Transporta-Transportation, Tallahassee, USA.

Lehnert ME, Romin LA, Bissonette JA. 1996; Mule deer-highway mortality in northeastern Utah: causes, patterns, and a new mitigative technique. In: Evink GL, Garrett P, Zeigler D & Berry J (eds) Trends in addressing transportation related wildlife mortality. Proceedings of the Transportation Related Wildlife Mortality Seminar, 105-112. Florida Department of Transportation, Tallahassee, USA.

Lovallo MJ, Anderson EM. 1996; Bobcat movements and home ranges relative to roads in Wisconsin. Wildlife Society Bulletin 24 (1): 71-76.

Mader H-J. 1984; Animal habitat isolation by roads and agricultural fields. Biological Conservation 29: 81-96.

Mader H-J, Schnell C, Kornacker P. 1990; Linear barriers to arthropod movements in the landscape. Biological Conservation 54: 209-222.

Malo JE, Suárez F, Díez A. 2004; Can we mitigate animal-vehicle accidents using predictive models? Journal of Applied Ecology 41: 701-710.

Mills LS, Allendorf FW. 1996; The one-immigrant-per-generation rule in conservation and management. Conservation Biology 10 (6): 1509-1518.

Munguira ML, Thomas JA. 1992; Use of road verges by butterfly and burnet populations, and the effect of roads on adult dispersal and mortality. Journal of Applied Ecology 29: 316-329.

Neal E, Cheeseman C. 1996; Badgers. T&AD Poyser Natural History, London, UK.

Opdam P. 1991; Metapopulation theory and habitat fragmentation: a review of holarctic breeding bird studies. Landscape Ecology 5: 93-106.

Opdam P, Foppen R, Vos C. 2002; Bridging the gap between ecology and spatial planning in landscape ecology. Landscape Ecology 16: 767-779.

Opdam PFM, Verboom J, Pouwels R. 2003; Landscape cohesion: an index for the conservation potential of landscapes for biodiversity. Landscape Ecology 18: 113-126.

Oxley DJ, Fenton MB, Carmody GR. 1974; The effects of roads on populations of small mammals. Journal of Applied Ecology 11: 51-59.

Paquet P, Callaghan C. 1996; Effects of linear developments on winter movements of gray wolves in the Bow River Valley of Banff National Park, Alberta. In: Evink GL, Garrett P, Zeigler D & Berry J (eds) Trends in addressing transportation related wildlife mortality. Proceedings of the Transportation Related Wildlife Mortality Seminar, Florida Department of Transportation, Tallahassee, USA.

Pfister HP, Keller V, Reck H, Georgii B. 1997; Bio-ecological functioning of green bridges across roads. Forschung Straßenbau und Straßenverkehrstechnik, Heft 756. Bundesministerium für Verkehr, Bonn-Bad Godesberg, Germany. [in German]

Reh E, Seitz A. 1990; The influence of land use on the genetic structure of populations of the common frog *Rana temporaria*. Biological Conservation 54: 239-249.

Rodríguez A, Crema G, Delibes M. 1996; Use of non-wildlife passages across a high speed railway by terrestrial vertebrates. Journal of Applied Ecology 33: 1527-1540.

Rost GR, Bailey JA. 1979; Distribution of mule deer and elk in relation to roads. Journal of Wildlife Management 43 (3): 634-641.

Ruediger B, Claar JJ, Gore JF. 1999; Restoration of carnivore habitat connectivity in the northern Rocky Mountains. In: Evink GL, Garrett P & Zeigler D (eds) Proceedings of the Third International Conference on Wildlife Ecology and Transportation, 5-20. Florida Department of Transportation, Tallahassee, USA.

Scheick BK, Jones MD. 1999; Locating wildlife underpasses prior to expansion of Highway 64 in North Carolina. In: Evink GL, Garrett P & Zeigler D (eds) Proceedings of the Third International Conference on Wildlife Ecology and Transportation, 247-251. Florida Department of Transportation, Tallahassee, USA.

Schippers P, Verboom J, Knaapen JP, van Apeldoorn RC. 1996; Dispersal and habitat connectivity in complex heterogeneous landscapes: an analysis with a GIS-based random walk model. Ecography 19: 97-106.

Servheen C, Waller J, Kasworm W. 1998; Fragmentation effects of high-speed highways on grizzly bear populations shared between the United States and Canada. In: Evink GL, Garrett P, Zeigler D & Berry J (eds) Proceedings of the International Conference on Wildlife Ecology and Transportation, 97-103. Florida Department of Transportation, Tallahassee, USA.

Simberloff D. 1998; Flagships, umbrellas, and keystones: is single-species management passé in the landscape era? Biological Conservation 83 (3): 247-257.

Singer FJ, Doherty JL. 1985; Managing mountain goats at a highway crossing. Wildlife Society Bulletin 13: 469-477.

Singleton P, Lehmkuhl JF. 1999; Assessing wildlife habitat connectivity in the Interstate 90 Snoqualmie Pass corridor, Washington. In: Evink GL, Garrett P & Zeigler D (eds) Proceedings of the Third International Conference on Wildlife Ecology and Transportation, 75-83. Florida Department of Transportation, Tallahassee, USA.

Singleton PH, Gaines W. 2001; Using weighted distance and least-cost corridor analysis to evaluate regional scale large carnivore habitat connectivity in Washington. In: Anonymous (eds) Proceedings of the International Conference on Ecology and Transportation, 583-594. North Carolina State Univesity, Center for Transportation and the Environment, Raleigh, USA.

Smith DJ. 1999; Identification and prioritization of ecological interface zones on state highways in Florida. In: Evink GL, Garrett P & Zeigler D (eds) Proceedings of the Third International Conference on Wildlife Ecology and Transportation, 209-229. Florida Department of Transportation, Tallahassee, USA.

Tremblay MA. 2001; Modeling and management of potential movement for elk (*Cervus elaphus*), bighorn sheep (*Ovis Canadensis*) and grizzly bear (*Ursus arctos*) in the Radium Hot Springs area, British Columbia. In: Anonymous (eds) Proceedings of the International Conference on Ecology and Transportation, 534-545. North Carolina State Univesity, Center for Transportation and the Environment, Raleigh, USA.

Van Bohemen HD, Bekker GJ, Veenbaas G. 2004; The fragmentation of nature by motorways and traffic, and its defragmentation. In: Van Bohemen HD (ed) Ecological engineering and civil engineering works. A practical set of ecological engineering principles for road infrastructure and coastal management, 109-157. Delft University of Technology, Delft, The Netherlands.

Van der Fluit N, Cuperus R, Canters K. 1990; Mitigation and compensation measures at main roads to improve ecological values, including elaborations for three Dutch regions. CML, Leiden, The Netherlands.

Van der Grift EA. 2004; Corridor Leusderheide. Benefit and need of the habitat linkage and recommendations for the design and placement of a wildlife overpass across Provincial Road N237. Alterra, Wageningen, The Netherlands.

Van der Grift EA. 2005; De-fragmentation measures in The Netherlands: a success story? GAIA 14 (2): 144-147.

Van der Grift EA, Pouwels R, Reijnen R. 2003b; Long-term programme for de-fragmentation – Bottleneck analysis. Alterra, Wageningen, The Netherlands. [in Dutch]

Van der Grift EA, Verboom J, Pouwels R. 2003a; Assessing the impact of roads on animal population viability. In: Irwin CL, Garrett P & McDermott KP (eds) Proceedings of the International Conference on Ecology and Transportation, 173-181. North Carolina State University, Center for Transportation and the Environment, Raleigh, USA.

Van Maanen FT, Jones MD, Kindall JL, Thompson LM, Scheick BK. 2001; Determining the potential mitigation effects of wildlife passageways on black bears. In: Anonymous (eds) Proceedings of the International Conference on Ecology and Transportation, 435-446. North Carolina State University, Center for Transportation and the Environment, Raleigh, USA.

Verbeylen G, De Bruyn L, Adriaensen F, Matthysen E. 2003; Does matrix resistance influence red squirrel (*Sciurus vulgaris* L. 1758) distribution in an urban landscape? Landscape Ecology 18: 791-805.

Verboom J, Foppen R, Chardon JP, Opdam PFM, Luttikhuizen PC. 2001; Introducing the key patch approach for habitat networks with persistent populations: an example for marshland birds. Biological Conservation 100 (1): 89-100.

Verboom J, Pouwels R. 2004; Ecological functioning of ecological networks: a species perspective. In: Jongman RHG & Pungetti G (eds) Ecological networks and greenways. Concept, design, implementation, 56-72. Cambridge University Press, Cambridge, UK.

Verboom J, Wamelink W. 2005; Spatial modelling in landscape ecology. In: Wiens JA & Moss MR (eds) Issues and perspectives in landscape ecology, 79-89. Cambridge University Press, Cambridge, UK.

Vos CC, Baveco JM, Chardon JP, Goedhart P. 1999; The role of landscape heterogeneity on the movement paths of the tree frog (*Hyla arborea*) in an agricultural landscape. In: Vos CC (ed) A frog's-eye view of the landscape. Quantifying connectivity for fragmented amphibian populations, 73-91. PhD-thesis. Wageningen University, Wageningen, The Netherlands.

Vos CC, Chardon JP. 1998; Effects of habitat fragmentation and road density on the distribution pattern of the moor frog *Rana arvalis*. Journal of Applied Ecology 35: 44-56.

Vos CC, Verboom J, Opdam PFM, Ter Braak CJF. 2001; Toward ecologically scaled landscape indices. The American Naturalist 183 (1): 24-41.

Vucetich JA, Waite TA. 2000; Is one immigrant per generation sufficient for the genetic management of fluctuating populations? Animal Conservation 3: 261-266.

Yanes M, Velasco JM, Suárez F. 1995; Permeability of roads and railways to vertebrates: the importance of culverts. Biological Conservation 71: 217-222.

CHAPTER 11: HABITAT AND CORRIDOR FUNCTION OF RIGHTS-OF-WAY

MARCEL P. HUIJSER & ANTHONY P. CLEVENGER

Western Transportation Institute – Montana State University, PO Box 174250, Bozeman, MT 59717-4250, USA.

1. Introduction

Roads, railroads and traffic can negatively affect plants, animals and other species groups (see reviews in Forman & Alexander 1998; Spellerberg 1998; van der Grift 1999). Transportation induced habitat loss, habitat fragmentation, reduced habitat quality and increased animal mortality can lead to serious problems for certain species or species groups, especially if they also suffer from other human-related disturbances such as large scale intensive agriculture and urban sprawl (Mader 1984; Ewing *et al.* 2005). Some species may even face local or regional extinction. However, other species or species groups can benefit from the presence of transportation infrastructure. Depending on the species and the surrounding landscape, the right-of-way can provide an important habitat or their only remaining functional habitat in the surrounding area. Rights-of-way may also serve as corridors between key habitat patches. The habitat and corridor function of rights-of-way can help improve the population viability of meta-populations of certain species in fragmented landscapes. This chapter aims to illustrate the habitat and corridor function of rights-of-way. While our focus is on roads, we include some examples of ecological benefits of railroads and railroad rights-of-way.

For this chapter we define the term "right-of-way" as the area between the edge of the road surface, which is usually asphalt, concrete or gravel, and the edge of the area that is not owned or managed by the transportation agency. In developed landscapes the latter usually coincides with a right-of-way fence and a change in land use, e.g. agricultural lands, gardens, or buildings (Figure 1). In undeveloped or less developed landscapes the far edge of the right-of-way usually coincides with a transition to native vegetation, e.g. native grasslands or forest (Figure 2). The vegetation in the right-of-way is usually disturbed as a result of road construction, alien soil, grading, seeding of native or non-native grasses and herbs to prevent soil erosion, trampling, dust and pollutants in air and water, mowing practices and the application of herbicides. The vegetated zone adjacent to the pavement is usually smooth and free from trees, shrubs, rocks or other large objects (e.g. higher then 10cm (4inches) above the ground) to allow drivers to regain control of their vehicle if they happen to run off the road. The width of this "clear zone" varies depending on the type of road, the speed limit and local conditions such as steep slopes that may not allow for the "ideal" clear zone width. However, in the United States the clear zone is often about 9-11m wide (about 30ft) (Forman *et al.* 2003). Depending on the climate, the clear zone may require regular mowing to prevent trees and shrubs from getting established. Regular mowing also allows drivers to see and read road signs. Furthermore, it improves the sight distance for drivers on the inside of curves and into the right-of-way so that they can see oncoming traffic, pedestrians or cyclists or large animals that may be present in the right-of-way (Rea 2003). A narrow zone immediately adjacent to the pavement may require a more intensive mowing regime, e.g. to prevent tall grasses and herbs from blocking reflectors that demarcate the road.

John Davenport and Julia L. Davenport, (eds.,) The Ecology of Transportation: Managing Mobility for the Environment, 233–254,

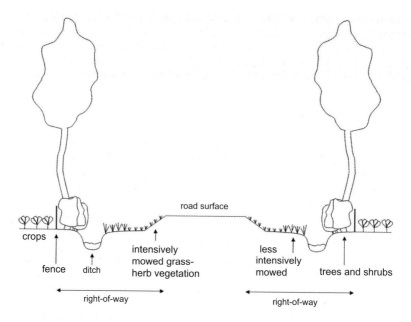

Figure 1. Cross-section of a road and its rights-of-way in a developed agricultural landscape.

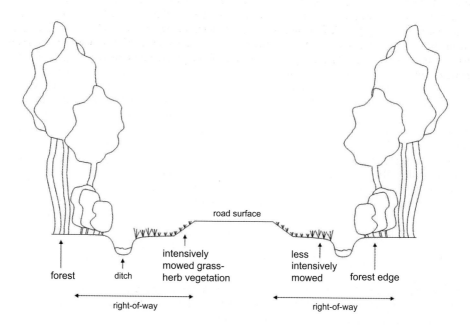

Figure 2. Cross-section of a road and its rights-of-way in an undeveloped forested landscape.

Rights-of-way vary greatly in width, but always run parallel to the road orientation by nature. Because most roads connect to other roads, right-of-ways can form extensive networks.

While there are many different types of rights-of-way and surrounding landscapes, we use the concept of rights-of-way in developed and undeveloped landscapes to illustrate different aspects of the habitat and corridor function of rights-of-way and how these may be valued.

2. Extent of road, railroad and rights-of-way networks

The density of the road and railroad network varies greatly between countries (Table 1). Not surprisingly, small and densely populated countries such as Belgium and The Netherlands have a relatively dense road and railroad network while larger countries with lower population density and vast regions with very low population density such as Russia and Australia have a relatively sparse road and railroad network. Nevertheless, the road density may be very high in densely populated regions of large countries as well, and the total length of road and railroad networks is enormous (Table 1).

The width of rights-of-way usually depends on the type of road, vehicle speed, and regional factors and may vary between just a few metres and a few hundred metres. For example, transportation corridors (road and rights-of-way combined) in some areas in Australia are up to 1,609m wide as they were originally designed for moving livestock from farms to water sources and towns (Spooner 2005a).

Estimates on the area covered by rights-of-way (excluding the road surface) are scarce, but they may cover 0.5-2.5% of a country or region's land area (Table 1). While these percentages seem small, they may be substantial when compared with the percentage land cover of nature areas in densely populated countries or regions. For example, nature areas, excluding multi-functional forests, only cover 3.9% of the land in The Netherlands (CBS 2000), and road reserves in New South Wales, Australia, were estimated to occupy 80% of the combined area of national parks (Bennett 1991). This means that the potential of rights-of-way to enhance habitat availability and connectivity for some species may reach or surpass that of designated natural areas.

Table 1. Road, railroad and rights-of-way statistics[*1]. Total road density is for paved and unpaved combined. Total road and railroad density is per km^2 of land (excluding water).

Country	Road length (km)		Railroad length (km)	Total road density (km/km^2)	Railroad density (km/km^2)	Road right-of-way area (ha) and % of land area
Belgium	paved:	116,687	3,518	4.90	0.12	
	unpaved:	31,529				
The Netherlands	paved:	104,850	2,808	3.44	0.08	50,000-70,000 (1.5-2.1%) (Schaffers 2000)
	unpaved:	11,650				
Japan	paved:	534,471	23,705	3.10	0.06	
	unpaved:	627,423				
France	total [*2]	1,565,669	32,175	2.87	0.06	
Germany	total[*3]:	656,182	46,039	1.88	0.13	
United Kingdom	total[*2]:	365,232	17,186	1.51	0.07	212,220 (0.9%) for England, Scotland and Wales (Way 1977)
Spain	paved:	657,157	14,268	1.33	0.03	
	unpaved:	6,638				
India	paved:	1,517,077	63,140	1.12	0.02	
	unpaved:	1,802,567				
United States	paved:	4,148,395	228,464	0.70	0.02	±4,856,400 (0.5%) (Forman *et al.* 2003)
	unpaved:	2,257,902				
Brazil	paved:	94,871	29,412	0.20	<0.01	
	unpaved:	1,630,058				
Canada	paved:	497,306	48,909	0.15	<0.01	
	unpaved:	911,494				
China	paved:	314,204	70,058	0.15	<0.01	
	unpaved:	1,088,494				
Australia	paved:	314,090	44,015	0.11	<0.01	±500.000 (2.5%) for Victoria (Straker 1998)
	unpaved:	97,513				
Russia	paved:	358,833	87,157	0.03	<0.01	
	unpaved:	173,560				

[*1] CIA (2005); [*2] Trocmé *et al.* (2003); [*3] FHWA (2000)

3. Habitat function of rights-of-way

In this section two categories of the habitat function of rights-of-way are distinguished:

   a. Partial habitat: individual animals may use the habitat in rights-of-way as part of their home range or they may use habitat in rights-of-way during part of their life cycle. Foraging, including mineral acquisition, mate searching, and reproduction are examples of the partial habitat function of rights-of-way.

   b. Complete habitat: individual plants or animals may spend their entire life in the right-of-way, and rights-of-way may support viable populations of these species.

We give examples of the partial and complete habitat function that rights-of-way can have for various plant and animal species and species groups. The examples are grouped based on the particular function of the right-of-way and the species or species groups concerned.

3.1 PARTIAL HABITAT

The partial habitat function of rights-of-way implies that the individuals are mobile. Therefore the examples relate to animal species rather than plant species.

In agricultural landscapes in France flowering vegetation in the right-of-way can be an important source of nectar for butterflies (Ouin et al. 2004). However, in this study butterflies did not tend to stay long in the rights-of-way; they only used it as partial habitat and rested in others.

In French and Spanish agricultural landscapes diurnal raptors and owl species selected rights-of-way and perches in the rights-of-way when hunting for small mammals or other prey (e.g. Fajardo et al. 1998; Meunier et al. 2000). Corvidae (South Africa, United Kingdom, Canada) and bald eagles (Haliaeetus leucocephalus) and turkey vultures (Cathartes aura) (Canada; Yellowstone National Park, United States) have been reported to scavenge on road- and train-killed animals (Wells et al. 1999; Gunther et al. 2000; Slater 2002; Dean & Milton 2003), and Corvidae, Columbidae, Anatidae have been observed eating grain spilled along a railroad (Wells et al. 1999). In forested landscapes edge and gap specialist bird species tend to be more abundant adjacent to roads than in forest interiors (Mumme et al. 2000; Laurance 2004). However, foraging on or near infrastructure also exposes birds to traffic which may result in high mortality (e.g. Fajardo et al.1998). In some cases the habitat along a road may even form a population sink (e.g. Mumme et al. 2000; Ramsden 2004).

Grizzly bear (Ursus arctos), black bear (Ursus americanus), elk (Cervus elaphus), white-tailed deer (Odocoileus virginianus), mule deer (Odocoileus hemionus), red squirrel (Tamiasciurus hudsonicus), Columbian ground squirrel (Spermophilus columbianus) and unidentified mice species have also been attracted to grain spills along railroads (Wells et al. 1999). Mammals are also known to be attracted to roads to feed on road- or train-killed animals. Species observed scavenging along roads in the United Kingdom include domestic cats, Eurasian badger (Meles meles), western polecat (Mustela putorius), red fox (Vulpes vulpes) and western hedgehog (Erinaceus europaeus) (Slater 2002). Grizzly bear, black bear, wolf (Canis lupus), coyote (Canis latrans), wolverine (Gulo gulo), and American marten (Martes americana) have all been observed scavenging on train-killed animals in Canada (Wells et al. 1999). Grizzly bears and coyotes have been reported to scavenge on road-killed animals in Yellowstone National Park, United States (Gunther et al. 2000). As with birds, mammals that spend time on or along roads or railroads run increased risk of being hit by vehicles or trains (e.g. Conover et al. 1995; Groot Bruinderink & Hazebroek 1996).

Several Cervid species are known to forage on the vegetation in rights-of-way. Roe deer (Capreolus capreolus) in Denmark were especially attracted to the vegetation along roads when there was little food available on the surrounding agricultural lands (Madsen et al. 2002). In Pennsylvania (United States) white-tailed deer were seen grazing or lying along roads year-round, but their numbers were especially high in spring and fall (Bellis & Graves 1971; Carbaugh et al. 1975). The spring and fall peak may be related to deer activity patterns (e.g. migration, dispersal, rut, hunting), but it has also been suggested that the deer are attracted to the right-of-way vegetation itself. The vegetation along roads may start to grow earlier in the season (light, partially sloped towards sun) than in the surrounding forested habitats and it may also remain relatively succulent in the fall (Bellis & Graves 1971; Feldhamer et al. 1986). Relatively green and abundant vegetation along roads has also been reported from Australia (Lee et al. 2004). Run-off from roads and relatively high levels of nitrogen deposition (Angold 1997) may also help explain the sometimes relatively abundant and attractive vegetation in rights-of-way.

In Pennsylvania the right-of-way vegetation is thought to be especially attractive in forested areas and less so in areas surrounded by agricultural lands (Carbaugh *et al.* 1975). Allen and McCullough (1976) suggested that white-tailed deer are mainly attracted to the vegetation in rights-of-way when foraging opportunities in the surrounding landscape are poor. Foraging of white-tailed deer on the vegetation in rights-of-way has also been reported from British Columbia, Canada (Kinley *et al.* 2003). Other species, including mule deer and black bear have also been reported foraging on the vegetation along roads and railroads (Lehnert & Bissonette 1997; Wells *et al.* 1999). Fabaceae (e.g. clovers (*Trifolium sp.*), alfalfa (*Medicago sativa*), vetches (*Vicia sp.*)) seem to be particularly attractive (Carbaugh *et al.* 1975; Wells *et al.* 1999).

Run-off of road salt used for de-icing may accumulate in low-lying areas along a road (Miller & Litvaitis 1992). This seems to be an important source of sodium for moose (*Alces alces*) in New Hampshire, United States, which are attracted to the salt in rights-of-way (Miller & Litvaitis 1992).

In the Midwest (e.g. Illinois, Iowa) of the United States most of the original prairie has been converted to intensively managed crops such as corn and soybeans. Rights-of-way are now among the few remaining open areas with a mixture of native and non-native grass and herb species. These rights-of-way have become very important to grassland birds for nesting and foraging (Warner 1992; Camp & Best 1993). However, traffic noise can cause bird species to avoid nesting close to roads (Reijnen *et al.* 1996) and remnant habitat strips away from a road may host more bird species than rights-of-way with similar vegetation (Bolger *et al.* 2001). Rights-of-way may also form nesting habitat for threatened or endangered mammal species in agricultural lands. In Europe hamsters (*Cricetus cricetus*) now often make their nests in rights-of-way, especially if the adjacent farmlands are ploughed deeply and frequently (Nechay 2000).

As illustrated above, various animal species and species groups use rights-of-way for nesting or foraging on vegetation, animal carcasses, minerals, or human caused food spills along transportation corridors. However, transportation corridors may also function as shelter. For example, in Alaska, caribou (*Rangifer tarandus*) walk, stand and run on gravel roads, apparently to seek relief from oestrid flies as the flies avoid the non-vegetated gravel roads (Noel *et al.* 1998).

3.2 COMPLETE HABITAT

The complete habitat function of rights-of-way implies that the individuals can be either sedentary or mobile. Therefore the examples relate to both plant and animal species.

Rights-of-way can be important relicts for plant communities or individual plant species if the surroundings are mostly developed and characterised by intensive large-scale agriculture or urban sprawl (Cousins & Eriksson 2001). The vegetation of interest is often a grassland community (e.g. Sýkora *et al.* 1993; Tanghe & Godefroid 2000; Tikka *et al.* 2000; Ries & Debinski 2001), but relatively wide rights-of-way, such as the ones in some parts of Australia, can contain substantial remnants of native forests (Bennett 2003; Spooner *et al.* 2004a). Rights-of-way can also be one of the last remaining growing sites for rare or endangered plant species (e.g. Godt *et al.* 1997; Yates & Broadhurst 2002; van Rossum *et al.* 2004).

Despite the conservation value that some plant communities in rights-of-way have, they may lack certain rare or indicative species when compared with natural or semi-natural habitats away from infrastructure. In addition, vegetation management (e.g. inappropriate mowing regime, herbicides), trampling, and air and water pollution can result in unfavourable conditions for the plant community or individual plant species (Liem *et al.* 1985; Tikka *et al.* 2000; Bryson & Barker 2002; Swaileh *et al.* 2004). On the other hand, the disturbance and specific environmental conditions in rights-of-way may allow other plant species to thrive (Dunnett *et al.* 1998). These may include invasive and non-native species (e.g. Wilcox 1989; Tyser & Worley 1992; Parendes & Jones 2000), and also species that are tolerant of, for example, trampling, road salt or heavy metals (e.g. Scott & Davison 1985; Sýkora *et al.* 1993; Yorks *et al.* 1997; Welch & Welch 1998).

Some animal species can also thrive in rights-of-ways. Invertebrate numbers were found to be highest within the first 5m away from a gravel road by Luce and Crowe (2001). Edge effects and a gradient in environmental conditions may help explain these high numbers, but it seems that the relatively high nitrogen levels along high volume roads play an important role too, as they lead to increased plant productivity (Port & Thompson 1980; van Schagen *et al.* 1992; Angold 1997). Defoliating larvae of moth species are even known to reach outbreak proportions along roads in the United Kingdom and Australia (Port & Thompson 1980; van Schagen *et al.* 1992). Other invertebrate species may benefit from other resources. For example, the species richness of ants in rights-of-way is higher than in adjacent rangeland, and rights-of-way also contain more rare species (Samways *et al.* 1997; Tshiguvho *et al.* 1999). Possible explanations include relatively low grazing pressure from large herbivores, relatively high moisture levels because of run-off from the road surface and greater variation in temperature because of slopes (Tshiguvho *et al.* 1999). However, the availability of food in the form of road-killed animals is also believed to be an important factor for ants (Samways *et al.* 1997; Tshiguvho *et al.* 1999). Depending on the vegetation and management practices, rights-of-way can also host a large and diverse butterfly population (Munguira & Thomas 1992; Ries *et al.* 2001). A certain mowing frequency and vegetation structure may not only promote plant species that provide nectar, but also create a habitat in which some butterfly species live and reproduce (Munguira & Thomas 1992; Ries *et al.* 2001).

In agricultural landscapes, right-of-way vegetation can form an important and complete habitat for small mammals (van der Reest 1992; Bellamy *et al.* 2000). The species richness and abundance of small mammals in rights-of-way may be similar to or higher than in similar habitat away from roads in agricultural fields (Adams & Geis 1983; Bellamy *et al.* 2000; Bolger *et al.* 2001). However, vegetation structure and mowing frequency influence the quality of the habitat (Adams 1984). Relatively wide and forested rights-of-way can provide a complete habitat for a wide range of mammal species (Downes *et al.* 1997).

4. Corridor function of rights-of-way

In this section three categories of the corridor function of rights-of-way are distinguished:

1. Home range movements: animal species may travel within rights-of-way as part of their movements within their home range.
2. Spread: plant or animal species may spread in rights-of-way over relatively short distances, e.g. through occupying growing sites or habitat adjacent to their original location.
3. Dispersal: plant or animal species may disperse in rights-of-way over relatively long distances. The seeds or animals may move in the rights-of-way but skip potential growing sites or travel many times the diameter of an average home range before reaching their final destination, either in the right-of-way or in a key habitat patch away from infrastructure.

We give examples of these three corridor functions for rights-of-way for various plant and animals species and species groups.

4.1 HOME RANGE MOVEMENTS

Home range movements imply the movements of individuals. Therefore the examples relate to animal species rather than plant species.

A butterfly species in a right-of-way in Iowa, United States, responded strongly to edges such as tree lines and tended to stay within the right-of-way (Ries & Debinski 2001). This suggests that the individuals moved mostly within the rights-of-way. However, the response to edges was reduced at low butterfly densities, and another butterfly species did not respond when edge habitat was encountered.

In Victoria, Australia, forested rights-of-way were part of the home range for southern brown bandicoot (*Isoodon obesulus*), long-nosed potoroo (*Potorous tridactylus*) and bush rat (*Rattus fuscipes*) (Bennett 1990). They used these linear landscape elements in addition to the larger forest patches. In south-east Australia, squirrel gliders (*Petaurus norfolcensis*) made their home ranges in forested rights-of-way (Van der Ree & Bennett 2003). Their home ranges were relatively small, indicating high quality habitat. In The Netherlands, road-killed hedgehogs (*Erinaceus europaeus*) were associated with road-railroad intersections (Huijser *et al.* 2000). This suggests that the hedgehogs travelled along railroads, either because they perceived the railroads as a barrier, or because they travelled in the vegetation along the railroads. In the United Kingdom, railroad rights-of-way are used by the red fox (*Vulpes vulpes*) (Trewhella & Harris 1990). The rights-of-way may influence red fox movements within their home ranges, but they appeared to have little effect on the distance or direction of dispersal movements. Bison (*Bison bison*) sometimes travel along roads in Yellowstone National Park, United States, but this type of use did not increase as a result of snow removal and grooming for snowmobiles (Bjornlie & Garrott 2001). Most movements along roads (61%) were less than 1km, but 12% were 5km or more.

Roads with low traffic volume may be attractive to predators as easy travel routes and provide greater access to prey (Thurber *et al.* 1994, James & Stuart-Smith 2000). In Nova Scotia, Canada, lynx (*Lynx canadensis*) followed road edges and forest trails for considerable distances (Parker 1981) and similar observations were made during winter for lynx in Washington State, United States, for roads less than 15m wide (Koehler & Brittell 1990).

4.2 SPREAD

A wide range of non-native plants were almost completely restricted to rights-of-way along roads, streams and clear cuts in Oregon, and along roads in Utah, United States (Parendes & Jones 2000; Gelbard & Belnap 2003). Disturbance, traffic and light were all associated with higher occurrence of these non-native plants. The spatial pattern suggested that the non-native plant species had spread in the rights-of-way. In New York, United States, the occurrence of a non-native plant, purple loosestrife (*Lythrum salicaria*), was investigated along a road (Wilcox 1989). Again, the spatial pattern suggested that purple loosestrife was spreading in the right-of-way, in this case from east to west. However, the species was believed to spread mostly through the transport of seeds in the water in the ditches in the rights-of-way.

In Australia non-native cane toads (*Bufo marinus*) were found to have relatively high density in rights-of-way, travelling on the road and in the right-of-way, especially along roads through rainforest (Seabrook & Dettmann 1996). The concentration of the cane toads in rights-of-way suggested that cane toads use rights-of-way to spread and expand their range.

The meadow vole (*Microtus pennsylvanicus*) expanded its range by 90-100km in about six years in Illinois, United States (Getz *et al.* 1978). The vole used the 5m wide dense grassy verges along recently constructed interstates. Bait removal rates along roads through forests on the island of Tenerife, Canary Islands, suggest that the non-native black rat (*Rattus rattus*) forages mainly along roads and that this may have enabled the species to spread (Delgado *et al.* 2001).

4.3 DISPERSAL

In a forested landscape in Finland, sites in rights-of-way that were several hundreds of metres to several kilometres apart were more similar than one would expect based on seeding of right-of-way vegetation and spatial autocorrelation (Tikka *et al.* 2001). The results suggest that grassland species dispersed along roads and railroads. Non-native and salt tolerant plant species in The Netherlands, Finland, United Kingdom, Canada and the United States are also known to have dispersed along roads and railroads (Scott & Davison 1985; Brunton 1989; Ernst 1998). Seeds in rights-of-way can be transported by water in roadside ditches (Wilcox 1989), crows, primates, cows or horses (Dean & Milton 2000; Campbell & Gibson 2001; Pauchard & Alaback 2004), mowing equipment (Strykstra *et al.* 1997), cars (Schmidt 1989; Lonsdale & Lane 1994) or trains (Brunton 1989; Ernst 1998). However, not all studies have been able to demonstrate spread or dispersal of non-native plants along roads (e.g. Harrison *et al.* 2001). In addition, depending on the dispersal capacity of the species and the width of the right-of-way, dispersal may be slow (van Dorp *et al.* 1997).

Dispersal distances for heathland carabid beetles in rights-of-way in The Netherlands were rather limited (Vermeulen 1994). The beetles dispersed up to 50-150m per year. In Germany, the cinnabar moth (*Tyria jacobaeae*) colonised a new area using linear landscape structures including rights-of-way along roads (Brunzell *et al.* 2004).

Multiple translocated hedgehogs in the United Kingdom favoured edge habitat and other linear habitats, including roads, when dispersing up to 3.8km from their release points (Doncaster *et al.* 2001). In Australia, dispersal (1.1km) between forest patches through a forested corridor along a road has been demonstrated for the long nosed potoroo and the bush rat (Bennett 1990). The Australian sugar glider (*Petauru breviceps*) was also found to disperse in a forested right-of-way, only several trees wide (Suckling 1984).

5. Factors affecting the quality of rights-of-way as a habitat or corridor

Right-of-way characteristics, both intrinsic and extrinsic, not only influence what species use rights-of-way, but they also have an effect on the quality of the habitat and corridor function of rights-of-way. These characteristics include traffic volume, the width of right-of-way habitat, whether the right-of-way is managed and how, the amount and type of disturbance to the right-of-way habitat, and the habitat adjacent to the right-of-way. In this section, we describe these features and how they may influence the quality of rights-of-way as habitat or corridor for a range of plant and animal species.

5.1 TRAFFIC VOLUME

Butterflies are known to be sensitive indicators of environmental change associated with natural and human-induced disturbances (Hogsden & Hutchinson 2004). Saarinen *et al.* (2005) studied butterfly and diurnal moth communities along Finnish roads with similar environmental conditions but varying in road size and traffic volumes. They found that species richness and total abundance of butterflies and moths were similar in each road type and not affected by traffic volume. Similarly, traffic levels had no apparent effect on butterfly and burnet (Zygaenidae) populations on rights-of-way in the United Kingdom (Munguira & Thomas 1992).

Traffic volume may affect the activity patterns of some sensitive raptor species. The occurrence of three raptor species along rights-of-way in southeastern Spain diminished on weekends associated with high traffic volumes compared with weekdays with relatively low traffic volumes. However, for six other raptor species the amount of traffic had no effect on occurrence. Potential explanations were noise, visual disturbance and prey concealment (Bautista *et al.* 2004). Similarly, little owls (*Athene noctua*) were found foraging in rights-of-way when traffic volumes were lowest (Fajardo *et al.* 1998). However, this is probably best explained by their nocturnal behaviour rather than a response to traffic levels.

5.2 WIDTH OF RIGHTS-OF-WAY

Wider rights-of-way result in larger areas that usually provide greater habitat and vegetation diversity, with more and diverse breeding habitats and food supply for certain animal species. There is a tendency for butterfly diversity to increase with the increasing width of the verge, from the narrow verges of rural roads to wider rights-of-way along highways. The highest number of meadow species was recorded along highways and the total abundance, particularly diurnal moths, decreased in line with decreasing road size along Finnish roads (Saarinen *et al.* 2005). Similarly, the density of adults and number of species were positively correlated with right-of-way width in the United Kingdom (Munguira & Thomas 1992). Broad corridors can also result in fewer losses of individuals to the surroundings and longer dispersal distances (Vermeulen & Opdam 1995).

Meunier *et al.* (2000) surveyed the relative abundance and activity of diurnal raptors along motorways and secondary roads in agricultural landscapes of western France. They found that kestrels (*Falco tinnunculus*) and buzzards (*Buteo buteo*) used rights-of-way for hunting and their abundance was directly related to the width of right-of-way habitat and availability of perch sites. Width of right-of-way habitat was also shown to influence positively the number of field voles (*Microtus agrestis*) and wood mice (*Apodemus sylvaticus*) along roads in the United Kingdom (Bellamy *et al.* 2000).

5.3 MOWING AND HERBICIDES

There are many reasons why roadsides are managed, e.g. to maximise human safety, enhance visual quality, control non-native species, enhance biodiversity, or reduce erosion and sediment flow (Forman *et al.* 2003). To manage right-of-way areas, transportation agencies often delineate roadside zones, each having a different management regime. In Washington State, United States, Zone 1 is closest to the road (usually <4m from the edge of the pavement) and usually kept bare and clear of vegetation unless the vegetation is not a fire hazard, does no damage to the pavement and does not obscure visibility. Zone 2 (usually <9-11m from the edge of the pavement), is often referred to as the "recovery zone" and is managed so that vehicles that have run off the road can recover. This zone is kept clear from objects >10cm (4inches) in height. Zone 3 is farthest from the road and normally abuts utility access and neighbouring land use (agricultural, residential, public etc). The management of these zones and the activities within them influence the quality of rights-of-way as habitat and corridors for plant and animal populations. Roadside mowing not only weighs heavily on maintenance budgets, but it also requires important decisions regarding time, space and intensity because of the ecological consequences. Regular mowing can affect normal periods of vegetation growth, flowering of plants, genetic diversity of certain plant species, attractiveness of plants to pollinators, pollination of flowering plants, nesting and denning opportunities for various animal species, and shelter from late or early frosts (Godt *et al.* 1997; Forman *et al.* 2003).

Less frequent mowing is less costly and can enhance plant species diversity, whereas frequent mowing usually favours a few grass species that out compete more diverse native plant species. On the other hand, frequent mowing (e.g. once in spring and once in autumn) in combination with hay removal can eventually deplete the soil of nutrients. In some rights-of-way in The Netherlands this has resulted in a reduction of the productivity and biomass of the right-of-way vegetation, allowing smaller plant species characteristic of relatively nutrient poor conditions to re-establish or increase in abundance (Sýkora et al. 1993). As the soil is depleted and right-of-way vegetation productivity reduced, mowing frequency can be decreased to less than it was before more intensive mowing began. Thus, in the long run the conservation value of the vegetation in the right-of-way can be increased and mowing frequency reduced. However, the timing of mowing needs to be carefully planned based on the phenology of the species of interest. Furthermore, it is important to remove the hay within one or two weeks of mowing to prevent leaching of nutrients back into the soil (Schaffers et al. 1998). However, immediate hay removal may not allow sufficient seeds to ripen and fall from the cuttings; this should be weighed against minimising the leaching of nutrients.

High nectar abundance has been shown to be the most important factor increasing the numbers of meadow butterflies along road verges (Munguira & Thomas 1992; Dover et al. 2000; Croxton et al. 2005; Saarinen et al. 2005). However, mowing and herbicides can suppress flowering and density of nectar plants. The most intensively mowed rights-of-way generally have the shortest vegetation and lowest amount of nectar, which together result in decreased butterfly abundance (Gerell 1997; Saarinen et al. 2005). Ries et al. (2001) demonstrated that roadside native prairie restoration involving vegetation management with restricted use of herbicides can benefit butterfly populations. They found species richness of habitat-sensitive butterflies was two to five times greater on restored sites and butterflies spent more time on restored rights-of-way and were also less likely to leave them, thus suggesting the rights-of-way were being or could be used as corridors. In order to enable the restoration and expansion of two reintroduced myrmecophilous butterflies in The Netherlands (Maculinea teleius, M. nausithous) road verges and canal borders were targeted and management practices changed to enhance the development of rough vegetation where the specific host ant species occurs (Wynhoff et al. 2000).

Road rights-of-way and central medians are often good habitat because of greater food and cover for herbivorous animals compared with neighbouring habitat outside of these areas. Cover is usually higher when the right-of-way and median habitat are not mowed but allowed to grow wild (Adams 1984; Meunier et al. 1999a). Road mitigation projects that include wildlife fencing can result in an effective "exclosure" for ungulates on the highway side of the fence. This can result in high quality habitat and enhancement of existing habitat through abundant forage and cover for smaller fauna (e.g. small mammals). In an extensive study carried out along the United States interstate highway system, Adams and Geis (1983) found that there were more small mammal species present and in higher densities on the right-of-way than in adjacent habitat. Their results also indicated that right-of-way habitat and its accompanying edge were attractive not only to grassland species but also to many generalist species that make use of the right-of-way and the edge with the adjacent habitat complex. At six interchanges along a highway in Ottawa, Canada, the density of woodchucks (Marmota monax) per hectare exceeded any density previously reported for this species in any habitat (Woodward 1990).

In Sweden, shrub and tree clearing in rights-of-way has reduced moose-vehicle collisions by 20% (Lavsund & Sandegren 1991). When trees and shrubs were removed along a railroad and sprayed with herbicides to prevent re-growth, this led to a 40-72% reduction in moose-train collisions (Jaren et al. 1991). Rea (2003) suggested that right-of-way shrub and tree cutting in the early season, just after their leaves have sprouted, would help to minimise right-of-way attraction to moose. Cutting later in the season promotes regrowth, which may be an attractant to moose.

5.4 SOIL DISTURBANCE AND BURNING

A number of studies focused on the effects of disturbance from heavy equipment on plant populations (Webb *et al.* 1983; Olander *et al.* 1998; Milchunas *et al.* 2000). Roadwork and the associated disturbance to the right-of-way and surrounding habitat are considered a major threat to native plants in roadside environments (see Godt *et al.* 1997). In addition, disturbance facilitates invasion of non-native plant species (Greenberg *et al.* 1997; Parendes & Jones 2000). Further, roadwork promotes spread and dispersal of non-native species by providing suitable linear habitat. Roads and railways are well known sites for non-native plant invasions (Borowske & Heitlinger 1981; Wilcox 1989; Tyser & Worley 1992; Gelbard & Belnap 2003; Hansen & Clevenger, 2005).

For some native plant species, soil disturbance from roadwork is analogous to periodic disturbance from a natural fire regime or other natural disturbance events. The importance of disturbance processes in shaping the spatial structure and temporal dynamics of ecological systems has been reviewed by White and Pickett (1985) and Hobbs (1987). The effects of natural and anthropogenic disturbances on plant populations also depend on complex interactions between the life history attributes of individual species, and the spatial and temporal structure of the disturbance regime (Spooner *et al.* 2004a). Anthropogenic disturbance from roadwork, in conjunction with historical changes in grazing pressure, are suggested as the main causes of increased recruitment for some *Acacia* species in rights-of-way in Australia (Spooner *et al.* 2004b). Ongoing management and disturbance regimes in roadside environments may even be critical to the persistence of some *Acacia* species and associated fauna habitat (Spooner 2005b). Frequent and intensive soil disturbance regimes appear to favour *Acacia* species that have strong re-sprouting ability, whereas *Acacia* species that are obligate seeders may be eliminated from roadside environments unless disturbance regimes are less frequent (Spooner 2005b).

Prescribed burning is one of the most important management tools in grassland areas (Engle & Bidwell 2001) and is commonly used to control the spread and establishment of non-native plants. However, fire can also increase non-native plants. In sclerophyll woodlands along highways in south-western Australia, fire enhanced the spread of weeds into the remnant right-of-way habitat (Milberg & Lamont 1995). They found that the number of weed species, their frequency, and cover increased after the fires, while the abundance of native species often decreased. Fire also caused the non-native Coolatai grass (*Hyparrhenia hirta*) to increase in south-eastern Australia (McArdle *et al.* 2004). In contrast, two rare *Acacia* species occur now mostly in Australian rights-of-way and depend on fire for germination (Yates & Broadhurst 2002).

Fires in grassland habitats like rights-of-way are known to affect bird abundance, distribution, nesting success and predation (Zimmerman 1997; Kirkpatrick *et al.* 2002). Burning can also cause changes in vegetative cover and arthropod abundance (Swengel 2001). Shochat *et al.* (2005) found that prescribed burning of right-of way vegetation increased bird nest success after the fire, possibly through an increased arthropod biomass on the re-growth. Similar results came from a study of bird density and nesting success on utility rights-of-way. Cool burns in early spring produced high structural diversity of herbs, shrubs and trees and supported a high density of birds and bird species (Confer & Pascoe 2003).

5.5 VEGETATION STRUCTURE & SURROUNDING LANDSCAPE

The composition of the landscape and vegetation structure along rights-of-way can have a strong effect on the quality of the right-of-way as habitat or corridors for many species. For example, tree lines along grassland habitat may be important for directing the movement of certain butterfly species (Ries & Debinski 2001). Saarinen *et al.* (2005) found high nectar abundance was the most important factor increasing meadow butterflies along road verges. However, for diurnal moths shelter provided by tall vegetation was the most important factor increasing their numbers (Saarinen *et al.* 2005), and the number of food plants is not always the most limiting factor for the presence of butterflies either. In The Netherlands, sections of motorway that were mowed twice a year appeared to have the highest diversity and density of nectar plants, but butterfly density was highest in right-of-way sections with a mowing frequency of only once every three years (Bak *et al.* 1998). High butterfly density occurred where there was relatively low diversity and density of nectar plants. The authors concluded that butterfly presence in their study area was most related to vegetation structure and favourable microclimate rather than food plants.

Meunier *et al.* (1999b) found that the structure of vegetation (trees and shrubs) was the most important factor influencing bird species richness along rights-of-way in France. In Central Amazonia, the height and density of forest re-growth markedly affected road-crossing movements by understorey rainforest birds (Laurance *et al.* 2004). In that study, clear differences were observed among different bird guilds, and species within the same guild often responded very differently to roads; even narrow dirt roads had large effects on some bird species movement. These authors recommended that land managers encourage forest regeneration along road verges and establish continuous canopy cover over road surfaces to facilitate movements of edge- and gap-avoiding species.

Small mammals on rights-of-way in the United Kingdom had affinity for tall vegetation, big hedges, and cover (Bellamy *et al.* 2000). Vegetation structure on road verges and adjacent habitat may also strongly influence road-crossing movements by mammals. Studies have shown that bears tend to prefer crossing roads at places where vegetation is most dense close to the road (Brandenburg 1996; Chruszcz *et al.* 2003).

Above we described how the structural complexity and type of vegetation along rights-of-way may influence the quality as habitat or corridors for many taxa. However, the habitat type and conditions adjacent to rights-of-way also strongly influence plant and animal distribution, abundance, and movements along or across roadway corridors. Thus, the surrounding landscape is also an important factor in assessing the habitat and corridor function of rights-of-way.

In Finland, the environment adjacent to the right-of-way had an effect on the species composition of butterflies (Saarinen *et al.* 2005). The authors studied three road types. Rights-of-way surrounded by cultivated fields were generally associated with low numbers of butterflies, whereas adjacent forests increased the total number of all species and favoured several butterflies inhabiting forest edges. Their results indicated that road verges should be considered as an important reserve for species that were dependent on semi-natural grasslands. The same was true in the United Kingdom, but here the abundance and diversity of butterflies was not correlated with adjacent habitats (Munguira & Thomas 1992). Beetle assemblages have also been found to depend on adjacent crops and have seasonal fluctuations in rights-of-way habitats (Varchola & Dunn 1999).

Shrubland bird density and nesting success on utility rights-of-way in forested habitat were high compared with other habitat types (Confer & Pascoe 2003). However, the composition of the surrounding landscape can also affect nest predation levels in road-side habitat (Bergin *et al.* 2000).

Meunier *et al.* (1999a) specifically investigated the effects of surrounding landscape type on the use of motorway rights-of-way by small mammals. In farmland landscapes they found that perturbations from agricultural activities (ploughing, harvesting etc.) largely explained the relative abundance of small mammals in rights-of-way. Other studies documented that mice inhabit crops for most of the year but take refuge in field margins and right-of-way habitats during intensive harvesting and when the fields are bare (Tew *et al.* 1994; Fitzgibbon 1997).

6. Potential problems

Throughout this chapter we have described the habitat and corridor function of rights-of-way and factors that influence the abundance and movements of a wide range of taxa. While rights-of-way can be beneficial to many species they also have the potential to lead to harmful effects as some animal species may experience consistently high rates of mortality, sometimes exceeding recruitment, and rights-of-way allow certain non-native species to establish themselves and spread along the right-of-way and into the surrounding landscape.

6.1 ROAD KILL AND POPULATION SINK

Several studies demonstrated that roadside territories can become population sinks and that young or inexperienced individuals are more vulnerable to road-related mortality as they tend to live closer to roads. In a 9-year study of Florida scrub jays (*Aphelocoma coerulescens*), Mumme *et al.* (2000) found that scrub jay habitat adjacent to highways was a demographic sink in which breeder mortality exceeded production of yearlings by a wide margin, and the scrub jays persisted only because of immigration from non-road territories. Road-naive immigrants that established roadside breeding territories suffered high annual mortality during their first two years as breeders, in many cases not surviving long enough to attempt nesting. The mortality of breeders and fledglings on road territories was caused by traffic and not by road-related habitat modifications.

However, two studies from The Netherlands showed that willow warblers (*Phylloscopus trochilus*), with greater experience or age, learned to avoid roads (Foppen & Reijnen 1994, Reijnen & Foppen 1994). They found that breeding territory densities were lower near busy highways and road zones were occupied primarily by first-year males that often bred unsuccessfully, and in subsequent years actively moved away from the highway.

Several populations of grizzly bears showed sex-related variation in habitat use adjacent to roads. Gibeau *et al.* (2002) found that sub adult male grizzly bears were found closer to highways than all other age and sex classes, when within or adjacent to high quality habitat near roads. He believed that social structure in grizzly bears was a large factor in explaining this result. Other grizzly bear studies have shown that cohorts of subordinate bears were found in poorer quality habitats near developments and displaced by more dominant classes, particularly adult males (Mattson *et al.* 1987; McLellan & Shackleton 1988). Thus, certain sex and age groups may be more vulnerable to traffic mortality than others.

Roads, their construction, and adjacent rights-of-way can create high-quality habitat where food resources are more abundant compared with adjacent areas. Lush forage along medians and verges created by exclusion fencing is attractive to herbivores. Locally abundant small mammal populations found in these habitats become targets for predators seeking easy and accessible prey, but the right-of-way habitat may also become a population sink for these predators (e.g. Ramsden 2004). In extensively forested areas the right-of-way created by the road corridor may be one of few open habitats around. This is the first place that grass will green up after winter (attracting ungulates) and is ideal for dandelions (*Taraxacum officinale*) and fruit-bearing shrubs that appeal to a variety of fauna, including bears. With herbivores grazing and predators hunting near the road, collisions with vehicles are inevitable, resulting in attractive carrion for avian and terrestrial scavengers, if carcasses are not removed promptly.

As opposed to what we have seen above, where road-side habitat is attractive to wildlife, there are some examples where animals are drawn to the actual road surface and the vehicles that travel on it. On roads requiring snow removal and the application of salt-based de-icing agents, problems arise during and after winter when mineral-deficient ungulates come to the road gleaning salt from the edge of the road and cracks in the pavement (Fraser & Thomas 1982). In warmer climes, reptiles that come to the road surface to bask during the day or thermoregulate at night quickly become road-kills and ultimately carrion for scavengers (Rosen & Lowe 1994; Kline & Swann 1998). Lastly, humans feeding wildlife from vehicles, food discarded from motorists while travelling, unsecured garbage containers along roads, or dead invertebrates on the grille or window of vehicles can be an easy, predictable source of food for wildlife such as coyotes, bears, corvids, grackles (*Quiscalus sp.*), house sparrows (*Passer domesticus*), squirrels, and raccoons. In 1997 there were more than 80 "bear jams" (traffic snarls caused by motorists stopping to look at bears) reported along roads in Banff National Park, Alberta, Canada (Pilkington 1997). This exposure to people and anthropogenic food sources results in bears becoming food-conditioned, habituated to humans, road-kills, or being removed from the area (Gibeau & Herrero 1998).

6.2 INVASIVE SPECIES

High concentrations of non-native species have been observed near roads in many ecosystems (Forcella & Harvey 1983; Tyser & Worley 1992; Hansen & Clevenger in press). There has been a growing interest in the effects of roads and other linear features on plant species composition (Angold 1997; Safford & Harrison 2001; Gelbard & Belnap 2003), and particularly the spread and establishment of invasive non-native species (Parendes & Jones 2000; Tyser & Worley 1992; Hansen & Clevenger in press). Studies have documented roads as suitable habitat and corridors for non-native plant dispersal (Forman *et al.* 2003; Gelbard & Belnap 2003; Watkins *et al.* 2003). Vehicle traffic on roads may aid in the invasion and dispersion of non-native species within road corridors (Clifford 1959; Wace 1977; Schmidt 1989; Lonsdale & Lane 1994). Heavy traffic can cause air turbulence and vehicles may act as vectors for spread of seeds and vegetative plant parts (Panetta & Hopkins 1991; Tyser & Worley 1992; Forman *et al.* 2003).

Road corridors also have enabled the dispersal of non-native fauna. Non-native beetles spread from populations of relatively high density located in the right-of-way into adjacent forest interior habitat in central Alberta, Canada (Niemela & Spence 1999). In Australia, cane toads (*Bufo marinus*), a species introduced to Australia, were more abundant in road corridors than in many surrounding habitats (Seabrook & Dettman 1996). Roads and rights-of-way assisted in extending their range and facilitated colonisation by toads of previously inaccessible areas. In general, however, the extent to which roads influence the distribution and abundance of exotic species, such as foxes, cats and dingoes, and the consequences for native fauna, are poorly known (May & Norton 1996; Forman *et al.* 2003).

The edge effect created by road construction, right-of-way habitat and the adjacent landscape matrix can benefit edge-foraging, generalist predators and nest parasites (Fagan *et al.* 1999). Edges that function as travel lanes for predators are considered to be key components of the ecological trap hypothesis, whereby fauna behaviourally favour edge habitat, but at the cost of high rates of mortality from edge-foraging generalist predators and nest parasites, e.g. brown-headed cowbird (*Molothrus ater*) (Gates & Gysel 1972; Angelstam 1986).

Both native and non-native small mammals frequently find optimal foraging conditions near road edges (Adams & Geis 1983; Downes *et al.* 1997). In Australian rain forests, edge generation creates opportunities for interspecific competition among rat species, perhaps leading to the local extirpation of one species (Laurance 1994). Goosem (2000) found that tropical rainforest roads did not affect community composition of small mammals, but roads did allow non-native mammal species to penetrate the forest interior from the right-of-way. High road densities that fragmented forests in the Canary Islands, Spain, and facilitated high non-native rat abundance along edges explained damaging effects on the native biota (Delgado *et al.* 2001).

7. Discussion and conclusion

This chapter has illustrated the habitat and corridor function of right-of-way habitat, including factors that influence the quality of these functions, and some of the potential problems. Species with large home ranges that are not tied to one particular habitat are more likely to use rights-of-way as only part of their home range than species that are sedentary, that have small home ranges and that are tied to a specific habitat. Therefore most examples of species that use right-of-way habitat but do not restrict their movements to these relatively narrow strips are birds and larger mammals.

While the habitat function of rights-of-way is generally well documented and understood, the corridor function is not. Dispersal, movements over relatively long distances, has rarely been documented. Most of the evidence is indirect; i.e. based on distribution patterns rather than the actual recording of dispersing individuals in the right of way. This might be expected as dispersal is typically a rare event and we would be fortunate to catch it in action in the right-of-way rather than simply observing that movement had happened and making the assumption that the individual had dispersed along the right-of-way rather than through the surrounding landscape. However, the corridor function has been suggested or demonstrated for various species, particularly for species that depend on specific habitat and conditions provided in the right-of-way which may not exist in, or that have disappeared from, the surrounding landscape. Depending on the species, disturbance, vegetation structure and the composition of the surrounding landscape seem to be among the most important variables. Nonetheless, we are only beginning to have some insight into the parameters that may influence animal movements in rights-of-way.

The habitat and corridor function of rights-of-way is not often discussed. Most studies address the negative effects of roads, including habitat loss, road mortality, reduced habitat quality and the barrier effect, and how we may mitigate these effects. When the habitat and corridor function of roads is discussed, there is often a marked difference in how we value these functions in different landscapes. In less developed landscapes, e.g. forested landscapes or landscapes dominated by natural grasslands, rights-of-way are usually perceived as disturbance zones that provide a habitat and corridor for non-native species. They may disrupt ecotones and ecosystem processes, and contribute to the habitat fragmentation caused by the actual road and the traffic that uses it. In some cases non-native species may not only spread or disperse along the transportation corridor itself, but they can also spread into the surrounding landscape and cause an additional threat to the integrity of the ecosystem. Reduction of disturbance and restoration of the native vegetation in the right-of-way, including transplanting large native plants, may reduce these problems (e.g. Harper-Lore & Wilson, 1999). In developed landscapes, for example in intensively managed agricultural landscapes, rights-of-way are often the only remaining natural or semi-natural habitat. They may form a refugium for certain native species in an otherwise hostile environment. Their linear shape and interconnectedness may provide an important habitat for such native species and they may help individuals move through the landscape, either for daily movements within their home ranges, gradual spread, or dispersal between larger habitat patches that are connected by rights-of-way (e.g. Viles & Rosier 2001). Even though right-of-way habitat and corridors are continuously exposed to disturbance from the roads and traffic, and even though they may be of lesser quality than habitat and corridors away from infrastructure, they can be important to the survival of some species, especially in developed landscapes.

References

Adams LW. 1984; Small mammal use of an interstate highway median strip. Journal of Applied Ecology 21:175-178.

Adams LW, Geis AD. 1983; Effects of roads on small mammals. Journal of Applied Ecology. 20: 403-416.

Allen RE, McCullough DR. 1976; Deer-car accidents in southern Michigan. Journal of Wildlife Management 40: 317-325.

Angelstam P. 1986; Predation on ground-nesting birds' nests in relation to predator densities and habitat edge. Oikos 47: 365-373.

Angold P. 1997; The impact of a road upon adjacent heathland vegetation: Effects on plants species composition. Journal of Applied Ecology 34: 409-417.

Bak A, Oorthuijsen W, Meijer M. 1998; Vlindervriendelijk wegbermbeheer langs de A58 in Zeeland. De Levende Natuur 99: 261-267.

Bautista LM, García JT, Calmaestra RG, Palacín C, Martín CA,. Morales MB, Bonal R, Viñuela J. 2004; Effect of weekend road traffic on the use of space by raptors. Conservation Biology 18: 726-732.

Bellamy PE, Shore RF, Ardeshir D, Treweek JR, Sparks TH. 2000; Road verges as habitat for small mammals in Britain. Mammal Review 30: 131-139.

Bellis ED, Graves HB. 1971; Collision of vehicles with deer studied on Pennsylvania interstate road section. Highway Research News 43: 13-16.

Bennett AF. 1990; Habitat corridors and the conservation of small mammals in a fragmented forest environment. Landscape Ecology 4: 109-122.

Bennett AF. 1991; Roads, roadsides and wildlife conservation: a review. In: Saunders DA & Hobbs RJ (eds.). Nature conservation 2: the role of corridors, 99-118. Surrey Beatty, Chipping Norton, NSW, Australia.

Bennett AF. 2003; Linkages in the landscape. The role of corridors and connectivity in wildlife conservation. IUCN, Gland, Switzerland.

Bergin TM, Best LB, Freemark KE, Koehler KJ. 2000; Effects of landscape structure on nest predation in roadsides of a Midwestern agroecosystem: a multiscale analysis. Landscape Ecology 15: 131-143.

Bjornlie DD, Garrott RA. 2001; Effects of winter road grooming on bison in Yellowstone National Park. Journal of Wildlife Management 65: 560-572.

Bolger DT, Scott TA, Rotenberry JT. 2001; Use of corridor-like landscape structures by bird and small mammal species. Biological Conservation 102: 213-224.

Borowske JR, Heitlinger ME. 1981; Survey of native prairie on railroad rights-of-way in Minnesota. Transportation Research Record 822: 22-26.

Brandenburg DM. 1996; Effects of roads on behavior and survival of black bears in coastal North Carolina. MSc thesis, University of Tennessee, Knoxville.

Brunton DF. 1989; The marsh dandelion Taraxacum section palustria Asteraceae in Canada and the adjacent USA. Rhodora 91: 213-219.

Brunzell S, Elligsen H, Frankl R. 2004; Distribution of the Cinnabar moth *Tyria jacobaeae* L. at landscape scale: use of linear landscape structures in egg laying on larval hostplant exposures. Landscape Ecology 19: 21-27.

Bryson GM, Barker AV. 2002; Sodium accumulation in soils and plants along Massachusetts roadsides. Communications in Soil Science and Plant Analysis 33: 67-78.

Camp M, Best LB. 1993; Bird abundance and species richness in roadsides adjacent to Iowa rowcrop fields. Wildlife Society Bulletin 21: 315-325.

Campbell JE, Gibson DJ. 2001; The effect of seeds of exotic species transported via horse dung on vegetation along trail corridors. Plant Ecology 157: 23-35.

Carbaugh B, Vaughan JP, Bellis ED, Graves HB. 1975; Distribution and activity of white-tailed deer along an interstate highway. Journal of Wildlife Management 39: 570-581.

CBS. 2000. Kerncijfers. Centraal Bureau voor de Statistiek. Available from the internet: URL: http://www.cbs.nl/ (accessed 05-06-2005)

Chruszcz B, Clevenger AP, Gunson K, Gibeau M. 2003; Relationships among grizzly bears, highways, and habitat in the Banff-Bow Valley, Alberta, Canada. Canadian Journal of Zoology 81: 1378-1391.

CIA. 2005. The world fact book. Central Intelligence Agency. Available from the internet: URL http://www.odci.gov/ cia/publications/factbook/index.html (accessed 05-06-2005)

Clifford W. 1959; Seed dispersal by motor cars. Journal of Ecology 47: 311-15.

Confer JL, Pascoe SM. 2003; Avian communities on utility rights-of-ways and other managed shrublands in the northeastern United States. Forest Ecology and Management 185: 193-205.

Conover MR, Pitt WC, Kessler KK, DuBow TJ, Sanborn WA. 1995; Review of human injuries, illnesses, and economic losses caused by wildlife in the United States. Wildlife Society Bulletin 23: 407-414.

Cousins SAO, Eriksson O. 2001; Plant species occurrences in a rural hemiboreal landscape: effects of remnant habitats, site history, topography and soil. Ecography 24: 461-469.

Croxton PJ, Hann JP, Greatorex-Davies JN, Sparks T.H. 2005; Linear hotspots? The floral and butterfly diversity of green lanes. Biological Conservation 121: 579-584.

Dean WRJ, Milton SJ. 2000; Directed dispersal of Opuntia species in the Karoo, South Africa: are crows the responsible agents? Journal of Arid Environments 45: 305-314.

Dean WRJ, Milton SJ. 2003; The importance of roads and road verges for raptors and crows in the Succulent and Nama-Karoo, South Africa. Ostrich 74: 181-186.

Delgado JD, Arévalo JR, Fernández-Palacios JM. 2001; Road and topography effects on invasion: edge effects in rat foraging patterns in two oceanic island forests (Tenerife, Canary Islands). Ecography 24: 539-546.

Doncaster CP, Rondinini C, Johnson PCD. 2001; Field test for environmental correlates of dispersal in hedgehogs Erinaceus europaeus. Journal of Animal Ecology 70: 33-46.

Dorp D, van Schippers P, van Groenendael JM. 1997; Migration rates of grassland plants along corridors in fragmented landscapes assessed with a cellular automation model. Landscape Ecology 12: 39-50.

Dover J, Sparks T, Clarke S, Gobbett K, Glossop S. 2000; Linear features and butterflies: the importance of green lanes. Agriculture, Ecosystems and Environment 80: 227-242.

Downes SJ, Handasyde KA, Elgar MA. 1997; The use of corridors in fragmented Australia Eucalypt forests. Conservation Biology 11: 718-726.

Dunnett NP, Willis AJ, Hunt R, Grime JP. 1998; A 38-year study of relations between weather and vegetation dynamics in road verges near Bibury, Gloucestershire. Journal of Ecology 86: 610-623.

Engle DM, Bidwell TG. 2001; Viewpoint. The response of central North American prairies to fire. Journal of Range Management 54: 2-10.

Ernst WHO. 1998; Invasion, dispersal and ecology of the South African neophyte Senecio inaequidens in the Netherlands: from wool alien to railway and road alien. Acta Botanica Neerlandica 47: 131-151.

Ewing R, Kostyack J, Chen D, Stein B, Ernst M. 2005; Endangered by sprawl: How runaway development threatens America's wildlife. National Wildlife Federation, Smart Growth America, and NatureServe. Washington, D.C., USA.

Fagan WF, Cantrell RS, Cosner C. 1999; How habitat edges change species interactions. American Naturalist 153: 165-182.

Fajardo I. 2001; Monitoring non-natural mortality in the barn owl (Tyto alba), as an indicator of land use and social awareness in Spain. Biological Conservation 97: 143-149.

Fajardo I, Pividal V, Trigo M, Jiménez M. 1998; Habitat selection, activity peaks and strategies to avoid road mortality by the little owl Athene noctua. A new methodology on owls research. Alauda 66: 49-60.

Feldhamer GA, Gates JE, Harman DM, Loranger AJ, Dixon KR. 1986; Effects of interstate highway fencing on white-tailed deer activity. Journal of Wildlife Management 50: 497-503.

FHWA. 2000; Highway Statistics 2000. Available from the internet: URL:
http://www.fhwa.dot.gov/ohim/hs00/index.htm (accessed 05-06-2005)

Fitzgibbon CD. 1997; Small mammals in farm woodlands: the effects of habitat, isolation and surrounding land-use patterns. Journal of Applied Ecology 34: 530-539.

Foppen R, Reijnen R. 1994; The effects of car traffic on breeding bird populations in woodland. II. Breeding dispersal of male Willow Warblers (Phylloscopus trochilus) in relation to the proximity of a highway. Journal of Applied Ecology 31: 95-101.

Forcella F, Harvey SJ. 1983; Eurasian weed infestation in western Montana in relation to vegetation and disturbance. Madrono 30: 102-109.

Forman RTT, Alexander LA. 1998; Roads and their major ecological effects. Annual Reviews Ecology and Systematics 29: 207-231.

Forman RTT, Sperling D, Bissonette JA, Clevenger AP, Cutshall CD, Dale VH, Fahrig L, France R, Goldman ChR, Heanue K, Jones JA, Swanson FJ, Turrentine Th, Winter ThC. 2003; Road Ecology. Science and Solutions. Island Press, Washington DC, USA.

Fraser D, Thomas ER. 1982; Moose-vehicle accidents in Ontario: relation to highway salt. Wildlife Society Bulletin 10: 261-5.

Gates E, Gysel LW. 1972; Avian nest dispersion and fledging success in field-forest ecotones. Ecology 59: 871-883.

Gelbard JL, Belnap J. 2003; Roads as conduits for exotic plant invasions in a
semiarid landscape. Conservation Biology 17: 420-432.

Gerell R. 1997; Management of roadside vegetation: effects of density and species diversity of butterflies in Scania, south Sweden. Entomologisk Tidskrift 118: 171-176.

Getz LL, Cole FR, Gates DL. 1978; Interstate roadsides as dispersal routes for Microtus pennsylvanicus. Journal of Mammalogy 59: 208-212.

Gibeau ML, Clevenger AP, Herrero S, Wierzchowski J, Wierzchowski S. 2002; Grizzly bear response to human development and activities in the Bow River watershed, Alberta. Biological Conservation 103: 227-236.

Gibeau ML, Herrero S. 1998; Roads, rails and grizzly bears in the Bow River Valley, Alberta. In: Evink G, Garrett P, Zeigler D, & Berry J. (eds.), Proceedings of the International Conference on Wildlife Ecology and Transportation, 104-108. FL-ER-69-98. Florida Department of Transportation, Tallahassee, Florida.

Godt MJW, Walker J, Hamrick JL. 1997; Genetic diversity in the endangered lily *Harperocallis flava* and a close relative, *Tolfeldia racemosa*. Conservation Biology 11: 361-366.

Goosem M. 2000; Effects of tropical rainforest roads on small mammals: edge changes in community composition. Wildlife Research 27: 151-163.

Greenberg CH, Crownover SH, Gordon DR. 1997; Roadside soils: a corridor for invasion by xeric scrub by nonindigenous plants. Natural Areas Journal 17: 99-109.

Grift EA van der. 1999; Mammals and railroads: impacts and management implications. Lutra 42: 77-98.

Groot Bruinderink GWTA, Hazebroek E. 1996; Ungulate traffic collisions in Europe. Conservation Biology 10: 1059-1067.

Gunther KA, Biel MJ, Robinson H.L. 2000; Influence of vehicle speed and vegetation cover-type on road-killed wildlife in Yellowstone National Park. In: Messmer TA & West B. (eds.). Wildlife and highways: seeking solutions to an ecological and socio-economic dilemma. September 12-16: Nashville, TN, USA.

Hansen M, Clevenger AP. 2005; The influence of disturbance and habitat on the frequency of non-native plant species along transportation corridors. Biological Conservation 125(2): 249-259.

Harrison S, Hohn Ch, Ratay S. 2001; Distribution of exotic plants along roads in a peninsular nature reserve. Landscape Ecology 16: 659-666, 2001.

Hobbs RJ. 1987; Disturbance regimes in remnants of natural vegetation. In: Saunders DA, Arnold GW, Burbridge, AA & Hopkins AJM. (eds) 233-240. Nature conservation: The role of remnants of native vegetation. Surrey Beatty, Chipping Norton, Australia.

Hogsden KL, Hutchinson TC. 2004; Butterfly assemblages along a human disturbance gradient in Ontario, Canada. Canadian Journal of Zoology 82: 739-748.

Huijser MP, Bergers PJM, ter Braak CJF. 2000; Road, traffic and landscape characteristics of hedgehog traffic victim sites. In: Huijser MP. Hedgehog traffic victims and mitigation strategies in an anthropogenic landscape, 107-126 PhD Thesis, Wageningen University, Wageningen, The Netherlands.

James ARC, Stuart-Smith AK. 2000; Distribution of caribou and wolves in relation to linear corridors. Journal of Wildlife Management 64: 154-9.

Jaren V, Andersen R, Ulleberg M, Pedersen PH, Wiseth B. 1991; Moose-train collisions: the effects of vegetation removal with a cost-benefit analysis. Alces 27: 93-99.

Kinley TA, Page HN, Newhouse NJ. 2003; Use of infrared camera video footage from a *wildlife protection system* to assess collision-risk behavior by deer in Kootenay National Park, British Columbia. Sylvan Consulting Ltd, Invermere, BC, Canada.

Kirkpatrick C, DeStefano S, Mannan RW, Lloyd J. 2002; Trends in abundance of grassland birds following a spring prescribed burn in southern Arizona. Southwestern Natualist 47: 282-292.

Kline NC, Swann DE. 1998; Quantifying wildlife road mortality in Saguaro National Park. In: Evink G, Garrett P, Zeigler D& Berry J. (eds.), Proceedings of the International Conference on Wildlife Ecology and Transportation, 23-31. FL-ER-69-98. Florida Department of Transportation, Tallahassee, Florida.

Koehler GM, Brittell JD. 1990; Managing spruce-fir habitat for lynx and snowshoe hares. Journal of Forestry 88: 10-4.

Laurance SGW. 2004; Responses of understory rain forest birds to road edges in central Amazonia. Ecological Application 14: 1344-1357.

Laurance SGW, Stouffer PC, Laurance WF. 2004; Effects of road clearings on movement patterns of understory rainforest birds in Central Amazonia. Conservation Biology 18: 1099-1109.

Laurance WF. 1994; Rainforest fragmentation and the structure of small mammal communities in tropical Queensland. Biological Conservation 69: 23-32.

Lavsund S, Sandegren F.1991; Moose-vehicle relations in Sweden: a review. Alces 27: 118-126.

Lee EU, Kloecker DB, Croft, Ramp D. 2004; Kangaroo-vehicle collisions in Australia's sheep rangelands, during and following drought periods. Australian Mammology 26: 215-226.

Lehnert ME, Bissonette JA. 1997; Effectiveness of highway crosswalk structures at reducing deer-vehicle collisions. Wildlife Society Bulletin 25: 809-818.

Liem ASN, Hendriks A, Kraal H, Loenen M. 1985; Effects of de-icing salt on roadside grasses and herbs. Plant and Soil 84: 299-310.

Lonsdale WM, Lane AM. 1994; Tourist vehicles as vectors of weed seeds in Kakadu National Park, northern Australia. Biological Conservation 69: 277-283.

Luce A, Crowe M. 2001; Invertebrate terrestrial diversity along a gravel road on Barrie Island, Ontario, Canada. The Great Lakes Entomologist 34: 55-60.

Mader H-J. 1984; Animal habitat isolation by roads and agricultural fields. Biological Conservation 29: 81-96.

Madsen AB, Strandgaard H, Prang A. 2002; Factors causing traffic killings of roe deer Capreolus capreolus in Denmark. Wildlife Biology 8: 55-61.

Mattson D, Knight R, Blanchard R, Blanchard B. 1987; The effects of developments and primary roads on grizzly bear habitat use in Yellowstone National Park, Wyoming. International Conference on Bear Research and Management 7: 259-273.

May SA, Norton TW. 1996; Influence of fragmentation and disturbance on the potential impact of feral predators on native fauna in Australian forest ecosystems. Wildlife Research 23: 387-400.

McArdle SL, Nadolyn C, Sindel BM. 2004. Invasion of native vegetation by Coolatai grass *Hyparrhenia hirta*: impacts on native vegetation and management implications.

McLellan BN, Shackleton DM. 1988 Grizzly bears and resource extraction industries: effects of roads on behavior, habitat use, and demography. Journal of Applied Ecology 25: 51-460.

Meunier FD, Corbin J, Verheyden C, Jouventin P. 1999a; Effects of landscape type and extensive management on use of motorway roadsides by small mammals. Canadian Journal of Zoology 77: 1-10.

Meunier FD, Verheyden C, Jouventin P. 1999b; Bird communities of highway verges: influence of adjacent habitat and roadside management. Acta Oecologica 20: 1-13.

Meunier FD, Verheyden C, Jouventin P. 2000; Use of roadsides by diurnal raptors in agricultural landscapes. Biological Conservation 92 (2000) 291-298

Milberg P, Lamont BB. 1995; Fire enhances weed invasion of roadside vegetation in southwestern Australia. Biological Conservation 73: 45-49.

Milchunas DG, Schulz KA, Shaw RB. 2000; Plant community structure in relation to long-term disturbance by mechanized military manoeuvres in a semiarid region. Environmental Management 25: 525-539.

Miller BK, Litvaitis J. 1992; Use of roadside salt licks by moose, *Alces alces*, in northern New Hampshire. Canadian Field Naturalist 106: 112-117.

Mumme RL, Schoech SJ, Woolfenden GE, Fitzpatrick JW. 2000; Life and death in the fast lane: demographic consequences of road mortality in the Florida scrub jay. Conservation Biology 14: 501-512.

Munguira ML, Thomas JA. 1992; Use of road verges by butterfly and burnet populations, and the effect of roads on adult dispersal and mortality. Journal of Applied Ecology 29: 316-329.

Nechay G. 2000; Status of Hamsters: *Cricetus cricetus, Cricetus migratorius,Mesocricetus Newtoni* and other hamster species in Europe. Convention on the conservation of European wildlife and natural habitats, Nature and Environment Series, No. 106, Council of Europe Publishing.

Niemela J, Spence JR. 1999; Dynamics of local expansion by an introduced species: *Pterostichus melanarius* III. (Coleoptera, Carabidae) in Alberta, Canada. Diversity and Distributions 5: 121-127.

Noel LE, Pollard RH, Ballard WB, Cronin MA. 1998; Activity and use of active gravel pads and tundra by caribou, *Rangifer tarandus granti*, within Prudhoe Bay oil field, Alaska. The Canadian Field-Naturalist 112: 400-409.

Olander LP, Scatena FN, Silver WL. 1998; Impacts of disturbance initiated by road construction in a subtropical cloud forest in the Luquillo experimental forest, Puerto Rico. Forest, Ecology and Management 109: 33-49.

Ouin A, Aviron S, Dover J, Burel F. 2004; Complementation/supplementation of resources for butterflies in agricultural landscapes. Agriculture Ecosystems and Environment 103: 473-479.

Panetta FD, Hopkins AJM. 1991; Weeds in corridors: invasion and management. In: Saunders D & Hobbs RJ (eds) Nature Conservation II: The Role of Corridors, 341-351. Surrey Beatty and Sons Pty Ltd, Chipping Norton.

Parendes LA, Jones JA. 2000; Role of light availability and dispersal in exotic plant invasion along roads and streams in the H.J. Andrews Experimental Forest, Oregon. Conservation Biology 14: 64-75.

Parker GR. 1981; Winter habitat use and hunting activities for lynx (*Lynx canadensis*) on Cape Breton Island, Nova Scotia. In: Chapman JA & Pursley D (eds) Worldwide Furbearer Conference Proceedings, 221-48. Frostburg, Maryland.

Pauchard A, Alaback PB. 2004; Influence of elevation, land use, and landscape context on patterns of alien plant invasions along roadsides in protected areas of south-central Chile. Conservation Biology 18: 238-248.

Pilkington R. 1997; Living with wildlife project report. Prepared for Friends of Banff National Park, Banff, Alberta, Canada. 19pp.

Port GR, Thompson GR. 1980; Outbreaks of insect herbivores on plants along motorways in the United Kingdom. Journal of Applied Ecology 17: 649-656.

Ramsden DJ. 2004; Barn owls and major roads. Results and recommendations from a 15 year research project. Barn Owl Trust, UK.

Rea RV. 2003; Modifying roadside vegetation management practices to reduce vehicular collisions with moose *Alces alces*. Wildlife Biology 9: 81-91.

Reest PJ van der. 1992; Small mammal fauna of road verges in the Netherlands: ecology and management. Lutra 35: 1-27.

Reijnen R, Foppen R. 1994; The effects of car traffic on breeding bird populations in woodland. I. Evidence of reduced habitat quality for willow warblers (*Phylloscopus trochilus*) breeding close to a highway. Journal of Applied Ecology 31: 85-94.

Reijnen R, Foppen R, Meeuwsen H. 1996; The effects of traffic on the density of breeding birds in Dutch agricultural landscapes. Biological Conservation 75: 255-260.

Ries L, Debinski DM. 2001; Butterfly responses to habitat edges in the highly fragmented prairies of central Iowa. Journal of Animal Ecology 70: 840-852.

Ries L, Debinski DM, Wieland ML. 2001; Conservation value of road-side prairie restoration to butterfly communities. Conservation Biology 15: 401-411.

Rosen PC, Lowe CH. 1994; Highway mortality of snakes in the Sonoran desert of southern Arizona. Biological Conservation 68: 143-148.

Rossum F Van, Campos De Sousa S, Triest L. 2004; Genetic consequences of habitat fragmentation in an agricultural landscape on the common *Primula veris*, and comparison with its rare congener, *P. vulgaris*. Conservation Genetics 5: 231-245.

Saarinen K, Valtonen A, Jantunen J, Saarnio S. 2005; Butterflies and diurnal moths along road verges: Does road type affect diversity and abundance? Biological Conservation 123: 403-412.

Safford HD, Harrison SP. 2001; Grazing and substrate interact to affect native vs. exotic diversity in roadside grasslands. Ecological Applications 11: 1112-1122.

Samways MJ, Osborn R, Carliel F. 1997; Effect of a highway on ant (Hymenoptera: Formicidae) species composition and abundance, with a recommendation for road side verge width. Biodiversity and Conservation 6: 903-913.

Schaffers AP. 2000; Ecology of roadside plant communities. PhD thesis, Wageningen University, Wageningen, The Netherlands.

Schaffers AP, Vesseur MC, Sýkora KV. 1998; Effects of delayed hay removal on the nutrient balance of roadside plant communities. Journal of Applied Ecology 35: 349-364.

Schmidt W. 1989; Plant dispersal by motor cars. Vegetatio 80: 147-152.

Scott NE, Davison AW. 1985; The distribution and ecology of coastal species on roadsides. Vegetatio 62: 433-440.

Seabrook WA, Dettman EB. 1996; Roads as activity corridors for cane toads in Australia. Journal of Wildlife Management 60: 363-368.

Shochat E, Wolfe DH, Patten MA, Reinking DL, Sherrod SK. 2005; Tallgrass prairie management and bird nest success along roadsides. Biological Conservation 121: 399-407.

Slater FM. 2002; An assessment of wildlife road casualties – the potential discrepancy between numbers counted and numbers killed. Web Ecology 3: 33-42.

Spellerberg IF. 1998; Ecological effects of roads and traffic: a literature review. Global Ecology and Biography Letters 7: 317-333.

Spooner PG. 2005a; On squatters, settlers and early surveyors: historical development of country road reserves in southern New South Wale Australian Geographer 36: 55-73.

Spooner PG. 2005b; Response of Acacia species to disturbance by roadworks in roadside environments in southern New South Wales, Australia. Biological Conservation 122: 231-242.

Spooner PG, Lunt ID, Briggs SV. 2004a; Spatial analysis of anthropogenic disturbance regimes and roadside shrubs in a fragmented agricultural landscape. Applied Vegetation Science 7: 61-70.

Spooner PG, Lunt ID, Briggs SV, Freudenberger D. 2004b; Effects of soil disturbance from roadworks on roadside shrubs in a fragmented agricultural landscape. Biological Conservation 117: 393-406.

Straker A. 1998; Management of roads as biolinks and habitat zones in Australia. In: Evink GL, Garrett P, Zeigler D & Berry J. (eds.) Proceedings of the International Conference on Wildlife Ecology and Transportation, 181-188. FL-ER-69-98, Florida Department of Transportation. Tallahassee, Florida, USA.

Strykstra RJ, Verweij GL, Bakker JP. 1997; Seed dispersal by mowing machinery in a Dutch brook valley system. Acta Botanica Neerlandica 46: 387-401.

Suckling GC. 1984; Population ecology of the sugar glider *Petauru breviceps* in a system of fragmented habitats. Australian Wildlife Research 11: 49-76.

Swaileh KM, Hussein RM, Abu ES. 2004; Assessment of heavy metal contamination in roadside surface soil and vegetation from the West Bank. Archives of Environmental Contamination and Toxiocology 47: 23-30.

Swengel AB. 2001; A literature review of insect response responses to fire compared to other conservation managements of open habitat. Biodiversity and Conservation 10: 1141-1169.

Sýkora KV, de Nijs LJ, Pelsma TAHM. 1993; Plantengemeenschappen van Nederlandse wegbermen. Stichting Uitgeverij Koninklijke Nederlandse Natuurhistorische Vereniging, Utrecht, The Netherlands.

Tanghe M, Godefroid S. 2000; Road verge grasslands in southern Belgium and their conservation value. Fragmenta-Floristica-et-Geobotanica 45: 147-163.

Tew TE, Todd IA, Macdonald DW. 1994; Field margins and small mammals. British Crop Protection Conference Monograph No. 58, pp.85-94.

Thurber JM, Peterson RO, Drummer TD, Thomasma SA. 1994; Gray wolf response to refuge boundaries and roads in Alaska. Wildlife Society Bulletin 22:61-8.

Tikka PM, Högmander H, Koski PS. 2001; Road and railway verges serve as dispersal corridors for grassland plants. Landscape Ecology 16: 659-666.

Tikka PM, Koski PS, Kivela RA, Kuitunen MT. 2000; Can grassland plant communities be preserved on road and railway verges? Applied Vegetation Studies 3: 25-32.

Trewhella WJ, Harris S. 1990; The effect of railway lines on urban fox (*Vulpes vulpes*) number and dispersal movements. Journal of Zoology London 221: 321-326.

Trocmé M, Cahill S, de Vries JG, Farrall H, Folkeson L, Fry G, Hicks C, Peymen J (eds.). 2003; Habitat fragmentation due to transportation infrastructure. The European review. COST Action 341, European Commission, Directorate-General for Research, Brussels, Belgium.

Tshiguvho TE, Dean WRJ, Robertson HG. 1999; Conservation value of road verges in the semi-arid Karoo, South Africa: ants (Hymenoptera: Formicidae) as bio-indicators. Biodiversity and Conservation 8: 1683-1695.

Tyser RW, Worley CA. 1992; Alien flora in grasslands adjacent to road and trail corridors in Glacier National Park, Montana. Conservation Biology 6: 253-262.

Van der Ree R, Bennett AF. 2003; Home range of the squirrel glider (*Petaurus norfolcensis*) in a network of remnant linear habitats. Journal of Zoology London 259: 327-336.

Van Schagen JJ, Hobbs RJ, Majer JD.1992. Defoliation of trees in roadside corridors and remnant vegetation in the Western Australian wheatbelt. Journal of the Royal Society of Western Australia 75: 75-81.

Varchola JM, Dunn JP. 1999; Changes in ground beetle (Coleoptera: Carabidae) assemblages in farming systems bordered by complex or simple roadside vegetation. Agriculture, Ecosystems and Environment 73: 41-49.

Vermeulen HJW. 1994; Corridor function of a road verge for dispersal of stenotopic heathland ground beetles Carabidae. Biological Conservation 69: 339-349.

Vermeulen HJW, Opdam PFM. 1995; Effectiveness of roadside verges as dispersal corridors for small ground-dwelling animals: a simulation study. Landscape and Urban Planning 31: 233-248.

Viles RL, Rosier DJ. 2001; How to use roads in the creation of greenways: case studies in three New Zealand landscapes. Landscape and Urban Planning 55: 15-27.

Wace, 1977; Assessment of dispersal of plant species – the car-bourne flora in Canberra. Proceedings of the Ecological Society of Australia 10: 167-86.

Warner RE. 1992; Nest ecology of grassland passerines on road rights-of-way in central Illinois. Biological Conservation 59: 1-7.

Watkins RZ, Chen J, Pickens J, Brosofske KD. 2003; Effects of forest roads on understory plants in a managed hardwood landscape. Conservation Biology 17: 411-419.

Way JM. 1977; Roadside verges and conservation in Britain: a review. Biological Conservation 12: 65-74.

Webb RH, Wilshire HG, Henry MA. 1983; Natural recovery of soils and vegetation following human disturbance. In: Webb RH & Wilshire HG (eds) Environmental effects of off-road vehicles: impacts and management in semiarid regions, 281-302. Springer-Verlag, New York.

Welch D, Welch MJ. 1998; Colonisation by *Cochlearia officinalis* L. (Brassicaceae) and other halophytes on the Aberdeen-Montrose main road in North-East Scotland. Watsonia 22: 190-193.

Wells P, Woods JG, Bridgewater G, Morrison H. 1999; Wildlife mortalities on railways: monitoring methods and mitigation strategies. In: Evink GL, Garrett P & Zeigler D (eds) Proceedings of the Third International Conference on Wildlife Ecology and Transportation, 85-88. FL-ER-73-99, Florida Department of Transportation, Tallahassee, Florida, USA.

White, P.S. & S.T.A. Pickett. 1985. Natural disturbances and patch dynamics. In: S.T.A. Pickett & P.S. White (eds.). The ecology of natural disturbance and patch dynamics. Pp. 3-13. Academic Press, San Diego, California USA.

Wilcox DA. 1989; Migration and control of purple loosestrife (*Lythrum salicaria* L.) along highway corridors. Environmental Management 13: 365-370.

Woodward SM. 1990; Population density and home range characteristics of woodchucks, *Marmota monax*, at expressway interchanges. Canadian Field-Naturalist 104: 421-28.

Wynhoff I, Oostermeijer JGB, van Swaay CAM, van der Made JG, Prins HHT. 2000; Herintroductie in de praktijk: het pimpernelblauwtje (*Maculinea teleius*) en het donker pimpernelblauwtje (*M. nausithous*) (Lepidoptera: Lycaenidae). Entomologische Berichten 60: 107-117.

Yates CJ, Broadhurst LM. 2002; Assessing limitations on population growth in two critically endangered *Acacia* taxa. Biological Conservation 108: 13-26.

Yorks TP, West NE, Mueller RJ. 1997; Toleration of traffic by vegetation: life form conclusions and summary extracts from a comprehensive data base. Environmental Management 21: 121-131.

Zimmerman JL. 1997; Avian community response to fire, grazing, and drought, in the tallgrass prairie. In: Knopf FL & Samson FB (eds) Ecology and Conservation of Great Plains Vertebrates, 167-180. Springer, New York.

CHAPTER 12: IMPACT OF ROAD TRAFFIC ON BREEDING BIRD POPULATIONS

RIEN REIJNEN[1] & RUUD FOPPEN[2]

[1]Wageningen University and Research Centre, Alterra, Landscape Centre,P.O. Box 47, 6700 AA Wageningen, Netherlands
[2]Dutch Centre for Field Ornithology, Rijksstraatweg 178, 6573 DG Beek-Ubber gen, Netherlands

1. Introduction

From a conservation perspective, human disturbance of wild life is important only if it affects demographic parameters and hence causes a population to decline (Gill et al. 2001). Transport corridors, particularly roads, may have a significant impact on animal populations at local, regional and national levels due to loss of habitat, fragmentation and isolation of habitats, increased mortality and degradation of habitats by pollution, disturbance (noise, visual stimuli, light) and increased human access (Findlay & Bourdages 2000; Spellerberg 2000; Forman et al. 2003).

This chapter deals with the direct effect on animal populations of roads when they are operational. Notwithstanding change of habitat conditions at the side of the road, many wildlife species are less common or absent near roads (e.g. Forman et al. 2003). Since these road-avoidance zones can extend to more than 1000m, this may effectively result in the loss of a population (Reijnen et al. 1997; Forman et al. 2003). Potential causes of these zones include all relevant influences of traffic, such as visual disturbance, vehicular pollution, road-kills, traffic noise and soil vibration.

Railway traffic might also be an important source of disturbance for animal populations (Tulp et al. 2002). Several potential causes are similar to those of road traffic, such as visual disturbance, noise and collisions, available data, however, are scarce.

This chapter focuses on breeding birds along roads because this species group has been the most extensively studied during the last decade and shows strong effects (Reijnen et al. 1997; Forman et al. 2003). A more thorough review of these effects is timely.

2. Effects on breeding densities

2.1 EVIDENCE FOR TRAFFIC AS THE MAIN CAUSE OF REDUCED DENSITIES NEAR ROADS

Effects on the breeding density of bird species near roads can be related to the presence of the road *per se*, such as favourable and unfavourable habitat conditions introduced by the presence of road infrastructure, and to the actual traffic using the road (van der Zande et al. 1980). Generally it is found that the effect of road presence could either reduce or increase breeding bird densities but effects of road traffic will reduce densities. The overall pattern inferred from 18 studies shows that negative impacts of road traffic on density of species by far outweigh the positive impacts (Table 1).

Most studies have separated the effects of the road *per se* and of the traffic upon it and attributed density-depressing effects to road traffic only. There is much evidence that road traffic is the main cause of the density-depressing effect. Significant relations between traffic-related factors and breeding density of species are shown by many studies (e.g. Raty 1979;

John Davenport and Julia L. Davenport, (eds.) The Ecology of Transportation: Managing Mobility for the Environment, 255–274,

van der Zande *et al.* 1980; Illner 1992; Reijnen *et al.* 1995, 1996; Reijnen & Foppen 1995; Weiserb & Jacob 2001; Forman *et al.* 2002). Furthermore, in studies that compare roads with different traffic densities the percentage of affected species along quiet roads is much lower than in studies along busy roads (Table 1, see also Figure1).

The few positive effects on the breeding bird density could be attributed to favourable habitat conditions introduced by road construction. In forested areas, edge effects introduced by the construction of the road are indicated as the main source of higher densities near the road (Ferris 1979; Clark & Karr 1979; Helldin & Seiler 2003), while in open areas other favourable habitat conditions introduced by road construction, such as ditches for water run-off, might be more important (Foppen *et al.* 2002). There are no indications that densities are higher because of the disappearance of competing species or because road kills are a source of food (Reijnen *et al.* 1995a).

It is likely that positive effects of favourable habitat conditions introduced by road construction will be more pronounced in quiet-traffic roads than in busy-traffic roads. Along quiet-traffic roads favourable habitat conditions introduced by road construction may outbalance the negative effects of traffic. This is shown by the figures ascertained by the studies: in five of the six studies that show positive effects, traffic densities were quite low.

Traffic is therefore the most important factor of the road/traffic complex that affects the density of breeding bird populations near roads.

2.2 HOW GENERAL IS THE EFFECT?

To determine the percentage of bird species showing reduced densities near roads studies that investigated 'all' species or a representative set of all species were used (Reijnen *et al.* 1995; 1996; Reijnen & Foppen 1995; Fernández-Juricic 1999; Milsom *et al.* 2000; Foppen *et al.* 2002; Helldin & Seiler 2003; Rheindt 2003; Peris & Pescador 2004). Within these studies the percentage of affected species varies greatly from 5-67%. The relationship between the percentage of affected species and traffic density was modelled as follows. First empirical logistic transformations were applied to the observed number of affected species and accompanying regression weights were calculated (see McCullagh & Nelder 1989). Nonlinear weighted regression was then used to model the transformed data as an exponential curve with traffic density as explanatory variable ($R^2=85\%$). Finally the exponential curve was back transformed to the percentage scale, giving the curve in Figure 1. For average traffic densities below 30,000 vehicles a day the differences are clearly related to differences in the maximum traffic density present in the studies. Above average traffic densities of ca. 30,000 vehicles a day the percentage of affected species does not increase further and remains around 55%. Apparently about 45% of breeding species are not affected by road traffic.

Table 1. Characteristics and main results of 18 studies that investigated the impact of roads and traffic on breeding bird densities.

Habitat: 1, open field (grassland, arable land, heathland); 2, woodland; 3, urban; 4, marshland. Traffic density: vehicles per day x 1000. Study sites: number of study sites (transects or plots); p refers to number of paired plots. N_t, number of species involved; S, number of species with either statistically negative or positive effect, numbers between brackets include possible effects. $\sum s$: combined effect of all species involved, negative effect (*), effect absent (-).

Reference	Habi-tat	Traffic density	Study sites	N_t	Negative effect		Positive effect	
					S	$\sum s$	S	$\sum s$
Ferris 1979	2	<10[1]	12+2	≥12	3	-	5	-
Clark & Karr 1979	1	0,3-3[1]	33	2	1		1	
Raty 1979	2	0,7-3	100	4	4		-	
Van der Zande et al. 1980	1	<1-54	4	4	2(3)		-	
Illner 1992	1	<1-60	10	1	1		-	
Reijnen et al. 1995	2	8-69	55p	43	26	*	-	-
Reijnen & Foppen 1995	2	40-52	16-18p	28	17	*	-	-
Reijnen et al. 1996	1	3-52	15p	12	8	*	-	-
Kuitunen et al. 1998	2	1-10	17p	55	3	*[5]	1	-
Fernández-Juricic 1999	3	city	search	14	3		-	
Green et al. 2000	1	4-34[2]	42	1	1		-	
Milsom et al. 2000	1	<5-30[3]	528	27		*		-
	1	<5-30[3]	528	10	4		-	
Weiserb & Jacob 2001	2	>50[2]	3	25		*		-
Foppen et al. 2002	1-4	8->100	>1000[4]	125	66		11	
Forman et al. 2002	1	3->30	84	5		*		-
Rheindt 2003	2	>50	2	26		*		-
Peris & Pescador 2004	½	<1-8,4[2]	3	20	3(6)	-	3	-
Helldin & Seiler 2003	1	0.5-17	12	1	*	-	-
	2	0.5-25	31	4	-	5	-

[1] indication of traffic densities is given by Dr. Richard T.T. Forman

[2] data on traffic densities are provided by the authors

[3] rough indication of traffic densities based on European datasets

[4] more than 1000 plots and national distribution data of The Netherlands

[5] effect if two common species are excluded

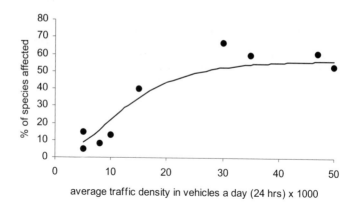

Figure 1. Relation between percentage of species that show a reduced density near the road and the average traffic density in vehicles per day based on data of nine studies (see table 1 and text). Selection of nine studies from Table 1 that investigated all occurring species in study area.

These data indicate that the density-depressing effect of road traffic is a common phenomenon in breeding birds. The affected species represent very different taxonomic groups and cover a wide range of habitat types, which is very well illustrated by the extensive study of Foppen *et al.* (2002) in The Netherlands (Figures 2 and 3). An important result of this study, from a conservation perspective, is that particularly species of national and international significance fared worse than the other species, 63% and 48% respectively were negatively affected by road traffic. The target species of international significance (in this case those on the European Bird Directive List) were particularly badly affected: 75% showed a negative impact of road traffic. Moreover, natural habitats (woodland, marshland and heathland) have on average a higher proportion of affected species (60%) than agricultural and urban habitats (40%).

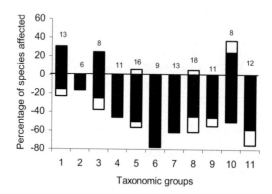

Figure 2. Percentage of breeding bird species in 11 taxonomic groups, showing negative or positive impact of road traffic on abundance and occurrence. The taxonomic groups are: 1, waterfowl; 2, raptors; 3, grouse, rails; 4, waders, gulls, terns; 5, pigeons, owls, woodpeckers; 6, larks, swallows, pipits; 7, thrushes; 8, warblers, flycatchers; 9, tits; 10, crows, starlings; 11, finches, buntings. Unshaded part of bar indicates near-significant impact; solid part of bar indicates significant impact. Above zero, positive impact; below zero, negative impact. Numbers above columns indicate total number of species analysed. (Foppen *et al.* 2002).

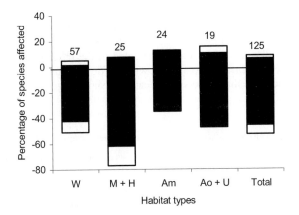

Figure 3. Percentage of breeding bird species on which road traffic has a negative or positive impact in terms of presence or abundance in the main terrestrial habitat types. The habitat types are: W, woodland; M, marshland; H, heathland; Am, wet agricultural meadows; Ao, other agricultural habitats; U, urban habitats. In habitat types that have been pooled, relatively few species have been investigated. Unshaded parts of bars, near-significant impact; black part of bars, significant impact. Above zero, positive impact; below zero, negative impact. Numbers above columns indicate total number of species analysed. (Foppen *et al.* 2002).

2.3 EFFECT SIZE

The size of the density-depressing effect of road traffic on breeding birds depends on the distance from the road at which a significant effect is detected (called effect-distance) and on the reduction of the density in the affected zone. Some available data on these effect-parameters in relation to the traffic density or other traffic-related factors are shown in Table 2.

2.3.1 *Effect distance*
Accurate data on effect-distances of many species of birds, inhabiting woodland and agricultural grasslands in The Netherlands, are presented by Reijnen *et al.* (1995,1996) and Reijnen and Foppen (1995). They quantified the relationship between road traffic and density of breeding birds by using a threshold regression model with traffic noise as the explanatory variable. The results are summarized and partly reconsidered in Reijnen *et al.* (1997). For single species the threshold value for traffic noise varied from 36-58dB(A) in woodland and from 43-60dB(A) in open grassland. For all species combined, the threshold value in woodland was 42-52dB(A) and in open grassland 47dB(A). Effect-distances are derived from these equations by transforming the threshold values in dB(A) into distance from the road (Reijnen *et al.* 1997). Estimated effect-distances at 50,000 vehicles a day vary from 75-930m for grassland birds and from 60-810m for woodland birds. For all species combined the effect-distance in grassland is 560m and in woodland up to 365m. For a traffic density of 10,000 vehicles a day the effect-distances are more than twice as low (Table 2).
Other data on effect-distances in forested areas refer to a few species or to estimates for some species combined. For woodland birds in Belgium, Weiserb and Jacob (2001) also used traffic noise as the explanatory variable. The estimated effect-distance of about 400m for all species

combined at a traffic density of 50,000 vehicles a day is quite similar to the effect-distance of 365m of Reijnen *et al.* (1997). Because the bird fauna in both studies is similar, this might indicate the same cause for the effect. The threshold values for the traffic noise however are measured in a different way and cannot be compared. Raty (1979) found depressed densities of tetraonids in woodland, a group not investigated in the other studies, at a distance up to 500m from roads. Since this involves roads with traffic densities varying from 700 to 3000 vehicles a day, it indicates that some species can also show large effect-distances along quiet-traffic roads.

Table 2. Maximum width of zone with depressed density (effect-distance) in relation to the traffic density.

Habitat: 1, open field (grassland, arable land, heathland); 2, woodland. Traffic density: vehicles per day x 1000. Negative effect: S, for single species (number); $\sum$s, for species combined (number). Effect-distance: S, for single species (range); $\sum$s: for species combined; -, no effect.

Reference	Habitat	Negative effect		Traffic density x 10	Zone with depressed density (m)	
		S	$\sum$s		S	$\sum$s
Clark & Karr 1979	1	1		0,3	250	
	1	1		3	≥500	
Raty 1979	2	4		0,7-3	250-500	
van der Zande *et al.* 1980	1	2		4.8	625	
				54	2000	
Illner 1992[1]	1	1		2.5	-	
				10	<100	
				60	300	
Reijnen *et al.* 1997[1,2]	2	33	28	10	38-305	125
				50	70-810	365
Reijnen *et al.* 1996[1]	1	8	12	10	30-360	120
				50	75-930	560
Green *et al.* 2000	1	1		6-16	3000	
Weiserb & Jacob 2001[1]	2		27	>50		400
Forman *et al.* 2002	1		5	3-8		-
				8-15		400
				15-30		700
				≥30		1200

[1]data are converted or inferred from threshold values in traffic noise based on the relationship between traffic noise and density of breeding birds.
[2]partly reconsidered data of Reijnen *et al.* (1995, 1996) and Reijnen & Foppen (1995).

In open field habitat several studies show that the effect-distance of bird species is greater at higher traffic densities (Table 2). In Germany, Illner (1992) estimated effect-distances for grey partridge (*Perdix perdix*) that are in the range of effect-distances presented by Reijnen *et al.* (1996). Van der Zande *et al.* (1980) found much higher effect-distances than those established by Reijnen *et al.* (1996, 1997) at similar traffic densities, for black-tailed godwit (*Limosa limosa*) and lapwing (*Vanellus vanellus*). However, Reijnen *et al.* (1996) argued that these large effect-distances might be over-estimated because of statistical flaws. Grassland birds in the United States seem to be more sensitive to road traffic then in The Netherlands (Forman *et al.* 2002). Effect-distances for all species combined at medium and high traffic densities are at least twice as high as in The Netherlands. Because the effects are dominated by two species, it is more appropriate to make a comparison with effect-distances of single species: then effect distances are rather comparable. Some species are much more sensitive to road traffic than others (Reijnen *et al.* 1996, 1997). The horned lark (*Eromophila alpestris*) showed effect-distances of 500m at a traffic density of 3000 vehicles a day (Clark & Karr 1979). For the stone curlew (*Burhinus oedicnemus*), an effect-distance of 3000m at a traffic density of 16,000 vehicles a day was established (Green *et al.* 2000).

2.3.2 *Reduction of Density*
The reduction of the density over the effect-distances for species varies greatly between species, from 25% to almost 100% (Table 3). Generally, density reductions increase when effect-distances become smaller (Reijnen *et al.* 1995, 1996; Reijnen & Foppen 1995). This is not surprising because effects at a short distance from the road can only be detected when the reduction of the density is great. For the total density of all species combined, the density reductions are similar and vary from 35-40%. Within the road-effect-zone the reduction of the density follows a log-linear relationship (Reijnen *et al.* 1997).

Table 3. Reduction of the density of species and of all species combined in the road effect zone.

Habitat: 1, open field (grassland, arable land,); 2, woodland.

Reference	Habitat	Reduction of density in the affected one (%)	
		Species	All species combined
Raty 1979	2	30-50	
van der Zande *et al.* 1980	1	46-66	
Illner 1992	1	27	
Reijnen *et al.* 1995; Reijnen & Foppen 1995	2	25-99	35
Reijnen *et al.* 1996	1	38-82	39
Green *et al.* 2000	1	30-40	
Weiserb & Jacob 2001	2		40

3. Probable causal factors and mode of action

3.1 INTRODUCTION

We have demonstrated that road traffic has negative impacts on the density of breeding birds in many parts of the world. What are the driving forces, and which mode of action leads to the observed effects? First, the traffic related aspects which can be considered as potential causal factors, must be identified. Then those aspects that are directly linked to the traffic source, and do not take into account aspects like habitat alteration caused by road construction, can be investigated. Potential causal factors e.g. factors such as noise, visual stimuli, reverberations, pollution and the direct effect of mortality due to road collisions are dependent on the type and extent of traffic on a particular road, and also are influenced by the environmental circumstances. For instance, the extent of visual disturbance strongly depends on the openness of the landscape and the visibility of the transport corridor and the traffic.

Visibility is affected by the height of the road relative to ground surface level and the presence of noise screens. To quantify traffic noise, models are available that predict the noise level depending on average traffic speed, type and number of vehicles (see e.g. Reijnen *et al.* 1995). Average noise levels, e.g. the 24-hour value of the equivalent noise level, appear to be a good indicator of traffic load and are often used as dose variables in effect studies (Reijnen *et al.* 1995). Studies on the type and extent of polluting substances associated with road traffic and their effect on wildlife are scarce. Birds in the immediate vicinity of a road show a slightly increased level of lead, but toxic effects could not be detected (Forman *et al.* 2003). Estimates on the extent of increased levels of various gases and other pollutants indicate that they reach background level within 50 m of the highway (Reijnen *et al.* 1995).

It is almost impossible under normal field conditions to study and distinguish between the potential traffic related causal factors because they are heavily correlated. However, to our best knowledge, there are no studies that use an experimental design to isolate and test the various factors one by one. To unravel the importance of these factors we therefore have to rely on circumstantial evidence. Below we describe the results of a number of studies. Together they construct a quite consistent view of the important factors, and shed some light on the probable mode of action behind the negative impact of traffic on breeding birds. We distinguish between two types of studies: studies looking for correlations between the impacts on breeding density and potential causal factors and studies that gain insight into the probable mode of action by showing effects on the individual or population level.

3.2 CORRELATIONS BETWEEN EFFECT ON BREEDING DENSITY AND CAUSAL FACTORS

Several studies have indicated that smaller effects on species along roads were found when the traffic load was relatively low. This supports the assumption that the presence of a road *per se* is not very important in affecting densities of breeding birds, but that we have to concentrate on the factors associated with the traffic load (Reijnen *et al.* 2002). All traffic-related factors, noise, visual stimuli, reverberations, pollution and the effects of collisions become less important at an increasing distance from the road. Reijnen *et al.* (1995) compared the range of the observed effects (distance to the road) on breeding densities with estimates of range of effect for the various causal factors (Figure 4). Their conclusion was that, in open landscapes, only noise and visual stimuli can explain the observed effects. In forests, at least beyond distances of 50 metres, only noise load is a potential causal factor (Forman 2003). This was supported by a study comparing the density effects in forest areas between sites with high and low noise load and equal visibility (Reijnen *et al.* 1995). The set with high noise load clearly showed more negative effects. Rearranging these study sites and comparing a set with and without visual traffic stimuli (sight of cars) and equal noise load did not show different effects on density.

In open landscapes the influences of other stimuli remain a potential causal factor. A few studies have been made of the effect that artificial light has on birds. It is known that light disturbs breeding and foraging behaviour of birds (Hill 1992). In an experiment with road light poles placed in the middle of a meadow area, black-tailed godwits avoided nesting close to the light sources in comparison with a situation with non-active poles (De Molenaar *et al.* 2000). They noted effects up to a few hundred metres from the road. Along most Dutch highways however no light poles are present, but the effects of car lights might have an equal effect. In a study of negative traffic impact on Stone Curlews, the authors hypothesize that vehicle headlights are the most probable causal factor and currently work is under way to test this (pers. com Rhys Green). In more open landscapes like lakes and heathlands, changes in plant growth have been observed that were induced by traffic exhausts (Seiler 2001). There are no studies linking these effects to breeding bird presence. Also in open landscapes some evidence exists that noise is an important causal factor. A German study on the Partridge (*Perdix perdix*) in open field areas showed that breeding densities along highways were lower even if the visual stimulus of traffic along the road was absent because the roads were bordered by hedgerows (Illner 1992). An American study relates species richness to ambient noise levels (not necessarily traffic noise). In an environment with a high ambient noise level, habitat and other explanatory variables being equal, less species appear (Stone 2000).

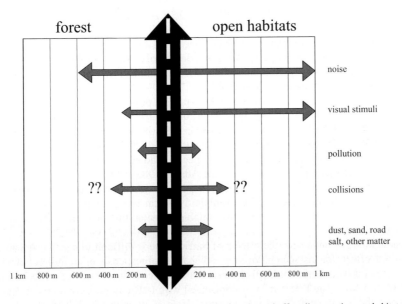

Figure 4. Schematic view of potential causal factors and their estimated effect distances in open habitat (e.g. grasslands) and forest (after Reijnen *et al.* 1995).

The effects of road collisions at first sight appear straightforward: road collisions cause an increased mortality along a road and this results in a lowered population level. The occurrence of road collisions can be determined by searching for road victims and many studies describe the large number of individuals and the wide variety of species found on roads (examples: Adams & Geis 1981; Hodson & Snow 1965; Füllhaas *et al.* 1989; Erritzoe *et al.* 2004). Other studies have concentrated on one species and the casualties are described as well as the consequences for the population dynamics being analysed. For instance, for the Barn Owl (*Tyto alba*), strong evidence indicates that road kills cause a significant decrease of the population size in several West-European countries including the UK (Ramsden 2003), France (Joveniaux 1985), Germany (Illner 1992b) and The Netherlands (Van den Tempel 1993). Also for the Nightjar (*Caprimulgus europeaus*) road kills are probably a major cause for the decline of the species in some areas (Kuitunen *et al.* 2003). A very convincing study was conducted on threatened Florida Scrub-Jay (*Aphelocoma coerulescens*) by Mumme *et al.* (2000): along a highway increased mortality was measured for adults and fledged young, probably as result of road kills. This increased mortality could not be compensated for by the reproductive output, consequently the territories along the road functioned as population sinks. Pied Flycatchers (*Ficdula hypoleuca*) breeding in nest boxes close to a highway showed a lower breeding success, possibly as a result of a higher parental mortality caused by road-kills (Kuitunen *et al.* 2003). A study on another small songbird the Willow Warbler (*Phylloscopus trochilus*) did not show significant differences in mortality rates between areas along roads and in control areas (Reijnen & Foppen 1994).

In conclusion, from these correlative studies it can be concluded that (traffic) noise and the effect of collisions appear to be most important causal factors for difference in breeding densities.

3.3 MODE OF ACTION: EFFECTS ON BEHAVIOUR AND POPULATION DYNAMICS

This section describes studies that give clues concerning the mode of action for the relation between traffic and the observed effects on breeding birds, concentrating on the two causal factors mentioned as most probable: noise load and road collisions. For both factors probable impacts on population dynamics are discussed.

In spite of the impressive numbers of road kills reported, poor evidence exists concerning the effect of road kills on bird population numbers (Forman 2003). This could be due to the fact that the majority of killed individuals are juveniles (Errritzoe *et al.* 2004) which would not contribute significantly to the recruitment rate because of their high natural mortality rate. Many authors therefore conclude that for many species road kills are not the cause of observed lower densities (Reijnen *et al.* 2002; Leedy & Adams 1982; Ellenberg *et al.* 1981; Forman 1995). The exceptions to this are long-lived species with a behaviour that makes them particularly vulnerable to collisions, for example owls that forage along the road verges and nightjars that roost on the roads. For most songbird species this does not offer a plausible explanation.

The possible mode of action for the effects of traffic noise is difficult to see. Noise can either cause a direct effect, for instance induced by a behavioural response, or an indirect effect. For the existence of direct, road avoiding responses (disturbances) little evidence exists and would be difficult to obtain. The fact that birds confronted with traffic noise have a raised heart beat rate suggests that noise leads to a stress reaction (Helb & Hüppop 1991). This stress reaction could lead to avoidance behaviour causing individuals to leave roadside areas. This is consistent with research into reactions of humans in response to noise. Long exposure to noise can induce psychological stress and physiological disorder in humans (e.g. Job 1996; Stansfeld *et al.* 1993).

In a willow warbler population along a highway there was some evidence that females did appear in territories along the road but were reluctant to stay, leaving the males unpaired (Reijnen & Foppen 1994). Consequently, in these roadside territories the total reproductive output *per capita* was negatively affected. That being said, the breeding success of males that could attract a mate was not different from territories in control areas. As a reaction to the poor reproductive output a high proportion of the males that occupied a territory along the road dispersed the following year and established territories at a greater distance from the road. They were replaced by first-year inexperienced breeders (Foppen & Reijnen 1994). The authors concluded that the territories along the road showed the characteristics of population sinks.

A number of studies address the impact of noise load on vocal communication behaviour of birds. Many bird species, particularly songbirds, rely heavily on song to attract mates and defend territories. There is a growing body of evidence that noise load hampers this communication and that this could lead to the observed effects on breeding densities. In reaction to traffic noise a number of songbird species start singing earlier in the morning before rush hour peaks and thus avoid unfavourable communications conditions (Bergen & Abs 1997). Other studies show that, as a response to noise load, species adapt their song. Chaffinches (*Fringilla coelebs*) in urban areas in Germany with high traffic noise use higher pitched sounds, probably to avoid the masking effect on their song (Bergmann 1993). Great tits (*Parus major*) show a similar response and also use higher pitched sounds in urban areas and adapt to a specific acoustic environment (Slabbekoorn & Peet 2003). Other studies indicate that not only is the frequency spectrum adapted as a response to a high ambient noise level, but also the amplitude of the song. The nightingale (*Luscinia megarhynchos*) responds to situations with a high traffic noise load by using higher sound levels (Brumm 2004) and this seems to hold true for other species (Madhusudan & Warren 2004).

If traffic noise induces a disturbance of communication one would expect a relationship between song frequency and the observed effects on density. Support for this comes from a German study (Rheind 2003). Breeding bird densities were compared between an area along a highway and an identical control area close by. A significant relation was found between effect size and the dominant frequency measured over the whole song. Species whose songs were generally at lower frequencies showed larger negative effects, meaning lower densities relative to control areas than species whose songs were generally at higher frequencies. This relation was irrespective of the size of the bird. Species with a dominant song frequency between 2000 and 4000Hz appeared particularly vulnerable. These are also the frequency ranges where traffic noise shows high sound pressure levels (Klump 2001). In a Dutch study traffic noise was used as dose variable for traffic load. Interestingly the observed threshold values (around 45dB(A)) are very comparable to predicted sound pressure levels that are expected to cause masking effects (around 47dB(A)). These predictions were derived from data on hearing capability of songbirds and information on the composition of traffic noise (Klump 2001).

We conclude that a large number of studies indicate that disturbance of vocal communication by traffic noise could be an important mode of action causing negative effects on breeding bird density along roads. Few studies show effects of traffic load on reproduction and probably these effects are indirect. A number of studies conclude that territories in the vicinity of roads function as population sinks (Reijnen & Foppen 1994; Kuitunen *et al.* 2003; Mumme *et al.* 2000). This means that the number and densities along roads depend on immigration. Supporting evidence comes from a Dutch study that demonstrates that, in years with higher population numbers the negative effects of road traffic on densities are smaller (Reijnen & Foppen 1995); probably because the influx from source areas towards sink areas is higher in these years.

Furthermore, this overview clearly demonstrates the need for studies with a proper experimental design in order to distinguish between the various potential causal factors. The studies on adapted song behaviour in relation to traffic noise show a very promising new line of study that will shed more light on the underlying mode of action. Figure 5 is a summary of the most probable causal relationships and underlying processes between road traffic and breeding birds.

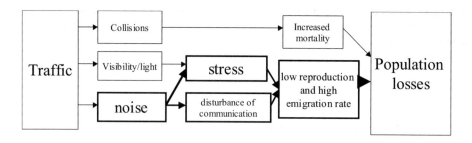

Figure 5. Probable relationship between road traffic and density of breeding birds.

4. Effects of road traffic on breeding bird populations at a regional scale: a case study for The Netherlands

4.1 INTRODUCTION

Extensive studies in The Netherlands have shown that road traffic depresses densities of half of the breeding bird species at distances over 1000m from busy highways and that affected species are spread over all taxonomic groups and all main habitat types (Reijnen *et al.* 2002; Foppen *et al.* 2002). Given the dense network of main roads with traffic densities up to 200,000 vehicles per day, it is argued that breeding density will suffer over large areas and lead to significant effects on the population size of species at the national scale (Reijnen *et al.* 2002).

Below are the results of an evaluation of the overall impact of road traffic on breeding bird populations in The Netherlands (Foppen *et al.* 2002), based on the quantitative relationships between densities and traffic related factors from Reijnen *et al.* (1995, 2002).

4.2 METHODS

The areas of the main Dutch terrestrial habitats were calculated using a digital map of habitat types (Reijnen *et al.* 2001). Traffic influence zones were based on the average and maximum (most sensitive species) threshold values for effects of traffic noise. Two different threshold values were available from previous studies, one for woodland birds and one for birds of wet agricultural meadows (Reijnen *et al.* 2002; see also Table 2). We assigned every species to one of these according to habitat preference and a simple decision rule: open (= meadow like) and closed (= forest like) habitats. With these thresholds and traffic information (intensity and velocity) we predicted the width of zones along roads that suffer from a lowered breeding bird density (Reijnen *et al.* 1995). The area of the habitat type affected by road traffic was estimated by using overlay-techniques in a GIS-system.

Effects on the national population size of species were calculated by multiplying the affected area of habitats with the reductions of the density for the average and the most sensitive species (Reijnen *et al.* 2002; see also Table 3). For one species, the black-tailed godwit (*Limosa limosa*), the loss of population was calculated in more detail. This species is the national icon for the wet agricultural meadows in The Netherlands, and the country harbours about 50% of the European population (Hagemeijer & Blair 1997). For the calculation the factors used were the affected area of agricultural wet meadows, an estimate of the percentage of population reduction within the traffic influence zones (47%, Reijnen *et al.* 1996) and an estimate of average breeding density in these areas in 5 x 5km squares calculated from the breeding bird survey (SOVON 2002), thus taking into account regional differences in the breeding density

4.3 RESULTS

The affected area of main terrestrial habitats in 1999 varied from 8% to 12% when the average threshold value was used and from 12% to 19% when the maximum threshold value was used (Figure 6). In woodland, marshland, heathland and a small part of the wet agricultural meadows, which together constitute most of the terrestrial habitats of the Dutch National Ecological Network, the pressure of road traffic was only slightly lower than it was in the other agricultural and urban habitat types. There were large regional differences, however: the total area of habitat types suffering from traffic impact varied from 26.5% in the centre of the country to 7.2% in the north (Figure 7).

Assuming that this reduction reflects a population loss, at national level this would mean a loss of 2-12% for an average species and 4-20% for the most sensitive species. For the black-tailed godwit the estimated population loss was about 10% or 5000 breeding pairs (40% of 5x5 squares affected, Figure 8).

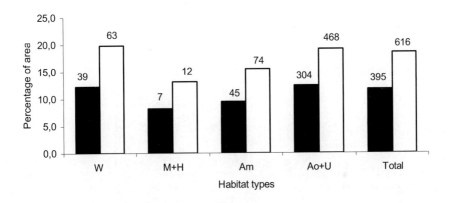

Figure 6. Estimated average (solid) and maximum (open) area (%) of main terrestrial habitat types in which breeding bird density is depressed by road traffic. The habitat types are: W, woodland; M, marshland; H, heathland; Am, wet agricultural meadows; Ao, other agricultural habitats; U, urban habitats. The numbers above the bars indicate area in ha x 1000.

R. REIJNEN & R. FOPPEN

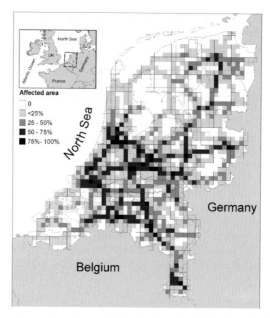

Figure 7. Distribution of area affected by road traffic, for the pooled habitat types found in The Netherlands.
Affected area in % per 5x5 km square.

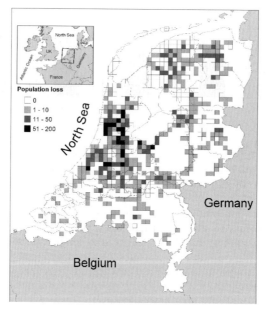

Figure 8. Distribution of traffic-induced population loss for the Black-tailed godwit (*Limosa limosa*) in The
Netherlands. Pattern of 5 x 5 km squares represents suitable habitat (wet agricultural meadows) of which 40% is
affected. Population loss in number of breeding territories per square.

4.4 DISCUSSION AND CONCLUSIONS

The assumption that the reduction of density reflects a population loss is supported by a demographic study of a population of willow warblers (*Phylloscopus trochilus*) along a highway showing that road traffic reduced the habitat quality: both breeding density and breeding success were lowered (Reijnen & Foppen, 1994; Reijnen & Foppen, 1995). The area affected by road traffic probably acted as a 'sink', which means that to survive, the population depends on immigrants from 'source' areas.

There are many indications that the size and persistence of breeding populations mainly depend on high quality areas (Wiens & Rotenberry, 1981; Bernstein *et al.* 1991). In consequence, the largest effects on the overall population size can be expected when a major loss of high-quality areas takes place. Also a further degradation of habitats with a low quality can have some effect, because these habitats may contribute significantly to the overall population size and act as a buffer for high-quality areas (Howe *et al.* 1991; Foppen *et al.* 2001). This is also supported by the study of Foppen and Reijnen (1994), who found that there is a quantitatively important breeding dispersal flow of male willow warblers from highway-induced low-quality habitat to high-quality habitat nearby.

On the other hand it is known that density is not always a good indicator of habitat quality and might be even misleading (Fretwell, 1972; van Horne, 1983). In several territorial bird species it has been shown that, when overall density is high, less-preferred habitat is more strongly occupied than when overall density is low (Kluyver & Tinbergen, 1953; Glas, 1962; O'Connor & Fuller, 1985). Similar relationships were found between habitats close to roads and habitats further away (Reijnen & Foppen 1995). This means that the size of zones adjacent to roads, that have a lower quality due to road traffic, can easily be underestimated when based on density data.

In The Netherlands, road traffic causes nationwide significant population losses for many species. The pressure of road traffic on nature areas in The Netherlands increased sharply from 1975 to 1999 and is mainly attributed to the increase of traffic intensity on existing roads (Foppen *et al.* 2002). Our findings, which can be extrapolated to many large areas in other industrialized countries (e.g. Forman 2000), illustrate the need for compensatory and mitigating measures. Such measures are particularly important since breeding birds in densely populated industrialized areas suffer from many other human impacts.

5. Practical implications for road planning and management

To minimize the ecological impacts in road planning one, can distinguish three successive steps (Reijnen *et al.* 1995; Cuperus 2004). The first step is to avoid ecological impacts by exploring route alignments that prevent conflicts. In the second step any unavoidable impacts are mitigated. Finally, in the third step ecological impacts that remain are compensated. The Netherlands is one of the countries currently applying the 'ecological compensation principle' to bring about a no-net-loss situation for nature when planning and constructing large-scale projects, such as highways (Cuperus 2004). The need to compensate for the impact of traffic has stimulated project initiators to develop alternative routes or route sections to avoid or reduce ecological impacts and also to apply coherent compensation and mitigation.

The impact of existing road networks on breeding bird populations is still receiving scant attention (Cuperus 2005). Given the large area affected mitigation is probably the most effective measure in reducing this impact.

5.1 AVOIDANCE

To avoid effects of road traffic on breeding bird populations, one should take into account a sufficient distance from important bird areas based on expected traffic densities. If extensive data on the relationship between breeding bird densities and traffic related factors are available, as in The Netherlands, accurate estimation of required distances to important bird areas is possible. Otherwise one can use the manual based on the results of Dutch studies that has become a standard reference in highway planning in The Netherlands and many other countries (Reijnen et al. 1995; see example in Figure 9). Generally estimated distances in relation to traffic density are similar amongst different studies. However, adjustments might be necessary, since some species absent in the Dutch studies exhibit much greater effect distances (Traonidae, Raty 1979; stone curlew, Green et al. 2000). In general, 1000m to either side of the road seems to be an adequate distance (Reijnen et al. 2002).

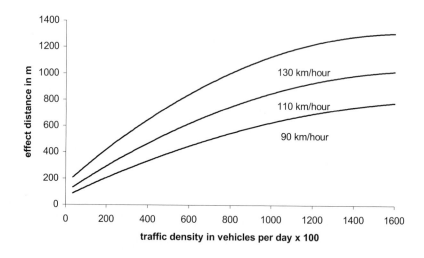

Figure 9. Prediction of the road effect zone for breeding birds in open field habitat based on data that describe the relationship between traffic noise and the total density of all species. To favour practical application traffic density and speed of vehicles are used to determine the effect distance (after Reijnen et al. 1995).

5.2 MITIGATION

Because traffic noise is probably the main cause of the density-depressing effect, noise-suppressing measures should be effective in reducing the impact on birds (Reijnen et al. 2002). Once in place experimental evidence and monitoring of their efficacy is desirable.
In the short term, the most achievable option is noise emission control by means of noise screens or earth walls, porous road surfaces and depressed or underground roads (Reijnen et al. 2002; Forman et al. 2003; Cuperus 2004); in the long term noise emission should be controlled by producing quieter vehicles (Sandberg 1991). To achieve a sufficient reduction of the noise level along roads the noise suppressing screens or walls should be high (Reijnen et al. 1995). However, in open areas, such screens or wall barriers may also act as a source of disturbance themselves, because many birds avoid the vicinity of hedgerows, wooded banks

and dikes up to several hundred metres (van der Zande *et al.*, 1980). So, application of high screens seems appropriate for only major problems (e.g. large disturbance distances in important areas for breeding birds). Moreover, in constructing noise screens for birds, one should take into account that they can hamper movements of other animal species. This may give rise to fewer problems if under- or overpasses for wildlife are present. On the other hand, noise screens and earth walls have the advantage that they will also reduce other traffic-related influences, such as visual disturbance.

5.3 COMPENSATION

To compensate for reduced densities adjacent to roads improvement of the conditions of existing habitats is more preferable than creating new habitats. The longer the time required for the habitat to develop, the more difficult it will be to compensate for the impacts (Anderson 1994). Generally, it is better to locate compensation sites outside the zone of road effects. Improvement of habitat conditions inside the road effect zone, as suggested by Cuperus (2004), probably has limited effect. For the willow warbler it was shown that, in an optimal habitat, both density and breeding success were lowered and that the road effect zone might act as a sink (Reijnen & Foppen 1994). This indicates that road effect zones make a minor contribution to the size and persistence of regional breeding bird populations (Reijnen *et al.* 2002). In view of this, it seems also better to compensate for the full densities in the affected zone rather than for the expected reduced densities. It is obvious that one should have low expectations of positive effects of roadside management for breeding birds in general. Contrarewise, in large agricultural areas, the occurrence of many species, including birds, may depend on roadside habitat (see e.g. Bennett, 1991).

Generally, compensation sites are located close to the affected habitats in the road effect zone. This approach has the disadvantage that it favours 'patchy' compensation and reduces overall ecological effectiveness (Cuperus 2004). To develop 'compensation banks', physically combining compensation for different projects and nature development programmes into larger contiguous areas is proposed as a more successful strategy (Cuperus 2004).

6. Conclusions

ROAD TRAFFIC AS A THREAT TO BREEDING BIRD POPULATIONS

- Road traffic is a general and widespread cause of decreased densities in breeding birds. More than half of the species of local and regional bird faunas spread over all taxonomic groups and in all main habitat types can be affected.
- The effects increase with growing traffic densities and extend outward for more than 1000m for busy highways. Population losses in the affected zone range from 30% to almost 100%.
- In countries or regions with dense road network traffic the effects can lead to significant population losses for many species. Examples for The Netherlands indicate effects on the population level of up to 20%.

PROBABLE CAUSAL FACTORS AND MODE OF ACTION

- Of the many potential traffic-related causal factors, noise is probably the most important at larger distances.

272   R. REIJNEN & R. FOPPEN

- For some specific species road-kills are an important causal factor for effects on the population level. Road kills directly influence survival and causes population losses.
- Available data point to stress and disturbance of communication as the most important mode of action.

PRACTICAL IMPLICATIONS

- In road planning ecological impacts can be minimised successively by avoidance, mitigation and compensation. The 'ecological compensation principle' implies that, for new road developments, there is a goal of a no-net-loss situation for nature. This principle is now a legal requirement in The Netherlands, and is beginning to be applied elsewhere.
- The impact of existing road networks on breeding bird populations which is still receiving scant attention shouldbe considered far more. Mitigation is probably the most effective measure to reduce this impact.
- To compensate for effects one should take into account the full densities in the affected zone rather than the expected reduced densities.
- Because traffic noise is probably the main cause of the density-depressing effect noise-suppressing measures are likely to be effective in mitigating the impact on birds

RESEARCH AGENDA

- Few studies have investigated the relationship between traffic density (or traffic-related factors) and the density of many breeding bird species in various habitat types. Such data are important particularly when compensation measures are needed. Study sites should be located along roads with a range of traffic densities from low to high. Combining this correlative approach with population studies for key species using capture-recapture methods greatly enhances appropriate interpretation of the results. The studies of Reijnen *et al.* (1995, 1996). Reijnen & Foppen (1994) and Foppen & Reijnen (1994) can serve as a model for this approach (Hill *et al.* 1997).
- There is also a need for studies with a proper experimental design in order to distinguish between various potential causal factors. Studies on adapted song behaviour in relation to traffic noise show a very promising new line of study that will shed more light on the underlying mode of action. Such data are important in testing the efficacy of mitigating measures.
- To understand the consequences of the impact of road traffic on population levels of species regional scale studies are necessary. The significance of such studies increases greatly when other threats are taken in account. This may be of great value in spatial planning at the regional scale.
- In planning new roads, long-term monitoring schemes are very useful in obtaining greater insight into the efficacy of measures to minimise the road impact.
- Similar studies should be undertaken for railways.

References

Adams LW, Geis, AD. 1981; Effects of highways on wildlife. Report No. FHWA/RD-81/067, Office of Research, Federal Highway Administration, U.S. Department of Administration, Washington.
Anderson P. 1994; Road and nature conservation; guidance on impacts, mitigation and enhancement. English Nature, Northminster House, Peterborough, England.

Bennett AF. 1991; Roads, roadsides and wildlife conservation: a review. In: Saunders DA & Hobbs RJ (eds) Nature conservation 2; the role of corridors, 99-117. Surrey Beatty & Sons, Chipping Norton, Australia.

Bergen F, Abs M. 1997; Etho-ecological study of the singing activity of the Blue Tit (Parus caeruleus), Great Tit (Parus maĵr) and Chaffinch (Fringilla coelebs). Journal für Ornithologie 138: 451-467.

Bergmann H.-H. 1993; Der Buchfink: neues über einen bekannten Sänger, Wiesbaden.

Brumm H. 2004; The impact of environmental noise on song amplitude in a territorial bird. Journal of Animal Ecology 73: 434-440.

Clark WD, Karr JR 1979; Effects of highways on Red-winged Blackbird and Horned Lark populations. Wilson Bulletin 81: 143-145.

Cuperus R. 2005; Ecological compensation of high impacts: negotiated trade-off or no-net-loss? Ph.D. thesis, University of Leiden, The Netherlands.

Ellenberg H, Müller K, Stottele T. 1981; Strassen-Ökologie. Broschürenreihe der Deutschen Strassenliga, Bonn.

Erritzoe J, Mazgajki TD, Rejt. L. 2004; Bird casualties on European roads- a review. Acta Ornithologica 38: 77-93.

Fernández-Juricic E. 1999; Avifaunal use of wooded streets in an urban landscape. Conservation Biology 14: 513-521.

Ferris CR. 1979; Effects of Interstate 95 on breeding birds in northern Maine. Journal of Wildlife Management 43: 421-427.

Findley CS, Bourdages J. 2000; Response time of wetland biodiversity to road construction on adjacent lands. Conservation Biology 14: 86-94.

Foppen R, Chardon JP, Liefveld W. 2001; Understanding the role of sink patches in source-sink metapopulations: Reed Warbler in an agricultural landscape. Conservation Biology 14: 1881-1892.

Foppen R, Reijnen R. 1994; The effects of car traffic on breeding bird populations in woodland. II. Breeding dispersal of male willow warblers (Phylloscopus trochilus) in relation to the proximity of a highway. Journal of Applied Ecology 31: 95-101.

Foppen R, van Kleunen A, Loos W.-B, Nienhuis J, Sierdsema H. 2002; Broedvogels en de invloed van hoofdwegen, een national perspectief. Onderzoeksrapport nr 2002/08, SOVON Vogelonderzoek Nederland, Dienst Weg- en Waterbouwkunde, Delft.

Forman RTT. 2000; Estimate of the area affected ecologically by the road system in the United States. Conservation Biology 14: 31-35.

Forman RTT, Reineking B, Hersperger AM. 2002; Road traffic and nearby grassland bird patterns in a suburbanizing landscape. Environmental Management 29: 782-800.

Forman RTT et al. 2003; Road ecology: science and solutions. Island Press, Washington.

Fretwell. 1972; Populations in a seasonal environment. Princeton University Press, Princeton.

Füllhaas U, Klemp C, Kordes A, Ottersberg H, Pirmann M, Thiessen A, Tschoetschel C, Zucchi H. 1989; Unterschuchungen zum Strassentod von Vögeln, Säugetieren, Amphibien und Reptilien. Beiträge zur Naturkunde Niedersachsens 42: 129-147.

Glas P. 1962; Factors governing density in the chaffinch (Fringilla coelebs) in different types of wood Archives Neerlandaises de Zoologie 13: 466-472.

Green RE, Tyler GA, Bowden CGR. 2000; Habitat selection, ranging behaviour and diet of the stone curlew (Burhinus oedicnemus) in southern England. Journal of Zoology 250: 161-183.

Helb H.-W, Hüppop O. 1991; Herzschlagrate als Mass zur Beurteilung des Einflusses von Störungen bei Vögeln. Ornithologenkalender 1992. Aula Verlag, Wiesbaden.

Helldin JO, Seiler A. 2003; Effects of roads on the abundance of birds in Swedish forest and farmland. In: IENE Conference 2003 Proceedings. Infra Eco Network Europe, Brussels, Belgium.

Hill D. 1992; The impact of noise and artificial light on waterfowl behaviour: a review and synthesis of available literature. British Trust for Ornithology. Tring, UK.

Hill D, Hockin D, Price D, Tucker G, Morris R, Treweek J. 1997; Bird disturbance: improving the quality and utility of disturbance research. Journal of Applied Ecology 34: 275-288.

Hodson NL, Snow DW. 1965; The road deaths enquiry, 1960-1961. Bird Study 12: 90-99.

Horne B van. 1983; Density as a misleading indicator of habitat quality. Journal of Wildlife Management 47: 893-901.

Howe RW, Davies GJ, Mosca V.1991; The demographic significance of 'sink' populations. Biological Conservation 57: 239-255.

Illner H. 1992a; Effects of roads with heavy traffic on grey partridge (Perdix perdix) density. Gibier Faune Sauvage 9: 467-480.

Illner H. 1992b; Road deaths of Westphalian owls: methodological problems, influence of road type and possible effects on population levels. In: Galbraith CA, Taylor IR & Percival S (eds) The ecology and conservation of European owls, 94-100. UK Nature Conservation 5, Peterborough.

Job RS. 1996; The influence of subjective reactions to noise on health effects of the noise. Environmental International 22: 93-104.

Joveniaux A. 1985; Influence de la misé en service d'une autoroute sur la faune sauvage. In: Bernard J-M, Lansiart M, Kempf C & Tille M. (eds) Actes du colloque Routes et Faune Sauvage, 211-228. SETRS, Colmar.

Klump GM. 2001; Die Wirkung von Lärm auf die auditorische Wahrnehmung der Vögel. Abgewandte Landschaftsökologie H. 44: 9-23.

Kluyver HN, Tinbergen L.1953; Territory and the regulation of density in titmice. Archives Neerlandaises de Zoologie 10: 265-289.

Kuitunen M, Rossi E, Stenroos A. 1998; Do highways influence density of land birds. Environmental Management 22: 297-302.

Kuitunen MT, Viljanen J, Rossi E, Stenroos A. 2003; Impact of busy roads on breeding success in pied flycatchers *Ficedula hypoleuca*. Environmental Management 31: 79-85.

Lowell LG, Best LB, Prather M. 1977; Lead in urban and rural song birds. Environmental Pollution 12: 235-238.

Madhusan K, Warren PS. 2004; Tits, noise and urban bioacoustics. Trends un Ecology and Evolution 19: 109-110.

McCullagh P, and Nelder JA. 1989; Generalized Linear Models, second edition. Chapman and Hall. London.

Milsom TP, Langton SD, Parkin WK, Peel S, Bishop JD, Hart JD, Moore NP. 2000; Habitat models of bird species' distribution: an aid to the management of coastal grazing marshes. Journal of Applied Ecology 37: 706-727.

Molenaar JG de, Jonkers DA, Sanders ME. 2000; Wegverlichting en natuur. III Lokale invloed van weg verlichting op een gruttopopulatie. DWW Ontsnipperingsreeks 38, Delft.

Mumme RL, Schoech SJ, Woolfenden GE, Fitzpatrick JW. 2000; Life and death in the fast lane: demographic consequences of road mortality in the Florida Scrub-Jay. Conservation Biology 14: 501-512.

O'Connor RJ, Fuller RJ. 1985: Bird population responses to habitat. In: Taylor K, Fuller RJ & Lack PC (eds), Bird census and atlas studies.197-211. British Trust for Ornithology. Tring.

Peris SJ, Pescador M. 2004; Effects of traffic noise in passerine populations in Mediterranean wooded pastures. Applied Acoustics 65: 357-366.

Ramsden D. J. 2003; Barn Owls and major roads: results and recommendations from a 15-year research project. The Barn Owl Trust, Ashburton.

Räty M. 1979; Effect of highway traffic on tetraonid densities. Ornis Fennica 56: 169-170.

Reijnen R, Foppen R. 1994; The effects of car traffic on breeding bird populations in woodland. I. Evidence of reduced habitat quality for willow warblers (Phylloscopus trochilus) breeding close to a highway. Journal of Applied Ecology 31: 85-94.

Reijnen R, Foppen R, ter Braak C, Thissen J. 1995; The effects of car traffic on breeding bird populations in woodland. III. Reduction of density in relation to the proximity of main roads. Journal of Applied Ecology 32: 187-202.

Reijnen R, Foppen R. 1995; The effects of car traffic on breeding bird populations in woodland. IV. Influence of population size on the reduction of density of woodland breeding birds. Journal of Applied Ecology 32: 481-491.

Reijnen R, Foppen R, Meeuwsen H. 1996; The effects of traffic on the density of breeding birds in Dutch agricultural grasslands. Biological Conservation 75: 255-260.

Reijnen R, Foppen R, Veenbaas G. 1997; Disturbance by traffic as a threat to breeding birds: evaluation of the effect and considerations in planning en managing road corridors. Biodiversity and Conservation 6: 567-581.

Reijnen R, Foppen R, Veenbaas G, Bussink H. 2002; Disturbance by traffic as a threat to breeding birds: evaluation of the effect and considerations in planning en managing road corridors. In: Sherwood B, Cutler D &. Burton JA (eds.) Wildlife and Roads, the ecological impact, 249-267. Imperial College Press, London.

Reijnen R, Veenbaas G, Foppen RPB. 1995; Predicting the effects of motorway traffic on breeding bird populations. Ministry of Transport, Public Works and Water Management, Road and Hydraulic Engineering Division, The Netherlands.

Rheindt FE. 2003; The impact of roads on birds: Does song frequency play a role in determining susceptibility to noise pollution? Journal für Ornithologie 144: 295-306.

Sandberg U. 1991; Abatement of traffic, vehicle and tire/road noise - the global perspective. Noise Control Engineering Journal. 49: 170-181.

Slabbekoorn H, Peet M. 2003; Birds sing at a higher pitch in urban noise. Nature 424, 267.

Spellerberg IF. 2002; Ecological effects of roads. Science Publishers, Inc. Enfield, USA.

Stansfeld SA, Sharp DS, Gallacher J, Babisch W.1993; Road traffic noise, noise sensitivity and psychological disorder. Psychological Medicine 23: 977-985.

Stone E. 2000; Separating the noise from the noise: A finding in support of the "Niche Hypothesis", that birds are influenced by human-induced noise in natural habitats. Anthrozoos 13: 225-231.

Tulp I, Reijnen MJSM(R.), ter Braak CJF, Waterman E, Bergers PJM, Dirksen S, Snep RPH, Nieuwenhuizen W. 2002; Effect van treinverkeer op de dichtheden van weidevogels. Rapport nr. 02-34, Bureau Waardenburg, Culemborg, The Netherlands.

Weiserbs A, Jacob J.-P. 2001 ; Le bruit engender par le traffic autoroutier-influence-t-il la repartition des oiseaux nicheurs? 2001. Alauda 69: 483-489.

Zande AN, van der, ter Keurs WJ, van der Weijden WJ. 1980; The impact of roads on the densities of four bird species in an open field habitat – evidence of a long-distance effect. Biological Conservation 18: 299-321.

CHAPTER 13: TOWARDS THE SUSTAINABLE DEVELOPMENT OF MODERN ROAD ECOSYSTEMS

LISA M.J. DOLAN,[1] H. VAN BOHEMEN,[2] PADRAIG WHELAN,[1] K.F. AKBAR,[3] VINCENT O'MALLEY,[4] GERARD O' LEARY[5] & P.J. KEIZER.[2]

[1] Department of Zoology, Ecology and Plant Science, University College Cork, Ireland;
[2] Dutch Road and Hydraulic Engineering Institute and Civil Engineering Section of Delft University of Technology, The Netherlands;
[3] Department of Botany, Government College, Sahiwal, Pakistan;
[4] National Roads Authority, Dublin, Ireland;
[5] Environmental Protection Agency, Dublin, Ireland.

1. Introduction

As road ecosystems are rapidly becoming the most familiar land-use type around the world, it is generally recognised that they will continue to remain an essential component of post-modern day living. Road ecosystems can extend for thousands of kilometres across several landscape types, regions, countries and continents, under estuaries and even seas. Road ecosystems are primarily built for socio-economic purposes, to facilitate commerce, including human economic service(s), and secondarily to provide access to landscapes for multiple values and leisure activities (Brown 2003). Examples of road ecosystems include: major arterial roads such as freeways, highways, and motorways; regional or local roads; forest roads; agricultural roads, and utility roads.

To accommodate economic services, road ecosystems must continue to remain efficient and competitive which generally demands larger and faster roads such as highways and motorways, while socially, transportation systems are required for day-to-living to reach employment centres, shopping centres, schools and recreational facilities. For these reasons roads must remain safe and accessible. In addition, transportation systems need to respect the natural environment including the landscape, its ecosystems and habitats.

Sustainable road ecosystem development is therefore concerned with integrating economic, social and environmental considerations into decisions regarding human mobility. In response to modern road ecosystem development, an increasing number of disciplines are becoming involved in the different phases of road development that comprise planning; design, construction, operation/use, maintenance and decommissioning. International best practice has demonstrated that modern road ecosystem development can improve the standard of living and quality of life for people, while simultaneously retaining and restoring landscape quality, ecological functions and services, contributing to biodiversity conservation and honouring local distinctive cultural and historical features of the landscape. Recent international literature and debate has recognised the need for a transdisciplinary and holistic focus on road ecosystem development that would bring together the broad range of formerly divergent disciplines now involved in the different phases of the road development process. This approach has the potential to merge what may currently be seen as the incompatible objectives of road development and biodiversity conservation and restoration, ultimately leading to further advancements in the field of road science and in best practice in the development of state-of-the-art sustainable road ecosystems.

However, a large body of literature exists concerning the negative effects and externalities of road ecosystems that continues to make it difficult for policy-makers to derive a coherent message about best practice, the potential positive effects of road ecosystem development and

John Davenport and Julia L. Davenport, (eds.), The Ecology of Transportation: Managing Mobility for the Environment, 275–331, how to adhere to sustainable development criteria. For these reasons, the first few sections of this chapter examine and summarise the large body of international literature available on all of the possible negative effects of road development. The remaining and majority of sections concentrate on the various aspects of international best practice which can retain and restore landscape quality, ecological functions and services, native vegetation amongst others. The sections also emphasise the important role of the roadside landscape in restoring elements to a landscape, where they have been lost due to past land-uses.

More specifically, this chapter establishes a chronological framework based on the phases of road ecosystem development which clarifies the various strategies and techniques available that can address sustainable development criteria, including a reduction in the consumption of natural resources at each phase of road development. The chapter also focuses on techniques that can be applied at the local level i.e. the road location or site scale, to address the dearth of techniques available at this level which address the principles of sustainable development. The chapter also contributes to the development of sustainable transportation systems as it facilitates the identification of possible solutions to environmental and technical problems posed by the various phases of road development. Most importantly, the framework identifies problems and solutions at the earliest possible development phase of a road scheme, and in advance of contractual commitment. Aspects of current policy and contractual obligations that provide for, and indeed, hinder the restoration of ecological function and landscape quality, pre- and post- road construction are also examined, as are the long-term benefits of the implementation of the European Union (EU) Directive on Strategic Environmental Assessment (SEA).

2. The 'Ecological Footprint' of modern road ecosystems

Material Flows Analysis (MFA) is increasingly being utilised by civil engineers and environmental scientists to monitor the flow of materials through urban, industrial and domestic systems or techno-systems. MFA identifies the various extraction methods, processes, inputs and outputs of the system in question, especially waste outputs, as these have to be avoided or minimized to reduce environment impacts. More specifically, the material flow chain includes: the extraction of the raw material or natural resource, processing, the supply of material, conversion into products, the placing of products on the market and their fate at end-of-life (Irish EPA 2002a).

In the literature, roads are increasingly being referred to as anthropogenic ecosystems, techno-systems or road ecosystems by ecologists, in recognition of the fact that roads exchange materials, including natural resources, with natural and semi-natural ecosystems in the surrounding landscape. This is the foundation of the ecosystem approach to road development. Lugo and Guncinski (2000), Dolan (2004) and Dolan and Whelan (2004b) described the interaction between roads and ecosystems, and how roads themselves have come to be defined as ecosystems.

Applying an ecosystem approach to roads is similar in principle to the MFA approach as it allows the various disciplines involved in road planning, design, construction, operation/use, maintenance and decommissioning to visualize the input, throughput and output of natural resources leading to a more dynamic and specific comprehension of how roads interact with semi-natural ecosystems.

It is now universally recognised that it is the application of the ecosystem approach to road development, which can explain the interaction between roads and adjacent ecosystems, landscapes and regions, ultimately providing a more accurate environmental assessment and reduction of the ecological footprint of road ecosystem development (CBD 2004). Ecological footprint analysis is a tool which estimates the natural resource consumption and waste assimilation of a defined human activity (Rees & Wackernagel 1996). Though roads are a necessary component of modern life, the establishment of road ecosystems can affect the integrity of natural and semi-natural ecosystems, as roads can have an adverse impact on natural resources, and on an ecosystem's ability to provide the flow of services within a landscape. During the process of land conversion for road development, natural resources can be heavily consumed such as land, water, biodiversity, air, light and energy, as well as the land itself. Consequently, the ecological functions and services, such as nutrient cycling and the regulation of water flow provided by semi-natural ecosystems may be severely compromised and previously self-sustaining ecosystems can be transformed into dependent ecosystems requiring mitigation, compensation and maintenance - where avoidance of adverse impacts is not possible (U.S. Council on Environmental Quality 1978; van Bohemen 1995).

Land, water and biodiversity are generally the most heavily consumed natural resources in road ecosystem development, as all three are linked with habitat loss and habitat fragmentation. Another comparatively minor impact, but critical from an ecological and sustainable development viewpoint, is the consumption loss of ecological functions and services of soil due to extensive soil 'sealing', movement and compaction within the road corridor.

Soil sealing refers to the covering of soil as a result of residential, industrial and transportation infrastructure with the result that soil is no longer able to perform the range of ecological functions associated with it (EEA 2002). Road ecosystem development creates corridors, and networks, of sealed soil due to the roadbed that supports the upper layer of asphalted road carriageway. Once sealed beneath the carriageway, or compacted by machinery, the ability of soil to provide habitat for flora and fauna, macrofauna, microfauna, and micro-organisms (bacteria, fungi, algae, actinomycetes and protozoa) is greatly reduced. Other ecological functions can be severely compromised, e.g. nutrient cycling (carbon, nitrogen, phophorus, sulphur), regulation and partition of water flow, microbial and biochemical transformations, (Irish EPA 2002b).

It should be noted that the sealing of soil is not adverse per se. Though it limits the extent of the ecological functions and services performed by soil, the integrity of the substrate can continue to remain intact. Rather it is the irreversibility of the sealing process (EEA 2002) and the soil movement and compaction associated with the construction of a road corridor, plus the increased rate in surface runoff that is more significant.

It must also be emphasised that residential infrastructure, industrial systems and road ecosystems also include land that is not actually sealed (EEA 2002). In fact it is the 'unsealed' roadside landscape that presents opportunities to mitigate and compensate for the negative effects of road development by the restoration of the essential ecological functions and services of soil and the retention, and restoration of landscape quality and native vegetation. Significantly the roadside landscape also provides a platform for restoring landscape quality, native vegetation and the integrity of soil where they have been previously compromised by past land-uses such as intensive agricultural management.

Best practice has demonstrated that, where appropriately planned and executed, roadside landscapes can evolve into self-sustaining linear habitats or corridors for wildlife, which exchange natural resources and a range of ecological functions and services with the adjacent "agro-ecosystems". However, self-sustaining characteristics cannot be achieved without specific sourcing, storage and manipulation of soil during the design and construction of roadside landscapes. Appropriate soil conditions are required to restore nutrient cycling and the regulation of water flow, and to provide a suitable substrate to support the restoration of the past native vegetation of an area.

In the construction of realignments and by-passes, there may also be a limited opportunity to restore the ecological functions of sealed soil, through the decommissioning of old road pavement which reverses the sealing process.

Apart from the loss of soil integrity, the consumption of natural resources in the construction, operation/use, maintenance and decommissioning of a road ecosystem may also contribute to the following impacts on semi-natural ecosystems: loss of habitat and native species, habitat fragmentation, invasions by non-native species, loss of landscape quality, decreased water quality; soil erosion; and, noise pollution (Southerland 1995; Lugo & Gucinski 2000; Dolan 2004). All of these impacts have their own specific avoidance, mitigation and compensation measures as required under international best practice.

2.1 HABITAT LOSS

In relation to road development, habitat loss and habitat fragmentation have been identified in the literature as having the most significant negative impacts on native vegetation, wildlife, ecosystem function and landscape quality (see Dolan 2004). Fahrig (1997) showed that the effects of habitat loss far outweigh the effects of fragmentation on wildlife populations. Loss of semi-natural habitat reduces the extent to which an ecosystem can support biodiversity and still provide ecological functions and services including the exchange of materials and natural resources.

2.1.1 *Construction phase and habitat loss*

The construction phase of road development is the most environmentally traumatic to adjacent ecosystems as it is primarily an agent of disturbance and habitat loss. Habitat loss can result from: (1) soil movement and compaction leading to a loss of or alteration to native vegetation during the initial construction of the road corridor (Forman 1995; Jaarsma & Willems 2002; Council of Europe 2001), (2) the inappropriate storage of soil and other materials during construction, (3) archaeological activities carried out pre-construction and during construction, (4) soil sealing as a result of the road bed and road carriageway, (5) inappropriate landscaping activities post-construction (Theobald *et al.* 1997; Dolan 2004), (6) rechannelisation or re-canalisation of streams and rivers (Figure 1).

Figure 1. Rock armouring, gabions and stream. (Dolan & Whelan)

During construction, vegetation and soil are removed as machinery moves through areas of high and low elevation, levelling the natural undulating topography of the landscape to form the new road corridor. The extent of habitat loss varies with the extent i.e. width and length of the road corridor which, in turn, depends on the type of road and its function (e.g. motorway, forestry road etc.), and the surrounding landscape type.

Additional habitat loss can occur in particular regions or landscapes that are known to be of significant archaeological interest since these areas can be subjected to extensive vegetation removal and soil movement to allow for archeological digs. Also, formerly unknown sites of archaeological interest may be uncovered during road construction that may require further investigation. In such circumstances, consultation is required with ecologists to identify vegetation of national, regional and local importance.

Further habitat loss can also occur, where existing native vegetation within the land-take is not incorporated into the re-vegetation of the roadside landscape, but is replaced by formal landscape design layouts utilizing non-native species. Where invasive species are overlooked or indeed planted during landscaping there can be a further loss of plant community structure and a loss of habitat for wildlife (see Dolan 2004).

Rechannelisation of streams, where inappropriately executed, can disturb native vegetation and the natural architecture of a streambed and bank that can result in loss of habitat, loss of feeding sites and reduced access to streams for wildlife.

2.1.2 *Operation/use and maintenance phases, and habitat loss*

Habitat loss can, however, also occur during the operation/use and maintenance phases of road ecosystem development. Noise pollution from traffic, and light pollution from street furniture, can lead to the creation of disturbance zones for birds and mammals including bats. Spillages from vehicles and surface water run-off can result in the pollution of aquatic systems such as watercourses, but also the degradation of soil and vegetation, which can make native vegetation cover in the vicinity of a road ecosystem uninhabitable. Prolonged maintenance activities can mimic construction, as resurfacing can alter roadside vegetation, along with frequent and inappropriately timed trimming, cutting and mowing of roadside vegetation. Such activities determine the wildlife and plant species found in an area (Theobald *et al.* 1997), ultimately contributing to habitat loss for certain species.

The alteration of native vegetation within the landscape can also affect the visual attributes of the landscape through loss of naturalness, rural structuring, variety and regional identity.

2.2 HABITAT FRAGMENTATION

Habitat fragmentation as a result of road ecosystem development is a complex phenomenon since it is brought about as a result of the combined effects of habitat loss and the indirect effects of construction, operation/use and maintenance phases of road ecosystems.

2.2.1 *Construction phase and habitat fragmentation*

As construction activities remove soil and vegetation within the road corridor, to make way for the road carriageway, the remaining habitat becomes bisected or fragmented which limits its ability to perform ecological functions and services, and to provide for the exchange of materials and natural resources within the landscape. This fragmentation is further compounded once the road surface and associated street furniture are in place i.e. when the road is fully operational and noise and light pollution (amongst others) arise.

Habitat fragmentation has been defined as the direct removal of, or the breaking up of habitat, leading to the fragmentation of remaining habitat into smaller parcels, isolated remnants or patches leading to more and smaller habitat patches. As a result, there is decreased connectivity between patches and complexity of patch shape leading to a higher proportion of 'edge' habitats which influences, the ability of organisms to migrate and disperse between ecosystems and across the landscape. Thus roads can significantly 'fragment' wildlife populations, reduce local population sizes and lead to local population extinctions (e.g. Forman 1995; Dolan 2004). The fear provoked by sudden and unusual noises, activities, or changes in the environment during initial construction can also create a barrier to wildlife, in the form of temporary 'disturbance zones' (Theobald *et al.* 1997).

Once in place the paved road surface can also act as a barrier to insects and other wildlife unwilling to cross the road pavement (Mader 1988).

The use of drainage culverts to divert a stream temporarily during construction may also block the passage of Otters (*Lutra lutra*) and amphibians. If the bottom of the culvert is too high above the base channel for fish to jump or where the gradients and velocities in the culverts are too steep and rapid for fish passage movement of juvenile and adult fish is inhibited (NCSAI 2001).

Best practice has shown that mitigation measures are available which can provide effective permeability for wildlife by providing for the safe passage of wildlife either under or over roads. Such measures include pipe culverts and green bridges.

2.2.2 *Operation/use phase and habitat fragmentation*

During operation/use, road ecosystems can create permanent barriers to the dispersal of wildlife as a result of chronic disturbances from traffic and other human activities, leading to the creation of almost permanent disturbance zones. Where such zones are present, fauna can maintain some distances from roads, ultimately dividing home ranges of species, which can have detrimental effects on populations, including local extinctions (Theobald *et al.* 1997). Rates of road use also select for certain types of organisms that can respond to road use (Lugo & Gucinski 2000), resulting in what is known as the 'filter effect' of road corridors (Forman 1995; Clevenger 2000). However, the main demonstrated impact of roads on wildlife to date, has been in terms of increased mortality due to road kills. Mortality numbers are high for certain species, especially: (1) during migratory periods, (2) during the breeding season (Forman 1995), (3) in highly populated areas (Bank *et al.* 2002), (4) where roadside food (Forman 1995) acts as an attractant, (5) where transparent noise barriers are placed on road schemes, (6) where damaged road surfaces provide watering holes for birds. Recent studies also suggest that vegetated centre medians can increase road mortalities of wildlife (Clevenger *et al.* 2003).

2.2.3 *Maintenance and decommissioning phases and habitat fragmentation*

Current maintenance practices can reduce the existing and potential floral and faunal diversity of roadside landscapes. Inappropriately timed mowing or cutting of grasslands and hedgerows, can remove or disturb habitats for birds, mammals, amphibians and invertebrates, through the noise associated with machinery and through the loss of vertical and horizontal structural differentiation of the vegetation. Where post-cutting removal of hay is not practised, the potential development of high species diversity within grasslands may not be realised (Parr & Way 1988; Perrson 1995). The use of herbicides should also be avoided.

Inappropriate activities during the decommissioning phase of a road ecosystem can lead to the disturbance of adjacent semi-natural habitats-especially sensitive habitats such as wetlands through noise disturbance, but also through the release of pollutants from the breaking up of the road surface. These adverse effects must be weighed against the fact that where old road pavement persists, it can continue to pose a barrier to wildlife dispersal.

2.3 NON-NATIVE INVASIVE SPECIES

The activities carried out during the construction, operation/use and maintenance of road ecosystems, are likely to aid in the dispersal and establishment of non-native invasive species, as a result of disturbance and the 'abrupt physical nature' or edge effect of the roadside landscape ecotone. In such circumstances the ability of the roadside landscape to exchange materials with semi-natural ecosystems can result in periodic invasions of non-native plant species into adjacent semi-natural ecosystems.

2.3.1 *Construction and operation/use phases and non-native invasive species*
Road construction can lead to a considerable amount of habitat disturbance and edge effects. This provides opportunities for non-native species to invade adjacent ecosystems, though they would not ordinarily have had a chance to survive in competition with a climax native plant community. Where such species invade the roadside landscape, they can lead to further alteration of plant community structure and composition (e.g. Saunders *et al*. 2002), and limit the community's ability to perform ecological functions and services.
Invasive species, where present within the proposed route corridor of a road ecosystem, can be disturbed by machinery, or they can be brought onto the construction site by machinery, either within soil loads, on the tyres or tracks of machinery or in the buckets of excavators. Roads also act as conduits for invasive plant species, therefore, once present within the land-take invasive species can be further distributed as construction activities progress along the route corridor. Invasive species can also be deliberately planted on roadside landscapes during landscaping activities (see Dolan 2004). Post-construction, it is the operation/use phase of road development that provides an opportunity for invasive species to invade roadside landscapes, as plant fragments and seeds can be carried on the tyres of vehicles (Wilcox 1989) and by wind dispersal.

2.3.2 *Maintenance and decommissioning phases and invasive species*
Road maintenance activities can mimic the effects of construction where prolonged. During maintenance the vegetation within the verge can be disturbed by mowing, trimming, repaving, and clearing of drainage ditches, all of which can contribute to the spread of non-native invasive species. Plant seeds and fragments can be carried on vehicle tyres or on the tailwinds of vehicles. The movement of soil during maintenance activities can also lead to further invasions of non-native species along a road corridor. Where invasive species are not managed prior to decommissioning, the breaking of the pavement can also provide an opportunity for non-native species to invade the old road corridor.
The use of de-icing salts during maintenance can also create ideal habitat for halophytic plants including the invasive plant species *Cochlearia danica* which has invaded roadside landscapes in The Netherlands.

2.4 LANDSCAPE QUALITY

Landscape quality can be defined as the dominant feature of a landscape, which is based on a particular combination of the subordinate visual and non-visual attributes of a landscape (see Dolan & Whelan 2004).
Native vegetation, soil, and elements of the natural terrain contribute to the visible appearance of the landscape and to the protection of archaeological and palaeontological sites (Irish EPA 2002b). Soil and parent bedrock provide: (1) the physical support within the landscape, for the road bed and carriageway, and the associated technical elements (including the street furniture), (2) the source of raw materials required in the construction of road ecosystems, of which there are two aspects.
Firstly, there are the extractions of raw materials required in the production of road construction materials and in the manufacture of street furniture, which usually takes place

elsewhere within the region or country, or indeed, the raw material or finished products may be imported from another country. Irrespective of origin, the extraction processes will involve quarrying and mining for the supply of rock, gravel, metals and minerals. The impacts of extraction of raw materials on landscape quality will vary with the type of raw material and landscape, and the quarrying and mining methods utilized.

At the local road ecosystem scale, there is the extraction and movement of raw materials within the land-take of the road. Road construction involves: (1) the extraction of soil, rock or scree, and the possible exposure of rock faces or scree slopes as a result of cuts into areas of local elevation, (2) localized soil movement and compaction, (3) filling in of areas of local depression and the aggregate base layers of the road bed with soil and rock fragments, (4) the subsequent sealing of soil as a result of the asphalted road carriageway.

Cuts and fills, in landscapes within an undulating topography, level the landscape to make way for the new road corridor. Such modifications to native vegetation, elements of the natural terrain and the topography of a landscape impact on peoples' perceptions and interpretations of a landscape i.e. their social and cultural relationship, or on the values which they place on a landscape (Nohl 2001). They may have a negative influence on one or more of the key visual attributes of the modern landscape: naturalness, variety, 'rural' structuring, regional identity and vista quality.

Inappropriate planning and design decisions can also affect these attributes, especially where large-scale technical elements (e.g. fly-overs, bridges, culverts) and the landscape design layout lead to the urbanisation of a rural landscape. The removal of vegetation during road construction can have a major effect on naturalness, regional identity, orientation and a sense of place, and 'rural' structuring for stakeholders. There can also be a loss of variety and visual landscape quality including vista and panoramic quality (Gulinck & Wagendorp 2001; Nohl 2001) and environmental aesthetics (both ecological and scenic aesthetics) as the driver and/or passenger's field of vision may be simplified, disturbed and narrowed down.

Inappropriately timed maintenance activities, such as cutting and mowing of vegetation can have an adverse effect on vegetation and consequently the quality of the roadside landscape. Associated or ribbon development can also have a knock on effect on landscape quality as it can lead to the 'rurbanisation' of a rural landscape.

The key non-perceptual landscape attributes of the modern landscape are: integrity, connectivity, function, coherence and continuity. Where these attributes are affected by construction, operation/use and maintenance, ecological services can be interrupted. Impacts may not be immediately visible to stakeholders, but can manifest over time. Where persistent, such impacts can even lead to soil erosion and loss of flood control within natural and semi-natural ecosystems.

Road decommissioning can, in part, reverse the negative effects of road ecosystem development on landscape quality and operation/use, but mitigation and/or compensation techniques are generally required.

2.5 POLLUTION

The activities within the four major phases of road development (construction, operation/use, maintenance, decommissioning) result in a myriad of waste products that can adversely impact the natural environment (Irish EPA 2002a). Where these waste products leave the road ecosystem, and enter natural and semi-natural ecosystems, they become pollutants.

Waste products can enter semi-natural ecosystems in the forms of; acute point, diffuse source and atmospheric pollution. The range of pollutant substances depends on: (1) the type of construction and street furniture materials selected during the design phase of road development, (2) the activities carried out during construction and subsequent landscaping of the roadside landscape, (3) the vehicle types utilised and the traffic loads carried during road operation/use, (4) the extent and type of maintenance required during the expected life-span of the road ecosystem.

Infrequently, indirect waste outputs can occur as a result of underground fuel tanks, long distance pipelines, filling and service stations, car wash facilities, roadside cafes, car scrap yards (Golwer & Sage 2005) and abandoned vehicles- all of which can substantially contribute to point source pollution of ground water.

2.5.1 *Construction phase and pollution*

As construction works progresses, acute point, chronic diffuse and atmospheric pollution, can be released in concentrated areas along the length of the new road corridor.

Following the removal of vegetation, the movement and compaction of soil during construction and soil sealing, the soil horizons, natural runoff flow-paths and percolation patterns of rain water can be disrupted, thus disrupting the natural regulation of water flow and generally increasing the rate of runoff. This makes water bodies, including ground water aquifers, more susceptible to pollution especially during periods of heavy rainfall (Golwer & Sage, in submission), as the combination of soil movement and large quantities of rainwater can result in pronounced soil erosion and the release of coarse particulates or sediment (such as mineral solids) with negative consequences for quality of drinking water, and indeed for salmonid fish and filter feeding invertebrate species. In contrast, during dry periods, fine soil particulates can form dust or dry spray that can find its way into watercourses, or can coat the ground, vegetation, adjacent buildings and vehicles.

Other forms of pollution which may arise during road construction include: (1) emissions and leakages from machinery, (2) spillages or leaks at refuelling depots and associated spillages along the road corridor which can enter runoff and can cause localised contamination, (3) sewage and waste water released from temporary quarters for construction workers (Golwer & Sage, in submission), (4) temporary noise disturbance zones for wildlife as a result of machinery and human activities, (5) left-over construction materials such as soil, piping, geotextile lining, spray cans, asphalt, gravel etc. Soil management measures must be put in place to reduce the extent of leftover soil post-construction.

2.5.2 *Operation/use phase and pollution*

The operation/use phase of road ecosystems results in chronic diffuse forms of pollution from vehicles, as a result of waste products from: combustion, component wear (engines, tyres, bearings, brakes), fluid leakage (oil and coolants) and corrosion of materials. These pollutants can enter the atmosphere as part of atmospheric emissions, or natural and semi-natural ecosystems, in wet or dry spray or as run-off. Infrequent acute point source pollution also contributes to road runoff during this phase e.g. accidental hydrocarbon leakages and/or chemical spillages from specialised transport vehicles. Due to the diversity of pollutants and the fact that the operation/use phase is longer in duration than any other phase, road ecosystems once operational, may release more pollution than any other phase of road development, and indeed the other phases combined.

Combustion processes release 1-3 buta-diene, acetaldehyde, formaldehyde, particulate matter (PM) e.g. PM 10, also carbon monoxide (CO), nitric oxide (NO), nitrogen dioxide (NO$_2$), volatile organic compounds (VOCs) including hydrocarbons (HCs) such as toluene and benzene (Irish EPA 2002a; U.S. EPA 2002) and carbon dioxide (CO$_2$) (though its effects are not apparent on a local scale).

Zinc (Zn), Cadmium (Cd) and Lead (Pb) may be released from the wear of tyre rubber. Copper (Cu) is derived from the wear of brakes and the corrosion of radiators; other heavy metals deposited have mixed origins (Oostergo 1997; Legret & Pagatto 1999; Weckwerth

2001; Sörme & Lagerkvist 2002). Tyres are vulcanized to provide elasticity through the use of a vulcanization accelerator such as Zn (Amari *et al.* 1999). Ahlbom and Duus (1994) reported 1.5–2% of Zn in the upper layer of the tyre rubber (Sörme & Lagerkvist 2002). It is believed that Cd is present in tyres as an impurity of Zn (see Ozaki *et al.* 2004).

Road surface or pavement degradation, from wear by tyres and weathering, can also release heavy metals such as Zn, Pb, and Cu depending on the stone constituent of asphalt. As the rock type varies within regions and between countries, great uncertainties are encountered when trying to identify the average heavy metal content of pavements, though granite, quartzite and porphyry rock are often utilized in road surfacing material. The use of studded tyres in particular increases wear and breakage of the road surface (Sörme & Lagerkvist 2002).

Porous asphalt may aid in minimizing direct run-off, but can also carry pollutants directly to vulnerable ground aquifers in places where appropriate drainage systems are not in place (Golwer & Sage 2005).

Road markings, usually of white and yellow paint, may also contribute to the pollution emitted from road ecosystems. In Japan, Ozaki *et al.* (2004) found that white road markings (which are employed most frequently internationally), were found to release Cd and arsenic (As) up to 0.510 and 2.52 μg/g dry weight, respectively. In the same study, yellow road markings were found to contain very high concentrations of Pb (up to 11.99mg/g dry weight) and high levels of chromium (Cr) (up to 820μg/g dry weight) indicating that yellow marking contains lead chromate, greatly contributing to the Pb and Cr pollution along the roadways, especially in urban areas where yellow paint is utilized more frequently (Ozaki *et al.* 2004).

Zinc is also released from street furniture that is manufactured from galvanized steel, such as crash barriers and lamp-posts, and to a lesser degree from road signs and road portals. There are some 70,000km of galvanized steel crash barriers and 50 million lamp-posts within road ecosystems across Europe (Ecosafe 1997). From a road safety viewpoint, galvanised steel crash barriers are successful at stopping cars but are less effective at stopping trucks and less safe for motor-cyclists (Ecosafe 1997). Galvanising extends the life-span of steel but corrosion of the metal means that crash barriers are a major source of diffuse Zn pollution. High levels of Zn have been found where such crash barriers are in place. The degree of pollution depends on the corrosion speed, which varies from 28-50g/m^2/yr. In 1996, corrosion speed was calculated at 63-82kg zinc per year per kilometer of highway in the Netherlands, though recent data show a decrease in corrosion speed over the last decade, in comparison with corrosion rates before the mid 1980s (see Korenromp & Hollander 1999).

Dead leaf-matter on a road carriageway can also be broken up by tyre action, where it builds up on the sides of roads, leading to enhanced and concentrated leaching of humic acids (Golwer & Sage 2005).

Drainage systems designed for storage and treatment of road run-off can themselves be significant sources of groundwater pollution through seepage or leakage, in the form of intermittent or continuous percolation (Golwer & Sage 2005).

The maintenance of vehicles during their lifetime also generates a considerable amount and range of waste products (e.g. tyres, batteries, oil, bulbs, parts etc.) Along with scrap vehicles or End of Life Vehicles (ELV's), such wastes need to be managed appropriately. Material flows analyses reveal that the ferrous material is sent for recycling to steel mills, the remaining waste is sorted into a 'heavy fraction' and the so-called Auto Shredder Residue (ASR). The heavy fraction, which contains aluminium, copper, brass, and some other non-metallic material such as rubber and glass can also be sent for recycling, while the remaining ASR (forming 25–30% of the car by weight) which contains the hazardous material, can't be recycled and is sent to landfill sites (Irish EPA 2002a).

Pollution of natural and semi-natural ecosystems by light or 'ecological light pollution' from road ecosystems includes chronic or periodically increased illumination from street lamps, lights on tunnel, flyover and bridge ceilings, unexpected changes in illumination from head lights and direct glare. All may impact on wildlife in a number of ways. Animals can experience increased orientation or disorientation from additional illumination and can be attracted to (phototactic) or repulsed (photonegative) by light, which affects foraging, reproduction, communication, and other critical behaviors. Increased illumination can extend diurnal or crepuscular behaviors into the nighttime environment by improving an animal's ability to orient its self e.g. many usually diurnal birds (Hill 1990) and reptiles (Schwartz & Henderson 1991) forage under artificial lights. In Galapagos, Yellow Crowned Night Herons sit under traffic lights to catch Painted Locusts (*Schistocerca melanocera*) that are attracted to the light, while Barn Owls hunt on the ground under streetlights. This has been termed the 'night light niche' for reptiles and seems beneficial for those species that can exploit it, but not for their prey (Schwartz & Henderson 1991). Moths are generally phototactic to lights that release ultraviolet wavelengths (such as high-pressure sodium lights), making them more susceptible to predators, possibly a factor in the recent decline in urban moth populations (Rydell 1992). Non-flying arthropods vary in their reaction to lights. Some nocturnal spiders are photonegative, whereas others will exploit light to locate easily available prey (Nakamura & Yamashita 1997). Artificial light can also disrupt intra-specific interactions that evolved from natural patterns of light and dark, so can have negative influences on community ecology e.g. street lighting may cause a false dawn that disrupt bird's behaviour. Some evidence suggests that artificial night lighting affects the choice of nest site of birds. De Molenaar *et al.* (2000) investigated the effects of road lighting on black-tailed godwits (*Limosa limosa*) in wet grassland habitats. When all other habitat factors were taken into account, the density of nests was slightly but statistically lower up to 300m away from the lighting on the road carriageway. The researchers also noted that birds nesting earlier in the year chose sites farther away from the lighting, while those nesting later settled in the remaining sites closer to the lights (De Molenaar *et al.* 2000). It is also known that illumination can disrupt the circadian rhythms (biological clock) of plants - affecting leaf fall and flowering times which are controlled by day length (Longcore & Rich 2004).

New road ecosystem development and road-widening expands the area exposed to elevated sound levels (Longcore & Rich 2001). Noise emissions from road ecosystems vary with traffic flow, speed and the proportion of heavy vehicles, but depend also on the topography, distance from the noise source and on the type of road surfacing material applied. The major contributors of noise from road ecosystems are those generated by engines and the complex interaction between tyres and the road surface material. At low speeds, it is the engine of the vehicle that dictates the noise level. When speeds exceed 40km/h, however, the properties of tyres and of the road surface are of greater significance (Irish EPA 2002a). Noise from vehicles moving along the road surface is generated by tyre vibration and the movement of air particles in the tread pattern that is reflected by dense or closed asphalt road surface (Golebiewski *et al.* 2003). Sandberg and Ejsmont (2002) calculated that traffic noise generated by tyres on the road surface peaks around 1000Hz and 75-80dB (A).

Research in The Netherlands has shown that the breeding density of many species is dramatically depressed near roads once a certain level or noise 'threshold' is attained. Two explanations are suggested for the decrease in density of breeding birds: (1) the disruption of vocal communication as male birds are perhaps unable to attract females when their songs cannot be heard, (2) birds avoid noisy areas because they are stressful. Increased stress because of difficult communication leads to an increase in emigration and decrease in reproduction (Ilner *et al.* 1992; Reijnen *et al.* 1995a; Reijnen *et al.* 1995b; Reijnen *et al.* 1997; Longcore & Rich 2001).

2.5.3 *Maintenance and decommissioning phases and pollution*

Road maintenance, such as resurfacing, can mimic the effects of the construction phase if prolonged. Road resurfacing, de-clogging and de-icing of asphalt, spraying of herbicides, washing of tunnel walls and road signs with detergents and water, are the main sources of pollutants that can be released into semi-natural ecosystems during road maintenance. However, pollution can also result from a lack of maintenance where, for example, interceptor drains, filters or oil separators are not cleared of blockages by grit and sediments.

Porous asphalt generally requires more frequent de-icing operations than closed asphalt, as it is more likely to develop black ice. De-icing salts mainly consist of sodium chloride and lower levels of calcium chloride and magnesium chloride. While chloride itself is not a health hazard, greatly increased chloride concentrations can increase the mobility of Zn and Cd through the formation of chloro-complexes (Bauske & Goetz 1993; Pagotto *et al.* 2000) leading to pollution and increases in the salinity of receiving waters. Chloride has been found to be toxic to certain vegetation (e.g. Beech, *Fagus sylvatica*, trees).

Where porous asphalt surfaces carry low traffic levels (e.g. in residential areas, emergency lanes on motorways) regular cleaning is required (in the absence of the pumping phenomenon) by means of high-pressure water jet and aspiration to release sediment that clogs the pores. These sediments along with the adsorbed heavy metals and hydrocarbons can contribute to run-off released by road ecosystems.

Sediments, hydrocarbons and heavy metals may also be released during the decommissioning phase, where porous pavement is broken up.

2.5.4 *Distribution of waste products and pollutants*

Waste products can be dispersed outside the road ecosystem (where they become pollutants) in various ways e.g. as run-off, in the form of wet and dry spray and deposition and via atmospheric emissions. Climatic conditions influence spray and deposition. Under dry conditions spray will dominate, while during wet periods road runoff will be important (van Bohemen & Janssen van de Laak 2004). In the case of dry spray, the particles are deposited as dust, while in the case of wet spray, they spread due to splashing and the vaporising rain water. In both cases a large part is deposited in the immediate vicinity of the roadside (see Boland 1995). Road runoff in a rural landscape will pass to the side of the road where it either travels directly or indirectly into watercourses, or infiltrates groundwater, unless intercepted by a sustainable drainage system.

A wide variety of measures can be put in place to avoid or reduce waste products at source and to mitigate the effects of road runoff and spray. Source-Oriented Measures can be divided into two sub-categories: (i) volume-oriented measures which avoid/reduce the waste products released by vehicles through road planning, (ii) technical design aspects which focus on reducing the source of waste products through vehicle design and manufacture and informed decision making by road users and the selection of alternative material for street furniture, crash barriers and road surfaces. Effect-Oriented Measures include the sustainable drainage systems put in place to reduce the effects of road runoff and spray.

2.6 ASSOCIATED OR RIBBON DEVELOPMENT

Apart from the direct and indirect impacts of road construction there are associated impacts due to ribbon development as new or upgraded roads provide greater mobility and access for further economic services leading to the construction of additional urban or industrial systems along the road corridor. The control of such development lies with planners, local authorities planning boards, politicians and stakeholders.

3. Strategic environmental assessment

Environmental Impact Assessment (EIA) is the process of evaluating the likely environmental impacts or the ecological footprint of a proposed development (see OECD 1997). It is now recognised that the commonest application of EIAs, at the project-level or the local ecosystem scale, has generally failed to determine or quantify the cumulative and large-scale impacts of road development. For this reason, the amended EU Directive on EIAs (97/11/EEC) called for a more detailed EIA for major road projects, to anticipate potential environmental problems from plans and projects (Council of Europe 2001; Bank *et al.* 2002); by focusing on a much broader or higher scale. More recently, the EIA Directive has been complemented by the introduction of the EU 'Strategic Environmental Assessment' (SEA) Directive (2001/42/EC) which calls for an SEA prior to the EIA process.

The basic steps of SEA are similar to the steps in EIA procedures, but the scope differs. By its nature, it covers a wider area, a wider range of activities often over a longer time span than an EIA and can be applied to an entire sector (e.g. national biodiversity plans) or to a geographical area, (e.g. a regional development plan) (CBD 2004). For this reason, SEAs complement the environmental impact assessment of road ecosystem development, since road ecosystems criss-cross landscapes or even regions, forming complex interactions and impacts on a very broad scale.

However, SEA does not replace or reduce the need for project-level EIA, rather it helps to streamline the incorporation of environmental concerns (including biodiversity) at an earlier and more strategic stage of the decision making process, often making project-level EIA more effective (CBD 2004).

The assessment of environmental impacts under an SEA requires a range of methods and tools such as Geographical Information Systems (GIS) that focus on the measurement of single impacts or methods such as Cost-Benefit Analysis, for easier assimilation by decision-makers. The significance of impacts is thus evaluated, and recommendations such as the preferred route, mitigation, compensation and monitoring measures are then proposed. Stakeholder participation is also an important requirement in SEA (Irish EPA 2002a). For this reason the process is on par with economic and social considerations and Local Agenda 21 criteria.

Typical environmental objectives for SEAs on transport activities can cover climate change, air, water and soil quality, biodiversity and noise, and can also include additional issues, such as the use of natural resources and land-use impacts (COMMUTE 1998). With regard to biodiversity considerations, an SEA can provide an early overview of the possible implications of transportation plans. Its application can reduce the potential risk of transport infrastructure conflicting with valued protected landscapes and habitats (Council of Europe 2001), therefore ensuring that a proposed new road development is compatible with international obligations to conserve protected habitats and their associated species (Treweek *et al.* 1998). The CBD recommends the ecosystem approach' (as described in decision V/6 of the Conference of the Parties) as an appropriate framework for the assessment of planned actions and policies in relation to transport (CBD 2004).

4. The ecosystem approach and sustainable road ecosystem development

Sustainable road ecosystem development is an expanding subject, prevalent in progressive engineering and ecological design professions, whereby all aspects of existing ecosystem function and services are weighed against the holistic environmental impact of establishing a new road ecosystem.

According to the literature, sustainable road ecosystem development needs to reflect the fundamental principle of sustainable development which is to seek the reduction of current inputs of natural resources in the planning, design, construction, maintenance, operation/use and decommissioning of a road scheme.

Sustainable road ecosystem development also needs to address Local Agenda 21 criteria fully and seek public participation in all aspects of the development, especially in planning and design.

It must be recognised that a transdisciplinary approach to road ecosystem development, where applied, could also lead to further advancements in the field of road science and the move towards more sustainable or state-of-the-art road ecosystems.

4.1 THE ECOSYSTEM APPROACH AND RESOURCE MANAGEMENT

The UN CBD (1992) imposes obligations for the conservation and sustainable management of biological diversity "and the ecological complexes" of ecosystems and natural habitats located within and outside protected areas. The CBD promotes the use of "the ecosystem approach" (Decision V/6 of the Conference of the Parties) as an appropriate framework for the assessment of planned actions and policies in relation to transportation infrastructure. The ecosystem approach provides a more accurate environmental assessment of the ecological footprint of road development. It also provides a methodology for the sustainable management of road ecosystems, since it provides a holistic, trans-disciplinary method for sustaining or restoring natural and semi-natural systems and their functions and values (Garrett & Bank 1995; Slocombe 1998) through the management and conservation of natural resources.

Resource conservation and management are central to a more sustainable approach to road ecosystem development. In accordance with the ecosystem approach, road managers must strive to manage the input, throughput and output of natural resources required in road ecosystem development. This can be achieved through appropriate planning, design, construction, operation/use, maintenance and decommissioning of a road ecosystem which can reduce the likelihood of potential impacts on adjacent semi-natural ecosystems whilst allowing both system types to interact with each other for their long-term mutual benefit. In order to achieve this, certain practices, procedures and techniques which result in the heavy consumption and depletion of natural resources must be avoided.

The first step to reduce the impact of a new road ecosystem is *avoidance* which is usually implemented during the planning phase. This often involves the avoidance of route selection options that would result in the heavy consumption and depletion of natural resources. Where avoidance is not possible, mitigation and/or compensation techniques are selected during the design phase to reduce the ecological footprint of road ecosystem development through the restoration of ecosystem function and services, native vegetation and landscape quality. Mitigation techniques are utilised to manage near natural levels of productivity and the flow of services between ecosystems as well as minimising the inputs of natural resources inputs e.g. land, soil, water, air, biodiversity (loss of habitat and road kills) and waste outputs such as road run off, spray and atmospheric pollution. Where avoidance is not possible and mitigation measures cannot compensate for the loss or potential loss of natural resources, compensation measures are required. For example, if disturbance or loss of habitat as a result of road ecosystem development is unavoidable, and mitigation measures cannot compensate for the loss, damage, or degradation of habitat, compensation in the form of habitat creation to achieve 'no net loss' (Iuell *et al.* 2003) may be the most appropriate response. Such an approach is consistent with the requirements of the EU Habitats Directive (CEC 1992). For this process to be sustainable, the establishment of mitigation and compensation measures must also be subjected to ecological footprint analysis.

In accordance with the ecosystem approach, the proper temporal and spatial scales of impacts should be determined, along with the type of adaptive mitigation and compensation techniques/measures required, because the restoration of ecosystem function and sustainable development is not achievable if managers only focus on the local ecosystem. Impacts on the landscape can only be identified by focusing on the landscape or regional scale, where the principal impacts of road ecosystems on adjacent semi-natural ecosystems are more discernible, such as habitat fragmentation (see Dolan 2004).

4.2 ECOLOGICAL ENGINEERING

Though road ecosystems are heavy consumers of natural resources, they also exhibit the ability to "adjust to conditions, blend with the landscape, and reach new ecological and hydrological states" (Olander et al. 1998) if designed appropriately and their resources managed efficiently.

It is the ecosystem approach that provides engineers with the necessary information to devise ecological engineering systems (see van Bohemen 2005) that take advantage of the remediation opportunities exhibited by semi-natural ecosystems. Ecological engineering refers to more sustainable design approaches in the management of onsite and offsite environmental impacts such as anthropogenic ecosystems, that integrate human and natural systems for their mutual benefit (Mitch & Jorgensen 1989). Such systems can be put in place at various spatial scales to reduce onsite and offsite environmental impacts by attempting to incorporate and/or reinstate ecological function and services in natural and semi-natural ecosystems. Examples of systems include: swales, detention basins, balancing ponds, retention ponds, constructed wetlands or helophyte filters. These systems can also store and/or treat road run off, restore landscape quality, as well as improving the quality of life for stakeholders.

Ecological engineering techniques are generally established within the roadside landscape that covers approximately 1ha or 2.5acres per km of new road in Europe. Along with providing locations for the establishment of ecological engineering techniques, roadside landscapes and certain elements within can (1) as refuges, contribute to the conservation of wildlife; wider roadside landscapes can harbour a higher diversity of edge and generalist flora and fauna species (even species associated with rare habitat may use them as supplementary habitat) especially in areas of intensive agricultural land use (Forman 1995), (2) contain the only extensively managed grasslands with the exception of a few nature reserves, (3) provide consistent microclimatic conditions which permit relictual species to persist, in essence establishing a biodiversity archive, (4) play a role as a conduit for certain species e.g. beetles and butterflies, as they provide fairly homogeneous conditions throughout the length of the corridor, uniting contrasting ecosystem types that interface with the road, thus increasing connectivity, (5) act as a buffer zone between the road pavement and adjacent native vegetation, (6) reduce noise disturbance, (7) act as a food source for raptors e.g. Barn Owls and Peregrine Falcons, (8) trap sediments and contaminants from road run-off, (9) provide possibilities for the recolonisation of native vegetation.

4.3 STAKEHOLDERS AND PUBLIC PARTICIPATION

The ecosystem approach, sustainable development and Local Agenda 21 criteria require the identification of peoples' tangible and intangible values for biodiversity and features within the landscape, which could be affected by a proposed road scheme, and the need for the participation of stakeholders in the decision-making process. Kemmis (1990) identified the less quantitative side of ecosystem-based management as being the link between management and peoples value's through a 'sense and management of place'. In ecosystem-based management m,anagers must provide for a diversity of peoples social and cultural values (Grumbine 1994 cited in Stein & Anderson 2002; Dolan 2004).

In Australia, Xuan Zhu *et al.* (2001) identified the need for a web-based decision support system for regional vegetation management. While scientific research has made valuable contributions to the subject of native vegetation management, the delivery of information and new knowledge to land managers, planners and the stakeholders has been slow. Meurk and Swaffield (2000) believe that if natural ecosystem processes and indigenous species do not enter into the consciousness of the majority of the population on a daily basis then there will be no ownership of, or support for it.

4.4 A TRANSDISCIPLINARY APPROACH

In the last fifty years road ecosystem development has evolved to meet the growing needs of society to extend transportation networks, societal concerns and legislative requirements for the prevention of, mitigation of, and compensation for the resultant adverse effects of road ecosystems on the surrounding landscapes. The number of agencies and disciplines involved in road ecosystem development has increased to meet these demands and in the recognition of the need for continual prudent commitment to sustained road ecosystems, including the responsible management and conservation of natural resources for future generations.

4.4.1 *Monodisciplinary*
Due to the pace of road ecosystem development, in the early part of the 20th century, transport planners were not ready or able to give due consideration to environmental considerations in road planning, design and construction (Box & Forbes 1992). This monodisciplinary approach led to the failure to recognise the adverse effects of road ecosystems on surrounding landscapes and ecosystems, and elements within.

In the late 1960s, attitudes of stakeholders towards construction and management of road ecosystems and their associated vegetation changed considerably, as a result of a growing awareness of the direct and indirect effects of such development e.g. pollution, loss of biodiversity and the urbanisation of rural landscapes. For this reason, the incorporation of environmental and social concerns, in the planning and design phases of road ecosystem development (which were earlier considered outside the traditional engineering framework) intensified. This change was the outcome of the large-scale acceptance of new concepts in resource management which advocated equal treatment of biological diversity and sustainable development, along with traditional aspects of resource management, in the planning process (Salwasser 1990).

4.4.2 *Multidisciplinary*
Road ecosystem development continues to be one of the most extensive and widespread activities carried out by local and government authorities. It has come to the forefront of political debate, and attracted the attention of National and local media as a result of: (1) the growing dependence on road ecosystems for daily travel, (2) as the footprint becomes greater, (3) as a result of the increasing awareness of the negative effects of road ecosystems on the landscape. There is a continuous need for governmental and non-governmental agencies and the media to adapt their focus, objectives and decision-making processes to address the principles of sustainable development in light of new advancements in the area of road science. Certain agencies face particular challenges, such as governments and local authorities in identifying an acceptable and workable definition for sustainable road ecosystem development and in adopting multi- or transdisciplinary approaches in the design and management of road ecosystems. In the UK, questionnaires were sent to local wildlife trusts

and local teams of English Nature in order to assess the contribution of local authorities to biodiversity assessments and the monitoring of roadside landscapes (see Akbar 1997). Only two English Nature teams and fifteen local wildlife trusts indicated that they had carried out or were involved with local authorities in studying and surveying roadside landscapes. Similarly, Alexander (1995) found that only 52% of Highway Authorities surveyed their verges and surveying was found to be informal and inadequate for proper management. Furthermore, the majority of the local wildlife trusts expressed their dissatisfaction over the development and management of roadside landscapes in their counties.

Meanwhile, the general public and the media must face up to the growing responsibility of recognising the more sustainable approaches to road ecosystem development that have come about as a result of the cumulative effort of stakeholders (i.e. the road users and local inhabitants of a road segment or network), road planners, road designers, civil engineers, structural engineers, environmental psychologists, soil scientists, geologists, hydrologists, landscape architects, nursery suppliers, road ecologists, wildlife rangers, archaeologists, social scientists and road managers who have worked in the area of road science over the past century.

4.4.3 *Transdisciplinary*

A transdisciplinary approach requires an awareness of, consultation with and a degree of integration between the various disciplines involved in the phases of road ecosystem development. It encourages an increased awareness of other professionals' roles in the road ecosystem development process.

For example the disciplines of ecology and civil engineering, and landscape architecture and landscape ecology have achieved a transdisciplinary status through the formation of landscape ecology and ecological engineering. Civil engineering has also adopted measures to reduce driver monotony and fatigue which emerged from the discipline of environmental psychology and to address the speed of vehicles entering urban landscapes through traffic calming schemes that are based on the concept of 'optical width' (NRA 1999), amongst others. The disciplines of road planning and design, and landscape ecology have further responsibilities to integrate social science in the landscape design process by providing opportunities for the participation of stakeholders at the local scale.

Most new road ecosystems, realignments and by-passes are developed in response to societal needs and pressures including the need for mobility, accessibility, connectivity and interaction. Since all of these activities have social dimensions, the justification for the inclusion of social scientists in road ecosystem development process is real. Social dimensions, where adequately incorporated into road ecosystem development, provide for: (1) a balanced and comprehensive decision making process, (2) fulfil government and local authority commitments on sustainability, such as Local Agenda 21 criteria, (3) reduce mistrust of local authority behaviour, (4) ultimately provide for the incorporation of stakeholders values and more environmental and ecologically sustainable road ecosystems.

According to Day (1998) social issues have not ranked high in the decision making criteria of road authorities because of: (1) the historical and current emphasis on cost-benefit analysis, and monitoring activities for engineering and more recently ecological purposes, (2) the kind of training imparted to those who have traditionally worked within road and transport authorities, (3) the problems (perceived and real) in using quantitative social data. (4) evaluating the representativity of advocacy groups that become involved in the proposed road ecosystem development, (5) the difficulties in understanding and adapting to social changes.

Monitoring of social issues is generally limited to demographic parameters such as development patterns, population trends, current and future patterns of employment, social equity, transport efficiency and public safety. Though these parameters play a major role in the planning of new road ecosystem development, they can fail to meet the day-to-day concerns of the potential road user and the local inhabitants of a new road scheme. More specifically, the route selection, design and construction phase of road ecosystem development are currently focused on the conservation of habitats or sites of historical or archaeological interest which are of regional, national or international importance. They fail on a local scale to incorporate stakeholders' cultural, social and aesthetic values for features within the landscape. Assessments of peoples' values and preferences can provide information leading to the identification, retention, restoration or creation of locally distinctive features within the road landscape. Locally distinctive features can be incorporated into the planning phase e.g. the route selection and design phase of road ecosystem development. For this reason road managers need to recognise the social context in which they are operating by involving social scientists to address sustainable development criteria in a more effective and realistic attempt to address stakeholder participation.

5. Planning phase

To reduce the ecological footprint of a new road ecosystem, planning, regional development strategies, and the design of road ecosystems need to be examined in accordance with the principles of sustainability (see Comhar 2002), to ensure that road ecosystems can meet our present needs without compromising the needs of future generations.

It is important, at this point, to differentiate between types of road ecosystems in relation to road ecosystem planning and design. There are obvious differences between urban roads and rural roads. However, there are also differences between local or regional roads and arterial roads or link roads that connect motorways. The bigger the road ecosystem, and the more rural the landscape, the greater the number of mitigating factors and opportunities to restore landscape quality, native vegetation and other elements of the landscape. Major arterial roads or highways, freeways or motorways lend themselves more to road ecology as the effects of such road ecosystems on landscapes, especially intensively managed landscapes is often very positive. It is much easier from an ecological viewpoint to address the principles of sustainable development when planning for major arterial roads, as there are more funds available to provide for ecological landscape design, wildlife crossing structures and traffic management measures which can funnel heavy traffic away from residential areas, or protected areas that contain sensitive flora and fauna.

The most effective planning strategy which can be put in place is avoidance of impacts, where zoning at the regional scale is utilised to designate sensitive social, ecological, hydrological and archaeologically areas, amongst others. For example; planners need to ensure that road safety requirements are augmented by the provision of a buffer zone, the retention of landscape quality and the needs of species moving through the landscape (Viles & Rosier 2001), especially where road ecosystems come into contact with protected areas or sites of significant archaeological or historical interest.

5.1 ROUTE SELECTION PROCESS

Route selection has been recognized as the most critical decision made in relation to road ecosystem development because the way in which a road network is positioned has enormous consequences for how it behaves and interacts with the landscape, thus determining how economically and environmentally costly (in terms of resource consumption) a road ecosystem will be (Lugo & Gucinski 2000). In the past the fitting, placing, positioning or aligning of road networks and segments in the landscape was usually considered an engineering and an economic problem (Forman 2001), especially relative to watercourses. However, today it is a key challenge for road planners and land-use managers (Swanson 2001) to combine both human mobility and environmental sustainability (e.g. a reduction in habitat fragmentation) in the route selection process.

Route selection is the main process by which adverse ecological impacts on landscapes, such as habitat loss and fragmentation, can be avoided. Aerial photography provides invaluable information during the route selection stage especially on matters relating to ecology. The concept of ecological infrastructure, which is based on aerial photography and satellite imaging, has been cited in the literature as the most effective approach to aid in the route selection and decision making process and has been applied in Holland, US and Ireland to reduce the habitat fragmentation effects of road ecosystems. The concept also complements Article 10 of the European Habitats Directive, as it encourages the establishment of corridors and other landscape features between protected areas.

The first step in the process involves focusing on the landscape scale, using the most effective landscape based techniques and analyses available to identify elements within the landscape that are essential in maintaining connectivity and meta-population dynamics. The process involves identifying ecological networks of core habitat areas and connectivity zones for a region in order to decrease the overall fragmentation of core habitat areas (Jackson & Griffin 2000).

For effective planning, the challenge and the solution is next to fit the road to the landscape ecologically by creating a map of the ecological network that depicts the distribution of the core habitat areas (and remnants thereof) together with elements of connectedness i.e. the major wildlife and water corridors, streams and wetlands, rare habitats and species and topographic sites. Once the road is superimposed onto the ecological network, bottlenecks or barriers to wildlife can be identified (van Bohemen et al. 1998; Forman 2001), and a practical strategy for mitigating the negative effects of a road ecosystem on wildlife selected during the design phase.

The route selection process can also restore the former native vegetation to particular landscape types, for example, where the road corridor passes through areas of low biodiversity value, such as intensive agricultural landscapes. Ecological landscape design techniques can strive to restore floral diversity within the roadside landscape by extending and restoring diversity within remnant hedgerow networks.

Another aspect of route selection process and the use of aerial photography is the identification of the land-take required for road ecosystem development.

5.2 VOLUME-ORIENTED MEASURES AND POLLUTION

Avoidance of pollution is an important objective in the establishment of a sustainable road ecosystem, and is usually achieved through: (1) the designation of floodplains and open space zones within a regional design plan as a means to avoid road run-off into hydrologically sensitive or protected areas. This can be achieved through route selection or traffic re-routing away from pollution 'hotspots' and/or sensitive areas, (2) the selection of traffic management measures and technical design aspects which can reduce the volume of waste products at source, (3) road closure and decommissioning.

At a smaller scale traffic management measures which can reduce the volumes of pollution at the road network or segment scale include: (1) environmental road or public transport pricing policies (to encourage modal shift), (2) environmental access control (to prevent access by a proportion of vehicles), (3) traffic signal control, (4) park and ride facilities, (5) parking control, (6) mass transit systems (Cloke et al. 1998; McCrae et al. 2000). In recent years demonstration projects of environmental access control measures have been carried out in

Paris, Athens and Rome, while entrance roads into the city of London have been tolled to reduce traffic loads. In certain cities in Germany vehicles that have not passed an emissions test may be banned from central city areas during severe air pollution episodes. In areas where pollution reaches dangerous levels, police set-up roadblocks marked with 'smog' signs along major corridors into cities.

Practical planning experience has also shown the necessity to involve drainage planners in the early stages of road ecosystem planning, as the extent of the land-take identified during this phase will determine the type of sustainable drainage system which can be put in place.

5.3 LAND-TAKE REQUIREMENTS

Aerial photography, the map of the ecological network and the flora and fauna surveys within the relevant EIA, can also be utilised by planners during the land-take decision making process, to identify areas along a proposed route that could potentially contribute to the mitigation of and/or compensation for habitat loss and habitat fragmentation, and which may make suitable candidate areas for the placement of drainage systems. These areas can include: (1) small sections of agricultural land left isolated as a result of road development, (2) pockets of disturbed or waste ground, (3) pockets of native vegetation such as wetland, grassland, marsh and woodland within the land-take of the road ecosystem. Where such areas are retained within the land-take of the road corridor, and are located adjacent to core habitat areas, they form ideal candidate areas to facilitate the extension of core habitats and connectivity zones between habitats, with possibilities for the integration of wildlife crossing structures into existing native vegetation within the landscape. However, an environmental cost-benefit analysis is required to ensure that environmental sustainability concerns in relation to the consumption of land as a resource are weighed against the requirement for mitigation and compensation measures selected during the design phase.

5.4 ROAD DECOMISSIONING

Planning for the upgrading of an existing road, through realignment or a by-pass, must also address the decommissioning of old road pavement. In many countries road decommissioning is not recognised as part of the general planning process for new road infrastructure. However, it is known that segments of old road pavement, where abandoned post construction, can continue to pose a barrier to the dispersal of wildlife. Decommissioned sections of old road pavement can provide valuable habitat for wildlife where the pavement is broken to facilitate the natural recolonisation of native vegetation. Such areas can compensate for past habitat loss while reducing the habitat fragmentation effects of the old road.

As part of the planning process, consideration can also be given to decommissioning part of the old road pavement, where the remainder is left paved, as a requirement for; vehicular access, footpaths, cycle-ways and greenways or green lanes. In such cases, decommissioning may also contribute to a reduction in habitat fragmentation but on a smaller scale (Iuell *et al.* 2003).

5.5 PLANNING FOR ROAD USERS

It is clear that road ecosystem development and zoning are controlled through planning, which delineates a route long before a definitive design can even be considered. This is why it is important, at an early stage of the planning process, for a new road scheme to form an impression of how a road should be designed, and to realize that decisions made at this stage on its alignment may later have a major influence on how the road is fitted or blended into the landscape and how the road user experiences the road aesthetically (Brown 2003; Dolan & Whelan 2004).

According to Brown (2003), it is the ability of a planner, to accept that road corridors are not just simply a path from Point A to Point B but also an expression of human values and perceptions of the landscape, that has the potential to transform the mission of those who plan, design and maintain road eco-systems (Dolan & Whelan 2004). As the ecosystem approach and Local Agenda 21 stress the importance of peoples' values and the need for the participation of stakeholders in the decision making process, planners and designers need to identify peoples' preferences for elements within the landscape and to incorporate these preferred elements into the road landscape as part of the planning and design process. Road planners and designers also need to integrate locally distinctive features and new large-scale technical elements (such as flyovers and bridges) into design layouts that aim to restore landscape quality and are sympathetic to the contextual framework of road users (see Houben 1999). The appropriate provision or management of such features can allow road users to retain social and cultural values and relationships with the landscape including regional identity and a sense of place (Brown 1984).

Currently, road planning and monitoring activity is orientated toward assessing and maintaining physical inventories for engineering purposes (e.g. road surface condition, structural integrity and hazard areas) because knowledge of these attributes is essential to road ecosystem functionality and safety. However, it is often the lack of social knowledge about peoples' relationship with a landscape that can generate road user conflict and mistrust of public transportation agency behaviour (Brown 2003; Dolan & Whelan 2004).

5.6 GREENWAYS

Greenways are multifunctional-use corridors that are usually established alongside road, rail or canal ecosystems where they enter or leave urban systems. Greenways provide significant opportunities to increase the ecological value and landscape quality of road ecosystems and urban ecosystems. The potential for establishing a greenway system must be identified during the planning phase of road development.

Up to the mid-1980s greenways were generally trail-orientated recreational routes that provided access to rivers, streams, ridgelines, rail beds and other corridors within an urban landscape and were usually vehicle free. More recently, greenways have evolved beyond recreation and beautification, to address: biodiversity conservation and connectivity needs for wildlife, urban flood damage control, improved water quality, outdoor education, historic preservation and other urban infrastructural objectives. However, in order to function as wildlife corridors, greenways need to be of a sufficient width, and provide sufficient cover for wildlife in the form of native vegetation to encourage the dispersal of wildlife (see Searns 1995; Dolan 2004). Greenways can also complement green lanes or modified flyovers (Dolan 2004).

6. Sustainable design

Decisions made during the design phase of road development will ultimately form the basis for the sustainable development of a modern road ecosystem. This is because the design phase determines the width, height and type of road carriageway, the extent of soil sealing, the type and amount of construction material, and mitigation and compensation measures required.

In accordance with the principles of sustainable development, design solutions must be selected which: (1) require a minimum consumption of natural resources, (2) mitigate habitat loss and fragmentation and restore ecological services, (3) restore landscape quality, (4) reduce waste outputs e.g. spray, road run-off and atmospheric emissions. Decisions made during this phase also determine the final content of the contractual commitment between the local authority and the contractor, including the design specifications for the various mitigation and compensation measures required. For this reason, variations to a contract are difficult to achieve post design phase, unless the required changes will have no added administrative and/or economic costs. This is unfortunate, given that the mitigation of the ecological impacts of roads requires a more flexible approach to design and construction of road ecosystems, because of: (1) unforeseen circumstances- where on site decisions may be required during and post construction phase, (2) prior to commencement of construction, additional surveys may be required to identify any wildlife which may have moved into the land-take of the road ecosystem since the original EIA surveys were undertaken (see Council of Europe 2001). At present, due to contractual limitations, any additional or previously unforeseen measures required to manage wildlife can rarely be put in place.

In addition to the application of ecological infrastructure analysis (see Section 5.1), aerial photography has several other important roles to play in the design phase. It can: (1) enable confirmation of existing and potential wildlife corridors and crossing points of the target species prior to construction, (2) assist in identifying the location of potential local provenance seed banks along a road corridor, (3) assist in identifying suitable areas for soil storage prior to the commencement of construction, (4) assist in the identification of features of archaeological interest, (5) document changes to the landscape pre- and post-construction.

6.1 HABITAT LOSS AND THE EXTENT OF LANDTAKE

Efforts to reduce the extent of the land-take for a road ecosystem are the ultimate means of reducing habitat loss as a result of road ecosystem development. However, a balance must be made between efforts to reduce the extent of land take and the extent of road corridor (i.e. of roadside landscape) required to facilitate mitigation measures to restore landscape quality and native vegetation, especially in intensively managed landscapes. With appropriate planning and through careful soil movement and storage, plus salvaging of plant material, habitats present within the road corridor can be avoided or translocated while other areas of less ecological value can be designed to provide additional or compensatory habitat for wildlife through ecological landscape design.

Slopes of 1:2 (vertical to horizontal) are increasingly being utilised in an effort to minimise offsite movement of soil (excess soil once removed from the road ecosystem is considered waste material), to maximise the areas available for access and for maintenance within the roadside landscape, while reducing the extent of the land-take. However such slopes may have associated stability issues. Significantly, it has been found that raised roads result in fewer road kills relative to level roads (see Clevenger *et al.* 2003).

A wide vegetated centre median with a soft landscape area fails to address sustainable criteria due to: (1) the extent of land-take required, (2) the extent of soil sealing, (3) associated maintenance of trees and shrubs, (4) specific demarcation and sight line maintenance regimes including mowing, (5) the 'habitat island' effect, (6) possibility of greater numbers of road kill, especially of birds. Alternatively, a more sustainable, ecological and cost effective design approach involves restricting the width to the minimum required for the crash barrier or other safety measure, and for the provision and access of services, without planting.

The extent of land-take, while it must address environmental sustainability issues, must also weigh up the requirement for areas that could potentially mitigate and/or compensate for habitat loss and habitat fragmentation and the provision of drainage systems. Sections of isolated agricultural land or pockets of disturbed or waste ground that do not contain vegetation of local ecological importance and are not required by the landowner can form ideal candidates for such measures.

6.2 HABITAT FRAGMENTATION AND WILDLIFE CROSSING STRUCTURES

Wildlife crossing structures put in place at particular locations along a road ecosystem allow for the undisturbed migration and dispersal of wildlife, and for the safe passage of wildlife either under or over the road carriageway. The type and size of crossing structures are to a large extent determined by the target species, the movement patterns and territories of the target species (Iuell *et al.* 2003), and by the identification of the wildlife crossing points where the wildlife corridor(s) of the target species and the road ecosystem intersect. These intersection points are termed 'bottlenecks' or 'barriers to the target species' and can be identified through on-the-ground surveys, the use of aerial photography and the map of the ecological network. It is the application of the concept of ecological infrastructure analysis and the designation of connectivity zones within the landscape that finally determines the positioning of crossing structures. On old roads, wildlife crossing structures may also be retrofitted where bottlenecks are identified post-construction i.e. where local peaks in road kills of wildlife have been found.

Once in place, crossing structures are integrated into the existing native vegetation in the surrounding landscape through the application of re-vegetation techniques which aim to reconnect severed wildlife corridors and fragmented core habitats. Re-vegetation techniques, such as the establishment of hedgerows, also provide food, shelter, habitat, and screening and can act as a noise barrier for wildlife attempting to utilise such structures. Fencing is also required to prevent wildlife crossing onto the road carriageway and to lead animals towards an overpass, or underground structures.

To date it is not known whether wildlife-crossing structures are successful in providing for long-term population survival, including the maintenance of sufficient genetic variability. It is also not known whether these structures are the most cost-effective method for wildlife conservation in relation to road ecosystem development. For these reasons the installation of such structures is sometimes considered a 'feel good' act undertaken by authorities or politicians, in light of the adverse effects associated with road ecosystem development.

6.2.1 *Target species*
The target wildlife species for crossing structures are identified by the flora and fauna survey within the relevant EIA. Where one or more target species are identified, consideration can be given to providing a multi-species use crossing structure e.g. an oversized culvert or an overpass with a variety of habitat types that may provide for the safe passage of several wildlife species. Identifying the target species is also an important basis for the planning of future monitoring procedures to evaluate the success of a structure.

Non-native species should not be target species for crossing structures, as they do not form part of the natural ecosystem. They can contribute to the spread of disease, affect inter-specific species interactions and alter community structure. Measures which could be put in place to prevent non-native mammal species from utilising wildlife-crossing structures include olfactory repellents, ultra sound and structural modifications to crossing structures (Iuell *et al.* 2003; Bank *et al.* 2002).

6.2.2 *Underground crossing structures*
Terrestrial underground wildlife crossing structures, beside the culvert itself, include fencing, earth berms or bunds, shrubs and trees all of which direct animals into the underground structure. Both van Apeldoorn *et al.* (1998) and Clevenger *et al.* (2003) recommend that

vegetative cover should be provided close to passage entrances to enhance animal use. Van Apeldoorn *et al.* (1998) described measures to be taken for local badger populations such the establishment of grasslands to provide a food source and the planting of woodlots and hedgerows that could provide habitat and serve as dispersal corridors, because dispersal is more likely to occur in the direction of increasing habitat quality.

Ramps or graded edges are required for aquatic crossing structures that contain mammal ledges (Bank *et al.* 2002), in order to allow easier access for wildlife (see Figure 2) e.g. oversized arched culverts and box culverts with mammal ledges.

Figure 2. Arched culvert with mammal ledge (Dolan & Whelan)

The use of additional paving material around the entrance of such structures is not required for structural support and is a non-sustainable use of construction material and natural resources. Similarly, rock armouring and gabions on streams should also be avoided. Nesting sites for bats and bird species can also be fitted to the ceiling of oversized arched or box culverts.

6.2.3 Overpasses

Landscape bridges or green bridges provide for the safe passage of wildlife over a road carriageway and can vary in width from 8.5-870 metres wide and can expand over several traffic lanes. Overpasses can be planted with hedges and clumps of bushes, connecting forested areas that can be utilised by a wide variety of species including deer, hares, bats and amphibians. They can also be fitted with ponds. In the Netherlands, tree stumps have been placed on such structures to provide cover and habitat for species attempting to cross. As guidance for soil depth for plants on overpasses in Germany, approximately 1-3m depth of soil is required for grass, 2-3m for shrubs and 1.5-2m for trees (Bank *et al.* 2002).

Flyovers for traffic can also be modified to provide a green lane for the safe passage of wildlife over the road carriageway. Where alterations in traffic patterns result in road decommissioning or reduced traffic loads, consideration can be given to the closure of flyovers and the creation of an overpass, or to the establishment of a green lane on the flyover, which would allow for wildlife to cross alongside traffic.

6.2.4. Technical features

Fences with one-way gates, reflectors and wildlife warning signals are technical elements that can be installed to increase the likelihood of safe passage for wildlife. Elements that can be put in place to prevent wildlife from crossing road carriageways include fencing, olfactory repellents, ultra-sound, population control, habitat modification and noise barriers (Bank *et al.* 2002).

6.3 INTERSECTIONS BETWEEN ROAD ECOSYSTEMS AND WATERCOURSES

Where rivers or streams intersect with the establishment of a new road ecosystem, measures can be put in place to facilitate the passage of the watercourse under the road carriageway and to avoid or mitigate interruptions to its natural architecture (flow and structure).

The first step in the process is to take measures to carry and/or divert the stream or river during initial construction through the use of temporary culverts. It is important that the placement of temporary structures should avoid blocking the passage of aquatic species by ensuring that the bottom of the temporary culvert is not placed too high above the base channel for fish to jump or, where the gradients and velocities in the culverts are too steep and rapid for fish passage (NCSAI 2001).

The permanent structures selected to carry watercourses under road carriageways include: concrete pipes or drainage culverts to carry small streams, and pre-cast concrete oversized-arched culverts and box culverts for larger streams or small river crossings. It is important that realignment reflects the turbidity of the former stream by reconstructing the natural architecture of a stream. This can help to adverse effects on fish and mammals such as otters (*Lutra lutra*).

It is important that the pre-cast structures have open bases that expose the stream bed. The silt and/or gravel material from the original stream bed can be stored during the initial earthworks for later use in the reconstruction of the new bed. Similarly, disturbed vegetation including riparian, marginal and emergent vegetation can be salvaged from the original stream bank and bed prior to the commencement of earthworks for later use in the re-vegetation of the newly realigned stream segment. Landscape designers need to ensure that non-native invasive species are not present within the salvaged material.

The design of the new river or stream realignment at the entrance and exit of a culvert should avoid steep sloping channelisation of the stream bank and bed, and hard engineering features such as rock armouring and gabions (see Figure1). A more sustainable design approach involves recreating the meandering or natural architecture of the stream, allowing for gently sloping banks stabilised through 'soft' engineering approaches such as the use of live cuttings of Willow (*Salix* spp) and the use of young Alder (*Alnus glutinosa*) trees that can stabilise the stream banks and reduce the likelihood of soil erosion. This architectural design facilitates natural recolonisation and allows a greater surface area for the establishment of emergent, marginal and riparian vegetation and also allows easier access to the stream for wildlife (see Simpson *et al.* 1982; Dolan & Whelan 2004). Efforts should also be made to reduce the amount of sediments released into watercourses during realignment and installation of culverts.

Bridges are selected for larger rivers and can also be fitted with mammal ledges or the span of the bridge can be expanded well beyond the width of the river into the adjacent upland habitat, so that a larger area of un-submerged land is bridged and available as a riparian corridor for the dispersal of wildlife along the riverbank (MacDonald & Smith 2000).

Oversized arched culverts, box tunnels and bridges can also be designed to accommodate nest boxes for birds and bats, especially those species whose habitat has been disturbed by road ecosystem development. The type of construction material selected can determine the extent to which bird species can build nests within such structures e.g. bridges constructed from steel girders or stone can provide more nesting sites for birds as opposed to concrete structures (Smiddy & O' Halloran 2004).

Viaducts and expansion bridges are selected for important river crossings and wetlands (Jackson & Griffin 2000). Such structures generally do not impact on the natural architecture of rivers, or result in the loss of habitat for wildlife.

6.4 SUSTAINABLE LANDSCAPE DESIGN

The idea that linear landscape elements, such as roadside landscapes, can provide possibilities for biodiversity conservation and the restoration of landscape quality is almost never realised without specific provisions being made within the design, construction and maintenance phases of road ecosystem development. It should be recognised that taking advantage of the habitat mitigation/compensation opportunities provided by road ecosystem development can even make a valuable contribution to the size and distribution of protected areas. This is significant because restricting nature conservation solely to designated nature reserves is insufficient. For this reason the restoration of habitat and landscape quality within the immediate vicinity of a new road ecosystem should be of continuous concern to Local Authorities and National road authorities (Sýkora *et al.* 1989). In light of this, the Irish National Roads Authority has recognised that the landscape design of Irish roadside landscapes is one of the biggest opportunities for wildlife conservation and woodland planting in the recent history of the state (see NRA 2005a).

6.4.1 *Ecological landscape design*
The management of natural resources, the use of native species and the restoration of ecological services and landscape quality are strongly promoted best practice policies that should underpin any approach to road landscape design. Ecological landscape design complements best practice as it offers the following main advantages. Firstly it moves the designer away from the viewpoint of the landscape that is dominated by visual attributes, towards a more comprehensive understanding of ecosystems, thus conserving biodiversity and ensuring environmental sustainability. Secondly, the selection of re-vegetation techniques develops from an understanding and appreciation of the historical idiosyncrasies of a site and the preferred features of the landscape, contributing to regionalism in the design process (Makhzoumi & Pungetti 1999; Makhzoumi 2000; Dolan & Whelan 2004).

6.4.2 *The approach*
Once appropriate wildlife crossing structures have been selected for the safe passage of target species, either over or under a road carriageway, the landscape designer must set about mitigating impacts on adjacent habitats and ecosystems, and on landscape quality.
The application of ecological landscape design to road ecosystem development complements the concept of ecological infrastructure analysis and the designation of connectivity zones, as it also recognizes that the real extent of impacts on the landscape (such as a loss of connectivity and naturalness, and rural structuring) are only discernible at the landscape scale, through the use of aerial photography and the Map of the Ecological Network. In accordance with this concept, re-vegetation techniques are applied within designated connectivity zones in order to: (1) integrate wildlife crossing structures, (2) facilitate the extension of existing wildlife corridors, (3) defragment core habitats including protected areas.
The approach also highlights the importance of incorporating existing vegetation within the land-take of the road scheme as a baseline feature for the design layout of the roadside landscape. Re-vegetation techniques can complement existing tree groups, tree-lines and hedgerows through the extension of such habitats. The landscape design layout can also extend core habitat areas present within the roadside landscape. Landscaping activities should not disturb or replace existing native vegetation as part of the design process.

Re-vegetation techniques are also put in place with an understanding of the resultant effects on wildlife. For instance, the planting of tree-lines and woodland areas within roadside landscape may naturally force birds to fly higher above roads when crossing between forest edges, as was shown in The Netherlands by Verkaar and Bekker (1991). Verges should be widened and landscape planting on bends or curves on a road should be set back, to improve visibility and to discourage crossings by wildlife (Clevenger *et al.* 2003). In contrast, along straight sections of road carriageway, vegetative cover should extend as close to the road as permitted by road construction and safety standards. Landscaping of centre medians, junctions, roundabouts or interchanges should be sensitive to wildlife. Berry-producing plants should not be selected as they may prove attractive yet fatal to wildlife due to the close proximity of traffic (Trocmé *et al.* 2003).

Ecological landscape design of road ecosystems also incorporates environmental sustainability issues by adapting the ecosystem approach to resource management. In accordance with best practice, the use of native planting material in the re-vegetation of roadside landscape should be underpinned by the minimisation of natural resource inputs in stock production and plant establishment and a reduction in the need for long-term maintenance. The landscape designer needs to ensure that the re-vegetated areas of the roadside landscape may develop into self-sustaining habitats that do not present a future hazard to the road user or require the use of fertiliser, herbicides and frequent cutting or mowing regimes. Appropriate soil preparation and management is central to the establishment of a self-sustainable roadside landscape. In 2000, The Netherlands adopted a policy on utilisation of chemical herbicides; they can now only be utilised where acute danger is posed to traffic and other road users.

Ecological landscape design needs to also consider the contextual framework of the mobile road user and the concept of dynamic scale that determines the view from within the vehicle. A reduction in driver monotony and fatigue, or improved conditions for driver safety should also be of key concern to the landscape designer.

Ecological landscape design addresses sustainable development issues such as Local Agenda 21 and the participation of stakeholders in the decision-making process. In accordance with best sustainable practice, Ireland has assessed road users' preferences for features found within the Irish landscape. Such features can be incorporated in to the landscape design layout of roadside landscapes.

Any approach to the ecological landscape design of road ecosystems, needs to incorporate the following: (1) aerial photography of the road scheme, (2) the Map of the Ecological Network, (3) the landscape assessment, and flora and fauna surveys within the relevant EIA, (4) a map of the past or potential native vegetation of an area, (5) dynamic scale, (6) the identification of the road users preferred landscape features, (7) a reduction in driver monotony and fatigue.

6.4.3 *Sustainable re-vegetation techniques*

Sustainable landscape design of roadside landscapes can be defined as that which minimises energy or physical and natural resource inputs (water, fertilisers, herbicides, pesticides etc.) in stock production, plant establishment and maintenance, is locally-appropriate (in terms of substrate type, choice of species and sources of plant material), maintains ecological integrity (see Hitchmough & Dunnett 1996) while incorporating the preferred landscape features of the various road user types.

6.4.3.1 *Soil preparation and management.* The acceptance of slower plant growth on less fertile soil (sub-soils, rubbles etc.) complements sustainable criteria including the reduction in the consumption of natural resources and the move away from the use of herbicides and fertilisers (Hitchmough & Dunnett 1996). Many authors have recommended the use of subsoil as opposed to topsoil (e.g. Mederake 1991; Melman & Verkaa, 1991; Box & Forbes 1992;

Hitchmough & Dunnett 1996; Sýkora *et al.* 2002; van Bohemen 2002, 2004; Dolan 2004) on roadside landscapes as it encourages species-rich wildflower meadows or grasslands. The abundance of topsoil in road construction in agricultural landscapes and its subsequent use in road landscaping activities, results in the establishment of highly productive vegetation types of low ecological and aesthetic value (Mederake 1991; Sýkora *et al.* 2002). Such vegetation types contain agricultural weeds and grasses and for this reason are heavy consumers of physical and natural resources requiring the use of herbicides and maintenance in the form of mowing.

During construction consideration should be given to covering topsoil with a minimum layer of 30cm of subsoil which is adequate for the root systems of wildflower meadows. Natural recolonisation, seed broadcasting and mulching can then be implemented. Where required trees and shrubs can be planted so that the root systems sit within the fertile topsoil, while the upper subsoil layer can act as a weed suppressant as well as providing the appropriate substrate for species rich wildflower meadows. To ensure an adequate supply of subsoil, its availability within the land-take of a road scheme must be assessed during the design phase, along with provisions for its appropriate storage and management of the leftover topsoil.

In terms of sustainability, the use of soil conditioners and fertilisers should also be avoided in the landscape design process, as should peat products and bark mulch from non-sustainable forests, and from those forests which do not practice green tree retention.

6.4.3.2 *Natural recolonisation.* Natural recolonisation, as opposed to planting trees and shrubs, addresses environmental sustainability issues as it requires comparatively little or no physical or natural resources inputs. It is generally recommended for small disturbed sites surrounded by native vegetation (the seed source) as such sites experience an easier and more direct succession towards vegetation similar (in species composition) to adjacent natural plant communities e.g. woodlands, wildflower meadows and wetlands. It can, however, be utilised on larger areas, especially on nutrient poor substrates such as rock cuttings and scree slopes where the establishment of planted material would not be environmentally sustainable as it would require a high input of resources for ground preparation and establishment, often with limited success. Natural recolonisation is also a suitable re-vegetation technique for: (1) realigned watercourses, (2) sustainable drainage systems where they are located close to existing wetlands, (3) decommissioned sections of roads, where a seed source such as a hedgerow or a woodland habitat is present in the vicinity of the old road.

For natural recolonisation to be successful, it is important that the surface geology of the proposed site is consistent with that upon which adjacent native vegetation has become established. The process of natural recolonisation can be complemented, and indeed quickened, through the use of soils and plant material (in the form of sods of turf) retained or salvaged from woodlands, hedgerows and wetland areas disturbed during initial construction of a road ecosystem. It may also be necessary to take measures to avoid the dispersal of existing non-native invasive species in the vicinity of proposed sites as disturbed soil can provide an opportunity for invasion by such species.

6.4.3.3 *Species choice and plant associations.* The basis of a sustainable approach to landscape planting must always be to choose native species that are well suited to a site. Not only will this remove the need for expensive manipulation of site conditions, and the consumption of natural resources in soil preparation and establishment (such as soil conditioners, fertilisers, irrigation and herbicides), but it may also avoid long-term maintenance requirements caused by the specification of unsuitable plants. The elements for design should be based on soil geographic factors, the past native vegetation of the area, and most importantly, should be based on the existing native vegetation in the vicinity of the new road ecosystem. The aerial photographs of the selected route and detailed floral surveys within the EIA contain information on the local plant communities, including the location of potential local provenance seed sources for wildflower meadow establishment, broad casting and mulching of shrub and tree seeds, and for instant planting of trees and shrubs.

6.4.3.4 *Wildflower meadow establishment.* The use of highly fertile topsoil is a limiting factor in the development of species rich wildflower meadows. Landscape designers can therefore utilise subsoil, to bury the seed bank of undesirable species and to smother existing weeds.
The natural recolonisation of wildflower meadows can be successfully facilitated where the roadside landscape possesses a surface geology or soil type consistent with that of the adjacent wildflower meadow. The soil type, however, can be altered to cater for the establishment of wildflower meadows). An alternative treatment to natural recolonisation and direct seed sowing is the use of hay from species-rich meadows. Hay strewing is a more sustainable technique in wildflower meadow establishment (apart from natural recolonisation or translocation of a seed bank) in comparison with the hydro-seeding of commercial wildflower seed. Hay is cut from the source meadow before the seed has fallen and is then spread thickly onto the soil surface. In the following spring the seed geminates under the protection of the hay covering. Where commercial wildflower seed mixes are selected for use, the seed should be of native local origin i.e. seed stock from local meadows containing local indigenous flora.

6.4.3.5 *Seed broadcasting and mulching.* Broadcasting and mulching of tree and shrub species is a more cost-effective and sustainable method than instant planting. It is also more predictable than natural recolonisation as it is not dependant on natural factors for the arrival of seed by wind and animal dispersal, or on the viability of seed. It is most suitable on soil cuttings and embankments, where there is no natural seed-source in close proximity (Luke *et al.* 1982). The environmental conditions at such sites may be severe, but advances have been made in the technology for the establishment of vegetation - both herbaceous and woody species (Wright *et al.* 1978; Colwill *et al.* 1979; Dunball 1985). The advantages of broadcasting and mulching include: (1) an increase in the number of woody species on roadsides, (2) the production of more natural assemblages of plants on a wide range of soil types, (3) an extension to the period of the year when plant material can be introduced onto roadside landscapes, (4) little or no soil preparation or use of physical and natural resources for establishment (Luke *et al.* 1982). Seed broadcasting and mulching can be carried out by hand, or through hydro-seeding. The former is a more sustainable approach to broadcasting seed and seed mulches.

6.4.3.6 *Instant planting.* Instant planting is the least sustainable approach to landscape design of roadside landscapes. Such planting is generally required where the landscape architect, ecologist, road manager or stakeholder want height and mass of plants at short notice (van der Sluijs & Melman 1991; Meunier *et al.* 1999) usually under the following circumstances: (1) where a road ecosystem enters an urban area (for aesthetic and traffic calming reasons), (2) for screening of large-scale technical elements and blending of the road into the landscape, (3) for soil stabilization, (4) as a bio-barrier for noise, (5) to integrate wildlife crossing structures in to the surrounding native vegetation, (6) to reconnect severed hedgerows and other wildlife corridors, (7) on sites where natural recolonisation cannot be facilitated i.e. where there is no seed source or the seed bank is not intact, or where there has been major soil disturbance, loss of soil structure and/or soil mycorrhizae.
Stock production and plant establishment associated with the use of mature plant material requires much energy or physical and natural resource inputs in the form of irrigation,

fertilisers, herbicides and pesticides. In descending order potted plants are the heaviest consumers of resources (and most costly) in comparison with root-balled and bare root plants. Root-balled plants can be wrapped in hessian or polyurethane bags. The former is a more sustainable option as it is constructed from a renewable resource and is biodegradable, while the manufacture of the latter releases pollutants and improper disposal can lead to littering.

Inappropriate selection of species, size or girth of plants, spacings or planting densities can have implications for long term maintenance requirements in the form of replacements, cutting, trimming, thinning and soil mycorrhizae inoculation which can be intensive and expensive (van der Sluijs & Melman 1991). It must also be emphasised that, in creating instant planting along roadsides, there is no question of an ecologically sound layout. The trees and shrubs that are planted have been grown elsewhere, therefore, the first few successional seres are skipped i.e. the natural development of the plants and associated fauna is not catered for (van der Sluijs & Melman 1991; Dolan 2004).

Baines (1994) estimated that up to 79% of the total cost of urban tree establishment (i.e. ground preparation, plant material and establishment) could be avoided through the use of natural recolonisation as opposed to planting of trees.

To ensure sufficient seed is available for broadcasting and mulching, and sufficient number of native plants of local or national provenance for the landscape design of road ecosystems, the tendering of the landscape design contract is required prior to commencement of earthworks in order to allow ample time for: (1) the collection of local provenance seed (including making provision for irregular supplies of seeds due to masting in the case of Oak trees), (2) seed stratification, (3) the production of stock of sufficient size or girth for aesthetics and/or successful establishment (in relation to instant planting).

An alternative to utilising potted plant material, root-balled or bare root plants is the use of live cuttings from trees and shrubs during the dormant season. This approach is more sustainable ecologically, and economically more sound than instant planting since it requires almost no natural resources and only physical resources in the gathering and planting of the cuttings. The use of cuttings also ensures that only local provenance plant material is utilised in landscape design layouts. This is important for re-vegetation of the roadside landscape in the vicinity of protected areas. Even though failure rates can be high, this can be counteracted by high density planting as the process is so cost-efficient. The failure rates of Willow (*Salix* sp.) cuttings can be dramatically improved if the cuttings are stored in coarse sand and water which encourages shoot and root formation.

The recycling of trees that have recolonised disturbed areas within the land-take of a road scheme, during the construction period, is another more sustainable alternative to potted, root-balled and bare root plants. Young Alder trees, as they are rapid self-seeders, lend themselves to this process and to the establishment of a wet woodland habitat in a short period of time. Root cuttings from certain species such as Aspen (*Populus tremula*) can also be utilised.

6.4.4 *The contextual framework of the road user*

As the growing urban population has increased mobility, a greater diversity of user groups, are now entering road landscapes and using the same road networks (Pauwels & Gulinck 2000).

Casual or tourist-type involvement with road ecosystems involves spontaneous sightseeing, with memories being the eventual trophies of travel. Such encounters are typically brief, emphasizing generalized landscape meaning versus individual object understanding (Steele 1981). While the experiences of such travellers might be considered superficial, they are far from being unimportant from a planning perspective. They provide indications of a region's tourism potential and as such, can be associated with extended economic benefits (Clay & Smidt 2004).

Long-haulage users distribute goods over long distances across countries and even between countries. For this reason they are susceptible to road monotony and fatigue. They may be very familiar with a particular route or routes along certain road networks. Due to mobility, encounters with roadside elements are brief but they may be revisited on a regular basis but with gaps of several days or weeks. Identification of features within the landscape that are of particular interest to this user type has the potential to combat road monotony, and fatigue, as the provision of such features could provide important stimuli and increase driver alertness. However, views of the road landscape from within trucks and buses are much different from those experienced by local road users within cars or vans due to the height above the ground and the angle of view from within the vehicle. Therefore road planners, designers and managers must acknowledge the specific design requirements of this road user type in the creation of road design plans.

Local users make short regular trips along the same segment or segments of a road network, therefore are very familiar with certain routes and with elements of the landscape within those routes which are frequently revisited on a regular, either daily and/or weekly basis.

Passengers may also be present within vehicles and their preferences for features of the roadside environment are not always taken into account in assessment of peoples' values for the road landscape. They may not be susceptible to road monotony and fatigue as a result of the activity of driving, but they can also find extensive stretches of road landscape which contain similar elements monotonous.

Pedestrians and cyclists are also mobile road users. Pedestrians take in larger swathes of landscape than a cyclist, as they are not involved in an activity that requires much of their attention.

Local inhabitants living and/or working within a road landscape may experience the road landscape in several contexts. Firstly, the local inhabitant frequents the road landscape as a mobile local user or passenger and secondly as a stationary local inhabitant gleans views of the road landscape from their home, work place or business.

6.4.4.1 *The View From Within a Vehicle.* The two fundamental types of views are the 'panorama' and the 'vista'. A panorama refers to a broad view with a good vantage point, while a vista refers to a view with moderate blocking of the environment e.g. a view restricted by bounding margins such as trees (Nelson 1997).

The relationship between the mobile road user and the roadside landscape is much more complex than for a person viewing the roadside landscape from a stationary position. Where a person is stationary, they generally only view a few features (e.g. a roadside art exhibit, a pond, or a group of trees) within a single landscape type. In contrast, through mobility, the driver and passengers can be greeted by multiple features and landscape types as they drive along a road corridor. It is important to note that some features can only be viewed at particular speeds and that views of features are also restricted or bounded by the confinements of the vehicle structure, while at the same time are opened from, or blocked by, vegetation, buildings and elements of the natural terrain (Nelson 1997; Clay & Smidt 2004; Dolan & Whelan 2004).

6.4.4.2 *Dynamic scale and the mobile road user.* It is the concept of dynamic scale that plays a pivotal role in understanding the opportunity presented to mobile road users to view features along a road corridor e.g. small scale objects such as ornamental shrubs cannot be experienced by the road user at speeds of 120km/h on motorways (Danish Roads Directorate 2002). Where the concept of dynamic scale is not addressed, re-vegetation techniques designed to restore environmental aesthetics can be rendered ineffectual as they may not be experienced by the road user. The speed at which the driver moves also determines how far ahead it is possible to focus on the landscape, and at what angle it is possible to see features. Vistas also need to be of a minimum width in order to be seen. For example, a 1 second duration at 100km/hr requires an opening of 27m, while at 42km/hr the space need only be 9.6m wide (Bell 1997). Further, the linearity of a road corridor reinforces the importance of spatial arrangement and visual sequencing of features within the roadside landscape (Magill & Schwartz 1989; Little 1990; Bell 1993; Landscape Institute 1995). For this reason, the landscape designer must be aware of the resultant affects of the design layout and sequencing of landscape layouts on the mobile road user e.g. the frequent use of a few relatively large blocks of planting containing only a few species, along a road corridor, can become repetitive leading to a loss of naturalness, rural structuring plus driver monotony and fatigue. The linearity of the road corridor also establishes a basis for discrete observations (Kent & Elliot 1995) of appropriately placed features en route, such as water features and roadside art exhibits (Dolan & Whelan 2004).

6.4.4.3 *Driver monotony and fatigue.* The landscape designer should also address the need to counteract monotony and fatigue for the driver through improved conditions for driver alertness, especially where extensive stretches of similar vegetation may exist, or where a potential loss of vista quality has been identified. Vistas serve to increase perception of interest as visually impenetrable margins increase the strength of hazard perception. Thus, features such as the partial blocking of a view by a copse of trees, the opening up of copse of trees to form a vista, particular landscape treatments including watercourses and water features, roadside art exhibits and kilometre pegs can all serve to stimulate or intrigue the driver by reducing the likelihood of fatigue (Nelson 1997; Thiffault & Bergerson 2002; Dolan & Whelan 2004).

6.4.4.4 *Stakeholder participation and preferred landscape features.* The participation of stakeholders in the decision-making process forms the basis for a sustainable approach to landscape design as it addresses Local Agenda 21 criteria.
Ecological landscape design promotes a thorough understanding of environmental aesthetics, including consideration that visual encounters with preferred features of the landscape can lead people to form emotional attachments to the land and develop a greater appreciation for sustainability goals (Parsons & Daniel 2002). The values associated with a feature of the landscape will vary from person to person, political group to political group, and from time to time (Forman 1990). Surveys are required to identify preferred features of the landscape that can be incorporated into the landscape design layout of roadside landscapes. A number of such surveys have been carried out in Ireland. In descending order of preference, the following features were highlighted by local inhabitants residing adjacent to Irish National road schemes: stone walls, water features (ponds or lakes), woodlands, panoramas (or open views), hedgerows and sites of historical or archaeological interest; while mobile road users identified: roadside art exhibits, water features (ponds or lakes), vistas, wildflower meadows, geological features (rock faces), a few mature trees in isolation and finally sites of historical or archaeological interest. The incorporation of such features can also promote regionalism, or regional identity in the design process through the retention of locally distinctive features (see Dolan & Whelan 2004).

6.4.4.5 *Water features.* The establishment of water features within roadside landscapes can complement environmental sustainability goals in several ways. These can: (1) treat and/or store road run off e.g. swales, attenuation/stilling/balancing ponds, extended detention or retention ponds and constructed wetlands or helophyte filters, (2) constitute a preferred landscape feature as they are aesthetically pleasing, (3) provide additional habitat for wildlife, (4) reduce monotony and fatigue by stimulating driver alertness (Nelson 1997; Thiffault &

Bergerson 2002), (5) benefit tourism, (6) induce a feeling of spaciousness and naturalness (Coeterier 1996).

6.5 RESTORATION OF LANDSCAPE QUALITY

In road ecosystem development, along with mitigating the impacts of habitat loss and habitat fragmentation, measures are required to restore landscape quality. Such measures need to address: (1) loss of variety: through the provision of aesthetically effective re-vegetation techniques, (2) loss of naturalness: through the provision of water features and techniques which may restore native vegetation and, integrate new large-scale technical elements such as flyovers, bridges and culverts, (3) loss of 'rural' structuring: through achieving agreement of scale, by using trees and hedgerows which may reduce pattern and orientation effects, (4) loss of regional identity: through the restoration of spatial arrangements which formed the unique individual character and appearance of the former landscape and the retention of locally distinctive features and views to the same, whilst screening large-scale technical elements through the manipulation of vegetation and scale, (5) loss of vista quality: through the creation of new vistas, by opening up and/or the planting of vegetation to create views with bounded margins (e.g. N11 Newtownmountkennedy by-pass, Co. Wicklow, Ireland). The use of noise barriers and other structures should also be sensitive to avoid simplifying and narrowing down the field of view.

6.5.1 Locally distinctive features
Locally distinctive features are features of a landscape that contribute to regionalism and regional identity (Nohl 2001). They may hold symbolic value for certain groups of people. Such elements often contribute to an (aesthetic) sense of the place which presupposes some history e.g. historical monuments and buildings, archaeological features, mature vegetation complexes, rock faces.
In traversing most landscapes, the establishment of a new road ecosystem will encounter areas of local elevation and depression therefore creating features such as cuttings and/or embankments. Cuttings that reveal rock faces and scree slopes can be of significant visual interest to mobile road users, as a locally motivated element, but they require sensitive treatment during construction. Rock faces are ideal surfaces for natural recolonisation, facilitating the natural recolonisation of rock cuttings bur require conditions that cater for the potential development of vegetation in crevices where organic matter may naturally accumulate over time. For this reason it is important to identify the potential for the exposure of a rock face or scree slope during the design phase to ensure that the rock cutting will expose an appropriate profile. Rock removed from such cuttings can also be utilised in the landscape design layout of urban areas.

6.5.2 Large-scale technical elements
Flyovers, bridges, junctions, interchanges and roundabouts are large-scale technical elements associated with road ecosystems, that may be visually dominating within the rural landscape as they can exhibit an over-scaled and 'urban' character. For this reason, sensitive selection of construction material and the management of vegetation and scale are essential to avoid a loss of naturalness and rural structuring, and to prevent the urbanisation of a rural landscape.
The design layout of flyovers can complement greenways, where consideration is given to providing a multifunctional-use corridor for cyclists and pedestrians including a green lane for wildlife that forms part of a larger greenway system. Existing two-lane flyovers in Holland and Belgium have been modified to create green bridges and green lanes.

A significant paved area is often placed at the base of the abutments of bridges and flyovers and around the entrances to arched culverts. Such paving constitutes a heavy consumption of natural resources and is generally not required for structural integrity. The pavement may also serve as a barrier to dispersal of grassland invertebrates.

Junctions and roundabouts are purposefully designed to be visually dominant, with elevational variation and illumination with merging and diverging traffic and associated over-bridges. Significant land-take may be enclosed by junctions and roundabouts. This land, however, is generally isolated from the surrounding landscape by the presence of access lanes and roundabouts. It should not be made attractive to wildlife through inappropriate planting e.g. berry producing trees and shrubs due to the risks associated with attracting wildlife to high levels of traffic activity. While the disturbance associated with traffic activity may deter animals from establishing in these locations, birds may still be attracted by particular planting schemes.

Illumination of such technical elements can create micro-climates for phototactic species of insects while deterring photonegative species. Careful consideration should be given to avoiding illumination in the vicinity of wetland and other sensitive ecosystems.

6.6 SOIL STABILISATION

The inclination of slopes, embankments and other graded areas within the roadside landscape plays a major role in determining the extent of coarse particulates or sediments released and the amount of natural and physical resources consumed by a road ecosystem.

Where cuts slopes of 1:2 slopes are present on a road scheme, hard engineering approaches may be required to achieve stabilisation e.g. rock armouring and gabions. The application of such measures involves a heavy consumption of natural and physical resources and can inhibit the restoration of native vegetation and impede wildlife. Where possible, consideration should be given to reducing the slope of cuttings from 1:2 (vertical to horizontal) to 1:5, in areas where the substrate is known to be potentially unstable.

Where rock armouring and gabions are applied over a large area, for example in the case of cuttings, there can be impacts on landscape quality (through a loss of 'rural' structuring and naturalness) along with the heavy resource consumption. In such cases modifying the road design layout to accommodate 1:5 slopes as opposed to 1:2, would make more sustainable, ecological, economic and aesthetic sense than design specifications which include a wide vegetated centre median .

Where 1:2 slopes are unavoidable, consideration can be given to utilising soft engineering or bioengineering methods to stabilise slopes e.g. locally sourced native grass seed mixes or sterile grasses with maximum life expectancy of one to two years. Live cuttings and recycled trees can also be utilised to stabilise soil slopes e.g. in the reconstruction of river, stream-banks and stabilisation of floodplains.

6.7 TECHNICAL DESIGN ASPECTS: THE ROAD SURFACE AND BED, AND STREET FURNITURE

The type of road surfacing material, road marking paint and street furniture selected during the design phase of road ecosystem development, determines the amount and type of pollutants released by these elements once a road is fully operational. For this reason, decisions made during the design phase, have a major role to play in the amount and type of pollutants released and deposited by a road ecosystem.

This section deals with technical design aspects that tackle waste products from road ecosystems at source.

Car manufacturers and service providers can make informed decisions to: (1) provide best available technology in vehicle manufacture to reduce the amount and type of emissions released by vehicles, (2) manage the waste outputs from the maintenance of vehicles, (3) provide for sustainable management of End of Life Vehicles.

6.7.1 Road surfacing material and marking paint

Pollution from road surfaces can occur due to the: (1) actual laying of the surface material or pavement during construction, (2) wear on the surface by weathering and the tyres of vehicles, during the operation/use phase, (3) maintenance of the surface, (4) the breaking up of old road pavement during decommissioning.

The type of road surfacing material applied on road carriageways determines the overall quantity and types of pollution released by a road ecosystem. This is because the manner in which pollutants are released (run-off, spray or atmospheric spread) depends on, (1) the sources and quantities of pollutants, (2) the quantity and distribution of rain water over time; (3) the type of the road surface material applied (van Bohemen & Janssen van de Laak 2004).

Porous road pavement contains a higher proportion of pores, >20% in comparison with closed asphalt. A transverse section through porous pavement reveals a 5 cm layer of porous asphalt placed on top of a layer of closed asphalt (Berbee *et al.* 2004). This design allows run-off to filter gradually through the top layer of the pavement and be discharged at the road edge, reducing or avoiding the phenomena of aquaplaning, water splashing and evaporation (Pagotto *et al.* 2000; Berbee *et al.* 2004). The risk of car accidents during rainfall events is also reduced as vehicle adhesion and visibility are improved (Pagotto *et al.* 2000).

Most importantly, from an ecological viewpoint, where porous asphalt is used, run-off, spray (due to wind dispersion) and atmospheric emission of pollutants are reduced. The quality of runoff from porous asphalt is much improved in comparison with closed asphalt surfaces. It is the retention of coarse particulate pollution and the adsorption and/or retention of certain dissolved forms of metals such as zinc and cadmium (RIZA 1996; Pagotto *et al.* 2000; van Bohemen & Janssen van de Laak 2004) within the pores during filtration that explains the reduction in the amount of hydrocarbons (HCs) and metals released from this type of surface. The discharge of heavy metal loads into the environment can be reduced from 20% (Cu) up to 74% (Pb), while coarse particulates are retained at a rate of 87% and HCs are intercepted at an even higher rate (90%). It has also been found that deposits by spray are much smaller on porous asphalt surfaces than in the case of closed asphalt (see Boland 1995) where 57-83% of the pollution from the latter is caused by dry and wet wind spray (van Bohemen & Janssen van de Laak 2004).

In addition to reducing the extent of pollutants released by road ecosystems, porous road surface absorbs rather than reflects (as in the case of closed asphalt) sound waves, thus reducing noise disturbance in the order of 3-4dB (Pichon 1993; Pagotto *et al.* 2000) and aggressiveness for local inhabitants and wildlife. Golebiewski *et al.* (2003), investigating noise reduction by porous asphalt, found a reduction in drive-by noise annoyance for all velocities investigated ($20<V<60$ km/h, where V = velocity). The majority of commercially available porous road surface materials are comprised of asphalt blended with reclaimed tyre rubber or other additives, in which the rubber component is e.g. 15% by weight of the total surface blend. Porous asphalt surfaces of this kind are referred to as drainage-, fluster-, or silent-pavements.

Porous asphalt, however, should only be selected for road ecosystems which are expected to carry a heavy traffic load as, in the absence of heavy traffic, the adherence of vehicles and the capacity of porous asphalt pavement can be reduced due to clogging (Pichon 1993) requiring regular maintenance in the form of washing. Thus, designers need to select the appropriate surface material to avoid the need for costly and frequent washing which could also lead to the concentrated release of sediments and pollutants (Pagotto *et al.* 2000; Berbee *et al.* 2004).

Apart from the potential release of pollutants during maintenance activities, the retention of heavy metals, HCs and sediments within the pores may also present a future hazard in road improvement works or in road decommissioning. In spite of these disadvantages, porous asphalt remains the preferred surface material, especially on motorways in temperate regions. Porous concrete surfaces are also under development but are not yet commercially available (Golebiewski *et al.* 2003).

Coloured road markings, can consist of a variety materials depending on the expected wear, abrasion resistance, skid resistance, texture and economic restrictions. The following materials are generally utilised: (1) Pigmented resin systems (epoxy, polyurethane, or thermoplastic) with coloured aggregate, (2) Pigmented slurry sealing, (3) Surface dressing with a suitable coloured aggregate, (4) Hot Rolled Asphalt (HRA) wearing course with a coloured pre-coat and clear resin coating. In the UK and in other countries, there are no national regulations that restrict the use of lead-based pigments or coatings in road markings materials. Therefore companies legally sell products containing lead-based pigments on the market for application on national roads. Even through environmental concerns have been voiced, these types of pigments continue to be utilized as they are less expensive than non-lead alternatives and because the characteristics found in lead-based pigments generally show a good balance of the required properties including the following: (1) They are good performers, retaining their colour during processing and in service for up to 3-5yrs or more, (2) They provide good stability to natural weathering, (3) Apart from good durability, they are heat resistant and are able to withstand relatively high temperatures during processing. Manufacturers are utilizing grades of pigments that have low levels of soluble lead compounds and have also adopted special coatings to seal the pigments so that there is minimum exposure of the pigment powder during processing in the factory and on site during application. Once the pigment is incorporated into the road-marking compound, the product is encapsulated by binders that seal in the pigment powder. Inevitably the seals or coatings applied can be removed due to abrasion. Some local authorities are now opting for lead-free products, despite their expense, for example in recognition of the possible end-of-life disposal problems of having to plane and remove markings.

6.7.2 Roadbed and fill

Fill for a roadbed consists of geotextile lining, sand, subsoil, gravels and crushed aggregate base material. Sensitive sourcing of aggregate material must be undertaken during the design phase. The principle of using local materials including local soil, local gravels and aggregate material for use in fill areas, embankments, the road bed and the road surface is essential. More sustainable alternatives to local quarried materials include the use of leftover stone from the blasting of rock faces in the aggregate base layer, or crushed recycled glass in the subsurface aggregate layer. The subsurface layers of the N2 Carrickmacross Road scheme in Co. Monaghan, Ireland contain a million crushed and recycled glass bottles instead of quarried rock. While only 10% of the stone in the subsurface layer was substituted with recycled glass, laboratory experiments show that substitution of up to 30% is feasible, with positive implications for road construction and waste management (NRA 2005b).

As the road surfaces receive high contact pressures from vehicle tyres, the various layers of material within the road bed are designed to dissipate the pressure down to a level that can be supported by the underlying soil. Over time, however, vehicle load pressure can cause subsoils to migrate into the aggregate base of the road bed, reducing the base thickness and the load-carrying capacity of the aggregate base, effectively reducing the life-span of the road surface material. The use of geotextile lining, such as Bentonite, in the design of road-beds, provides separation between the different layers of material that make up the road-bed, thus, preserving the integrity and extending the life-span of the road surface material. Filtration is another function widely performed by geotextiles. The filtration function has two concurrent objectives, to permit a certain quantity of water to pass through the plane of the lining and the retention of filtered coarse soil particulates. These functions prevent the release of silt and sediments into groundwater aquifers and adjacent aquatic systems (NAFS 2005).

6.7.3 *Street furniture*

Zinc originates from tyres and asphalt, but also from street furniture such as crash barriers and lamp posts, and to a lesser degree from road portals and road signs. Some 3,500km of crash barriers and two million new lamp-posts are placed on European road ecosystems each year (Ecosafe 1997). However, new technology means that alternative materials can now be selected to avoid or reduce zinc deposition, such as fibre reinforced plastic (FRP), wood and even recycled tyres.

FRP crash barriers and lamp-posts are inherently corrosion-resistant, having a longer life span than galvanized steel and without the polluting galvanizing process. FRP as an alternative to galvanized steel and concrete has been found to be the safest alternative for all road-users. In contrast to galvanized crash barriers that are less effective at stopping trucks and unsafe for motor-cyclists, or concrete crash barriers, which are more capable of stopping trucks but less suitable for small cars (Ecosafe 1997), the energy absorbing capacity of FRP material means that it can be tailor-made to absorb the impact of different vehicle types (Ecosafe 1997). The same applies to lamp-posts made from FRP, which can be designed so that they snap-off in the event of collision. Other advantages of FRP include its lightweight nature and the optimization of the design that make installation easier and quicker, resulting in lower maintenance costs and fewer traffic delays for maintenance. Other functions can also be integrated into the design, such as anti-dazzle shields or information on speed limits (Ecosafe 1997).

Wood is a desirable building material because it is relatively inexpensive and widely available. Inexpensive wooden crash barriers can be designed to absorb the energy from an impact and protect the more expensive object or vehicle hitting it. The crash barrier can then be easily and cost-effectively replaced or left to biodegrade if required (Heyliger *et al.* 2004).

Recycled whole tyres can also be utilised as crash barriers due to their physical form, resilience to impact, and longevity (Amari *et al.* 1999). FRP and recycled plastic also provide alternatives to timber or concrete fence posts for boundary demarcation. The use of local or vernacular material is recommended for the design and reconstruction of stone walls for land boundaries and noise barriers, that would retain regional identity and reduce the urbanisation of the rural landscape.

At the road location or site scale, barriers put in place to reduce noise levels include: timber fencing, stone walls, concrete structures, bio-barriers (which consist of rock and soil bunding and/or landscaping) and glass panels. Timber is the most popular material selected for noise barriers in road ecosystems. From a sustainability view point timber should be sourced locally, from sustainable or renewable forests including green tree retention forests and should be pressure treated using eco-friendly wood preservatives. Alternative to wooden barriers are bio-barriers constructed from soil and/or rock bunds planted with trees and shrubs. Such structures can serve as additional habitats and corridors for wildlife, particularly small mammals and bats. Where design specifications require a bio-barrier, care must be undertaken to select locally sourced rock and plant material that is based on an understanding of the resultant effects on wildlife. Transparent noise barriers that have been utilised in The Netherlands for aesthetic reasons have led to greater numbers of road kill (Keizer pers. comm.) and therefore should be avoided. The Dutch road ministry has ordered vertical stripes to be placed on existing barriers to reduce the likelihood of collisions (see Iuell *et al.* 2003).

The positioning of noise barriers and landscape planting also has knock on effects on the deposition and settlement of pollutants from road ecosystems as they can also serve to concentrate pollutants directly behind them (e.g. Erisman *et al.* 1999 cited in van Bohemen & Janssen van de Laak, 2004). Therefore strategic assessment and monitoring is required where these structures are present in the vicinity of sensitive aquatic ecosystems.

The design of road signage, and other street furniture, should be consistent in colour shape and size throughout a road corridor in order to reduce contrasts of association or harmony. Street furniture can also be fitted with nesting boxes for wildlife e.g. for kestrels on United Kingdom (UK) motorways.

Positioning of street lighting needs to be sensitive to wildlife and wildlife habitats, especially in the vicinity of protected areas. The extent of light pollution can be reduced by a number of simple design and positioning measures: (1) light fittings can be designed to reduce the amount of light emitted upwards, (2) proper positioning of street lighting and directing it downwards i.e. shielded lighting, (3) using only the necessary amount of lighting, (4) switching off unnecessary lighting, particularly decorative floodlighting and advertising lighting, late at night and in the early morning (UK EPA 2005). Consideration should also be given to the use of low-pressure sodium lights that do not produce ultraviolet light and will not attract moths (Rydell 1992).

6.8 EFFECT ORIENTED MEASURES

Effect-orientated measures usually consist of drainage systems which are implemented to reduce the imminent impact of diffuse and acute point pollution on environmental quality at the road segment or road location or site scale. For this reason they have less of an effect than approaches which reduce the amount of pollutants at source. Traditional rural drainage systems involved the use of kerbs and gullies or interceptor drains. More recently there has been a move away from the use of such systems in rural environments towards the use of French drains and swales. This has resulted from the pressure to rethink conventional drainage systems, and to address sustainable development and a minimum consumption of natural resources, as well as ecological criteria.

The sustainable drainage approach is currently an expanding subject area, prevalent in progressive engineering design, that has brought drainage management potentially much closer to nature than the traditional engineering approach. In sustainable drainage management, all aspects of the road hydrological cycle are weighed against the potential holistic environmental impact. Hence, the storage and treatment of road runoff is considered along with interrelated factors such as water use, adjacent land uses and run-off, existing wastewater treatment processes, effluent discharges and receiving water quality (INTERURBA II 2001).

Sustainable drainage systems are required for both rural and urban landscapes. Sustainable Urban Drainage Systems (SUDS) once restricted to urban landscapes are now increasingly being utilised in rural landscapes. There are several types of drainage systems, including: (1) French drains, (2) swales, (3) bio-swales, (4) filter strips/level spreaders, (5) attenuation, balancing/stilling ponds or basins, (6) extended detention or retention ponds, (7) constructed wetlands or helophyte filters, (8) landscaping (BCAFAP 2003; Revitt & Ellis 2001). Swales, French drains and attenuation, balancing/stilling ponds or basins and extended detention or retention ponds are utilised in both rural and urban landscapes while filter strips/level spreaders and constructed wetlands or helophyte filters are generally restricted to urban areas. Drainage systems in both the rural and urban environments can maintain or restore environmental quality, and with the exception of French drains, can provide additional wildlife habitats and enhance aesthetic aspects of the roadside landscape.

6.8.1 *Permeable conveyance systems*
There are two main types: underground systems such as French drains and bioswales and surface water features such as swales (SEPA 1999).

6.8.1.1 *French (or filter) drains.* French drains are underground systems comprised of a trench, filled with gravel wrapped in a geotextile membrane into which runoff water is led, either directly from the drained surface or via a pipe system. The system moves runoff water slowly towards a receiving watercourse, allowing storage, filtering and some loss of runoff water through evaporation and infiltration before the discharge point. The gravel in the filter drain provides some filtering of the runoff, trapping organic matter and oil residues that can be broken down by bacterial action through time. Runoff velocity is slowed, and storage of runoff is also provided. Infiltration of stored water through the membrane can also occur and some filter drains need not lead to a watercourse at all. Filter drain systems have been widely used by road authorities. Hybrid infiltration systems and filter drains have been used for a variety of developments, including both residential and industrial systems (SEPA 1999).

6.8.1.2 *Swales.* Swales are grassed depressions that lead surface water overland from the drained surface to a storage or discharge system, typically using the green space of a roadside landscape. When compared with a conventional ditch, a swale is shallow and relatively wide, providing temporary storage for storm water and reducing peak flows. They are appropriately close to source and can form a network within a development scheme, linking storage ponds and wetlands. A swale is dry during dry weather, but during a rainfall event water flows over the edge and slowly moves through the grassed area. The flow of surface water is retarded and filtered by the grass. Sediment is deposited and oily residues and organic matter retained and broken down in the top layer of soil and vegetation. Swales can be lined below the soil zone where necessary, to protect the underlying aquifer. During a rainfall event a proportion of the runoff can be lost from the swale by infiltration, and by evaporation and transpiration. If necessary, overflows can be placed at high level to prevent flooding in times of exceptionally heavy rainfall. Swales should be designed to be dry between storm events to enhance their pollutant removal capability. Swales work best with small gradients both for their side slopes and longitudinally. Performance can be enhanced by placing check dams across the swale to reduce flow velocities, which in turn reduces the risk of erosion in a swale. Even where swales discharge directly to a watercourse, a considerable reduction in pollution load can be achieved. In addition, where runoff is conveyed via surface channels, wrong connections become obvious and can be fixed without the need for the expensive surveys that are typically required with traditional piped systems. Swales also require a minimum consumption of resources as the need for expensive piping, roadside kerbs and gullies and ongoing maintenance is avoided. They also reduce risk to amphibians such as toads and newts, which are often trapped in interceptor drains or gully pots. Some regular maintenance is required to keep a grassed swale operating correctly, chiefly mowing during the growing season. The optimum grass length is around 150mm (SEPA 1999; BCAFAP 2003).

6.8.1.3 *Bioswale.* A bioswale is a more advanced swale system as it includes an infiltration trench, a perforated underdrain and an optional geotextile lining, designed to capture and temporarily store road run-off, allowing the removal of pollutants while attenuating flows. Bioswales can be applied at the road network or segment scale, and can be retrofitted to existing road ecosystems in both urban and rural landscapes.
Bioswales can be planted with native grasses and forbs that can enhance filtration thus improving water quality and preventing sealing of subsoils (BCAFAP 2003). More specifically bioswales can: (1) reduce impervious run-off volumes and rates, (2) recharge

groundwater and sustain base flows, (3) reduce sediment and nutrient runoff, (4) increase biological oxygen demand (BOD) control, (5) reduce detention needs, (6) improve aesthetics, (7) act as temporary habitat for wildlife.

Several design considerations are required in the establishment of a bioswale including: (1) size and design must account for the catchment or drainage area and soil types, (2) filtration can be improved by planting native deep-rooted marsh vegetation, (3) infiltration storage should be designed to drain in 24 hours to prevent sealing of subsoils, (4) frequent and effective ditch maintenance is required, (5) soil can be amended with compost and/or sand to improve organic content for filtering and to achieve adequate infiltration rates, (6) direct entry of flood runoff into the infiltration trench should be prevented to protect groundwater quality (BCAFAP 2003).

6.8.2 *Passive permeable systems*

Passive permeable systems include filter strips, ponds, basins and constructed wetland ecosystems (SEPA 1999). Such systems are designed to store and treat large quantities of road run-off (with the exception of filter strips) by attempting to incorporate natural hydrologic processes within a constructed system. Such systems represent a radical paradigm shift in the philosophy of drainage management as they encourage above-ground storage close to the point of runoff through raised outlets (as opposed to conveying it) i.e. the runoff is visible whilst at the same time being in a constructed system. Ponds, basins or wetlands can be inserted into roadside landscapes (and even retrofitted in some situations) where they have the capability to improve the quality of road run-off significantly, providing additional wildlife habitat as well as improving the quality of life for stakeholders.

6.8.2.1 *Filter strips.* Filter strips are vegetated surface features utilized in urban landscapes that provide infiltration facilities to treat run-off from car parks and rooftops.

A filter strip is an area of dense native vegetation cover that reduces the volume and rate of runoff from road pavement through absorption and infiltration over a large area. Filter strips form part of a level-spreader system, which consists of multiple narrow lateral inlets or trenches laid on the contour, in between the vegetated filter strips, that distribute runoff over the filter strips. Filter strips can: (1) recharge groundwater and sustain base flows, (2) reduce sediment and nutrient runoff, (3) de-concentrate runoff from flood overflow pipes and detention basins, (4) dissipate energy, (5) reduce potential scour or erosion, (6) reflect the former natural rainwater runoff flow paths and percolation patterns into the receiving watercourse or aquatic system, (7) reduce detention needs. Filter strips/level-spreader must be designed and sized to account for the drainage area, contours or slopes and soil types, as chronic hydraulic overloading of filter strips can cause erosion. Infiltration storage within the level-spreader trench should be designed to drain within 24 to 48 hours to prevent sealing of subsoils. Filtration can be improved by planting native deep-rooted vegetation, minimizing the slope and by reducing compaction of the soil through the use of leaf compost and coarse sand to improve, not only filtration but also, infiltration and plant establishment. Runoff should be diverted away from filter strips during construction activities until vegetation has become established (BCAFAP 2003).

6.8.2.2 *Attenuation, balancing, settling and stilling ponds or basins.* Attenuation, balancing, stilling and detention ponds/basins are used to store road run-off temporarily and release it at a rate allowed by ordinances. Such systems are effective at managing run-off at the road network and segment scales, and can be retrofitted to improve the water quality of aquatic ecosystems in the vicinity of existing road ecosystems.

Investigations on German highways by Lange *et al.* (2003) found low efficiencies for the removal of heavy metals by concrete settling basins, but determined much better results for earth basins planted with native vegetation, such as reeds (67-84% for SS, COD and metals, 96% for PAH).

The benefits of basins and ponds include: a reduction in runoff rates (even during flooding periods), effective removal of sediment, nutrients and associated pollutants, BOD control and the provision of temporary wildlife habitats and aesthetic features within the landscape. The following specific design considerations are required in the establishment of such features: (1) the pond should be sized to control the release of run-off at an allowable rate, (2) the size should reflect the volume of run-off in the conveyance systems e.g. swales, (3) water level fluctuations should be limited to 3-4 feet to maximize plant diversity, (4) shallow water entry angles minimize shoreline erosion, improve water quality and species diversity.

Detention basins are planted floodwater systems designed to provide conveyance, infiltration and a reduction in coarse particulates and nutrients in run off for a typical 10-year storm event.

6.8.2.3 *Extended detention or retention ponds.* Extended detention basins or retention ponds and constructed wetlands retain a certain amount of water at all times and can be categorised as systems that are designed to store and treat floodwater prior to releasing it at an appropriate rate once the peak flow has passed.

Extended detention basins specifically provide flow attenuation and detention for up to 24h, allowing the removal of suspended solids through settlement and biodegradation processes. The incorporation of aquatic macrophytes enables additional treatment through biofiltration, adsorption and biological uptake of solids (Revitt *et al.* 2004).

The pollutant removal efficiencies of detention basins have been shown to be dependent on residence time with suspended solids removal decreasing from a maximum of 70% to 20% as containment time reduces from 48h to 2h (Stahre & Urbonas 1990). The removal efficiencies of hydrocarbons, BOD and metals (Zn and Pb) were reduced by similar factors. Hares and Ward (1999) have indicated removal efficiencies in excess of 84% for a range of 11 metals in a 500m^2 detention pond receiving run-off from a major motorway.

In an extensive study of retention ponds in the Florida area, Yousef *et al.* (1996) reported average sediment accumulation rates of 1.3, 13.8 and 6.9 kg/ha/yr for Cu, Pb and Zn, respectively. Similar metal accumulation rates have been observed in French studies of retention basins (Lee *et al.* 1997).

6.8.2.4 *Constructed wetlands or helophyte filters.* Purposefully designed constructed wetlands or helophyte filters are ecological engineering systems that take advantage of the natural processes or remediation opportunities exhibited by semi-natural wetland ecosystems, such as temporary storage and bio-filtration, adsorption and biological uptake of pollutants (nutrients, heavy metals etc.) through the selective use of wetland plant species such as aquatic macrophytes and trees e.g. Willows. Constructed wetlands that manage road runoff are generally utilized in urban areas and treat road runoff to the same degree as retention ponds, but over shorter periods of time, often at the expense of incorporating an equivalent storage capacity (Revitt *et al.* 2004). An indirect consequence of such systems is the provision of extensive wildlife habitat. Wetland systems can also be created solely for biodiversity conservation where appropriate conditions exist.

Extensive data sets are available for constructed wetlands which treat urban storm-water with removal efficiency ranges of 67–97% for TSS, 25–98% for Ntot, 5–94% for Pbtot and 10–82% for Zntot (Strecker *et al.* 1992). The variability in performance was attributed to a number of factors including shortcircuiting, short detention and contact times, pollutant remobilisation and seasonal vegetation effects (Revitt *et al.* 2004). Detention ponds and constructed wetlands can also contribute to the restoration of landscape quality and ecosystem function in adjacent semi-natural ecosystems as well as to improve the quality of life for stakeholders.

6.8.3 *Re-vegetation, additional habitat and landscape quality*

Landscape design with native species, as a sustainable drainage management system, stands alone as a category due to the importance of vegetation in the treatment of pollution, soil erosion and flood control, the restoration of landscape quality and the provision of habitat for wildlife. The use of native hydrologically and ecologically appropriate plants from wetlands and grassland ecosystems can provide: (1) high quality road run-off management through the infiltration and cleansing properties of the systems, (2) a reduction in runoff volumes, (3) an increase in infiltration rates, (4) an increase in the permeability of compacted soils, (5) a reduction in fertilization requirements, (6) soil stabilization and a reduction in sediment release, (7) wildlife habitat, (8) a minimal consumption of natural resources and waste outputs such as the use of fossil fuels and air pollution relative to grassed or turfed areas that require regular maintenance and mowing. For these reasons native landscaping often forms an important and complementary component of sustainable drainage systems (BCAFAP 2003). The technology of wetland creation is well advanced. The most environmentally sustainable re-vegetation technique for such areas post-construction is the facilitation of natural recolonisation. Though wetland species generally have good dispersal mechanisms, they can also be introduced into a constructed system by salvaging plant material and riverbed material from watercourses or wetlands of similar characteristics disturbed by road construction (in such cases measures must be undertaken to avoid the introduction of invasive species). Natural recolonisation can also be supplemented through seed broadcasting or mulching while live cuttings or recycling of trees can provide instant cover.

Sustainable drainage systems (with the exception of French drains), especially basins/ponds and constructed wetlands have the ability to contribute to the restoration of landscape quality post construction as they restore naturalness and a feeling of spaciousness (see Coeterier 1996) together with providing additional habitat for wildlife.

Surveys of flora and fauna within such systems have shown rapidly increasing plant species numbers and diversity as the systems mature, along with the presence of otters (*Lutra lutra*) and nesting sites of bird species such as Swans, Ducks and Herons, providing evidence that they can contribute to biodiversity conservation.

Systems can also be utilised and modified to provide a specific type of habitat known to attract particular bird species. The Royal Society for the Protection of Birds (RSPB, UK) has developed sophisticated design criteria for bank and water margin patterning, depth and substrate, to the point where it is almost possible to guarantee that a given constructed system will attract a particular bird species, or at least allow strong predictions concerning the likely species that may benefit from such a system. The type of maintenance regime put in place will also determine the flora and fauna species to be found within a system.

SUDS and native species can also improve the quality of life for stakeholders. In Scotland surveys of public perception of sustainable drainage systems ponds have shown a clear belief that sustainable drainage systems such as ponds are aesthetically pleasing and add to the amenity of an area provided they appear to resemble a natural pond (Apostolaki *et al.* 2003). From the public perception studies the following design recommendations were made: (1) the provision of shallow flowing natural contours and softer shore slopes or shallow edges to allow wildlife to access pond and to allow natural recolonisation, (2) the provision of additional habitat and cover for wildlife conservation, (3) the design of an aesthetically pleasing pond. Water features including ponds and constructed wetlands can also reduce driver monotony and fatigue by stimulating driver alertness.

The need for regular inspection and maintenance of these systems cannot be underestimated, especially for systems where the accumulation of heavy metals is high (Revitt et al. 2004) since such systems can be the source of significant groundwater pollution. Other ecological concerns include the identification of a suitable overflow destination (van Bohemen & Janssen van de Laak 2004); the fact that amphibians and other species can get caught in interceptor drains; and, the recognition that inappropriately timed maintenance regimes can disturb wildlife.

7. Sustainable construction

Road construction is often most environmentally traumatic to adjacent ecosystems as it is primarily an agent of disturbance and change (Lugo & Gucinski 2000) and is a heavy consumer of natural resources. To address sustainable development issues, activities undertaken during this phase of road ecosystem development must aim to minimise: (1) the consumption of natural resources through appropriate soil management and the sustainable use of construction material, (2) noise and light pollution, amongst others, (3) the management of non-native invasive species. Environmental specifications for contractual documents for the contractor, should cover the management of work forces, machinery (speed, noise, and traffic), and the prevention of erosion and pollution incidents during construction.

Construction activities must also be sensitive to landscape quality where they encounter locally distinctive features such as features of historical and archaeological interest, but also unprotected features such as old stone walls, water features, old mill and farm buildings, and particular vegetation complexes such as mature trees in isolation or hedgerows that are among the preferred features of the roadside landscape.

Where archaeological features are uncovered either before construction or during initial construction of the road corridor, construction activities may be postponed. Care must be taken to determine the cost-benefit analysis of any excavations. Archaeologists can consult with ecologists and the relevant EIA to determine whether any further investigatory excavations outside the land-take of the road corridor, and indeed the boundary limits of the EIA, may disturb vegetation of ecological importance especially in the vicinity of protected areas. Additional ecological surveys may be required while hydrologists and soil scientists may need to determine the extent of any adverse indirect effects. For example, soil movement in the vicinity of wetland areas can potentially lead to soil erosion and loss of flood control.

Where significant finds are uncovered as a result of an archaeological excavation, road managers may decide to select an alternative route in order to conserve archeological finds of national and/or international importance. For example, an alternative route was identified for a section of the N25 Waterford By-Pass, in southeast Ireland when a former Viking settlement of international importance was uncovered prior to commencement of construction.

7.1 SOIL MOVEMENT AND STORAGE

Very large quantities of soil may be displaced during the construction of a new road ecosystem, especially where the road corridor passes through areas of local elevation. For this reason, suitable soil storage sites such as depots and borrow pits, should be carefully selected to minimise disturbance, further loss of vegetation and waste outputs. Similarly, the location of site offices and compound storage facilities for construction materials must be carefully managed. Suitable sites for soil storage and site offices can be identified through the use of aerial photography and the flora and fauna survey within the relevant EIA, which aid in selecting sites that will limit further and avoidable disturbance to habitats of ecological importance within the land-take. Soil, construction and demolition waste must not be located within identified buffer zones for watercourses, especially in relation to salmonid rivers. Strict site restoration and re-vegetation programs can be implemented post construction that can aim to restore or establish native vegetation or return sites to their former land-use.

The retention and storage of subsoil for use in wildflower meadow establishment and soil from disturbed ecosystems such as woodlands and wetlands should also be carefully carried out during initial construction activities and should be stored separately for the duration of construction activities, until the landscape designer and nursery supplier are ready to commence the re-vegetation of the roadside landscape.

7.2 ROCK CUTTINGS

Cuttings which reveal rock faces and scree slopes can be of significant visual interest, where they receive sensitive treatment during construction. Construction should expose an appropriate profile that will allow numerous small ledges and a varied micro-topography. Where blasting of rock face is undertaken, explosives must be handled with extreme sensitivity and care, especially in the vicinity of protected areas or locally important sites for bird species. The rock removed from faces can be utilized elsewhere within a road ecosystem to reduce the amount of construction material or natural resources required in the construction of a road ecosystem examples include the construction of stone walls for boundary demarcation or noise barriers, bio-barriers and in the aggregate base layer of the road. The stone can also be utilized as a feature within the landscape design layout of the roadside landscape especially in the urban environment e.g. in roundabouts. The use of such material can also reduce the abrupt transition between the urban and rural environment, as in the urban area of Tralee town in Co. Kerry, Ireland, by bringing locally distinctive features into the urban landscape.

Rock cuttings may also reveal significant palaeontology finds which may require further investigation. Fossils can be removed from the site or can be recorded and left *in situ*.

7.3 DISTURBANCE OF VEGETATION

On-site decisions can be made by engineers, ecologists and landscape architects to retain mature vegetation complexes within the land-take of a new road-ecosystem during construction. Trees can also be retained within the soft landscape area of junctions, interchanges and roundabouts. Efforts should also be made to avoid unnecessary damage to trees and shrubs and other vegetation during construction. Damage to trees includes damage to lower areas of the trunk, limb loss and root damage as a result of machinery operation or as a result of soil degradation from oil or fuel spillages, which can result in ill-health and the death of trees. Preventative measures include the flagging of trees and fencing in of trees, while hedgerows can be lifted by excavator buckets and pushed out of the path of construction activities. Regular checks and maintenance of construction machines and refuelling depots can reduce the likelihood of on-site spillages or leakages from vehicles.

7.4 SALVAGING PLANT MATERIAL

The careful removal and storage of plant material during road construction must also be exercised. Such material can play a central role in the re-vegetation of roadside landscapes especially in mitigation and compensation measures that aim to restore disturbed or lost habitat. Plant material which can be salvaged from the road corridor includes: (1) marginal and emergent aquatic plant material from watercourses, (2) sods, turf and soil from meadows, wetlands, woodlands and hedgerows, (3) hedgerows can be pushed out of the path of construction activities, (4) cuttings and young trees from wetlands, woodlands and hedgerows. Suitable short-term storage facilities must be identified for live plant material, most critically for aquatic species, in order to avoid desiccation.

7.5 CONTROL OF INVASIVE PLANT SPECIES

Invasive species can take advantage of the opportunities provided by perturbation regime (disturbance and soil movement) during construction. Prior to the commencement of earthworks, non-native invasive species need to be adequately managed to avoid the spread of plant material along the road corridor. EIAs must highlight the presence of non-native invasive species and recommend the flagging of contaminated areas, prior to the commencement of earthworks and landscaping activities. Proposed removal methods need to reflect best practice. Regular spot checks of soil loads and tyres and tracks of vehicles leaving the vicinity of these areas should also be carried out. Informative brochures, posters and transdisciplinary identification keys are required and should be made available to supervisory staff and engineers on-site for early identification and easier access to reference material and detail on best available technology for removal of non-native invasive species.

7.6 CONTROL OF POLLUTION

Site offices, as they are temporary facilities, may not be subject to the same level of scrutiny or environmental impact assessment as permanent installations. Siting of such compounds should be subject to pollution risk assessments and adequate prevention and mitigation measures should be provided (Golwer & Sage 2005).

Site offices and site compounds may contain flood lights, for safety/security reasons. Such lighting may not be subjected to the same level of scrutiny or risk assessment as permanent street lighting proposed for the road corridor. For this reason, the illumination of site offices and compounds should be subjected to assessments and adequate measures implemented such as shielded lighting and censored lighting to reduce the possibilities of ecological light pollution, especially in the vicinity of protected areas.

Sudden and unusual noise emissions from construction machinery can create temporary disturbance zones for wildlife as work progresses in concentrated locations along the length of a new road corridor. Such noise may also disturb local residents in the vicinity of the road scheme. During construction, decisions can be made to put in place, temporary wooden noise barriers in place for local residents until more permanent installations can be put in place. Maximum permissible noise levels during construction are generally set between 65-70dB during daytime and 60dB at nighttime (see NRA 2004).

Regular maintenance and proper use of machinery should also be ensured for optimum operating conditions and to minimize noise emissions. Mitigation options during construction include: (1) temporary noise barriers, (2) prioritizing the use of low-noise machinery, (3) the use of noise reduction devices, (4) the rearrangement of construction schedules to reduce noise impacts at particular times. For example, high-noise machinery should not operate from 22.00h to 06.00h.

To fulfil contractual commitments, construction work may extend into night-time, which may not have been foreseen by a risk assessment. For this reason adequate measures must be put in place to reduce associated noise and light pollution during night time hours to avoid any disruption to the foraging behaviours and other critical behaviours of nocturnal animals.

Risk assessments are also required in relation to the possibility of soil erosion and/or sediment or coarse particulate pollution. During construction activities, engineers and fishery inspectors should strategically place percolation wells and silt traps to remove sediments before they can enter an aquatic system. Straw bales, silt fences and mulching can also be utilised as temporary measures to reduce soil erosion from disturbed soil (e.g. on unstable cuts and fills) and to prevent the release of sediments into streams.

Where required geotextiles, must be appropriately placed within the road bed to ensure the adequate filtration of coarse particles and the separation of grade materials so that the expected life-span of the road surface material can be realised.

Refuelling and associated spillages often cause localised contamination, and in sensitive areas construction of such refuelling depots/areas should be assessed for leakage and spillages (Golwer & Sage 2005). Emergency procedures and regular maintenance of interceptor drains, filters and oil separators are required in advance of oil and fuel spills. Petrol/oil filters can be utilised to prevent fuel from entering watercourses and aquifers. Assessments must also be undertaken to identify the risk of groundwater pollution from lubricants and hydraulic oil released by construction machinery.

7.7 SUSTAINABLE USE OF CONSTRUCTION MATERIAL

Sustainable development also involves the minimum consumption of natural resources through the sensitive use of construction material, and in the sourcing and extraction of material required in the construction of the road ecosystem. Implementing the principle of sourcing and utilizing local materials, from within the land-take in fill areas, embankments, the road-bed and the road surface material is essential to reduce impacts on landscape quality, and to reduce inputs of natural resources in road ecosystem development.

Wastes from construction, in the form of littering and construction waste should be minimised (e.g. soil, concrete and plastic piping, geotextile lining, road marking spray cans etc.). Soil, construction and demolition waste can form the main waste outputs from road ecosystem development. Efforts should be made to retain such materials within the land take of a road ecosystem. Such wastes can be buried in borrow-pits within the roadside landscape to avoid the cost of transportation and disposal off-site. In such circumstances, however, the environmental quality of the land is reduced, no matter how deeply the waste is buried

8. Sustainable operation/use

8.1 ROAD USERS

With respect to sustainability issues, road users can make informed decisions with respect to the choice of tyres, batteries, bulbs, fuel and vehicles, the recycling of the latter and the distance they travel per day, month or year. Car-pooling is one measure that can reduce the distance travelled per unit of time as is the use of public transportation. Road users also need to refrain from littering of the roadside landscape and the illegal dumping of domestic refuse and construction and demolition wastes within road ecosystems. Road users also have a responsibility to maintain their vehicles regularly to reduce the likelihood of car parts dropping from vehicles, to be left as litter within the roadside landscape.

8.2 TECHNICAL DESIGN ASPECTS AND VEHICLES

As vehicles are the main source of diffuse and acute pollution from road ecosystems, continuous efforts are being employed by scientists and manufacturers in the pursuit of best available technology to achieve improved air quality, fuel efficiency, the conservation of natural resources in vehicle manufacturing and the recovery of energy and materials.

Technical design aspects that focus on reducing emissions of dangerous substances at source include engine modifications, catalytic converters and changing the composition of fuel to capitalise on the potential of alternative fuels. The management of waste streams that provide for the optimal recovery of materials from vehicle maintenance, through reuse and energy release are discussed below as are waste streams from End of Life Vehicles.

8.2.1 *Type of tyres*
In relation to noise pollution, it is generally accepted that further technological advances in engine design will only result in marginal improvements (SOU 1993). However, some scope for noise reduction does exist in the area of tyre manufacture. For example, if speed limits were reduced tyres could be made narrower and of softer rubber leading to less noise (Irish EPA 2002a; Alenius & Forsberg 2001).

The constituents of tyres selected by road users can determine the range and amount of heavy metal pollutants released by the wear of the tyre on the road surface. Road users can make informed decisions to select tyres which contain the least amounts of Zn and Cd, while in colder climates road users can select tyres with lightweight studs to reduce wear and breakage of the road surface (Sörme & Lagerkvist, 2002). More wear resistant pavements have also been developed (Lindgren 1998).

8.2.2 *Alternative Fuel Vehicles*
It is recognised that Alternative Fuel Vehicles (AFVs) release fewer, less damaging emissions than petrol and diesel-fuelled vehicles. Ford are at the forefront of AFV research and technology, and have the widest range of alternative fuels in comparison with any other manufacturer e.g. compressed natural gas, liquefied petroleum gas (LPG), methanol, ethanol and electricity (FORD 2004). The AFVs include Fuel Cell, Ethanol and Hybrid Electric Vehicles.

A Fuel Cell Vehicle (FCV) is an electric vehicle that utilizes a catalyst combined with oxygen and hydrogen (or methanol fuel) to produce electricity, instead of utilizing a battery. FCVs are more than twice as efficient as petrol/gasoline vehicles, and hold the most promise to produce a new generation of vehicle which is efficient, silent, producing zero emissions, except for water and heat.

Ethanol is a renewable fuel source for AFVs. It is usually produced from maize, but can be made from other starch feed stock such as sugar cane, wheat, or barley. Ethanol Vehicles are also called Flexible Fuel Vehicles (FFVs) as they generally operate on a combination of two fuels (85% denatured ethanol and 15% petrol/gasoline) called E85 or indeed can operate on either conventional petrol/gasoline or ethanol. E85 has been used by local government vehicles, buses, and trucks in many countries for several years and is growing in popularity.

Hybrid Electric Vehicles are powered by a conventional engine with an electric motor added for enhanced fuel performance thus halving the consumption of fuel compared to that of a petrol/diesel engine, helping to conserve world limited petroleum resources. This type of vehicle also has reduced emissions through increased average engine efficiency and regenerative brakes which recapture energy to recharge the battery (FORD 2004) and is much quieter.

8.2.3 *Vehicle maintenance and End of Life Vehicles*
As mentioned above, the maintenance of vehicles during their lifetime also generates a variety of waste products which can be sent to disposal facilities where varying degrees of reuse and recycling are achieved.

Old lead-acid batteries from vehicles are generally shredded, baled and sent to lead recovery facilities, where the acid is neutralised and the resultant precipitated 'cake' is sent to landfill

(Irish EPA 2002a). With regard to engine oil, however, significant amounts of oil are generally not recovered partly due to owner home maintenance, with subsequent disposal to soil, watercourse or drains but also due to a lack of car maintenance resulting in atmospheric emissions from the fraction burnt in poorly maintained engines. Waste engine oil can be utilised in 'space heaters' in motor garages. Spent oil filters contain significant quantities of oil and can also be recycled.

In many countries the disposal of tyres is a growing concern due to the ever-increasing number of vehicles moving within road ecosystems. A variety of tyre disposal alternatives are currently being practised internationally. Materials recovery (such as reuse options, scrap tyres and crumb rubber), and fuel and energy recovery are the principal options. Reuse options, including retreading, remoulds and sale of part-worn tyres, offer the most resource efficient strategy for used tyre recovery, saving both material and energy. However, the remould market has been declining due to the poor public perception of the quality of the tyres and reduced profitability compared with new tyres.

Beside reuse, scrap tyres can be recycled whole or size-reduced for civil engineering applications, agricultural uses and for composite materials such as landfill engineering, crash barriers and marine dock-fenders. Tyres have been used in silage pits and artificial reefs. In more recent years, granulated or 'crumb rubber' has been used in asphaltic concrete mixes for road surfaces, and as a shock absorbent surface for playgrounds and sports tracks. Crumb rubber from tyres can also be added to other polymers (rubber or plastic) to extend or modify properties of thermoplastic polymeric materials (Amari *et al.* 1999).

Tyres can also be used as fuel either in shredded form (Tire Derived Fuel (TDF)), or whole, depending on the type of combustion furnace. Pyrolysis has proven to be a feasible way of converting waste tyres into economically valuable products of liquid and gaseous HCs. As a fuel, tyres are superior to coal in specific energy content and the environmental burden of residues, as they result in heat content 10-16% higher than that of coal, lower sulphur content than the ash from burnt coal and lower NOx emissions when compared with many US coals (Ohio Air Quality Development Authority 1991; Amari *et al.* 1999). TDF recovery from waste tyres is utilised to a large extent in cement production in rotary fuel-fired kilns, (Amari *et al.* 1999). While the use of tyres as fuel may not represent the optimal strategy for value recovery, the environmentally responsible combustion of tyres for energy is preferable to the health and aesthetic problems resulting from their accumulation in landfills.

The EU Directive (2000/53/EC) concerning End of Life Vehicles has had far-reaching implications for manufacturers, dismantlers and recycling industries, as the choice of materials used in vehicles, and their overall design, now has to focus on the new reuse, recovery and recyclability requirements of the Directive. Technologies are now being developed whereby the majority of generic materials and fluids [both hazardous and non-hazardous materials (steel, heavy metals, plastics, glass, rubber, natural fibres,) several of which are intimately bound together into complex compounds] from vehicles can be separated and then reused or recycled by appropriate processes (ECRIS 1999). Since 2002, ELVs have been officially categorised as hazardous waste.

9. Maintenance phase

The decisions made during the design phase of road ecosystem development should minimise the need for long-term maintenance requirements. In general Local Authorities are responsible for maintaining road ecosystems.

Lugo and Gucinski (2000) identified two aspects of road maintenance that require consideration of disturbance. First are the effects that disturbances might have on the structure and functioning of road ecosystems (at the landscape scale) and second, maintenance activities can approximate the effects of building activities through the generation of local irregular but concentrated disturbance zones along the road corridor (Theobald *et al.* 1997). Such maintenance influences the state of a road ecosystem upon which wildlife can become dependent and can favour the establishment and survival of non-native species where maintenance activities allow them to extend their range through disturbance.

The design phase can reduce the subsequent frequency and time periods of maintenance regimes, and indeed traffic hold-ups, through the selection of lightweight materials such as FRP for use in street furniture, which is long-lasting, easily installed and also safer for the road user.

9.1 ROAD SURFACE MATERIAL

Maintenance of the road surface can involve road works including road widening and resurfacing with concrete, concrete asphalt mixes, asphalt or bitumen. Regular and strategic maintenance of localised damaged to a road surface may reduce the need for major resurfacing activities at a later stage in a road's development.

The advantages of utilizing porous asphalt are described in Section 6.7.1. However, where porous road pavement proves inefficient in the event of (1) black ice, (2) clogging (Pichon 1993), and (3) weed growth more frequent maintenance activities are required with associated release of pollution. For instance, in the event of black ice, more frequent de-icing operations are required on porous surfaces. Porous surfaces with low-traffic levels areas (e.g. residential areas, motorway hard-shoulders and emergency lanes.) require regular maintenance to (1) Release clogging sediments through the use of high-pressure water jets and aspiration, (2) Remove weeds that may establish themselves, mainly within hard-shoulders. It has been found that such weeds can be difficult to remove from the porous surface. For these reasons, and because porous asphalt also costs double that of non-porous asphalt, informed decision making must be undertaken when selecting the type of road asphalt surface for particular segments of road ecosystems.

During maintenance activities consideration must be given to the appropriate storage of deicing agents, herbicides and cleaning agents for road signs or tunnel walls. Assessments are also required to evaluate whether good practice is utilized in the handling and application of such agents, whether these are regularly checked and verified, and whether their application is minimised to ensure no excess is available for pollution (World Bank Group 1997).

9.2 SUSTAINABLE DRAINAGE SYSTEMS

Regular maintenance and monitoring of sustainable drainage systems must be implemented to prevent the possibility of groundwater pollution through seepage from such systems (Golwer & Sage in submission). Maintenance includes unblocking of interceptor drains, filters or oil separators from grit and sediments, more frequent and effective ditch management (Southerland 1995; LaFayette *et al.* 1996) and the appropriate management of vegetation (Lugo & Gucinski 2000; Swanson 2001).

The maintenance of balancing ponds, extended detention, retention ponds and constructed wetlands needs to be timed to permit the associated flora and fauna to remain in an undisturbed part of the system (van Bohemen & Janssen van De Laak 2004).

9.3 VEGETATION

In the past, the main goal of managing roadside vegetation has been for transport safety (*e.g.* Parr & Way 1988; Persson 1995) in relation to specific demarcation and line of sight issues.

Where transport safety and traffic engineering are primary functions of verges, cutting and/or mowing may be required more frequently, for example at corners and bends, where visibility is essential or in strips containing marking and signposting equipment (van der Sluijs & van Bohemen 1991) or other street furniture. However, it is now recognised that regular maintenance of roadside landscapes selects for particular wildlife and plant species that can take advantage of disturbance, while the overall species diversity can remain quite low. Cutting and trimming of large vegetation complexes, for example, needs to be timed to allow for the nesting periods of birds.

Where frequent and inappropriately-timed mowing of grasslands or meadows extends into the roadside landscape beyond the area required for line of sight issues, the potential area for the development of wildflower meadows is limited, while associated maintenance costs and physical and natural resource inputs are also unnecessarily high and not sustainable. The Netherlands has produced a guide to cutting regimes for the management of verges on national highways within each province (van der Sluijs & van Bohemen 1991; Sýkora *et al.* 2002), where the cutting frequency and period is adapted to the existing (and sometimes potential) vegetation (van der Sluijs & van Bohemen 1991). It is carried out after the flowering periods, and includes aftermath removal of hay and usually takes place twice a year (Dolan 2004). Such maintenance regimes require minimal consumption of natural resources and production of waste outputs. It is also recognized within the literature that intensive horticultural practices including the use of fertiliser and general broad-leaved weed control should be avoided, as such activities only serve to reduce potential species diversity of wildflower meadows. The application of fertiliser is considered counter-productive, as it encourages the development of agricultural grass/weeds species, increasing the need for increased mowing frequency and herbicide use (Hitchmough & Dunnett 1996).

Indeed, minimal maintenance and natural resource inputs, promotes sustainable processes such as natural recolonisation and nutrient cycling that can restore ecological integrity to the roadside landscape (see Figure 12). However, in the absence of mowing, wildflower meadow communities can disappear due to natural succession towards closed tall-herb communities and encroaching tree and shrub species which may oust species-rich low vegetation (Sýkora *et al.* 2002). Where large vegetation complexes are not maintained, plant succession can reach unpredictable states and influence road function and safety (Lugo & Gucinski 2000). Ecotope boundaries between the roadside landscape and adjacent land-uses can disappear (Pauwels & Gulinck 2000).

Sensitive use of de-icing salts must be exercised to minimise damage to existing native plant communities in the vicinity of road carriageways and to prevent colonisation by saline-tolerant plants species, and indeed plant communities, within terrestrial ecosystems (Lugo & Gucinski 2000).

9.4 INVASIVE PLANT SPECIES

Early identification and effective management of non-native invasive species can reduce long-term maintenance and removal costs. However, maintenance activities can also contribute to the spread of such species, for example, trimming and hedge cutting, releases plant fragments that can be carried on blades, tyres or wind. International best practice indicates that an integrated removal approach is the most successful form of management.

To prevent invasions by non-native species, maintenance can cater for the development of mature vegetation and the promotion of vegetation complexity at the ecotone between the road ecosystem and adjacent vegetation (Lugo & Gucinski 2000).

9.5 GREEN LANES AND MODIFIED BRIDGES

Green lanes have been established in The Netherlands, where reduced traffic loads allowed one lane of a two-lane flyover, to be closed to traffic and be transformed into a green lane for wildlife. The asphalt on the closed lane of the flyover was replaced by a 300mm layer of soil in order to establish vegetation while tree stumps were placed to provide cover, habitat and a corridor for wildlife (Bank *et al.* 2002). The green lane crosses the road carriageway parallel to the remaining traffic lane. More recently in Belgium an entire flyover was closed to traffic for the provision of a cycle lane and a green lane for wildlife. Such modifications can complement the establishment of multi-use greenways.

Replaced, repaired and modified bridges provide an opportunity to re-establish habitat connectivity. For example, extending the span of the bridges just beyond the banks of the river water into the adjacent habitat, so that a larger area of un-submerged land is bridged allows a larger area to be utilised as a riparian corridor by riverine species (Jackson & Griffin 2000; MacDonald & Smith 2000). Alternatively, bridges can be retrofitted with mammal ledges through on-site casting of concrete ledges, or the instalment of wooden ledges on existing bridges and culverts to provide for the safe passages of wildlife above flood level. Jackson and Griffin (2000) also suggest the selective use of viaducts at important stream or river crossings.

10. Road decommissioning phase

In the course of road planning and design, sections of old road pavement can be abandoned due to (1) the establishment of a new road ecosystem, (2) the realignment of an existing road, (3) the By-Pass of traffic 'hotspots', (4) required road closure for environmental reasons. At a regional level the effects of road ecosystems on hydrologically sensitive sites within catchments and on protected areas can be minimised through closure of a segment of a road ecosystem (Lugo & Gucinski, 2000). Where old road pavement persists it can continue to (1) inhibit the ecological functions and services of the soil sealed beneath the carriageway and the road bed, (2) pose a barrier to the dispersal of wildlife, (3) release pollutants from surface run-off. It is for these reasons that the process of road decommissioning is carried out.

In general the extent of decommissioned roads is small, but occasionally, it can be large enough to significantly extend the vegetation within the existing roadside landscape and indeed within adjacent core habitat areas. Road decommissioning can therefore: (1) provide a compensatory habitat, (2) reinforce the ecological network, (3) contribute to the restoration of landscape quality in the vicinity of a new road ecosystem.

Native vegetation can be re-established on decommissioned roads, through natural recolonisation, where they are located adjacent to hedgerows and woodlands. If the pavement is generally broken up and mixed with the underlying soil or covered with suitable soil (from the disturbance of a hedgerow elsewhere on a road scheme), the resulting decommissioned section of road shows a rapid recovery as vegetation successional processes can more easily recolonise the road segment resulting in valuable additional habitat for wildlife, especially birds. Where roads are abandoned without decommissioning (i.e. the pavement is not broken up and soil is not utilised) the process of natural recolonisation is much slower.

11. Conclusion and recommendations

Identifying activities and techniques that address the principles of sustainable development has received much attention since the Earth Summit and the Rio Declaration on Environment and Development in 1992. National policies, regional development plans and spatial planning strategies have been adapted worldwide to address sustainable road development across a diversity of regions, landscape and ecosystem types.

Sustainable road ecosystem development requires a multi- and transdisciplinary approach in order to implement national policies and regional development plans at the local ecosystem scale. Informed decision-making during the relevant phase of road ecosystem development i.e. in advance of contractual commitment and commencement of construction and soil movement, is required to ensure that the principles of sustainable development can be addressed.

Participants in the various disciplines involved in road ecosystem development have a responsibility to make informed decisions with respect to the management of natural resources including land (soil), water, biodiversity air, light and energy. This can only be achieved through an awareness of the roles and contributions of the various professionals currently involved in road science. Informed decision-making will ultimately form the basis for the restoration of ecological function and services such as nutrient cycling and the regulation of water flow in semi-natural ecosystems.

In local ecosystem development, participation is seen as the key to equality, inclusiveness and sustainability (Comhar 2002). The identification of peoples' relationships and values for features within the road landscape acknowledges the importance of participation, co-operation and consensus of all actors in society, especially the importance of local/indigenous knowledge, as it actively seeks participation of the stakeholders involved in the development of a road ecosystem. Gran (1987 cited in Roseland 2000) stated that, for people to prosper anywhere, they must participate as competent citizens in the decisions and processes that affect their lives, to ensure that social and environmental as well as economic dimensions are included in the development process.

Acknowledgments

The primary author would like to acknowledge the contributions of the various professionals who provided valuable insights into the transdisciplinary requirements of sustainable road ecosystem development. The primary author would also like to acknowledge Enterprise Ireland and the Embark Initiative, a Government of Ireland Scholarship funded by the Irish Research Council for Science Engineering and Technology (IRCSET) under the Irish National Development Plan 2002-2006.

References

Ahlbom J, Duus U. 1994; New Wheel Paths – A Product Study of Rubber Tyers, Kemikalieinspektionen,

Akbar KF. 1997; Aspects of the Ecology and Conservation Value of Roadside Vegetation. Ph.D. Thesis. University of Bradford, UK.

Alenius K, Forsberg B. 2001; Consideration of health aspects in environmental impact assessments for roads. National Institute of Public Health, Sweden.

Alexander L. 1995; The Roadside Verge Report. The Wildlife Trusts National Office. Edinburgh. UK.

Amari T, Themelis NJ, Wernick IK. 1999; Resource recovery from used rubber tires. Resources Policy (25): 179-188.

Baines C. 1994; The Way Forward. In: Chambers K & Sangster M (eds) A Seed in Time. Proceedings of the Third International Conference on Urban and Community Forestry, 1993. Forestry Commission, Edinburgh. UK.

Bank FG, Irwin CL, Evink GL, Gray ME, Hagood S, Kinar JR, Levy A, Paulson D, Ruediger B, Sauvajot RM, Scott DJ, White P. 2002; Wildlife Connectivity Across European Highways. Federal Highway Administration, U.S. Department of Transportation 1-34.

Bauske B, Goetz D. 1993; Effects of de-icing-salts on heavy metal mobility. Acta hydrochim. hydrobiol. 21: 38-42.

BCAFP 2003; Best Management Practices: Appendix A. Blackberry Creek Alternative Futures Analysis Project, Illinois.

Bel. S. 1997; Design for Outdoor Recreation. E & FN Spon, London. UK.

Bell S. 1993; Elements of Visual Design in the Landscape. E & FN Spon, London. UK.

Berbee R, Vermij P, van de Laak WJ. 2004; Policy development for the reduction of pollution caused by traffic experiences from The Netherlands. Water Science and Technology 49(3): 183-188.

Boland M. 1995; Microverontreinigingen langs rijkswegen: een evaluatie [Micro pollutants along government roads: an evaluation]. DWW 95-734. Delft.

Box JD, Forbes JE. 1992; Ecological considerations in the environmental assessment of road proposals. Highways Transport April: 16-22.

Brown G. 2003; A method for assessing highway qualities to integrate values in highway planning. Journal of Transport Geography 11: 271-283.

Brown T, 1984;. The concept of value in resource allocation. Land Economics 60(3): 231-246.

CBD 2004;Convention on Biological Diversity: http://www.biodiv.org/decisions/default.asp?lg=0&m=cop-05&d=06 (accessed 05-05-05).

CEC 1992; Council Directive 92/43/EEC of 21 May 1992 on the conservation of natural habitats and of wild fauna and flora. Official Journal European Community 35(L 206): 7-49.

Clay GR, Smidt RK. 2004; Assessing the validity and reliability of descriptor variables used in scenic highway analysis. Landscape and Urban Planning 33: 239-250.

Clevenger AP. 2000; Effects of highway and other Linear Developments on Wildlife Populations. A Literature Collection. Columbia Mountains Institute of Applied Ecology.

Clevenger AP, Chruszcz B, Gunson KE. 2003; Spatial patterns and factors influencing small vertebrate fauna road-kill aggregations. Biological Conservation 109: 15-26.

Cloke J, Boulter P, Davies GP, Hikman AJ, Layfield RE, McCrae IS, Nelson PM. 1998; Traffic management and air quality research programme. TRL Report 327.

Coeterier JF. 1996; Dominant attributes in the perception and evaluation of the Dutch landscape. Landscape and Urban Planning 34: 27-44.

Colwill DM, Thompson JR, Rutter AJ. 1979; The impact of road traffic on plants. Department of Transport, Crowthorne, Berks, United Kingdom.

Comhar 2002; Principles for Sustainable Development. The National Sustainable Development Partnership, Dublin, Ireland.

COMMUTE (Common Methodology for Multi-modal Transport Environmental Impact Assessment). 1998; Final report. Commission of the European Communities – 4th Framework funded project.

Council of Europe 2001; Code of Practice for the introduction of biological and landscape diversity considerations into the transport sector. Bureau of the Committee for the activities of the Council of Europe in the field of biological and landscape diversity.

Danish Roads Directorate, 2002; Beautiful Roads. A Handbook of Road Architecture. Danish Road Directorate, Copenhagen, Denmark.

Day A. 1998; Meeting the needs of the community: Social issues in road and transport planning. Lecture No.3: Main Roads Lecture Series. Main Roads. Western Australia.

De Molenaar JG, Jonkers DA, Sanders ME. 2000; Road illumination and nature. III. Local influence of road lights on a black-tailed godwit (*Limosa l. limosa*) population. Wageningen, The Netherlands: Alterra.

Dolan LMJ. 2004; Roads as Ecosystems: a more sustainable approach to the design of Irish rural roadside verges. In: Davenport J & Davenport JL (eds.) The Effects of Human Transport on Ecosystems: Cars and Planes, Boats and Trains, 15-62. Dublin: Royal Irish Academy.

Dolan LMJ, Whelan P. 2004a; Ecologically Friendly Highways. The Engineers Journal 54 (1): 43-49. The Institution of Engineers of Ireland, Dublin.

Dolan LMJ, Whelan P. 2004b; Sustainable Road Landscapes I: an ecological landscape design approach to Irish rural roadside verges including design for the driver within the vehicle. IENE Conference: "Habitat Fragmentation due to Transportation Infrastructure" 13-15th Nov, 2003, Brussels, Belgium.

Dunball A. 1985; The importance of urban roadside planting in the United Kingdom. In: Karnosky DF & Karnosky SL (eds) Improving the Quality of Urban Life with Plants 28-34. Proceedings of the International Symposium on Urban Horticulture held in Bronx, N.Y., June 21-23 1983.

Ecosafe 1997 Consortium to design safer crash barriers Industry News. Reinforced Plastics July/August 1997 co financed by the European Union (EU) under its BRITE/EURAM programme.

ECRIS 1999; ECRIS – A Research Project in Environmental Car Recycling 1994-1998.

EEA 2002; (European Environmental Agency) Environmental Signals; Soil Sealing http://reports.eea.eu.int/environmental_assessment_report _2002_9/en/signals2002-chap13b.pdf. (accessed 05-05 2005).

Fahrig L. 1997; Relative effects of habitat loss and fragmentation on population extinction. Journal of Wildlife Management, 61: 603-610.

FORD 2004; Environmental Vehicles. www.ford.com/en/support/faq/environmentalVehicles.htm (accessed 05-05-2005)

Forman RTT. 1990; Ecologically sustainable landscapes: the role of spatial configuration. In: Zonneveld IS & Forman RTT (eds), Changing landscapes: an ecological perspective 261-278. Springer-Verlag, New York.

Forman RTT. 1995; Land Mosaics: the ecology of landscapes and regions. Cambridge University Press.

Forman RTT. 2001; Spatial Models as an Emerging Foundation of Road System ecology and a handle for transportation planning and policy. In: Evink GL, Garrett P & Zeigler D (eds) Proceedings of the Third International Conference on Wildlife, Ecology and Transportation 119-124. Florida Dept. of Transportation, Tallahassee, Florida.

Garrett PA, Bank FG. 1995; The Ecosystem Approach to Transportation Development. U.S. Department of Transportation. Federal Highway Administration.

Golebiewski R, Makarewicz R, Nowak M, Preis A. 2003; Traffic noise reduction due to the porous road surface. Applied Acoustics 64: 481-494.

Golwer, Sage. 2005: Traffic and transport: Information needs. In: World Health Organisation (2005) Protecting groundwater for health: managing the quality of drinking-water sources.

Gran G. 1987; An annotated guide to global development: capacity building for effective social change, University of Pittsburgh Economic and Social Development Program, Pittsburgh.

Grumbine RE. 1994; What is ecosystem management? Conservation Biology 4(1): 27-38.

Gulinck H, Wagendorp T. 2001; References for fragmentation analysis of the rural matrix in cultural landscapes.

Hares RJ, Ward NI. 1999; Comparison of heavy metal content of motorway stormwater following discharge into wet biofiltration and dry detention ponds along the London Orbital (M25) motorway. Science of the Total Environment 235:169-178.

Heyliger P, Kienholz J, McLean C, Ramirez F. 2004; Highly Flexible Crash Barriers. Department of Civil Engineering, Colorado State University, Fort Collins CO 80523.

Hill D. 1990; The impact of noise and artificial light on waterfowl behaviour: a review and synthesis of the available literature: British Trust for Ornithology Report No. 61. Norfolk, United Kingdom.

Hitchmough J, Dunnett N. 1996; Sustainable Landscape Planting. Landscape Design,

Houben FMJ. 1999; Ingenieurskunst en mobiliteitsesthetiek. Stedebouw en Ruimtelijke Oreening 1999:3.

Illner H. 1992; Effect of roads with heavy traffic on grey partridge (*Perdix perdix*) density. Gibier Fuane Sauvage 9: 467-480.

INTERURBA II, 2001; Interactions between sewers, treatment plants and receiving waters in urban areas – INTERURBA II. Proceedings of the 2nd International Conference in Lisbon, Portugal, 19–22 February, 2001.

Irish EPA. 2002a; Scope of Transport Impacts on the Environment (2000-DS-4-M2) Final Report Environmental RTDI Programme 2000-2006. Environmental Protection Agency, Ireland.

Irish EPA. 2002b; Towards Setting Environmental Quality Objectives for Soil; Developing a Soil Protection Strategy for Ireland. A discussion document. Environmental Protection Agency, Ireland.

Iuell B, Bekker H, Cuperus R, Dufek J, Fry G, Hicks C, Hlavác V, Keller V, Rosell C, Sangwine T, Tørsløv N, Wandall B. 2003; Wildlife and Traffic. A European Handbook for Identifying Conflicts and Designing Solutions. Cost 341 Habitat Fragmentation due to Transportation Infrastructure.

Jaarsma CF, Willems GPA. 2002; Reducing habitat fragmentation on minor rural roads through traffic calming. Landscape and Urban Planning 62: 1-11.

Jackson SD, Griffin CR. 2000; A strategy for mitigating highway impacts on wildlife. In: Messmer TA & West B (eds) Wildlife and Highways: seeking solutions to an ecological and socio-dilemma. 7th Annual Meeting of the Wildlife Society, Tennessee.

Kemmis D. 1990; Community and the politics of place. University of Oklahoma Press, Norman.

Kent RL, Elliot CL. 1995; Scenic routes linking and protecting natural and cultural landscape features: a greenway skeleton. Landscape and Urban Planning 33: 341-355.

Korenromp RHJ, Hollander JCTh. 1999; Diffuse emissions of zinc due to atmospheric corrosion of zinc and zinc coated (galvansied) materials. TNO-MEP Report Order No. 28924.

LaFayette RA, Pruitt JR, Zeedyk WD. 1996; Riparian area enhancement through road design and maintenance. In Neary D, Ross KC & Coleman S (eds.) National Hydrology Workshop 85-95. USDA Forest Service Gen. Tech. Rep. RM-279. Rocky Mountain Forest and Range Research Station, Fort Collins, CO.

Landscape Institute 1995; Guidelines for Landscape and Visual Impact Assessment. E & FN Spon, London,

Lange G, Grotehusmann D, Kasting U, Schütte M, Dieterich M, Sondermann W. 2003; Wirksamkeit von Entwässerungsbecken im Bereich von Bundesfernstraßen. Forschung Straßenbau und Straßenverkehrstechnik, 861, Bonn.

Lee PK, Touray JC, Bailif P, Ildefonse JP. 1997; Heavy metal contamination of settling particles along the A-71 motorway in Sologne France. Science of the Total Environment 201: 1-15.

Legret M, Pagatto C. 1999; Evaluation of pollutant loadings in the runoff waters from a major rural highway. The Science of the Total Environment 235: 143-150.

Lindgren A. 1998; Road construction materials as a source of pollutants. Dissertation. Lulea University of Technology, Sweden, 1998.

Little C. 1990; Greenways for America. Johns Hopkins University Press, Baltimore, MD.

Longcore T, Rich C. 2001; A Review of the Ecological Effects of Road Reconfiguration and Expansion on Coastal Wetland Ecosystems. The Urban Wildlands Group Inc. California.

Longcore T, Rich C. 2004; Ecological Light Pollution. A Review. Front Ecol Environ 2(4): 191-198.

Lugo AE, Gucinsk, H. 2000; Function, effects and management of forest roads. Forest Ecology and Management 133: 249-262.

Luke AGR, Harvey HJ, Humphries RN. 1982; The Creation of Woody Landscapes on Roadsides by Seeding. A comparison of Past Approaches in West Germany and the United Kingdom. Reclamation and Revegetation Research 1: 243-253.

MacDonald LA, Smith S. 2000; Bridge Replacements: An opportunity to improve habitat connectivity. Defenders of Wildlife. Washington D.C.

Mader HJ. 1988; The significance of paved agriculture roads as barriers to ground dwelling arthropods. In Schreiber KF (ed) Connectivity in Landscape Ecology, 97-100. Muntersche Geographisce Arbeiten 29. Ferdinand Schoningh, Paderborn, Germany.

Magill AW, Schwartz CF. 1989; Searching for the Value of a View. Research Paper PSW. 193. Pacific SW Forest and Range Experiment Station, Forest Service, USDA, Berkeley, California.

Makhzoumi J, Pungetti G. 1999; Ecological Landscape Design and Planning: The Mediterranean Context. E & FN Spon, London.

Makhzoumi JM. 2000; Landscape ecology as a foundation for landscape architecture: application in Malta. Landscape and Urban Planning 50: 167-177.

McCrae IS, Green JM, Hickman AJ, Hitchcock G, Parker T, Ayland N. 2000; Traffic management during high pollution episodes. TRL Report 459.

Mederake R. 1991; Vegetationsentwicklung und Standortsbedingungen von Strassenbegleitflächen bei unterschiedlicher Pflege. – Thesis Georg-August-Univ. Göttingen.

Melman PJM, Verkaar HJ. 1991; Layout and management of herbaceous vegetation in road verges. In: van Bohemen HD, Buizer DAG & Little A (eds) Nature Engineering and Civil Engineering Works 63-78. Centre for Agricultural Publishing and Documentation (PUDOC), Wageningen, Netherlands.

Meunier FD, Corbin J, Verheyden C, Jouventin P. 1999; Effects of landscape type and extensive management on use of roadsides by small mammals. Canadian Journal of Zoology 77: 108-117.

Meurk CD, Swaffield SR. 2000; A landscape ecological framework for indigenous regeneration in rural New Zealand-Aotearoa. Landscape and Urban Planning 50: 129-144.

Mitch WJ, Jorgensen SE. (eds). 1989; Ecological Engineering: an introduction to ecotechnology. John Wiley & Sons, New York.

NAFS 2005; www.nafs.fed.us.access.roads.stream.htm) (accessed 02-01-2005).

Nakamura T, Yamashita S. 1997; Phototactic behavior of nocturnal and diurnal spiders: negative and positive phototaxes. Zoological Science 14: 199-203.

NCSAI Forest Watershed Task Group, 2001 Forest Roads and Aquatic Ecosystems: A Review of Causes, Effects and Management Practices. http://www.ncsai.org/forestry/research/watershed/road.pdf (accessed19/03/03).

Nelson TM. 1997; Fatique, Mindset and Ecology in the Hazard Dominant Environment. Accident Analysis and Prevention. 29(24): 409-415.

Nohl W. 2001; Sustainable landscape use and aesthetic perception-preliminary reflection on future landscape aesthetics. Landscape and Urban Planning 54: 223-237.

NRA 1999; Guidelines on Traffic Calming for Towns and Villages on National Routes. National Roads Authority, Ireland.

NRA 2004; Guidelines for the Treatment of Noise and Vibration in National Road Schemes. National Roads Authority, Ireland.

NRA 2005a; Landscape Treatments for National Road Schemes in Ireland. National Roads Authority, Ireland.

NRA 2005b; Use of Recycled Glass Bottles in Road Pavement Subsurface Layers. http://www.nra.ie/ Environment/ ResearchInitiatives/ (accessed 05-05-2005).

OECD 1997 'Reforming Environmental Regulation in OECD Countries', OECD, Paris, France.

Ohio Air Quality Development Authority, 1991; 50 W. Broad Street, Suite 1718, Columbus, Ohio 43215.

Olander LP, Scatena FN, Silver WL. 1998; Impacts of disturbance initiated by road construction in sub-tropical cloud forest in the Luquillo Experimental Forest, .

Oostergo H. 1997; Milieudata, Emissies van metalen en PAK door wegverkeer [Emission of metals and PACs by road traffic]. Ministry of VROM, The Hague.

Outen AR. 1998; The Possible Ecological Implications of Artificial Lighting. Hertfordshire Biological Records Centre, Hertford, UK.

Ozaki H, Watanabe I, Kuno K. 2004; Investigation of the Heavy Metal Sources in Relation to Automobiles. Water, Air and Soil Pollution 157: 209-223.

Pagotto C, Legret M, Le Cloirec P. 2000; Comparison of The Hydraulic Behaviour and the Quality of Highway Runoff Water According to the Type of Pavement Water Research 34(18): 4446-4454.

Parr TW, Way JM. 1988; Management of roadside vegetation: the long-term effect of cutting. Journal of Applied Ecology 25: 1073-1087.

Parsons R, Daniel TC. 2002; Good looking: in defense of scenic landscape aesthetics. Landscape and Urban Planning 910: 1-14.

Pauwels F, Gulinck H. 2000; Changing minor road networks in relation to landscape sustainability and farming practices in West Europe. Agriculture, Ecosystems and Environment 77: 95-99.

Persson TS. 1995; Management of roadside vegetation: vegetation changes and species diversity. Department of Ecology and Environmental Research, Swedish University of Agricultural Sciences, Report 82.

Pichon A. 1993; Clogging of porous media in civil engineering: case study of French North motorway porous asphalt (in French). Ph.D. thesis, University of Paris VI.

Rees W, Wackernagel M. 1996; Our Ecological Footprint: Reducing Human Impact on the Earth. New Society Publishers, Gabiola Island, BC.

Reijnen R, Foppen R, Meeuwsen H. 1995; The effects of traffic on the density of breeding birds in Dutch agricultural grasslands. Biological Conservation 75: 255-260.

Reijnen R, Foppen R, ter Braak C, Thissen J. 1995; The effects of car traffic on breeding bird populations in woodland. III. Reduction of density in relation to the proximity of main roads. Journal of Applied Ecology 32: 187-202.

Reijnen R, Foppen R, Veenbaas G. 1997; Disturbance by traffic of breeding birds: evaluation of the effect and considerations in planning and managing road corridors. Biodiversity and Conservation 6: 567-581.

Revitt DM, Ellis JB. 2001; Drainage, runoff and groundwater. Guidelines for the environmental management of highways. Institution of Highways and Transportation, London.

Revitt DM, Shutes RBE, Jones RH, Forshaw M, Winter B. 2004; The performances of vegetative treatment systems for highway runoff during dry and wet conditions. Science of the Total Environment 335: 261-270.

Rio Declaration on Environment and Development 1992. The United Nations Conference on Environment and Development. Rion de Janeiro.

RIZA 1996; Treatment of runoff from highways. Report nr. 96.017 (in Dutch) Institute for Inland Water Management and Waste Water Treatment, Lelystad, The Netherlands.

Roseland, M. 2000; Sustainable community development: integration environmental, economic and social objectives. Progress in Planning 54(2): 73-132.

Rydell J. 1992; Exploitation of insects around streetlamps by bats in Sweden. Functional Ecology 6: 744-50.

Salwasser H. 1990; Gaining perspective: forestry for the future. Journal of Forestry 88: 32-38.

Sandberg U, Ejsmont JA. 2002; Tyre/Road Noise Reference Book. Informex, SE-59040 Kisa, Sweden (www.Informex.Info).

Saunders SC, Mislevits MR, Chen J, Cleland DT. 2002; Effects of roads on landscape structure within nested ecological units of the Northern Great Lakes Region, USA. Biological Conservation 103: 209-225.

Schwartz A, Henderson RW. 1991; Amphibians and reptiles of the West Indies: descriptions, distributions, and natural history. Gainesville, FL: University of Florida Press.

Searns RM. 1995; The evolution of greenways as an adaptive urban landscape form. Landscape and Urban Planning 33: 65-80.

SEPA 1999; Sustainable Urban Drainage Systems. Scottish Environmental Protection Agency, Edinburgh.

Simpson PW, Newman JR, Keirn MA, Matter RM, Guthrie PA. 1982;. Manual of stream channelization impacts on fish and wildlife. U.S. Fish and Wildlife Service, Dept. of Interior. Contract No. 14-16-0009-80-066. Washington, D.C.

Slocombe. SD. 1998; Lessons from experience with ecosystem-based management. Landscape and Urban Planning 40: 31-39.

Smiddy P, O' Halloran J. 2004; The Ecology of River Bridges. In: Davenport J &. Davenport JL (eds) The Effects of Human Transport on Ecosystems: Cars and Planes, Boats and Trains, 15-62. Royal Irish Academy, Dublin.

Sörme L, Lagerkvist R. 2002; Sources of heavy metals in urban wastewater in Stockholm. The Science of the Total Environment 298: 131-145.

SOU 1993; 65 Handlingsplan mot buller. (Plan of action against noise). Miljo- och naturresursdepartementet, Goteborg (in Swedish).

Southerland MT. 1995; Conserving biological diversity in highway development projects. The Environmental Professional. 17: 226-242.

Stahre P, Urbonas B. 1990; Stormwater detention for drainage, water quality and CSO management. Prentice Hall, New Jersey.

Steele F. 1981; The Sense of Place. Massachusetts, CBI Publishing Company, Inc.

Stein TV, Anderson DH. 2002; Combining benefits-based management with ecosystem management for landscape planning: Leech Lake watershed, Minnesota. Landscape and Urban Planning 912: 1-11.

Strecker EW, Kersnar JM, Driscoll ED, Horner RR. 1992; The Use of Wetlands for Controlling Stormwater Pollution., The Terrene Institute, EPA, US.

Swanson FJ. 2001; Road Systems Interacting with the Land. In: Roads Ecology Seminar 16-17. The International Conference on Ecology and Transportation (IOCET) 2001, Keystone, Colorado.

Sýkora, KV, Kalwij JM, Keizer P-J. 2002; A phytosociological and floristic evaluation of a15-year ecological management of roadside verges in the Netherlands. Preslia 74: 421-436.

Sýkora KV, Pelsma TAHM, de Nijs LJ. 1989; The vegetation of Dutch road side verges. Ges. f. Ökol. Verhandl. 19: 149-150.

Theobald DM, Miller JR, Thompson Hobbs N. 1997; Estimating the cumulative effects of development on wildlife habitat. Landscape and Urban Planning 39: 25-36.

Thiffault P, Bergerson J. 2002; Monotony of the environment and driver fatigue: a simulator study. Accident Analysis and Prevention 848: 1-11.

Treweek JR, Hankard DBR, Arnold H, Thompson S. 1998; Scope for strategic ecological assessment of trunk-road development in England with respect to potential impacts on lowland heathland, the Dartford warbler (*Sylvis undata*) and the sand lizard (*Lacerta agilis*). Journal of Environmental Management 53(2): 147-163.

Trocmé M, Cahill S, De Vries JG, Farrall H, Folkeson L, Fry G, Hicks C, Peymen J. (eds) 2003; COST 341 Habitat fragmentation due to transportation infrastructure. Office for Official Publications of the European Communities, Luxembourg.

UK Environmental Protection Agency 2005; Light Pollution. Http://Www.Environment-Agency.Gov.Uk/Yourenv/Eff/Pollution/152227/? Version=1&Lang=_E: (accessed 11-02-2005).

UN CBD 1992; Proceedings of The United Nations Convention on Biological Diversity, Earths Summit, Rio De Janeiro.

US EPA 2002; National Air Quality Status Report, United States Environmental Protection Agency. www.epa.gov (accessed 11-02-2005).

U.S. Council On Environmental Quality, 1978; Implementation Of Procedural Provisions, Final Regulations. Federal Register, V.43, No.230. Nov.29, 1978, 55978-56007.

Van Apeldoorn RC, Knaapen JP, Schippers P, Verboom J, Van Engen H, Meeuwsen H. 1998; Applying ecological knowledge in landscape planning: a simulation model as a tool to evaluate scenarios for the badger in The Netherlands. Landscape and Urban Planning 41: 57-69.

Van Bohemen HD. 1995; Mitigation and compensation of habitat fragmentation caused by roads: strategy, objectives and practical measures. Transportation. Research Record 1475: 133-137.

Van Bohemen HD. 1998; Habitat Fragmentation, infrastructure and ecological engineering, Ecological Engineering 11: 199-207.

Van Bohemen HD. 2002; Infrastructure, ecology and art. Landscape and Urban Planning 59: 187-201.

Van Bohemen HD. 2004; Ecological Engineering and Civil Engineering Works. Dissertation Thesis Road and Hydraulic Engineering Institute of the Directorate-General of Public Works and Water Management, Delft The Netherlands.

Van Bohemen HD. 2005; Ecological Engineering; Bridging between Ecology and Civil Engineering, Aeneas, Technical Publishers.

Van Bohemen HD, Janssen van de Laak WH. 2004; The Influence of Road Infrastructure and Traffic on Soil, Water and Air Quality. Environmental Management 31(1): 50-68.

Van der Sluijs J, Melman PJM. 1991; Layout and management of planted road and canal verges. In: van Bohemen HD, Buizer DAG & Little A (eds) Nature Engineering and Civil Engineering Works 79-85. Centre for Agricultural Publishing and Documentation (PUDOC), Wageningen, Netherlands.

Van der Sluijs J, van Bohemen HD. 1991; Green elements of civil engineering works and their (potential) ecological importance. In: van Bohemen HD, Buizer DAG & Little A (eds) Nature Engineering and Civil Engineering Works 21-32. Centre for Agricultural Publishing and Documentation (PUDOC), Wageningen, Netherlands.

Verkaar HJ, Bekker GJ. 1991; The significance of migration to the ecological quality of civil engineering works and their surroundings. In: van Bohemen HD, Buizer DAG & Little A (eds) Nature Engineering and Civil Engineering Works 79-85. Centre for Agricultural Publishing and Documentation (PUDOC), Wageningen, Netherlands.

Viles RL, Rosier DJ. 2001; How to use roads in the creation of greenways: case studies in three New Zealand landscapes. Landscape and Urban Planning 55: 15-27.

Weckwerth G. 2001; Verification of traffic emitted aerosol components in the ambient air of Cologne (Germany), Atmospheric Environment 35(5): 525-536.

Wilcox DA. 1989; Migration and control of purple loosestrife (*Lythrum salicaria* L.) along highway corridors. Environmental Management 13: 365-370.

World Bank Group 1997; Roads and the Environment: A handbook. Edited by Tsunokawa K, & Hoban C. World Bank Technical Paper No. 376. The World Bank, Washington D.C.

Wright DL, Blaser RE, Woodruff JM. 1978; Seedling emergence as related to temperature and moisture tension. Agron. Journal 70:709-712.

Xuan Zhu, McCosker J, Dale AP, Bischof RJ. 2001 Web-based decision support for regional vegetation management. Computers, Environment and Urban Systems 25: 605-627.

Yousef YA, Baker DM, Hvitved-Jacobsen T. 1996; Modeling and impact of metal accumulation in bottom sediments of wet ponds. Science of the Total Environment 189/190: 349-54.

CHAPTER 14: ENVIRONMENTAL IMPACTS OF TRANSPORT, RELATED TO TOURISM AND LEISURE ACTIVITIES

JOHN DAVENPORT[1] & T. ADAM SWITALSKI[2]

[1]*Department of Zoology, Ecology and Plant Science. University College Cork, Cork, Ireland*
[2]*Wildlands CPR, Missoula, MT. USA*

1. Introduction

Mass tourism is a modern phenomenon, stemming primarily from the introduction of personal vehicles and motorised mass transport from the late 19th Century onwards, accelerating particularly after 1945 by the development of passenger airlines. Initially, mass tourism was a short-range phenomenon largely within nation states, but is now global with tourists from developed countries visiting almost all parts of the globe.

The greatest ecological threats that mass tourism poses undoubtedly lie in the infrastructure and transport arrangements required to support it, particularly in situations where numbers of tourists are subject to little control. Physical development of resorts, consumption of fuel by buildings, aircraft, trains, buses, taxis and cars, overuse of water resources, pollution by vehicle emissions, sewage and litter all contribute to substantial, often irreversible environmental degradation, as well as to dramatic social consequences. The first part of the chapter focuses on the transport-related aspects of these large-scale problems.

The second part of the chapter is aimed at assessing the ecological impact of individual leisure transport. This has developed from a simple matter of walking or horse riding that sufficed for centuries, through bicycle touring and leisure boating that burgeoned at the end of the 19th Century, to the use of high-powered four-wheel drives, snowmobiles and jet skis by 21st Century leisure consumers.

2. Mass tourist transport

Travel and tourism together form the world's fastest-growing economic sector, and this has been the case for much of the past half century. Worth around US$ 3.5 trillion per annum and employing 200 million people at the end of the 20th Century, the growth rate for national and international travel/tourism has averaged some 3-4% (i.e. more than general growth) for many years. Europe is still the most frequently visited tourist destination in world terms (roughly 60% of international tourism), but many developing countries gain significant (sometimes dominant) income from the trade. This is particularly true of islands or countries with substantial coastal tourism: in these cases tourism is often a major proportion of the gross domestic product – Caribbean countries are four times more dependent on tourism than any other area in the world.

Tourism brings economic benefits to countries, but there are usually substantial socio-economic and environmental costs associated with it. Such costs can be overwhelming for small island resorts. Social scientists in the 1980s developed the 'self-destruct theory of tourism' that applies particularly to such resorts. Wiese (1996) gives a good account of irreversible environmental and socio-economic degradation on the island of Cancún (Mexico).

John Davenport and Julia L. Davenport, (eds.), The Ecology of Transportation: Managing Mobility for the Environment, 333–360,
© 2006 Springer. *Printed in the Netherlands*

2.1 RAIL AND ROAD TRANSPORT INFRASTRUCTURE

Tourist resorts require effective transport links. The explosion of car- and coach-based tourism in the 20th Century contributed heavily to the development of extensive road networks throughout the developed world, increasing habitat loss to tarmac and augmenting habitat fragmentation. Tourist resorts are also generally characterised by extensive car-parking facilities, taking yet more land. For example, in Italy 50% of the 8000km of coastline has been built over in the past 3-4 decades and in the US it is estimated that 16 million hectares are covered by roads and car parks.

More eco-friendly coastal tourism developments are feasible. In May 2004 a development by a commercial concern in partnership with WWF started in Portugal to build a resort (Mata de Sesimbra) south of Lisbon that will involve dramatic savings in water usage, and a design that fosters wildlife corridors, restores forests, constructs 6000 sustainable homes and avoids all transport other than walking, cycling or use of non-petrochemical vehicles (website 1). However, although this is a significant improvement on the resort that might have been built, it will still bring a large number of extra tourists, who will use road, rail and air transport to travel, to Portugal.

Post 1945, international mass tourism developed most spectacularly in southern Europe (averaging 6-10% growth rate per annum), though the pattern has since been repeated all over the warm regions of the globe. The coastal areas of the Mediterranean Sea have been impacted by human presence for millennia and have long been affected by deforestation, intensive agriculture, irrigation and the resulting land erosion. However, during the past half century tourism has burgeoned, especially in areas with sandy beaches. Coastal road construction, tourist resorts and car parks have replaced natural habitats with concrete, tarmac and golf courses, while hotel, marina and street lighting now fringe most of the Mediterranean coast and its island systems. Beaches themselves are tramped and occupied by millions of people, while promenades and walkways often replace dune or rocky systems. Ecological effects have been dramatic. Wetlands have disappeared, taking their fauna and flora with them. Disturbance and habitat fragmentation have reduced biodiversity. Some vulnerable species have been driven close to extinction. Sea turtles provide good examples. Two centuries ago there were substantial numbers of green (*Chelonia mydas*), loggerhead (*Caretta caretta*) and leatherback (*Dermochelys coriacea*) turtles within the Mediterranean, all of which sustained breeding populations on the sandy beaches of southern Europe, Mediterranean islands (e.g. Corsica, Sicily, Malta) and North Africa. Green turtle breeding is now limited to Cyprus, while Loggerhead populations (declining by as much as 10% per year) are limited to small areas of Greece and Turkey. Leatherback breeding is now virtually unknown. Turtle life history is particularly badly affected by coastal transport-related development. Female turtles naturally dig nests close to vegetational fringes at the top of beaches. These areas have often been replaced by roads and are also frequently subject to additional planting of vegetation that overshadows the nesting area, thus lowering the temperature in the nest and influencing the sex of the offspring. Trampled, compacted sand is often too difficult to dig into, while tourist activity can destroy nests. Night-time lighting and disturbance inhibits beach crawls by nesting females, while hatchlings leaving the nest at night are programmed to seek out the lightest part of the horizon, which will be over the sea in natural systems, but the nearest brightly-lit coastal road, hotel or car park along much of the present-day Mediterranean. Hatchling mortality is naturally high, but made worse by tourist litter traps and the building of concrete paths that provide 90° edges that hatchlings can't climb (Cheng 1995).

2.2 FERRIES AND CRUISE SHIPS

Large passenger vessels showed a decline in importance during the 1960s and 1970s after being the premier mode of transoceanic transport for about a century. However, the last 30 years has seen a renaissance of passenger traffic, with the increasing success of cruise ships. By 2003 about 250 cruise ships carrying some 12 million passengers per annum had entered service. The Mediterranean and Caribbean are presently major destinations, but polar waters attract the more adventurous tourists. Cruise ships create a number of ecological problems. Illegal discharge of substances (mainly discharge of oil or other hydrocarbons) is common. Cruise ship anchoring in tropical waters has been associated with severe long-term damage to coral reefs, while dredging channels for the larger vessels causes increased turbidity that is damaging to both corals and sea grass beds. The turbulence produced by their propulsive screws also stirs up sediment (e.g. Ward 1999), smothering nearby communities and thus reducing biodiversity. Cruise ships are effectively mobile villages or towns (the largest vessels carry up to 5,000 passengers and crew), and produce substantial quantities of waste water and sewage that are often discharged untreated into pristine marine habitats. A typical cruise ship discharges around 1 million litres of 'black water' (sewage) during a 1 week voyage (United States Environmental Protection Agency 2000). A relatively unappreciated problem is that cruise ships also disembark large numbers of people onto remote sites that could not be reached by other means; this can result, for example, in high numbers of snorkelers who can damage reefs (see personal transport section below), or generate large local demand for personal water craft (q.v.). Finally, since ship-generated solid waste can no longer legally be dumped at sea, solid waste is often dumped in landfill sites at tourist destinations, thereby contributing to pollution and habitat loss.

Cruise ships in polar waters usually concentrate on eco-tourism (q.v.) and generally their operators seek to minimise environmental impacts. However, polar waters are inherently dangerous, and concerns centre around problems of oil spills, should a vessel run aground or strike ice. Fuel oil spillages are a particular problem as heavy fuel oil is more toxic than crude oil; also, petrochemicals break down much more slowly in cold, ice-covered water than they do in the tropics (e.g. Patin 1997).

Cruise ships are also popular in some freshwater systems, particularly the Rhine, Danube, Amazon, St Lawrence and Nile. Specific studies have been rare, but Ali et al. (1999) noted that the shallow waters of the Nile in Upper Egypt were adversely affected by cruise ship wash, with sediments being coarsened and aquatic plant diversity consequently being reduced.

2.3 ECOTOURISM TRANSPORT

In the past few decades there has been a substantial growth in ecotourism as tourists have demanded access to wildlife in as non-destructive fashion as possible. Whale and dolphin watching, bird watching, glass-bottom boat excursions and 'safari' holidays are all increasingly common. Policed properly, much of such tourism is either inoffensive or has potentially beneficial effects – exemplified by the preservation of large tracts of wetland for bird watching, and of conservation of forest areas for giant pandas and great apes. Problems tend to be caused by too-heavy demand. Intensive whale and dolphin watching activities by boats and swimmers can disturb marine mammal behaviour and acoustic activity (e.g. Constantine 2001; Van Parijs & Corkeron 2001), while too close contact between humans and great apes, risks catastrophic disease transmission (Woodford et al. 2002). Seabird and seal colonies can have their breeding activities disrupted by tourists (e.g. Yorio et al. 2001).

2.4 SIGHT-SEEING AIR TOURISM

Separately from mainstream commercial aviation, short-range, low-altitude sight-seeing flights by fixed wing aircraft or helicopters also have an impact. Mostly these have been associated with noise pollution of various US National Parks such as the Grand Canyon and the Volcanoes National Park of Hawaii (e.g. Stokes *et al.* 1999), but such flights occur in many wilderness areas around the world where scenery is spectacular (e.g. Milford Sound, New Zealand; Hunt 1999). Public interest has centred around the nuisance of noise to humans, but noise impacts upon mammals and birds are likely to be similar, and exacerbated by the predator-like appearance of low-flying aircraft. The US National Park Service reviewed the many impacts to wildlife which include: direct collision with aircraft, flushing of birds, alteration of movement and feeding patterns, increased heart rates, and decreased calf survival (USDOI 1994).

3. Individual leisure transport

Increasing prosperity in developed countries has created a worldwide demand for individual leisure transport, from simple walking and swimming to modern phenomena such as off-road vehicles (ORVs), snowboarding, SCUBA, jet skiing and paragliding. It is interesting to note that ORVs are amongst the fastest growing forms of recreation on public lands in the United States. In 2001, 42 million Americans were reported using ORVs on US National Forests or Grasslands an 110% increase on the 1982 figure (Cordell *et al.* 2004). An additional 13.5 million used snowmobiles and 5.5 million used personal watercraft (e.g. jet skis) on public access areas. While day hikers (76 million) and bird watchers (73 million) far exceed the number of people using motorised recreation, ORVs are having a growing impact on the environment.

This part of the chapter considers the ecological impact of land, water and aerial personal transport. Broadly speaking, serious impacts are associated with localised pressure of numbers of people indulging in these activities and with the development of new technologies that have so far escaped the regulations associated with conventional mass transport mechanisms, or which have greatly enhanced human mobility in challenging environments.

3.1 LAND TRANSPORT

3.1.1 *Walking, hiking and trekking*
Recreational walking was well established by the 19th Century in the form of long distance way-marked walks in Germany, Sweden, the US and UK. The availability of footpaths and trails is now a major tourist attraction throughout the world. Most walkers undertake short journeys. A recent survey of the adult population in Scotland showed that 1 million people had gone walking in the previous year, but only 50,000 (5%) had achieved a 50+ km walk in that period. Environmental pressure by walkers tends therefore to be localised (and often extremely intense).

Solo or small-group wildlands walkers trekking in low altitude remote areas subject to occasional use seldom do significant ecological damage, though interactions with ground-breeding birds (e.g. terns in coastal areas, grouse on moors) can cause problems. Mostly such walking literally leaves no footprint, provided that the participants behave responsibly, leave no litter and dispose of waste safely. There are a few exceptions associated with especially vulnerable habitats such as sand dune vegetation, cyanolichens and cyanobacterial flora, sphagnum bog, tussock grass, cushion plants and tundra vegetation. The Dry Valley of Antarctica is a dramatic example; the natural supply of nitrogen to soils is extremely limited and the biota mostly microbial. It is accepted practice that scientists working on foot in the Dry Valley collect and bag all their urine and faeces to avoid damaging the environment by eutrophication. In contrast, at low levels of intensity, in more robust areas trampling may even be beneficial in biodiversity terms as some trample-resistant plant species benefit, while the creation of open spaces favours breeding of some insects (e.g. wasps, bees and ants) or promotes basking and foraging in reptiles (website 2). Widespread ecological problems associated with walking arise from pressure of numbers, plus repeated walking over tracks (Figure 1).

a) path eroded by repeated walking   b) closeup of path, showing soil and vegetational damage

Figure 1. Walker damage to heather moorland on Isle of Cumbrae, Scotland, UK. Photographs: John Davenport

Numbers of people walking in rural areas inevitably disturb wildlife, particularly mammals and birds (e.g. ground-dwelling species such as red grouse, *Lagopus lagopus*). Heavily-walked areas have a lowered diversity of these groups, while popular walked paths can contribute to habitat fragmentation and population isolation. The effects on soil and vegetation depend largely on repetition and duration. A recent study by Thurston and Reader (2001) showed that 500 passes over 1 metre-wide lanes substantially (<100%) reduced vegetational stem density and species richness, as well as increasing the area of exposed soil by <54%. However, a year later these treatment effects were undetectable, so the damage was reversible if the activity ceased.   Sustained repeated walking over tracks causes rutting; walkers walk on verges, widening tracks and extending damage to vegetation. The creation of multiple, deepened, tracks ('braiding') leads to soil erosion and concentration of drainage - the paths become small stream beds that drain water from the surrounding soil, altering its composition and vegetation. Places subject to extreme walking pressure show marked signs of soil erosion, vegetational loss and reduced biodiversity. Management and trail maintenance minimise environmental damage, but are expensive (Figure 2). Severe damage involving total soil loss is essentially irreversible over human timescales.

In Australia, bush walking and horse-riding (q.v.), together with the camping implicit in spending several days in such activity has been implicated in creating physical deleterious effects by track formation, soil erosion or compaction, and greatly increased risk of fires.

Littering and water pollution are localised effects, but walkers have been implicated in the spread of plant and soil pathogens, as well as in weed dispersal (Sun & Walsh 1998).

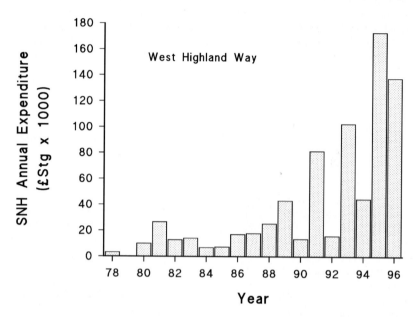

Figure 2. Grant expenditure by Scottish Natural Heritage on a single long distance walk (1978-96). Source of data: SNH Policy Paper - Long Distance Routes in Scotland (1997)

High altitude trekking and expeditions pose severe problems. Mountaineering has become an increasingly popular pastime and has been revealed in recent years to cause serious ecological damage, particularly in the Himalayas, but also in glacial areas of every continent. In the 1950s a handful of climbers reached the high mountains of Tibet and Nepal: now there are thousands each year. Expeditions invariably leave behind garbage; over time the non-biodegradable portion (plastics and metal) has been left in such quantities that vegetational growth has been slowed and topsoil rendered unfit for plant growth in some cases. Metal and plastics do not reflect sunlight as much as snow, so local albedo is decreased and snow melting is caused. Expeditions also take fuelwood from the limited vegetation available and this has led to reduced vegetational cover, heightened local temperatures and enhanced glacial retreat. To exemplify the scale of such deforestation, a single 20-person expedition consumed 290kg of rhododendron fuel wood (Kuniyal 2002). Pack animals put further pressure on limited vegetation and a number of mammals, including snow leopard (*Panthera uncia*) have become increasingly uncommon in mountainous parts of Asia as a result of disturbance and habitat loss.

3.1.2 *Intertidal trampling/collection disturbance*
In many parts of the world tourists walk in rocky or sandy intertidal zones. Simple walking overlaps with curiosity-driven exploration of boulders and rock pools, and informal exploitation of the resources of the intertidal zone for food and fishing bait. Impacts can be severe. Liddiard *et al.* (1989) worked on two extensive rocky shores near Swansea in South Wales, UK that are exploited for 'peeler' crabs by anglers. At least 3,000 rocks were

overturned each day during low tide periods at both sites. Bell *et al.* (1984) indicated that 90% of all boulders at one site were overturned within 2 weeks, and some boulders might be turned 40-60 times during a single summer. Rocks were rarely replaced in their original position. Boulders have markedly different fauna and flora on their upper and lower surfaces, so this human activity causes degraded habitat stability and reduced biodiversity. The destruction of large algae caused by boulder-turning also removes their understorey habitats, which are important for the shelter provided to small algae and invertebrates. Even simple walking in the rock intertidal zone causes damage; Brosnan and Crumrine (1994) showed that such walking causes significant declines in foliose algae and barnacles. Damage to mussel beds was also long-lasting (up to 2 years after trampling ceased), and community structure changed as mussel cover was replaced by algal turf. There is substantial qualitative evidence that many rocky shores in the United Kingdom, for example, that are extensively walked by tourists and school/university educational visitors have lower levels of biodiversity than they did in the 19th Century when such usage was negligible.

3.1.3 *Mountain biking*

For more than a century from the bicycle's invention, riding was limited to paved roads or broad tracks. In the past 30 years, recreational cycling has been significantly displaced from roads to paths and tracks, often shared with hikers and horse riders. The development of broad-tyred mountain bikes with low gearing has given the cyclist almost as much mobility off-road as walkers. Whether mountain bikers inherently cause any more environmental damage than hikers is debateable. Short-term studies suggest that their effects on vegetation and soil are similar (Thurston & Reader 2001), though Cessford (1995) noted that there was some extra damage caused when skidding downhill, or as a result of torque-induced wheelspin when riding up steep, wet slopes. However, mountain bikers can also cover much more ground (by a factor of 5-10) in a given time than walkers, especially downhill, and a subset of mountain bikers are devoted to cutting new trails in ever-more remote areas, thereby spreading damaging effects. The general consensus is that mountain biking is less damaging than often claimed by walkers or environmental managers, but is still a significant cause of environmental damage, particularly where mountain bikers illicitly build tracks, ramps and ditches to foster competition (Woodland Trust 2002; see also Figure 3).

a) Hillside eroded by constructed mountain bike paths and ramps. Note loss of vegetational cover in foreground resulting from repeated riding

b) Constructed ramps. Note loss of shrub cover on left

Figure 3. Damage to pasture and woodland by mountain bike tracks and ramps in Douglas, Cork, Ireland
Photographs: John Davenport

3.1.4 *Horse riding and pony trekking*

Horse riding and pony trekking, using privately – or commercially owned – horses, are major leisure transport activities, mainly in temperate areas, throughout the world. In extensive wildlands areas, particularly in the US, South America and Australia, riding animals are often accompanied by pack animals that carry camping gear. Worldwide it is likely that recreational horse riding in rural areas is far more common than at any time in the past.

Horses have surprisingly heavy environmental impacts, far greater than walkers or mountain bikers (see Figures 4 & 5). Cole and Spildie (1995) conducted careful studies in Montana forests demonstrating that horses (with shoed hooves that can deliver 1,000 tonnes/m^2) damaged vegetation far more than human walkers or llamas (whose impacts could not be distinguished from that of humans). Horses did particular damage to herbs, though this was quickly reversible, while shrub damage was more long lasting. This data has been used in the US to encourage use of llamas rather than horses as camping pack animals. Horse traffic, particularly when repeated along paths, causes severe damage, with significant soil erosion, soil compaction and vegetation damage occurring in as few as 20-30 horse passes in alpine habitats (Harris 1993).

Figure 4. Comparison of impact of boot and foot of shod horse on grass verge, Isle of Cumbrae, Scotland, UK
(Photograph: John Davenport)

It has long been known that trails used by horses eat deeper into the soil than trails used by hikers alone (Dale & Weaver 1974). This often causes tracks used by horses to feature muddy quagmires – in consequence hikers tend to skirt round them, contributing to track widening and braiding. In this respect horses appear to be as damaging as off-road motorcycles. Only dry grassland appears to be relatively immune to horse damage. In addition, horses cause ecological damage by browsing, manuring soils and pools, and by transmitting weeds on feet or in droppings.

Tethering horses to trees can damage tree bark to the extent that trees are killed (Cole 1983). Horse riding is so damaging that it is recommended that it be limited to hard tracks in sensitive alpine and sub-alpine habitats such as those of the Tasmanian World Heritage Area, where substantial effects on shrubland, herbfield and bolster heath were seen in as little as 20-30 passages by horses (Whinam & Comfort 1996).

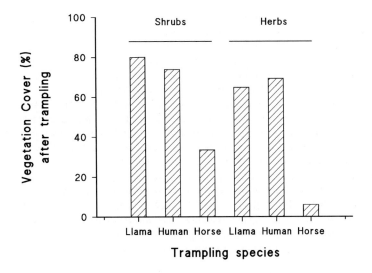

Figure 5. Comparison of trampling effects of humans, horses and llamas. In each case the low shrubs or erect herbs were trampled by 150 passes through previously undisturbed forest understorey in Montana, USA. Source of data: Cole and Spildie (1998)

3.1.5 *Skiing and snowboarding*

There are two basic types of skiing. Cross country skiing involves minimal technology and is similar in principle and effects to walking over trails or virgin snow; it is a popular pastime in Scandinavian countries, but is generally a low intensity sport that has not shown the explosive growth of other snow sports in recent decades. There appears to be little concern about environmental impact, which is unlikely to exceed that associated with walkers. Alpine ('downhill') skiing is an extremely popular mass sport (currently worth 10% of world tourism financial turnover) that has grown enormously in popularity over the past 50 years, spreading across alpine areas throughout the world. There is constant development of new resorts and most countries with high mountains either have such facilities (e.g. Chile, Kazakhstan) or intend to acquire them. The building and use of ski-lift stations and restaurants as well as the use of specialised machinery produces waste, oil and grease which pollute the surrounding soil and water. In recent years alpine snowsports have diversified with the invention of snowboards and a variety of 'radical' skis, but all rely on the same type of infrastructure. Most skiers/snowboarders use very broad tracks ('pistes') cut from summits or glaciers through tree cover and alpine meadows down to villages or towns. The pistes fragment forest and a combination of skiing (by thousands of individuals) and management of snow by machinery ('piste bashing') severely damages vegetation and accelerates the erosion of hillsides (see Ruth-Balaganskaya & Myllynen-Malinen 2000 for review). Heavily skied runs show very low diversities of plants by comparison with nearby natural or semi-natural vegetation (e.g. Klugpumpel & Krampitz 1996) and this is not helped by commercial operators often sowing inappropriate lowland seed taxa to 'repair' the damage. Noise and visual disturbance by people, ski lifts, gondolas and cable cars mean that skiers displace many animals, so that large areas of habitat are lost to them. There is a general tendency for chamois and mountain goats to retreat from open grasslands to forests, where they damage trees by browsing (SFSO/SAEFL 1997).

As skiing has become easier to learn, by virtue of improvements in technology combined with more frequent access to the sport, general competence and adventurousness of skiers has increased. This has led to the development of 'off-piste' skiing and snowboarding. At the simplest level this involves minor diversions from main pistes, but most hillsides in popular resorts are now crisscrossed by thousands of trails that weave through scrub, spreading vegetational and soil damage to wider areas. Off-piste skiing is frowned upon by resort managers because participants are far more vulnerable to avalanches. However, the practice is essentially impossible to control. Until recently, alpine skiing, whether on or off-piste, was essentially limited to the area of the resort, which controlled damage to some extent. Recent developments in technology, particularly mountain skis that allow (extremely fit) participants to climb up hills before they ski down, mean that this degree of control is beginning to disappear. Helicopter skiing, in which participants are taken to high altitude sites and then ski off-piste to the bottom of the mountains, is an ominous new trend. The industry is currently responding to global climate change that is already making some low altitude resorts uneconomic; restoration of damaged meadow and forest systems is attracting increasing attention. This positive response is offset by plans to move European alpine skiing to higher altitudes, thus spreading the influence of the industry into hitherto pristine glacial areas.

3.1.6 Off-road vehicles (ORVs) and all terrain vehicles (ATVs)
Since ORVs and ATVs first gained popularity in the 1970s, scientists have documented the physical and ecological consequences of their use. The impacts have been documented in hundreds of research articles, and several literature reviews and books (e.g. Joslin & Youmans 1999; Schubert & Associates 1999; Havlick 2002; Sutter 2002; Gaines *et al.* 2003; Gilbert 2003; Hulsey *et al.* 2004). ORVs are able to traverse land, snow, and water and are found on all continents and ecosystem types. Technological advances have given ORVs more power and control, allowing even novice riders into some of the most remote places on Earth.

ORVs include a variety of four-wheeled drive 'sports utility vehicles' (SUVs; rugged passenger vehicles that are also usable on normal roads), specialist ATVs that can have three, four or six wheels, and off-road motorcycles. ATVs were originally military, forestry or agricultural vehicles that had limited effects on pristine environments. Recreational ATVs first debuted as a modified motorcycle with three balloon wheels. Manufacturers quickly added a fourth wheel and the modern ATV was born. Sales of ATVs have now surpassed those of street legal motorcycles in the USA. The development of more powerful and sophisticated machines has permitted their widespread use in wildlands areas of North America, North Africa and many parts of Australia. Accurate position finding by GPS and reliable communication by radios and cell phones allows ever deeper penetration into wildlands areas. ATVs are able to traverse almost any landform, many riders are going off-trail and riding cross-country. The result of cross-country travel is the creation of user-created trails, also known as renegade or non-system routes. There are hundreds of thousands of these unplanned user-created routes on public lands across the US. In large countries with relatively low population densities it is difficult to police ATV activity even within nature reserves. In Florida there are around 37,000 km of trails (mostly illegal) cut by ATVs in the Big Cypress National Reserve (website 3). There is significant use of ATVs in most European countries, though it is generally illegal for ATVs to be used in designated wildlife parks and vehicle insurance often does not cover the use of private ATVs more than a few metres from paved roads. The densely populated nature of most European countries means that detecting illegal use is easier.

The millions of ATVs that now roam lands around the world have had an enormous impact on soil, vegetation, and wildlife. Inland they have been implicated in the rapid deterioration of white roads and paths, and in damaging trees and other vegetation (website 2). As with walkers and trekkers, erosion created by the ruts of pioneer ATV users (Figure 6) tends to force subsequent drivers to avoid the centre of paths, gradually widening the area affected. Extensive creation of rutted paths channels surface water as well as directly causing soil and vegetational degradation. The first comprehensive studies on the impacts of ATVs took place in the mid-1970s. Most of the work focused on the influx of ATVs in arid regions (e.g. Nakata *et al.* 1976; Snyder *et al.* 1976; Vollmer *et al.* 1976; Wilshire & Nakata 1976) and coastal regions (e.g. Liddle & Greig-Smith 1975; Godfrey *et al.* 1978). These studies found that ATV use resulted in significant soil compaction – often after just a couple of passes. Compaction is measured as soil bulk density (mass of soil relative to volume). An increase in soil bulk density decreases soil porosity, permeability and infiltration capacity resulting in soil erosion (Misak *et al.* 2002). Soil erosion from ATV use can be severe. A recent study by Sack and deLuz (2003) found erosion rates as high as $0.11m^3/m^2$ over the course of a riding season in Appalachia, USA. This equates to a loss of over 200 kg of soil each year on just a 60 m section of trail. Significant soil erosion like this can greatly reduce soil fertility and add sediment to streams, decreasing water quality and reducing habitat quality for stream flora and fauna.

Figure 6. Ruts produced by pioneer ATV passes on an hillside, Isle of Cumbrae, Scotland. Photograph: John Davenport

A number of factors are related to the degree of soil compaction and subsequent erosion. Compaction is generally higher in wetter, poorly drained soils than in well-drained soils (Willard & Mace 1970; Burde & Renfro 1986). Similarly, forests that receive higher precipitation are more susceptible to erosion than drier forests (e.g. Cole 1983; Burde & Renfro 1986). However, wet forests may recover faster if ATVs are removed, while desert systems can take centuries to recover (Belnap 2003). Soils with fine texture are more erosive than coarse or heterogeneous soils (e.g. Bryan 1977; Welch & Churchill 1986). Trails at high elevations generally experience more erosion than lower elevations (Willard & Marr 1970; Marion 1994), and trails on slopes are more susceptible to erosion than on flatter ground (Welch and Churchill 1986). The impacts on soil by ATVs can also influence vegetation by decreasing nutrient uptake, reducing root growth, and weakening plant stability (Blackburn & Davis 1994).

Desert areas have been particularly subject to ATV influence, especially in North America, Australia and the Middle East. Desert soils are particularly sensitive to disturbance and vehicle tracks are visible for very long periods. Soil compaction by repeated ATV passage inhibits germination of seedlings, which cannot penetrate the soil (e.g. Lovich & Bainbridge 1999). A wide variety of desert animals only survive desert temperatures by living in burrows that buffer thermal extremes (Figure 7).

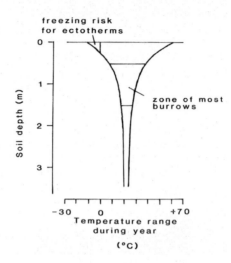

Figure 7. Changes in temperature with desert soil depth (information from Davenport 1985)

Use of ATVs compacts soil (making burrowing difficult) and collapses existing burrows e.g. the burrow of the fringe-toed lizard (*Uma notata*) (Stebbens 1974). It is likely that ecological damage is widespread, but relevant systematic studies are rare. However, in the Mohave Desert endangered desert tortoises of the genus *Gopherus* have attracted particular attention. These animals spend the day in their burrows, which are extensive and often dug by successive generations of these long-lived (>70 years) animals. ORVs have been implicated in the 90% decline in populations of gopher tortoises over the past few decades. Tortoises have been crushed beneath ATV wheels, or their burrow systems destroyed so that they suffocate or cannot find shelter from the desert sun (e.g. Burge 1983).

Repeated driving over-land by ATVs results in the crushing, breaking and overall reduction of vegetative cover. For example, Misak *et al.* (2002) found that vegetation cover was twice as high in protected areas as in sites open to ATVs. Vehicle passes can also result in indirect effects including damaging germinating seeds, and weakening plants making them more susceptible to disease and insect predation. Furthermore, such passes result in the simplification of vegetation and changes in plant species composition. Communities of shrubs are generally replaced by forb and grass communities following ATV use (Leininger & Payne 1971; Stout 1992). Many sensitive species have the potential to go locally extinct in areas of high ATV use (Stensvold 2000; Brown & McLachlan 2002).

Some plant communities can shift to non-native invasive species following ATV use. With knobbly tires and large undercarriages, non-native 'weedy' species' seeds can be unintentionally taken deep into wildlands by ATVs. The spread of invasive weeds has been cited by the Chief of the US Forest Service as one of the four 'great issues' facing these lands today (Bosworth 2003). Several recent studies have documented how ATVs are a major vector for spreading invasive exotic species (Montana State Extension Service Bulletin 1992, Gelbard & Harrison 2003; T. Rooney in prep.). For example, on a 10 mile ATV course in Montana, 2,000 spotted knapweed (*Centaurea biebersteinii*) seeds were dispersed in just one trip. In Wisconsin, exotic species were found on 88% of the segments sampled along ATV trails (T. Rooney in prep).

The effects of ATVs are synergistic and can have cascading effects throughout an ecosystem. For example, the damaging and reduction of vegetative cover coupled with the changing of entire plant communities can greatly affect the food and cover requirements of wildlife. The result is often a reduction in native plant and wildlife populations (Bury 1980). For example, on an intensively used ATV trail in Idaho, native shrubs, bunch grasses and microbiotic crust were greatly reduced close to the trail and replaced with non-native cheat grass (*Bromus tectorum*) and rabbitbrush (*Chrysothamnus* spp.; Munger *et al.* 2003). Because of these habitat changes, fewer reptiles were found alongside the trail than 100 m away.

The negative effects of ATVs on wildlife are well documented. Most studies cite habitat loss as a primary concern, although there are a number of direct and indirect effects as well. While the scope of impacts is still debated, most agree that the four fundamental impacts of ATVs on wildlife are harvesting or killing, disturbance, habitat modification and pollution (Gutzwiller *et al.* 1994).

Collisions and impact-induced mortality are the most obvious effects of ATVs on wildlife populations. Most species can fall victim to collisions with an ATV, but small animals are the most susceptible. Small mammals can be easily crushed when run over by ATVs (Bury 1980; Wilkins 1982; Rosen & Lowe 1994). Birds can be maimed and killed by ATVs. Piping plovers (*Charadrius melodus*) are a threatened bird species that nest on beaches. Their eggs and young have been trampled by ATVs, proving to be a significant threat to the viability of their populations (Strauss 1990). For larger animals, ATVs can impact populations by allowing increased harvesting and/or poaching. ATVs are used for hunting game species such as deer (*Odocoileus spp.*) and elk (*Cervus elaphus*) in the US. Most ATV hunters use them to access remote hunting grounds and to transport their kill. While legal in many areas, ATV hunting is coming under increased scrutiny by 'traditional' hunters on foot. For example, the Idaho Department of Fish and Game has banned game retrieval by ATVs on large sections of public land. Additionally, ethical concerns have been raised over other uses of ATVs for hunting. Of great concern is the use of ATVs to herd animals and chase them down. While certainly not legal, there have been increased reports of ATV hunters taking 'flock' shots at running pronghorn antelope (*Antilocapra americana*) from long ranges (Canfield *et al.* 1999).

Carnivores are also at risk from poaching or over harvest. Increased access on ATVs can increase the trapping vulnerability of pine marten (*Martes americana*), fisher (*Martes pennanti*), and wolverine (*Gulo gulo luscus*; Weaver 1993). Wolves (*Canis lupus*) often travel on ATV trails because of ease of movement. However, wolves risk increased poaching pressure. Boyd and Pletscher (1999) reported that 21 of 25 human-caused wolf mortalities in the central Rockies occurred less than 200 m from human linear features. Another federally protected species, the grizzly bear (*Ursus arctos horribilis*), is also at risk from poaching on ATV routes and bears avoid open roads. Mace *et al.* (1996) found that most bears avoided even little travelled roads throughout the year. The study resulted in the decommissioning of hundreds of miles of old logging roads and greatly limited ATV use within the range of this species. The Big Cypress National Reserve, Florida (website 3) is highly diverse and contains endangered species such as manatees, but is particularly associated with the critically endangered Florida panther (*Felis concolor coryi*). ATVs have harassed panthers to the extent that they avoid preferred hunting areas and home ranges; ATVs also carry hunters who shoot the pigs and deer on which the panthers feed (website 3).

ATVs can have a number of indirect effects on wildlife as well. Noise and disturbance from ATVs can impact on a number of species resulting in increased stress (Nash *et al.* 1970), loss of hearing (Brattstrom & Bondello 1979), altered movement patterns (Nicola & Lovich 2000; Veira 2000; Wisdom *et al.* 2004), avoidance (Janis & Clark 2002), and disrupted nesting activities (Hooper 1977; Fyfe & Olendorff 1976; Strauss 1990). These behavioural changes can result in increased energy expenditure and loss of productivity in wildlife.

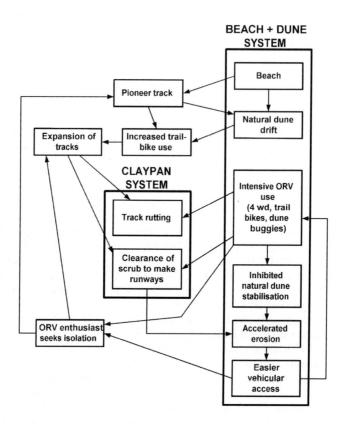

Figure 8. Simplified model of feedback relationships between off-road vehicle use, access track network development and dune/claypan systems in South Australia (modified from Gilbertson, 1981)

Coastal areas are also affected by ATVs that often use dune and beach systems. There is much evidence that beach use by ATVs is damaging. Dune and sand vegetation is vulnerable to crushing and supports specialised insect communities. ORV tracks are deep enough in beaches to act as traps that prevent turtle hatchlings reaching the sea, while intense use of off-road vehicles was long ago shown to decimate ghost crab populations in Virginia by collapsing their burrows (Fialka 1975). Interactions with breeding birds are a problem (Watson *et al.* 1996), with ground nesting birds especially vulnerable. A detailed study of impacts on a coastal barrier system with varied soil types, was conducted near Adelaide, Australia by

Gilbertson (1981). This involved study of several types of ATVs. Overall he found that ATVs caused increased soil erosion and destabilised dunes. Scrub and sand-binding grass damage, plus collapsing of dune ridges were responsible for enhanced dune mobility, while compacting of clay pans behind dunes led to reduced water penetration, surface waterlogging and enhanced water-driven erosion. This study also led to a simple positive feedback model of ATV use, access track development and dune drift (Figure 8). Overall, ATV use had done more damage to the barrier system than previous centuries-long pastoralism.

3.1.7 Snowmobiles

From 1982 to 2001 snowmobile use increased by125% in the US. Today, over 13 million Americans use snowmobiles as a form of wintertime recreation. As snowmobiling gained popularity in the US during the 1970s, several studies documented their negative ecological impacts. These researchers could not have anticipated the exponential increase in snowmobile popularity and technology. The power and mobility of today's snowmobiles has greatly increased the ability of snowmobilers to go off-trail and into very steep terrain. Snowmobilers' access is only restricted by regulations and the risk of avalanches cascading upon them. Many still take the risk confident that their machines can 'outrun' avalanches and the law.

Yellowstone National Park (USA) is subject to many tourist pressures throughout the year. The winter period is now characterized by an extraordinary dominance of snowmobiles (Table 1). In 2000 almost 60% of winter visitors entered the Park on snowmobiles, which are extremely difficult to manage as they can easily circumvent the barriers and gates that may be used to regulate cars and coaches. It has been claimed that snowmobiles do less damage to soils than off-road vehicles do in the summer. However, snowmobiles can crush and damage vegetation, particularly woody plants. Neuman and Merriam (1972) found 78% of saplings were injured after just one pass of a snowmobile. Nearly 28% of those saplings were predicted to have a high probability of mortality. In Minnesota, Wanek (1971) reported damage to 47% of Norway pines (*Pinus resinosa*) and 55% of white spruce (*Picea glauca*) following snowmobile use. In a marsh in Iowa, snowmobiling resulted in a 23% decrease in cattail (*Typha spp.*) density, 12% decrease in cattail height, and a 44% increase in sedge (*Carex spp.*) density (Sojda 1978).

Table 1. Numbers of people entering Yellowstone Park in Winter 2000
(Source: Idahonews)

Mode of transport	Numbers of tourists
Cross country skiing	512
Snow coaches	10,779
Cars	40,727
Snowmobiles	76,271

The trampling effects of snowmobiles can be especially damaging when travelling on steep slopes or thin snowpack. On steep slopes, shallow roots and rhizomes are at an increased risk of damage (Boucher & Tattar 1975). In areas where wind has blown the snow away, or during years of low snowpack, the amount of damage to vegetation can be much greater. For example, Wanek and Schumacher (1975) found 93% of Norway pines were damaged in a year that received below average snowfall. Of these damaged pines, 45% received heavy damage while 8% died. Additionally, as snowmobiles become increasingly powerful, the increased torque creates a potential for greater impact on vegetation.

Snowmobiles can also indirectly influence vegetation through compaction. Compaction of the snow reduces the insulating air spaces and conducts cold air to the ground. These colder temperatures lower plant density and composition (Neumann & Merriam 1972), reduce productivity and growth (Wanek & Potter 1974; Wanek & Schumacher 1975), delay seed germination and flowering (Keddy *et al.* 1979), as well as affecting decomposition rates, humus formation, and microbial activity (Neumann & Merriam 1972; Rongstad 1980).

In Minnesota and Michigan snowmobiles have been found to kill off populations of small mammals that inhabit the subsnow ('subnivean') habitat (Rongstad 1980). This has adverse secondary consequences to birds of prey and mammalian predators such as foxes (e.g. Brander 1974). It is not uncommon in wildlands areas for snowmobile riders to drive their vehicles off patchy snow onto vulnerable soils, particularly in tundra regions. In Alaska snowmobiles have been found to compact tundra and permafrost ecosystems, causing permafrost to melt prematurely (website 5), thus exacerbating permafrost retreat due to global climate change.

Like jet-skis (q.v.), snowmobiles are powered by simple, rugged engines that are very noisy and release substantial quantities of pollution. In America's first National Park, Yellowstone, photos of Park Rangers wearing gas masks at entrance gates have prompted several recent studies to address snowmobile toxicity (e.g. USDOI 2000; Bishop *et al.* 2001; Sive *et al.* 2003). In Yellowstone, snowmobiles account for 68-90% of the total hydrocarbon (HC) and 35-68% of the total carbon monoxide (CO) emissions depending on how emissions are estimated (USDOI 2000). This equates to about 20 tonnes of HC and 54 tonnes of CO released in the park on a peak day (USDOI 2000). Automobiles in comparison omit 2.5 and 17.9 tonnes of HC and CO respectively on a busy summer day (USDOI 2000).

Up to 30% of the fuel is unburned in a two-stroke engine. With an average of 60,000 snowmobile visits per year (USDOI 2003), the equivalent of three tanker trucks of gas/oil mixture is being released into the air and snowpack each year. Furthermore, it is estimated that running a two-stroke engine for seven hours produces more air pollution than running a modern car for 160,000 km on a road (CEPA 1999). The ecological consequences of the fall out and eventual snowmelt into streams has not been well explored (Adams 1975), but warrants investigation. The park has begun to phase out the more polluting two-stroke snowmobiles in favour of quieter four-stroke snowmobiles and snow coaches which release two to twenty times less emissions and are much quieter (Sive *et al.* 2003). Unfortunately, the new regulations are also allowing increased numbers of snowmobiles into the Park, thus cancelling any perceived ecological benefits.

With freezing temperatures and deep snow, winter can be the most trying time of the year for wildlife in temperate regions. Disturbance from snowmobiles can add additional stress at this critical time of year. Noise pollution can be extreme; at 11 of 13 sites sampled within Yellowstone Park during daytime, snowmobiles could be heard for 70% of the time, at 8 of the sites for 90% of the time and at 1 site for 100% of the time. Modern snowmobiles are often equipped with headlights, which allow wildlands areas to be subject to both noise and light pollution at night. Animals, particularly mammals, show evidence of disturbance of normal activity patterns. The effects of snowmobiles on herbivores have received the greatest attention in the literature. While a few studies have found only small impacts resulting from snowmobiles travelling on trails (Bollinger *et al.* 1972; Richens & Lavigne 1978), most studies have found otherwise. Snowmobiles have been shown to disturb and create an energy cost for caribou (*Rangifer tarandus*; Fancy & White 1985), muskoxen (*Ovibos moschatus*; McLaren & Green 1985), bison (Bjornlie 2000), deer (Kopischke 1972; Moen *et al.* 1982), and elk (Creel *et al.* 2002). Even when mammals do not show visible signs of disturbance they may be stressed: Moen *et al.* (1982) found increases of up to 250% in heart rate in white-tailed deer when snowmobiles approached, even though they neither got up nor ran.

Some of the older literature suggests that unpredictable and erratic disturbance such as off-trail cross-country skiing creates a greater response in herbivores than do snowmobiles (e.g. Freddy *et al*. 1986). However, in those studies, snowmobiles remained on trails where wildlife became accustomed to their movements. Today's snowmobilers are not restricted to trails but, crisscross entire drainages in search of 'fresh powder'. While these newer, more powerful and far-reaching snowmobiles are probably having a greater impact, no study has yet quantified the extent of their intrusion.

Carnivores can also be negatively affected by snowmobiles. Carnivores such as wolves, grizzly bears and lynx (*Lynx canadensis*) have received increased attention lately because many are considered threatened or endangered in the US. While these species will use snowmobile trails because of ease of movement, they can also suffer from stress from snowmobiles (Creel *et al*. 2002), and have a higher risk of being illegally shot or run over (Claar *et al*. 1999). Additionally, carnivores may follow snowmobile trails into less remote human-dominated landscapes where there is a greater risk of conflict.

Snowmobile trails can also change species composition resulting in competition between species in places where they would not naturally co-occur. Recent concern has been expressed over coyotes using snowmobile trails to access lynx habitat. Lynx are well adapted for deep snow conditions, but are easily out competed in snow-free areas. Bunnell *et al*. (2004) found coyotes (*Canis latrans*) used snowmobile trails over 70% of the time potentially displacing lynx in the Uintah Mountains of Utah. Another species of special concern is the grizzly bear it is endangered in much of its range in the US. While den abandonment by grizzly bears has been rarely documented because of snowmobiles (Linnell *et al*. 2000), as the off-trail capabilities of snowmobiles has increased, so has the potential for disturbance (Claar *et al*. 1999; Hilderbrand *et al*. 2000). Copeland (1996) studying wolverines (*Gulo gulo*) in Idaho concluded that snowmobiles displaced wolverines from potential denning habitat (wolverines construct sub-snow tunnels and dens).

At the time of writing, moves to ban snowmobiles from National Parks in the USA are in progress, but are strongly opposed by vested interests. Generally in Europe snowmobile use is substantially restricted, especially in Norway, though neighbouring Finland advertises snowmobile riding as a tourist attraction. Manufacturers are introducing cleaner, quieter snowmobiles, but it will take many years for the existing snowmobile fleet to be replaced – and these changes will not affect damage to vegetation and subnivean habitats.

3.2 WATER TRANSPORT

3.2.1 *Swimming and surfing*
Swimming and surfing are among the most innocuous of pastimes in themselves, though they usually require a panoply of coastal tourist infrastructure to support them. However, in some countries, notably Australia and the USA, the installation of shark netting off popular swimming beaches has locally reduced numbers of rays, dolphins and turtles due to entanglement.

3.2.2 *Snorkelling and SCUBA*
Snorkelling and SCUBA are among the fastest growing pastimes in the world. Participants are drawn to the most attractive diving sites, in particular warm-water sites with high levels of biodiversity, especially coral reefs. Many dive sites are in marine protected areas (MPAs). Generally SCUBA divers are carried to dive sites by motor vessels whose impacts are described below. In former times snorkellers and SCUBA divers caused substantial impacts through their recreational fishing, principally by use of spear guns, but also by collecting crabs, lobster and octopus. These activities are now largely frowned upon by national and international bodies that organise SCUBA clubs, as well as by tour operators. Spearfishing is

illegal in many areas. However, there is still some residual leisure fishing in most parts of the world. In Madeira, for example, snorkellers deplete limpet populations as these are a local delicacy. Since limpets control seaweed growth, there is probably a resultant increase in macroalgal abundance, though there is no available scientific evidence to confirm this as before and after data were not collected. There is still some collection of coral and living shells by divers, though this is probably dwarfed globally by deliberate semi-industrial illegal collecting of such material by dynamiters who supply the curio trade.

Most concerns about diving are now concentrated upon physical damage by divers and ecological disturbance by divers who feed large fish (e.g. Badalamenti et al. 2000). Physical damage is a particular problem for coral reefs as most coral species are slow to regenerate. Snorkellers (e.g. Alison 1996) and SCUBA divers inadvertently damage corals by clambering over them, by kicking them accidentally with their fins, or by stirring up silt that suffocates them. Even in the MPAs of the Cayman Islands, where highly restrictive and well-policed management practices have been in place for more than ten years, it is still the case that substantial damage is being done (Tratalos & Austin 2001). This appears to reflect sheer numbers of divers. Currently about 350,000 divers per year visit the Cayman reefs. Even if a tiny proportion inflict damage, this is cumulatively significant. Hawkins et al. (1999) suggested that abrasion and tissue loss in corals due to diver damage may potentiate disease transmission, particularly in large, massive corals. These workers, who studied Caribbean coral communities at Bonaire, also found that dive sites showed altered species' composition with a trend towards dominance by branching corals that grow more quickly than massive non-branching corals. Similar results have been found at reefs in the northern Red Sea, with heavily-dived sites having significantly higher levels of coral damage and algal overgrowth than more pristine areas. Zakai and Chadwick-Furman (2002) studied SCUBA impact at Eilat in Israel and reported about 10 reef contacts per person dive; they concluded that the present level of diving (250,000 dives per annum on 12km of coast) was unsustainable as damage exceeded regeneration capacity.

Feeding of large fish by SCUBA divers also causes ecological problems. Feeding attracts predators that scare off smaller fish, reducing local biodiversity (probably reversibly over short periods). It also attracts sharks that may damage corals by abrasion while seeking prey (Tratalos & Austin 2001).

3.2.3 Yachting and motorboats

Small sailing vessels (sailing dinghies, sailboards, motorless small yachts) appear to cause little or no environmental impact. Larger sailing vessels with auxiliary engines, or motorboats with outboard or inboard motors have been implicated in several types of environmental degradation. Historically, private yachts and motorboats were the prime users of tributyl tin (TBT)-based antifouling paints that had profound effects on molluscs such as oysters and whelks in the 1970s and 1980s before such paints were withdrawn from use. Marinas where large numbers of vessels are moored tend to be hotspots of pollution by HCs, sewage and antifouling compounds. In consequence there are often steep gradients of biodiversity between marinas and neighbouring pristine habitats, though usually over fairly small distances.

Leisure craft also create problems at more dispersed anchorages. Boat anchors and moorings create 'halos' in sea grass meadows as the chains drag along the sea bed and boat position shifts with tide, current and wind (e.g. Ward 1999). Benthic animals are often damaged by anchoring as well, though a recent study by Backhurst and Cole (2000) found that, although anchoring scars persisted for several months, and anchors killed the bottom dwelling mussel Atrina zelandica, this damage was reversed within a year or so. Such rapid recovery does not occur when anchoring takes place at coral reef sites, where serious damage is inevitable if anchors touch corals themselves. Damage to coral is likely to take decades to repair.

Leisure motor vessels are also implicated in the introduction of pest species, just as is the case for larger commercial vessels. For example: in Ireland it is believed that leisure craft have had a role in the introduction and spread of zebra mussels (*Dreissena polymorpha*) in the Shannon watershed, while in Bermuda the wildlife reserve of Nonesuch Island has been recently invaded by cane toads (*Bufo marinus*) almost certainly carried in the bilges of motor boats (Davenport *et al.* 2001).

When they are moving, most ecological effects of motorboats tend to be associated with wake formation, impact with air-breathing vertebrates and with visual/acoustic disturbance. Wake formation is a particular problem in freshwater systems where wakes erode riverbanks, damaging riparian vegetation and adversely affecting mammals and birds that live at or near the water's edge.

Powerboats are usually (but not always) detected and avoided by fast-swimming marine mammals and seals. However, two groups of air breathers seem particularly vulnerable to injury or death as a result of collisions with boats: sea turtles and manatees. Amongst sea turtles, green turtles (*Chelonia mydas*) typically browse in near-shore sea grass habitats where boat traffic is intense. They are slow swimmers and propeller or impact damage is common (Figure 9). Freshwater Florida manatees (*Tricechus manatus latirostris*) have often suffered death or injury from powerboat propellers, though management procedures are in place to minimize the impact of boating activities (e.g. Buckingham *et al.* 1999).

Moving motorboats create visual and acoustic disturbance for birds. Burger (1998) studied flight behaviour of terns and reported that motorboats caused significant disturbance to common tern colonies, especially during the breeding season. Motorboats have also been implicated in acoustic disturbance of a number of marine mammals such as dolphins (e.g. Van Paris & Corkeron 2001), particularly during breeding periods.

3.2.4 *Personal Watercraft*

Personal watercraft (PWC) such as jet skis and WaveRunners have gained popularity in the past decade. For example, in Montana, PWCs dramatically increased from 560 to 4,470 (700%) between 1990 and 1998 (Youmans 1999). Managers across the US have begun to restrict or outright ban PWC use in order to mitigate water contamination, wildlife harassment and increasing conflicts with non-motorised recreationists.

As with snowmobiles, PWCs use inefficiently burning two-stroke engines, so release as much as 30% of the gas/oil mixture unburned into the water and atmosphere (CEPA 1999). There are no legal restrictions on exhaust emissions for jet skis. The scope of this pollution is remarkable. There are more than 1.2 million jet skis in the US alone (1998 figure), discharging some 200 million litres of oil and petrol annually to water bodies and atmosphere. For perspective, it was estimated that two-stroke engines (including PWCs) on Lake Mead National Recreation Area (Arizona/Nevada) discharge over 100,000 litres of unburned fuel into the lake per day on a peak summer weekend (USDI 2002). Even more striking is the finding that, on an average weekend day in California, jet skis emit 50% more HCs and nitrogen oxides than all of California's cars combined (website 6). Drinking water can also be quickly contaminated by PWC use. For example, operating a single two-stroke engine for one hour makes 11,000 m^3 of water undrinkable (Jüttner *et al.* 1995).

The most serious contaminants released while riding PWCs are methyl tertiary butyl ether (MTBE) and polycyclic aromatic hydrocarbons (PAHs). MTBE is an oxygenate added to gasoline and PAH compounds are a by-product of the combustion process. Few studies have investigated the impacts of MTBE on aquatic organisms, but an ecological risk analysis by Johnson (1998) considered potential adverse effects on rainbow trout (*Oncorhynchus mykiss*) and insect communities. More is known about PAH compounds, at least three of which are probably carcinogenic (USDI 1999). It is probable that hydrocarbons are accumulated up inshore marine food chains, and in the food chains of lakes and reservoirs. They are also likely to contaminate shellfish reared in shallow confined areas. PAHs have been found to impact fish negatively. One study found that PAHs interfered with the biological function of rainbow trout; exposure to two-stroke engine exhaust resulted in disruption of enzyme activity and carbohydrate metabolism resulting in abnormal growth (Tjärnlund *et al.* 1995, 1996).

The ability of PWCs to travel at high speeds in very shallow water has a great potential to disturb wildlife. Outboard-powered boats are not able to enter these shallow areas that once served as a refuge for wildlife. Jet skis are very manoeuvrable, but have very long stopping distances (80-150m) and their high frequency noise does not give warning of their approach until they are very close to human or animal swimmers. As with mountain bikes and snow-mobiles they exemplify exciting, rapidly developing technologies aimed primarily at young consumers with the subtext of freedom from regulation.

Most public concern about jet skis has centred on noise pollution and accidents (they cause 40% of all boating injuries in the US) in which swimmers have been injured or killed, usually by blunt impact trauma. Jet skis also undoubtedly kill a range of surface-dwelling air breathing vertebrates, including seals, dolphins, and turtles (see Figure 9).

a) Turtle decapitated by propeller of power boat   b) Turtle killed by blunt trauma impact with power boat
                                                        or jet ski

Figure 9. Bermudan Green Turtles (*Chelonia mydas*) killed by power boats or jet skis (Photos: Jennifer Gray, Bermuda Aquarium, Museum & Zoo)

Recent studies have documented significant impact on wildlife from PWCs (Burger 1998; Nowacek *et al.* 2001; Rodgers & Schwikert 2002). The most commonly reported impact is the disturbance of feeding and nesting areas of birds.

Rodgers and Schwikert (2002) compared the flush distance of a number of bird species in response to PWCs and outboard-powered boats. Flush distances ranged from 19.5m (Least Tern, *Sterna antillarum*) to 49.5m (Osprey, *Pandion halietus*) for PWCs while outboard-powered boats flushed birds at greater distances ranging from 23.4m (Forester's Tern, *S. forsteri*) to 57.9m (Osprey). Burger (1998), however, found that PWCs actually caused significantly more flight behaviour in common terns (*S. hirundo*) than motorboats. She hypothesized that the large vertical and horizontal spray of PWCs may make them appear larger and created the greater flush distances than more traditional outboard motorboats. Nowacek *et al.* (2001) found that PWCs caused behavioural changes in bottlenose dolphins (*Tursiops truncates*) more often than outboard boats. In Ireland in 2001 when three Orcas

(*Orcinus orca*) spent some weeks of the summer in Cork Harbour there were initial problems with jet ski riders harassing the whales by riding very close to them until the Naval Service intervened. In California sealion and elephant seal colonies have been repeatedly disturbed, with disruption of rest patterns and occurrences of stampedes that separate pups from mothers. The erratic operations of PWCs and the ability to intrude into shallower water where the marine mammals can escape normal boat traffic was cited as the cause for their disturbance.

The smooth hulls of jet skis even allow them to be driven up and over low banks in lakes, rivers and salt marshes, damaging vegetation and disturbing or killing ground-nesting birds (Burger 1998). Additionally, anecdotal reports suggest that PWCs use resulted in nest abandonment of common loons (*Gavia immer*) in Montana (Hamann *et al.* 1999), and disturbance of brown pelicans (*Pelecanus occidentalis*) and osprey in Florida, and swans (*Cygnus* spp.) in Alaska (Jenkins 2002). Further accounts have seen PWCs separate harbour seal (*Phoca vitulina*) pups from mothers, stampede seals and sea lions (*Zalophus californianus*), and harass sea otters (*Enhydra lutris*; Jenkins 2002). PWCs and motorboats also create waves that cause shoreline erosion and loss of marsh habitat (Dayton & Levin 1996). These wakes can affect semi-aquatic mammal habitat by swamping and eroding muskrat (*Ondatra zibethicus*) canals and houses, beaver (*Castor canadensis*) caches and lodges, and river otter (*Lutra canadensis*) latrine sites (Waller *et al.* 1999).

3.3 AIR TRANSPORT

3.3.1 *Hang gliding and Paragliding*

Sometimes grouped together as 'foot launched flight', hang gliding and paragliding are relatively young sports, having developed in the early 1970s and late 1980s respectively. Their ecological impact appears limited and takes two forms. Take off areas ('launches'), usually on hill summits or cliff tops, tend to become devoid of vegetation and physically eroded by repeated running by the participants as they take off. There appears to be little difference from the impact of walkers in this respect – the summits of hills accessible to hikers are often denuded in similar fashion.

The second impact that hang gliders and paragliders have stems from their size, shape and flight style. Visually they resemble large raptors – indeed many participants regularly fly with vultures and eagles, which appear not to be disturbed by them. There is some (mainly anecdotal) evidence that they displace small and medium sized mammals from areas regularly used by the sports.

3.3.2 *Kitesurfing, Kitejumping and Kitebuggying*

In the late 1990s even newer forms of 'extreme sports' based on parafoil kites arose very rapidly. Large kites, some 5-8m^2 in area and flown at heights of around 30m are used to tow participants, either at sea on small boards rather like miniature sailboards (kitesurfing), or on land-towing vehicles similar to sand yachts (kitebuggying). Some participants dispense with boards or vehicles and fly in a series of giant hops (kitejumping), while attempts have also been made to combine the activity with skiing. The sport demands great skill and experts demonstrate spectacular displays. Typically these activities take place on or near flat beaches and sand dune systems. Because the large, highly mobile and colourful kites are visible from a great distance, they are far more obtrusive than sailboards or sand yachts. There are also sudden loud noises associated with kites hitting the water. Already there is ample evidence of a conflict between these kite users and birds, particularly wading birds, terns and gulls. Kitesurfers disturb the nearshore areas where terns and other birds feed on shoals of sand eels. They and kite buggies also interfere with feeding by waders at the water's edge, and with feeding trips to nests in the dunes. In the Dee Estuary of the United Kingdom, severe disturbance of oystercatchers and redshank have already been detected at West Kirby, which is one of the most important areas in Western Europe for wetland birds, even though relatively few kite surfers were involved (website 8).

3.3.3 *Private aircraft*

Fixed-wing private aircraft (including microlight aircraft) and privately owned helicopters are still relatively few in number and their use is heavily circumscribed by national and international legislation. Most concerns are similar to those that apply to commercial aircraft and are related particularly to noise pollution. In wilderness or national park areas aviation noise can sometimes be extreme (e.g. in Hawaii). Since the noise is sufficient to annoy humans, it is likely that it has an impact on wildlife too.

4. Conclusions

This review reveals common trends for both mass tourist transport and individual leisure transport:

- Leisure transport and the technologies to support it provide substantial employment and economic activity. Once a community, area or industry becomes dependent on leisure transport, powerful vested interests oppose change or regulation; such vested interests can usually rely on the support of the activity's participants to back their opposition. Participants, in turn, cite individual freedom, often in association with a long history of the activity (albeit usually at much lower intensity), to avoid regulation.

- Many of the ecological problems associated with leisure transport of all types are created by excessive numbers of participants combined with open access. This situation has been created by rising prosperity (generating demand) and perceptions of 'rights' to automatic access to countryside and wildlands areas, both terrestrial and aquatic. Superimposed upon these are two further factors, a quest for adventure and the technological development of transport technologies providing that excitement.

- Serious, sometimes irreversible, ecological damage often takes place before control measures are put in place. The power of vested interests is often such that a brake is not put on activities until tourists themselves recognize that their environment is degraded to an unacceptable level. Their response, and that of developers, is often to widen the area affected (i.e. to enter more pristine neighbouring habitats), or simply to transfer activities to 'more attractive' areas elsewhere.

- Individual leisure transport has become increasingly mechanised and associated with high speed. Particularly worrying has been the explosive emergence of two technologies (snowmobiles and jet skis) that are highly polluting, very rapid and extremely noisy. Noise pollution is already a severe human problem in urban areas. Newer, vehicle-based leisure transport technologies such as ORVs and jet skis transfer noise stress to wildlands ecosystems – with effects that have been poorly studied. A problem with all of the newer technologies has been that they have not been evaluated for environmental impact before widespread promotion in the marketplace, nor have they automatically been subject to the same regulations as related existing technology.

- *A priori*, leisure transport should take place within a sustainable framework that minimises ecological deterioration. At the very least, it should be possible to limit damage to fairly small areas, and the damage should ideally be repairable over relatively short periods. This requires control over access. Where environments are particularly vulnerable or valuable, access may have to be prohibited at least to sanctuary areas that are large enough to be biologically meaningful and sustainable.

Acknowledgments

JD acknowledges the help of Julia L. Davenport in literature searching. He also thanks the Higher Education Authority of Ireland for financial support in the form of a PRTLI3 grant for research scholarship under the heading *'Ecology of Transport'*. TAS thanks Tom Peterson and Chad Dear for reviews of the manuscript.

References

Adams ES. 1975; Effects of lead and hydrocarbons from snowmobile exhaust on brook trout (*Salvelinus frontinalis*). Transactions of the American Fisheries Society 2: 363-373.
Ali MM, Murphy KJ, Langendorff J. 1999; Interrelationships of river ship traffic with aquatic plants in the River Nile, Upper Egypt. Hydrobiologia 415: 93-100.
Alison WR. 1996; Snorkeller damage to reef corals in the Maldive Islands. Coral Reefs 15(4): 215-218.
Backhurst MK, Cole RG. 2000; Biological impacts of boating at Kawau Island north-eastern New Zealand. Journal of Environmental Management. 60: 239-251.
Badalamenti F, Ramos AA, Voultsiadou E, Lizaso LJS, D'Anna G, Pipitone C, Mas J, Fernandez JAR, Whitmarsh D, Riggio S. 2000; Cultural and socio-economic impacts of Mediterranean marine protected areas. Environmental Conservation 27 (2): 110-125.
Bell DV, Odin N, Austin A, Hayhow S, Jones A, Strong A, Torres E. 1984; The impact of anglers on wildlife and site amenity. Report by the Department of Applied Biology UWIST Cardiff.
Belnap J. 2003; The world at your feet: desert biological soil crusts. Frontiers in Ecology and the Environment 1(5): 181-189.
Bishop GA, Morris JA, Stedman DH. 2001; Snowmobile contributions to mobile source emissions in Yellowstone National Park. Environmental Science and Technology 35(14): 2874-2881.
Bjornlie DD, Garrott RA. 2000; Ecological effects of groomed roads on bison in Yellowstone National Park. Abstract and presentation The Wildlife Society 7th Annual Conference Nashville Texas.
Blackburn J, Davis MB. 1994; Off-road vehicles: fun or folly. Livingston Kentucky ASPI Publications 56p.
Bollinger JG, Ronstad OJ, Soom A, Larson T. 1972; Snowmobile noise effects on wildlife. Final Report Engineering Experiment Station University of Wisconsin Madison Wisconsin.
Bosworth D. 2003; Managing the National Forests System: Great Issues and Great Diversions. Speech presented to the San Francisco Commonwealth Club and Berkeley University on Earth Day April 22.
Boucher J, Tattar T. 1975; Snowmobile impact on vegetation. Forest Notes 121: 27-28.
Boyd DK, Pletscher DH. 1999; Characteristics of dispersal in a colonizing wolf population in the central Rocky Mountains. Journal of Wildlife Management 63: 1094.
Brander RB. 1974; Outdoor recreation research: applying the results: ecological impacts of off-road recreation vehicles. North Central Forest Experiment Station USDA Forest Service St Paul MN General Technical Report NC-9.
Brattstrom BH, Bondello MC. 1979; The effects of dune buggy sounds on the telencephalic auditory evoke response in the Mojave fringe-toed lizard *Uma scoparia*. Unpublished report to the US Bureau of Land Management California Desert Program Riverside California.
Brosnan DM, Crumrine LL. 1994; Effects of human trampling on marine rocky shore communities Journal of Experimental Marine Biology and Ecology 177: 79-97.
Brown AC, McLachlan A. 2002; Sandy shore ecosystems and threats facing them: some predictions for the year 2025. Environmental Conservation 29(1): 62-77.
Bryan RB. 1977; The influence of soil properties on degradation of mountain hiking trail at Grovelsjon Geografiska. Annaler 59A: 49-65.
Buckingham CA, Lefebvre LW, Schaefer JM, Kochman HI. 1999; Manatee response to boating activity in a thermal refuge. Wildlife Society Bulletin 27 (2): 514-522.

Bunnell KD, Flinders JT, Wolfe ML, Bissonette JA. 2004; Quantifying the impacts of coyotes and snowmobiles on lynx conservation in Utah and the West. The 84th American Society of Mammalogy Annual Meeting Abstracts .

Burde JH, Renfro JR. 1986; Use impacts on the Appalachian Trail Pages 138-143 in RC Lucas editor Proceedings: National Wilderness Research Conference: Current Research USDA Forest Service Intermountain Research Station General Technical Report INT-212 Ogden Utah USA.

Burge BL. 1983; Impact of Frontier 500 off-road vehicle race on desert tortoise habitat. Proceedings of the 1983 Annual Symposium of the Desert Tortoise Council pp. 27-38.

Burger J. 1998; Effects of motorboats and personal water-craft on flight behaviour over a colony of Common Terns. The Condor 100 (3): 528-534.

Bury RB. 1980; What we know and do not know about off-road vehicle impacts on wildlife. In: Andrews RNL & Nowak P (eds) Off-road vehicle use: A management challenge (University of Michigan Extension Service) Michigan League The University of Michigan School of Natural Resources USDA The Office of Environmental Quality.

Canfield JE, Lyon LJ, Hillis JM, Thompson MJ. 1999; In: Joslin G & Youmans H (coordinators) Effects of recreation on Rocky Mountain wildlife: a review for Montana, 61-625. Committee on Effects of recreation on Wildlife Montana Chapter of The Wildlife Society.

CEPA 1999; California Environmental Protection Agency 1999; Fact Sheet – New Regulations for Gasoline Engines http://www.arb.ca.gov/msprog/marine/marine.htm (accessed 19-01-2005).

Cessford GR. 1995; Off-road Impacts of mountain bikes. Science & Research Series No 92 Wellington New Zealand: Department of Conservation.

Cheng I-J. 1995; Tourism and the green turtle in conflict on Wan-An Island Taiwan. Marine Turtle Newsletter 68: 4-6.

Claar JJ, Anderson N, Boyd D, Cherry M, Conrad B, Hompesch R, Miller S, Olson G, Ihsle Pac H, Waller J, Wittinger T, Youmans H. 1999; Carnivores, 71-763. In: Joslin G & Youmans H. (coordinators) Effects of recreation on Rocky Mountain wildlife: A Review for Montana Committee on Effects of Recreation on Wildlife Montana Chapter of the Wildlife Society.

Cole DN. 1983; Campsite conditions in the Bob Marshall Wilderness Montana. US Department of Agriculture Forestry Service Intermountain Forest and Range Experiment Station Research Paper INT-312 Ogden UT.

Cole DN, Spildie DR. 1998; Hiker horse and llama trampling effects on native vegetation in Montana. USA Journal of Environmental Management 53(1): 61-71.

Constantine R. 2001; Increased avoidance of swimmers by wild bottlenose dolphins (Tursiops truncatus) due to long-term exposure to swim-with-dolphin tourism. Marine Mammal Science 17: 689-702.

Copeland JP. 1996; Biology of the wolverines in Central Idaho. Thesis, University of Idaho, Moscow, Idaho, USA.

Cordell HK, Betz CJ, Green GT, Mou S, Leeworthy VR, Wiley PC, Barry JJ, Hellerstein D. 2004; Outdoor Recreation for the 21st Century America, A Report to the Nation: The National Survey on Recreation and the Environment. Venture Publishing, Inc., State College, PA.

Creel S, Fox JE, Hardy AR, Sands J, Garrot B, Peterson RO. 2002; Snowmobile activity and glucocorticoid stress responses in wolves and elk Conservation Biology 16(3): 809-14.

Dale D, Weaver T. 1974; Trampling effects on vegetation of the trail corridors of North Rocky Mountain forests. Journal of Applied Ecology 11: 767-772.

Davenport J. 1985; Environmental Stress and Behavioural Adaptation. London: Croom Helm.

Davenport J, Hills J, Glasspool AF, Ward J. 2001; Threats to the endangered endemic Bermudian skink Eumeces longirostris. Oryx 35(4): 332-339.

Dayton PK, Levin LA. 1996; Erosion of cordgrass at Kendall-Frost Mission Bay Reserve. California Sea Grant. Biennial Report of Completed Projects 1992-1994. California Sea Grant College Program, University of California. La Jolla, California.

Fancy SG, White RG. 1985; Energy expenditures by caribou while cratering in snow. Journal of Wildlife Management 49(4): 987-993.

Fialka J. 1975; Running Wild. National Wildlife 13(2): 36-40.

Freddy DJ, Bronaugh WM, Fowler MC. 1986; Responses of mule deer to disturbance by persons afoot and snowmobiles. Wildlife Society Bulletin 14 (1): 63-68.

Fyfe RW, Olendorff RR. 1976; Minimizing the dangers of nesting studies to raptors and other sensitive species. Canadian Wildlife Service Information, Canada Catalogue Number CW69-1/23 Ottawa Canada.

Gaines WL, Singleton PH, Ross RC. 2003; Assessing the cumulative effects of linear recreation routes on wildlife habitats on the Okanogan and Wenatchee National Forests. General Technical Report PNW-GTR-586 Portland OR: US Department of Agriculture Forest Service, Pacific Northwest Research Station. http://www.fs.fed.us/pnw/pubs/gtr586.pdf (accessed 19-01-2005).

Gelbard JL, Harrison S. 2003; Roadless habitats as refuges for native grasslands: interactions with soil, aspect, and grazing. Ecological Applications 13(2): 404-415.

Gilbert BK. 2003; Motorized access on Montana's Rocky Mountain Front: a synthesis of scientific literature and recommendations for use in revision of the travel plan for the Rocky Mountain Division. The Coalition for the Protection of the Rocky Mountain Front. http://www.wildmontana.org/gilbertreport.pdf (accessed 19-01-2005).

Gilbertson DD. 1981; The impact of past and present land use on a major coastal barrier system. Applied Geography 1: 97-119. (Figure 8 reprinted with permission from Elsevier, copyright 1981).

Godfry PJ, Leatherman SP, Buckley PA. 1978; Impact of off-road vehicles on coastal ecosystems. Coastal Zone 2: 581-600.

Gutzwiller KJ, Wiedenmann RT, Clements KL, Anderson SH. 1994; Effects of human intrusion on song occurrence and singing consistency in subalpine birds. Auk 111: 28-37.

Hamann B, Johnston H, McClelland P, Johnson S, Kelly L, Gobielle J. 1999; Birds. Pages 31-334 in Joslin G & Youmans H (coordinators) Effects of recreation on Rocky Mountain wildlife: A Review for Montana Committee on Effects of Recreation on Wildlife. Montana Chapter of the Wildlife Society 307p.

Harris J. 1993; Horse riding impacts in Victoria's Alpine National Park. Australian Ranger 27: 14-16.

Havlick DG. 2002; No Place Distant: Roads and Motorized Recreation on America's Public Lands. Foreword by Mike Dombeck. Island Press Washington DC 297p.

Hawkins JP, Roberts CM, Van't Hof T, de Meyer K, Tratalos J, Aldham C. 1999; Effects of recreational scuba diving on Caribbean coral and fish communities. Conservation Biology 13(4): 888-897.

Hilderbrand GV, Lewis LL, Larrivee J, Farley SD. 2000; A denning brown bear *Ursus arctos* sow and two cubs killed in an avalanche on the Kenai Peninsula Alaska. Canadian Field-Naturalist 114(3): 498.

Hooper RG. 1977; Nesting habitat of common Ravens in Virginia. Wilson Bulletin 89: 233-242.

Hulsey B, Hopkins E, Olson E, Burg E, Carlson M. 2004; Highway Health Hazards. Published by the Sierra Club San Francisco, CA. http://trborg/news/blurb_detailasp?id=4033 (accessed 19-01-2005).

Hunt M. 1999; Management of the environmental noise effects associated with sightseeing aircraft in the Milford Sound area New Zealand. Noise Control Engineering Journal 47(4): 133-141.

Jenkins D. 2002; Hostile Waters – The impacts of personal watercraft use on waterway recreation. American Canoe Association 52p http://bluewaternetwork.org/reports/rep_pwc_ACAreport.pdf (accessed 19-01-2005).

Johnson ML. 1998; Ecological risk of MTBE in surface waters. John Muir Institute of the Environment, University of California, Davis, California. http://tsrtp.ucdavis.edu/mtberpt/homepage.html.

Joslin G, Youmans H (coordinators). 1999; Effects of recreation on Rocky Mountain wildlife: A review for Montana Committee of Effects of Recreation on Wildlife Montana Chapter of the Wildlife Society 307p. http://www.montanatws.org/pages/page4a.html (accessed 19-01-2005).

Jüttner F, Backhaus D, Matthias U, Essers U, Greiner R, Mahr B. 1995; Emissions of two- and four-stroke outboard engines-II. Impact on water quality. Water Resources 29(8): 1983-1987.

Keddy PA, Spavold AJ, Keddy CJ. 1979; Snowmobile Impact on Old Field and March Vegetation in Nova Scotia Canada: An Experimental Study. Environmental Management 3(5): 409-415.

Keenan HT, Bratton SL. 2004; All-terrain vehicle legislation for children: a comparison with and without a helmet law. Pediatrics 113(4): 330-334.

Klugpumpel B, Krampitz C. 1996; Conservation in alpine ecosystems: the plant cover of ski runs reflects natural as well as anthropogenic environmental factors. Bodenkultur 47(2): 97-117.

Kopischke EB. 1972; Effects of snowmobile activity on the distribution of white-tailed deer in South-Central Minnesota. Minnesota Department of Natural Resources, Game Research Project Quarterly Report. 32(3): 139-142.

Kuniyal JC. 2002; Mountain expeditions: minimising the impact. Environmental Impact Assessment Review. 22(6): 561-581.

Leininger WC, Payne GF. 1979; The effects of off-road vehicle travel on rangeland in southwestern Montana. Agricultural Experiment Station, Montana State University. Research Report 153. 47p.

Liddiard M, Gladwin DJ, Wege DC, Nelson-Smith A. 1989; Impact of boulder-turning on sheltered sea shores. Report to the Nature Conservancy Council. School of Biological Sciences, University College of Swansea. NCC CSD Report 919.

Liddle MJ, Greig-Smith P. 1975; A survey of tracks and paths in a sand dune ecosystem. Journal of Applied Ecology 12: 909-930.

Linnell JDC, Swenson JE, Andersen R, Brain B. 2000; How vulnerable are denning bears to disturbance? Wildlife Society Bulletin 28(2): 400-413.

Lovich JE, Bainbridge D. 1999; Anthropogenic degradation of the southern California desert ecosystem and prospects for natural recovery and restoration. Environmental Management 24: 309-326.

Mace RD, Waller JS, Manley TL, Lyon LJ, Zuuring H. 1996; Relationships among grizzly bears roads and habitat in the Swan Mountains. Montana Journal of Applied Ecology 33: 1395-1404.

Marion JL. 1994; An assessment of trail conditions in Great Smoky Mountains National Park. US Department of the Interior National Park Service Southeast Region Research/Resources Management Report Atlanta GA.

McLaren MA, Green JE. 1985; The reactions of muskoxen to snowmobile harassment. Arctic 38(3): 188-193.

Misak RF, Al Awadhi JM, Omar SA, Shahid SA. 2002; Soil degradation in Kabad area, southwestern Kuwait City. Land Degradation & Development 13(5): 403-415.

Moen AN, Whittemore S, Buxton B. 1982; Effects of disturbance by snowmobiles on heart rate of captive white-tailed deer. New York Fish and Game Journal 29(2): 176-183.

Montana State University Extension Service Bulletin. 1992; Controlling knapweed on Montana rangeland. Circular 311, February 1992.

Munger JC, Barnett BR, Novak SJ, Ames AA. 2003; Impacts of off-highway motorized vehicle trails on the reptiles and vegetation of the Owyhee Front. Idaho Bureau of Land Management Technical Bulletin 03-3: 1-23.

Nakata JK, Wilshire HG, Barnes GG. 1976; Origin of Mojave Desert dust plumes photographed from space. Geology 4(11): 644-648.

Nash RF, Gallup jr. GG, McClure MK. 1970; The immobility reaction in leopard frogs (*Rana pipiens*) as a function of noise induced fear. Psychonometric Science 21(3): 155-156.

Neumann , Merriam. 1972; The ecological effects of snowmobiling. Canadian Field Naturalist 86: 207-212.

Nicola NC, Lovich JE. 2000; Preliminary observations of the behavior of male flat-tailed horned lizards before and after an off-highway vehicle race in California. California Fish and Game 86(3): 208-212.

Nowacek SM, Wells RS, Solow AR. 2001; Short term effects of boat traffic on bottlenose dolphins *Tursiops truncatus* in Saratosa Bay Florida. Marine Mammal Science 17(4): 673-688.

Patin SA. 1997; Ecological Problems of Oil and Gas Resource Developments on Marine Shelf Moscow: VNIRO 1997 350 p (In Russian).

Richens VB, Lavigne GR. 1978; Response of white-tailed deer to snowmobiles and snowmobile trails in Maine. The Canadian Field Naturalist 92: 334-344.

Rodgers JA, Schwikert ST. 2002; Buffer-zone distances to protect foraging and loafing waterbirds from disturbance by personal watercraft and outboard-powered boats. Conservation Biology 16(1): 216-224.

Rongstad OJ. 1980; Research needs on environmental impacts of snowmobiles. In: Andrews RNL & Nowak P (eds) Off-road vehicle use: A management challenge. US Department of Agriculture Office of Environmental Quality. Washington, DC.

Rooney TP. In Prep; Off-road vehicles as dispersal agents for exotic plant species in a forested landscape. Submitted to Environmental Management.

Rosen PC, Lowe CH. 1994; Highway mortality of snakes in the Sonoran desert of southern Arizona. Biological Conservation 68:143-148.

Ruth-Balaganskaya E, Myllynen-Malinen K. 2000; Soil nutrient status and revegetation practices of downhill skiing areas in Finnish Lapland – a case study of Mt Ylläs. Landscape & Urban Planning 50(4): 259-268.

Sack D, daLuz S. 2003; Sediment flux and compaction trends on off-road vehicle (ORV) and other trails in an Appalachian forest setting. Physical Geography 24(6): 536-554.

Schubert and Associates. 1999; Petition to enhance and expand regulations governing the administration of recreational off-road vehicle use on National Forests. Published by Wildlands CPR, Missoula, MT 188p http://www. wildlandscpr.org/orvs/ORVpetition.doc.

Sive B, Shively D, Pape B. 2003; Spatial variation of volatile organic compounds associated with snowmobile emissions in Yellowstone National Park. A research report submitted to the National Park Service, United States Department of the Interior. 85p.http://www.nps.gov/yell/technical/planning/winteruse/plan/sive_report.htm.

Snyder CT, Frickel DG, Hadley RE, Miller RF. 1976; Effects of off-road vehicle use on the hydrology and landscape of arid environments in central and southern California. US Geological Survey Water-Resources Investigations 76-99. 45p.

Sojda jr. R. 1978; Effects of snowmobile activity on wintering pheasants and wetland vegetation in northern Iowa marshes. Thesis. Iowa State University, Ames, IA, USA.

Stebbins RC. 1974; Off-Road Vehicles and the Fragile Desert. The American Biology Teacher Part 1 36(4): 203-208

Stensvold MC. 2000; The conservation status of Ophioglossaceae in southern Alaska. Proceedings of Botany 2000. August 6-10. Portland OR.

Stokes JB, Leese DJ, Montgomery SL. 1999; Citizens get relief from recreational noise: the case in the skies from Hawaii. Noise Control Engineering Journal 47(4): 142-146.

Stokowski PA, LaPointe CB. 2000; Environmental and social effects of ATVs and ORVs: an annotated bibliography and research assessment. School of Natural Resources, University of Vermont. 31p. http://www.anr.state.vt.us/anr/atv_nov20_final.pdf.

Stout III BM. 1992; Impact of off-road vehicle use on vegetative communities of northern Canaan Valley, West Virginia. Final Report of The Canaan Valley Task Force. Wheeling Jesuit College, Department of Biology, Wheeling, WV.

Strauss EG. 1990; Reproductive success, life history patterns, and behavioral variation in a population of piping plovers subjected to human disturbance. Dissertation. Tufts University.

Sun D, Walsh D. 1998; Review of studies on environmental impacts of recreation and tourism in Australia. Journal of Environmental Management 53 (4): 323-338.

Sutter PS. 2002; Driven Wild: How the Fight Against Automobiles Launched the Modern Wilderness Movement. Foreword by William Cronon. Weyerhaeuser Environmental Books series. Seattle: University of Washington Press. 343p.

Swiss Federal Statistical Office (SFSO)/ Swiss Agency for the Environment, Forests and Landscape (SAEFL.) 1997; The Environment in Switzerland 1997: Facts, Figures, Perspectives. Neuchâtel: Berne.

Thurston E, Reader RJ. 2001; Impacts of experimentally applied mountain biking and hiking on vegetation and soil of a deciduous forest. Environmental Management 27(3): 397-409.

Tjärnlund U, Ericson G, Lindesjöö E, Petterson I, Åkerman G, Balk L. 1996; Further Studies of the Effects of Exhaust from Two- Stroke Outboard Motors on Fish. Marine Environmental Research 42: 267-271.

Tjärnlund U, Ericson G, Lindesjöö E, Petterson I, Balk L. 1995; Investigation of the Biological Effects of 2 -Cycle Outboard Engines' Exhaust on Fish. Marine Environmental Research 39: 313-316.

Tratalos JA, Austin TJ. 2001; Impacts of recreational SCUBA diving on coral communities of the Caribbean island of Grand Cayman. Biological Conservation 102(1): 67-75.

United Nations Environment Programme (UNEP). 2001; Environmental Impacts of Tourism URL. http://www.unepie.org/ps/tourism/sust-tourism/environment.htm (accessed 18-2-2003).

United States Environmental Protection Agency. 2000; Cruise Ship White Paper.

US Department of the Interior (USDOI). 1994; Report to Congress: Report on the effects of aircraft overflights on the National Park System. National Park Service. http://www.nonoise.org/library/npreport/intro.htm (accessed 19-01-2005).

US Department of the Interior (USDOI). 1999; Water quality concerns related to personal watercraft usage. Draft internal literature review. Water Resources Division, National Park Service, Fort Collins, Colorado. 11pp.

US Department of the Interior (USDOI). 2000; Air quality concerns related to snowmobile usage. Yellowstone National Park: Air Resources Division, National Park Service. 32p.

US Department of the Interior (USDOI). 2002; The draft Environmental Impact Statement for Lake Mead National Recreation Area/Lake management plan. National Park Service, Pacific West Region.

US Department of the Interior (USDOI). 2003; Winter use plan for the Yellowstone and Grand Teton National Parks and John D. Rockefeller, Jr. Memorial Parkway, Idaho, Montana, and Wyoming (Final Supplement to the EIS). http://www.nps.gov/grte/winteruse/fseis/intro.htm.

Van Paris SM, Corkeron P. 2001; Boat traffic affects the acoustic behaviour of Pacific humpback dolphins Sousa chinensis. Journal of the Marine Biological Association of the United Kingdom 81: 533-538.

Vieira MEP. 2000; Effects of Early Season Hunter Density and Human Disturbance on Elk Movement in the White River Area Colorado. MS Thesis. Fort Collins CO: Colorado State University.

Vollmer AT, Maza BG, Medica PA, Turner FB, Bamberg SA. 1976; The impact of off-road vehicles on a desert ecosystem. Environmental Management 1(2): 115-129.

Waller AJ, Sime CA, Bissell GN, Dixon B. 1999; Semi-aquatic Mammals. Pages 51-525 in Joslin G & Youmans H (coordinators) Effects of recreation on Rocky Mountain wildlife: A Review for Montana Committee on Effects of Recreation on Wildlife. Montana Chapter of the Wildlife Society 307p.

Wanek WJ. 1971; Observations on snowmobile impact. The Minnesota Volunteer 34(109): 1-9.

Wanek WJ, Potter. 1974; A continuing study of the ecological impact of snowmobiling in northern Minnesota. Final Research Report for 1973-74. The Center for Environmental Studies, Bemidji State College, Bemidji, MN. 54p.

Wanek WJ, Schumacher LH. 1975; A continuing study of the ecological impact of snowmobiling in northern Minnesota. Final Research Report for 1974-75. The Center for Environmental Studies, Bemidji State College, Bemidji, MN. 34p.

Ward JAD. 1999; Bermuda's coastal seagrass beds as habitat for fish. MSc Thesis. University of Glasgow, UK.

Watson JJ, Kerley GIH, McLachan A. 1996; Human activity and potential impacts on dune breeding birds in the Alexandria Coastal Dunefield. Landscape and Urban Planning 34: 315-322.

Weaver J. 1993; Lynx, wolverine and fisher in the western United States: Research assessment and agenda. USDA Forest Service Intermountain Research Station, Contract Number 43-0353-2-0598, Missoula, MT.

Welch DM, Churchill J. 1986; Hiking trail conditions in Pangnirtung Pass, 1984, Baffin Island, Canada. Parks Canada Report, Ottawa, Canada.

Whinam J, Comfort M. 1996; The impact of commercial horse riding on sub-alpine environments at Cradle Mountain, Tasmania, Australia. Journal of Environmental Management 47 (1): 61-70.

Wiese PV. 1996; Environmental Impact of Urban and Industrial Development A Case History: Cancún Quintana Roo, Mexico (http://www.unesco.org/csi/wisc/cancun1.htm) (accessed 12-10-2004).

Wilkins KT. 1982; Highways as barriers to rodent dispersal. Southwest Naturalist 27(4): 459-460.

Willard BE, Marr JW. 1970; Effects of human activities on alpine tundra ecosystems in Rocky Mountain National Park Colorado. Biological Conservation 2: 257-265.

Wilshire HG, Nakata JK. 1976; Off-road vehicle effects on California's Mojave Desert. California Geology 29(6): 123-132.

Wisdom MJ, Preisler HK, Cimon NJ, Johnson BK. 2004; Effects of off-road recreation on mule deer and elk. Transactions of the North American Wildlife and Natural Resource Conference 69. http://bluewaternetwork.org/reports/rep_atv_forestservice.pdf (accessed 19-01-2005).

Woodford MH, Butynski TM, Karesh WB. 2002; Habituating the great apes: the disease risks. Oryx, 36; 153-160.

Woodland Trust. 2002; Urban woodland management guide. I: Damage and misuse. (http://www.woodland-trust.org.uk/policy/publications.htm). (accessed 19-01-2005).

Yorio P, Frere E, Gandini P, Schiavini A. 2001; Tourism and recreation at seabird breeding sites in Patagonia, Argentina: current concerns and future prospects. Bird Conservation International 11: 231-245.

Youmans H. 1999; Project overview. Pages 11-118 in Joslin G & Youmans H (coordinators) Effects of recreation on Rocky Mountain wildlife: A Review for Montana. Committee on Effects of Recreation on Wildlife. Montana Chapter of the Wildlife Society. 307p.

Zakai D, Chadwick-Furman NE. 2002; Impacts of intensive recreational diving on reef corals at Eilat, northern Red Sea. Biological Conservation 105 (2): 179-187.

Websites
1.   http://www.bioregional.com/programme_projects/opl_prog/portugal/portugal_hmpge.htm (accessed
     05-05-2005)
2.   http://www.defra.gov.uk/wildlife-countryside/access/appraise/13.htm (accessed 05-05-2005)
3.   http://www.wildlifeadvocacy.org/programs/panthercypress.htm (accessed 03-02-2005)
4.   http://www.naturaltrails.org/issues/factsheets/fs_wildlife.html (accessed 07-02-2003)
5.   http://www.acousticecology.org/wildlandvehicles.html (accessed 28-01-2005)
6.   http://www.iwla.org/reports/pwcrep.pdf (accessed 28-01-2005)
7.   http://www.naturaltrails.org/vroomreport/1_28_03.html (accessed 07-02-2003)
8.   http://www.deeestuary.co.uk/decgks.htm (accessed 11-01-2005)

CHAPTER 15: CONTAMINANTS AND POLLUTANTS

DAGMAR B. STENGEL[1], SARAH O'REILLY[2] & JOHN O'HALLORAN[3]
[1]Department of Botany, Martin Ryan Institute and Environmental Change Institute, National University of Ireland, Galway
[2]Biochemical Toxicology Laboratory, Department of Biochemistry, University College Cork
[3]Department of Zoology, Ecology and Plant Science, University College Cork, Cork, Ireland

1. Introduction

The desire by human kind for increased mobility and access to goods and services, recreation and exploration has led to an explosion in the number and extent of movements. The world has become a small place; easily accessible transport can reach all parts of the world in a few hours. Enlarged economic and social networks have also led to an increased demand in transport systems. However, this increase in transportation has led to a significant increase in infrastructure related to, and discharges from, the different modes of transport.

Aviation now plays a very important role in the world's transportation networks. In spite of the predictions of a collapse in the air transport network following the September 11 2001 terrorist attack on the Twin Towers, New York, some 4 years later the number of civilian and military aircraft movements has increased. It is estimated that during 2001-2002 British Airways had nearly 1,500 flights daily (Somerville 2003). It is believed that the growth in air transportation will increase by at least 5% per year, though increased fuel costs and congestion may lead to some levelling off in this growth rate. The world's airlines currently carry about 1.6 billion people and 30 million tonnes of freight each year. Whitelegg and Cambridge (2004) estimate that the number of kilometres flown is expected to triple and aircraft numbers double by 2025. Shipping activity worldwide has also increased significantly over the past 15 years. Over 2,000 major ports worldwide handle 80% of trade with origins or destinations in developing countries (Bailey & Solomon 2004), freight carried by ship in the US doubled between 1990 and 2001 and is expected to further increase in the next few years.

Transportation, along with most other human activities, is dependent on either the manufacture of compounds or the exploitation of natural resources. During the mining and processing of fuels and/or the manufacturing of transport vehicles there will be waste, discharges, and emissions. Fossil fuel consumption (in both the manufacture and use of transport vehicles) and subsequent emissions has led to changes in air-quality with impact on human health, agriculture, aquaculture, sensitive organisms, ecosystem function and global climate.

This chapter sets out to explore the sources of pollution and contaminants that result from transportation, their effects, and possible remediation steps. Because of the breadth of this topic, this chapter is intended to be introductory in nature and further reading and references are listed to guide readers to the more specialist literature.

'Contaminants' are defined as inputs of alien and potentially toxic substances into the environment; not all contaminants cause pollution, as their concentrations may be too low. 'Pollutants' are defined as anthropogenically-introduced substances that have harmful effects on the environment. Sometimes the distinction between contaminants and pollutants is not simple, since concentrations at which contaminants become pollutants cannot always be defined; also long-term damage to organisms or systems may occur that is not evident initially.

John Davenport and Julia L. Davenport, (eds.), The Ecology of Transportation: Managing Mobility for the Environment, 361–389,

2. Sources of contaminants and pollutants from transport

2.1 EMISSIONS FROM THE COMBUSTION OF FOSSIL FUELS

Fossil fuels are hydrocarbons that, if they were pure and burnt in a plentiful supply of oxygen, would produce carbon dioxide and water. However, naturally occurring pollutants and artificial additives are also present in fuels and these, combined with incomplete combustion (due to inefficient engines), cause other pollutants and contaminants to be released into the atmosphere. Transport systems that use a variety of fuels for combustion, cause the emission of a wide range of compounds. Some of these compounds are likely carcinogens, or will influence the health of humans, flora and fauna through impacts on cardiovascular, photosynthetic, or reproductive processes. The United States Environmental Protection Agency (USEPA) has identified 21 atmospheric pollutants specifically associated with on-road and non-road mobile sources (US Government 2001).

2.1.1 Railways
As the emission of atmospheric pollutants is considered one of the major problems associated with traffic and air transport, railways are sometimes considered the preferred option when planning transport, both for transport of freight and passengers.
Historically, high emissions of CO, CO_2, NO_x and SO_x occurred from coal and steam engines that have been mostly abandoned, except in China and some parts of Eastern Europe (Carpenter 1994). Today, compared with road traffic, emission of air pollutants by railways is generally low. However, a large proportion of trains are powered by diesel engines which emit CO, CO_2, NO_x, SO_x, and particulates, and thus have potentially significant environmental impacts. Emissions of atmospheric pollutants from old diesel engines are high, while modern diesel engines are equipped with catalytic converters, similar to diesel road vehicles. Discharges from diesel locomotives are low when calculated on a per-person-per-kilometre or a per-tonne-per-kilometre basis and compared with other modes of land transports, except for large trucks and buses (for detailed review see Carpenter 1994). For example, in 1995/6, 0.9% of the total emissions of CO_2 in the UK came from rail traffic, compared with 17.1% from road traffic, 3.3% from air traffic and 0.5% from shipping (Colvile *et al.* 2001).
In many countries, most trains are now electric. The environmental impacts of electric engines depend on the energy source used for electricity production. This may be oil or coal, natural gas, hydroelectric, wind powered or nuclear power stations. Their environmental effects should be considered when assessing the energy efficiency and 'greenness' of railways. However, even if fossil-fuelled power stations supply the energy for railways, their energy production may be more environmentally friendly than diesel engines, since most modern power stations are equipped with SO_2, NO_x and particulate filtration devices.

2.1.2 Ships and ferries
While emissions from land-based transport and industry have been monitored and reduced significantly since the 1960s, emission inputs into the atmosphere from ships and ferries have been largely neglected until recently. Since the 1990s, monitoring of ship emissions, both port-based and during normal operation, has intensified globally (International Maritime Organisation 1998; European Commission 2002), but in many countries, including the US, ships are excluded from emission reduction programmes and policies, and will therefore continue to contribute significantly to riverine, coastal, off-shore and global air pollution. In addition, ships at berth in ports, which are often located in the vicinity of urban settlement, generate emissions that can significantly decrease urban air quality (Isakson *et al.* 2001).

Emissions from ships and boats arise from the combustion of fuels, and are similar to those from land-based transport as most engines run on diesel. Air pollutants from ships include SO_2 and NO_x, both of which can contribute to acidification of soils and water via precipitation; in addition, they can form airborne particles that represent a risk to human health. In 2000, emissions from the European shipping fleet were estimated to be 2.6 million tonnes of SO_2, just under half of the total emission from all land-based and domestic sea traffic (European Commission 2002, European Environmental Agency 2002). Also, significant emissions of polycyclic aromatic hydrocarbons (PAHs), CO, CO_2 and particulates have been recorded both when in berth (e.g. Isakson *et al.* 2001; Cooper 2003) and during transit (European Commission 2002). As might be expected, emissions are related to fuel type, fuel consumption and uses of different types of engines (Cooper 2003).

2.1.3 *Aviation*
Whitelegg and Cambridge (2004) showed that the world's airlines burn 205 million tonnes of kerosene and produce about 300 million tonnes of greenhouse gases per annum. The greatest concerns about gaseous emissions from aircraft are focused on water vapour and CO_2. For every tonne of kersosene burned, 1.23 tonnes of water vapour are released into the atmosphere. At altitude in the upper troposphere and lower stratosphere, under conditions of low temperature and relatively high humidity, mixing with the ambient air cools the exhaust stream, and water vapour condenses to form a visible trail. Minute particles in the exhaust stream, mostly soot and aerosol particles produced during combustion, provide the nuclei for condensation (Green 2003). Emissions also include CO_2, methane, ozone and various nitrogenous compounds. These gases are of concern in relation to their potential to influence climate change (Intergovernmental Panel on Climate Change (IPCC) 2001a, 2001b). The emissions from aircraft, however, have been reduced considerably in the past number of decades, particularly since 1996 when new international standards for gaseous emissions became effective. Although aircraft kerosene combustion contributes as little as 2.5% of the total global carbon dioxide emission from burning fossil fuels (Somerville 2003), Whitelegg and Cambridge (2004) note that since these gases are injected at a relatively high levels in the atmosphere and that they have a radiative force of 3, i.e. are about 3 times more damaging in terms of climate change than if they had been emitted at ground level. Therefore while the total contribution may be small, the fact that their radiative force is so great means that any increase in discharges may be significant.

In addition to the discharge of CO_n derivatives and NO_x, aircraft also emit Volatile Organic Compounds (VOCs) (Table 1).

Table 1. Some of the VOCs emitted during one cycle of engine use (Data from: EEA 2001)

VOC	% total VOC mass
Ethylene	19.2
Formaldehyde	16.6
Siliceous compounds	13.1
Propene	5.7
Acetaldehyde	5.1

2.2 RUNOFF

2.2.1 *Road Runoff*
The main contaminants found in road runoff are suspended solids, heavy metals, hydrocarbons such as PAHs, de-icing salts and nutrients (Buckler & Grenato 1999). These contaminants can accumulate on road surfaces due to road maintenance operations and wear and tear of car

components. Heavy metal and PAH levels in runoff tend to raise the most concern because of their potential toxicity and persistence in the environment. However, other contaminants such as high levels of suspended solids can also have adverse impacts if they enter streams and rivers.

The most common metals found in runoff are zinc, lead, copper, cadmium and nickel. Vehicular component wear is a source of chromium, zinc, iron and aluminium deposition on the road surface. Wear of car brakes deposits copper, lead, chromium, manganese and zinc. Engine wear and fluid leakage is also a source of aluminium, copper, nickel and chromium deposition on roads (Sansalone et al. 1996). A number of studies have also begun to investigate the platinum group elements (platinum, palladium and rhodium) which are associated with catalytic converters in exhausts (Buckler & Grenato 1999).

The discharges from land vehicles can be emitted into the air, or directly onto the road, but in most cases they are washed off the road surface by the combined effect of rain and traffic movement. Metals in water can be in the form of free metal ions, metal adsorbed onto organic or inorganic complexes and metal bound to particulate matter. The pH, hardness and dissolved organic matter concentration also affects metal speciation (Sriyaraj & Shutes 2001). Some metals such as cadmium, zinc and nickel are found mainly in the dissolved phase, whereas other metals like lead, iron and aluminium tend to be associated with particles (Legret & Pagotto 1999). The more soluble forms of metals are more bioavailable and more likely to bio-accumulate. A study completed in Northwest England found that the particulate phase of runoff contained >90% of the inorganic lead, ~70% of the copper and ~56% of the cadmium (Hewitt & Rashed 1992).

PAHs are semi-volatile organic compounds that consist of a number of aromatic rings. Sixteen of these compounds were classified by the USEPA as priority pollutants. Seven of these have also been classified as probable or possible human carcinogens. PAHs are present in fuels and lubricating oils and are released during combustion processes. Release of PAHs from exhausts and leakage of fuel and oil from cars are the main sources of PAH deposition on roads (Bomboi & Hernandez 1991). Due to the fact that PAHs are semi-volatile, they are highly mobile throughout the environment and combustion is one of the main sources of PAHs in the environment. Low-molecular weight PAHs e.g. naphthalene, are emitted mainly in the gas phase and become dispersed in the atmosphere, whereas higher molecular weight compounds are emitted in particulate form (Maltby et al. 1995a). PAHs are also hydrophobic and have a tendency to adsorb to particulate matter. Up to 80% of semi-volatile compounds in runoff are attached to suspended solids. This can lead to the transportation of PAHs on suspended solids in runoff and eventual accumulation in freshwater or marine sediments.

Other contaminants in runoff that have the potential to cause impacts are de-icing salts such as sodium chloride or calcium chloride, which are used during winter periods. Loads and effects of nutrients from runoff can also be substantial. Suspended solids, as mentioned previously, can act as a carrier of contaminants. Many contaminants in runoff tend to associate with particulate material and therefore can become accumulated in stream sediments.

2.2.2 Rail Runoff

Railway runoff has received increased attention recently, in Europe in particular in view of the European Commission Drinking Water Directives Standards of 'zero tolerance' (<0.1µg/L) for herbicides and pesticides in drinking water, as well as the recent EU Water Framework Directive.

Harmful substances have in the past been (and are to some extent still) used in the construction and maintenance of railways. Some persistent pollutants have accumulated in the underlying soils and are slowly leaching into groundwater.

Metals are not only emitted from trains through fuel combustion but also dispersed through corrosion and wear during normal railway operation, e.g. through physical contact on contact lines, where copper and silver are released into the environment at significant amounts. Heavy metals, such as lead, mercury, and cadmium, as well as polychlorinated biphenyls (PCBs) are also used during railway construction.

Railway sleepers, poles and other timber products are treated with wood preservatives containing oils, tar and chromated copper salts, phenols and creosote. PCBs and PAH-containing substances, toxic to animals and humans, leak into soils and groundwater. Also, oils and lubricants are directly used in general railway operations and maintenance (Broster et al. 1974), and eventually leach into the underlying substrates and soils, affecting groundwater and therefore flora, fauna and humans.

For transport safety reasons, herbicides and pesticides, toxic to biota, are usually generously applied to railway tracks, banks and terraces over long distances. Track drainage is therefore severely contaminated with pesticides and herbicides and poses an environmental threat. For example, German Railways use 250tonnes of herbicides each year to clear vegetation from over 70,000 km of railway lines; this represents the country's biggest single use of herbicides. Herbicides such as bromacil and diuron were used until recently and have reportedly caused significant soil and water contamination. Because of growing scientific evidence of the environmental damage, but also increased public awareness, in most countries the use of soil-acting herbicides such as diuron and triazine along railways has been abandoned, and foliar herbicides such as glyphosate are used instead. Although more efficient spraying techniques have been developed and the total amount of herbicide applied along railways has been reduced, it is still significant compared with other transport systems.

Construction and building work sometimes requires movements of soils and materials which are often contaminated from long-term and/or recent application of chemicals. Some railway routes have existed for over 150 years, and soils have gradually accumulated large amounts of chemicals and preservatives over extended periods. Such soils must be considered as hazardous wastes and need to be disposed of or treated accordingly.

Train accidents involving hazardous substances are rare but their consequences can be very serious if toxins spill into soils and water bodies.

2.2.3 Runway runoff

One of the major water-quality issues relating to aircraft transportation is the use of de-icing fluids. Bielefeldt et al. (2004) reported that of 50 airports in the United States, 45 use anti-icing chemicals such as propylene glycol. They estimate that in excess of 11 million gallons are used for de-icing aircraft in the US each year. Whilst much of the fluid is composed of glycol and water (98%), a small fraction is made up of surfactants, flame-retardants and corrosion inhibitors that have known toxicity to aquatic organisms. Coupled with de-icing, runway runoff from tyre wear, hydrocarbon combustion and particle deposition will be similar to that produced by road transport.

2.3 HUMAN WASTES AND SERVICING

Solid and liquid wastes produced from human activity are not unique to transport. Wastes from trains, including organic and solid wastes from toilets are disposed of along the tracks, but they are not considered a serious ecological problem. In developing countries, and also remote areas (e.g. along the Trans Siberian Railway), solid wastes, including plastics and other non-degradable wastes, are still dumped along the tracks. While organic wastes decompose eventually, contaminants are released directly into the environment. Plastics are not only unsightly, but also threaten wildlife in otherwise unspoilt environments, by ingestion when scavenging through the rubbish.

Wastes and solids are disposed of into the sea along commercial shipping routes, from leisure boats, and in harbours and marinas. Although one of the more visible pollutants in the oceans, the disposal of litter along busy shipping routes is less obvious to most people. Litter, and in particular plastics can travel long distances in the sea and be washed ashore thousands of miles from their place of entry into the sea, and the source of origin is not always traceable. It has been estimated that about 2 kg of rubbish are accumulated per person per day on board of ships, and most of this is disposed at sea. Even though waste disposal from ships at sea has been banned since 1988 (Clark 2003), 6.5 million tonnes of plastics are still discarded from ships annually.

2.4 OIL POLLUTION

Oil consists of a mixture of hydrocarbons, including chained and cyclic (aromatic) molecules, and up to 20-25% of non-hydrocarbons containing heavy metals and sulphur, and vanadium. The composition and size of hydrocarbons and non-hydrocarbons varies according to the origin of the oil and the degree to which it has been processed; large, long-chained molecules are not easily broken down by bacteria and are therefore very persistent in the environment.
While small quantities of oil are spilled on land and enter the marine system through riverine inputs, major amounts enter the sea directly through oil spillages from tanker and other shipping accidents, tanker cleaning and discharge of bilge and ballast water, as well as indirect (but still largely transport-related) incidents such as spillages during regular production procedures at off-shore or coastal refineries. Although all shipping activity can potentially lead to the release of contaminants into the marine environment, shipping incidents and accidents that occur due to groundings, fires, storms and collisions, are the greatest sources of oil pollution. Estimated annual inputs vary between 20,000 and 600,000 tonnes. Analysis of oil spill data shows that, despite the introduction of the Oil Pollution Act in 1990, oil spills are likely to become more frequent, but the rate of increase is likely to decrease (Ketkar 2002). The volume of oil spilled increases with the volume of petroleum imports, but decreases with the amount of domestic movement. In addition, risks of spillage increase with recent and predicted increases in marine traffic (Ketkar 2002).
Improved technology and practices reducing the accidental release of oils during normal tanker operations, dry-docking, ballast storage, in compartments separate from oil, have reduced inputs significantly recently. It is thought, however, that atmospheric inputs from oil evaporation during tanker transport may be as high as 3.5 million tonnes annually (Clark 2003).
While coastal habitats and flora and fauna on exposed shores often recover quickly from oil pollution, hydrocarbons accumulated in sediments can result in the slow release of hydrocarbons over long periods. In many cases, mitigation procedures and clear-up attempts using toxic chemicals as dispersing agents have caused more ecological damage than the oil. Toxins released from oil and dispersants that are accumulated in sediments may prevent recovery of affected shores for up to 15 years (Southward & Southward 1978).
In freshwater systems, accidental spillage of oil and fuel, in particular in inland marinas can have severe local effects. Oils spills on rivers may affect drinking water quality and water needed for crop irrigation. In busy tourist regions with high leisure boat activity, small oil spills may be common, often in combination with sewage-driven eutrophication due to high population densities in the tourist season.

2.5 ANTIFOULING AGENTS

Sessile animals and plants attached to ships and boats cause drag, increase weight, significantly decrease speed, increase fuel consumption and require regular expensive removal. To avoid such adverse effects, antifouling agents are used to kill and remove bacteria, eukaryotic uni- and multicellular algae, barnacles and other sessile animals from ships hulls. In addition, piers and ferry terminals are regularly treated with pesticides and herbicides that are slowly released into the water to facilitate easy and safe access for vehicles and foot passengers.

In the 1960s organo tin-containing (in particular tributyl tin, TBT) paints were introduced worldwide, as they killed marine fouling organisms more effectively than any other antifouling agent. The use of TBT was further extended to fish farms and cages, in particular Atlantic salmon cages and trestles for the cultivation of the Pacific oyster (*Crassostrea gigas*). TBT and TBT-derived breakdown products are readily accumulated into lipids of marine organisms, including organisms of commercial importance farmed for human consumption. The subsequent deformation of shells and reproductive organs of oysters, dogwhelks and other shellfish resulted in substantial losses of farmed and natural populations.

There is a worldwide ban on the application of TBT-containing paints to vessels under 25m, and all vessels in the European Union, the US, Australia and New Zealand. Other organo-metals have been used, in particular copper-based paints. In California alone, some 180 tonnes of copper from antifouling paints were estimated to enter the sea annually, before tin-containing paints were introduced (Clark 2003). Like tin, however, copper-containing compounds are accumulated, by means of biomagnification, in the marine food chain. Only about 15% of all copper is present in soluble form in seawater, with most copper bound to organic particles. Only ionic copper species are available for uptake by algae, but particulate copper may be accumulated by animals. A range of fouling organisms, in particular filamentous algae such as *Ectocarpus siliquosus* have developed copper-resistant strains, e.g. by avoiding cellular uptake of copper by exclusion (Hall *et al.* 1979) or detoxifying mechanisms. In the future since more organisms are likely to exhibit this response and as more detrimental long-term cumulative effects of antifouling agent release become known, new less damaging antifouling techniques will need to be found.

2.6 SHIPPING ACCIDENTS

Besides oils and fuels, other pollutants originating from accidents and incidents during marine transport include toxic chemicals, pesticides, herbicides, fertilisers, acids, alkalis, solid and nuclear wastes. Shipping accidents are particularly common on rivers prone to changing water levels caused by drought and flooding, when pressures to keep shipping routes open overrule safety regulations.

Parts of ships may be lost at sea; also large scale scrapping of ships, often of first world origin, occurs in several places in South East Asia. While parts are re-used and recycled, large quantities of waste materials, as well as oil remains, are dumped and washed out to sea.

Ship wrecks themselves probably only have minor environmental impacts and are rapidly occupied by sessile marine organisms, forming artificial substrata for settlement similar to artificial reefs, however, fuels or toxic cargo leaching from wrecks potentially have long-term impacts.

3. Consequences

The consequences of contaminant or pollutant discharge or emissions from transportation are varied. Few pollutants are specific to transportation, but transportation adds to other sources of pollutants discharged from either discrete or diffuse sources into the environment.

Thousands of studies have been undertaken worldwide, which have examined the link between environmental pollution and health and ecosystems. The extent to which environmental pollutants affect human health depends on socio-economic factors, levels of epidemiological surveillance and the extent of environmental contamination. While there are many uncertainties as to some of the precise mechanisms involved in connecting human health impacts and pollutants, a number of studies have demonstrated strong links. Among the best-described impacts is the impact of airborne lead on human health. Lead was originally added to petroleum products (now banned in most parts of the world as a petrol additive) to reduce engine knocking and increase combustion efficiency. The consequence of this was that lead became widely distributed around the world, even to the Arctic and Antarctic ice caps. This lead, which has no known physiological function, resulted in a low-level exposure for most of the human populations of the industrial world. For those exposed to higher burdens of lead, anaemia, memory loss and reduced reaction times and behavioural abnormalities resulted (Bellinger 2004). The effects of NO_2 on human health include asthma and lung disease, cardiovascular diseases and increased mortality these are reviewed in detail by Colvile *et al.* (2001). In combination with volatile organic carbons and other pollutants, NO_2 is also a precursor for ozone, in particular ground-level ozone, which is damaging to humans as well as vegetation, including commercial crops. Other vehicle discharges include known toxic and genotoxic compounds such as benzene and PAHs.

3.1 ATMOSPHERIC CHANGES

3.1.1 *The formation of ozone*
Complex photochemistry involving NO_x and VOCs results in the formation of ozone. Within the low troposphere ozone causes a problem for public health. Much of these emissions emanate from urban combustion of fuels, but there is increasing evidence to suggest that aircraft emission during take-off and landing, often associated with the use of reverse thrust to slow the aircraft, may also contribute to this low troposphere ozone. Some workers (e.g. Rice 2002) have suggested that as much as 10% of regional NO_x in metropolitan areas may be due to aircraft emissions. In one of the few studies of the impact of aviation on the lower troposphere, Pison and Menut (2004) report that the maximum impact of air emission on ozone occurs during the night due to the fast titration with NO.
By contrast, many studies have been undertaken to examine the impact in the upper troposphere. These studies showed, for example, that about 60% of aircraft NO_x are emitted between 10-12 km above the ground, 93% of which were in the northern hemisphere. They also suggest that aircraft emissions may represent up to 60% of NO_x emissions in the upper troposphere and could lead to an increase of a few percent of global ozone concentrations in this atmospheric layer (Pison & Menut 2004). The most positive impact on ozone was during the day in remote areas and at altitude, where the atmosphere is not saturated with NO. It is clear, therefore, that atmospheric emission for aircraft can have both positive and negative impacts on air quality.
Whilst technology has improved such that emissions from individual aircraft have declined, the increase in the number of flights has resulted at best in a modest reduction in pollution levels. Nonetheless the impact of aviation *per se* on air quality seems to be much lower than other forms of transportation (Somerville 2003).

3.1.2 *Global dimming*
One interesting phenomenon recently linked to emissions from aircraft is global dimming (Stanhill & Cohen 2001). This phenomenon appears to have been reported many decades ago, but most scientists have ignored its significance. Research has shown that the amount of solar

radiation hitting the earth's surface has declined in recent decades, but its significance remained unclear. However, following the almost complete suspension of air traffic in North America following the September 11 2001 terrorist attack on the Twin Towers in New York, scientists noted that the skies were clearer and that temperatures were about $1°$ C warmer. It has since been deduced that the clear weather and warmer temperatures on those days following September 11 2001, were due to the lack of air pollution from aircraft.

While the exact mechanisms of global dimming have yet to be resolved, it is believed that pollution dims sunlight in two ways. Some light bounces off soot particles (from incomplete combustion) in the air and is reflected into outer space, and pollution also causes more water droplets to condense out of air, leading to thicker, darker clouds, which also block more light. For that reason, the dimming appears to be more pronounced on cloudy than sunny days. Some less polluted regions have had little or no dimming.

Stanhill and Cohen (2001) review the evidence for a widespread and significant reduction in global radiation and discuss its probable causes and possible agricultural consequences. They indicate that a 10–20% decrease in solar radiation reaching the surface of the earth, if unaccompanied by other climatic changes, would probably have only a minor effect on crop yields and plant productivity. Experimental studies indicate that where, or when crop productivity is limited by water, small decreases in solar radiation, especially if accompanied by increases in the diffuse radiation component, will either have no impact on productivity, or may actually increase it. This moisture limitation will often occur in tropical latitudes and in the arid and semi-arid regions of other latitudes. In wet climates with low radiative heat load on the plants, any decrease in solar radiation is likely to be accompanied by a small decrease in productivity (Stanhill & Cohen 2001).

3.1.3 Climate change

Transportation involving the combustion of fossil fuels is responsible for over 20% of world CO_2 emissions (Table 2).

Table 2. Anthropogenic carbon dioxide emissions from energy use (average figures from various sources)

Sector where energy used	Percentage of total CO_2 emissions from energy use
Industry	42%
Buildings	32%
Transport	20%
Agriculture	6%

Total global CO_2 emissions are projected to rise, with the percentage attributable to transport remaining at 20% (Schipper & Fulton 2003). CO_2 contributes to the greenhouse effect with likely impacts on flora and fauna. The impacts of climate change (which also involve other greenhouse gases e.g. ammonia and CFCs) include a reduction in snow, retreat of polar ice-caps and statistically significant changes in the distribution and magnitude of precipitation (IPCC 2001a). These changes are predicted to involve shifts in species distributions and significant coastal flooding following changes in global temperature.

The impact of climate change induced by rising greenhouse gas levels (CO_2 methane etc.) and associated seas level rise on flora, fauna and humans have been described and modelled in a large number of studies world-wide. Although it is not possible here to deal with each in detail, it is worth considering some of the major impacts. The effects of climate change can be divided into direct (increases in sea level, sea surface temperatures and UV-B penetration) and indirect effects (changes in nutrients, circulation, etc.). It seems that only the direct effects can be predicted with any degree of confidence (Harrison et al. 2001).

The impact of climate change on ecology has been intensively studied over the past decade, and while there is certainty in some processes, there are uncertainties about the potential impacts. A recent review of the observed impacts of global climate change in the US sets out 10 major impacts (Parmesan & Galbraith 2004). They report that average US temperatures have increased by 0.6°C over the past century, and precipitation also increased slightly. The extent of the change is variable spatially. Their findings are based on more than 40 studies with observed ecological changes summarised below. Many other studies in Europe and elsewhere have observed similar patterns of change in phenology, geographical ranges, and ecosystem processing. These stressors are adding to already vulnerable ecosystems and habitats.

Of those species studied, it appears that most shift their distributions or become extinct when responding to climatic changes. In other words, they generally do not evolve in response to changes in temperature, but their distributions shift to track optimum conditions for survival and reproduction. What is interesting is that in many species, the timing of important events has already altered because of recent climate change. In Britain, for example, the timing of breeding in a number of species of birds has shifted, with many now breeding earlier (Crick *et al.* 1997; Crick 2004). It is likely that other seasonal events such as the timing of migration, flowering and germination may also be significantly influenced by climate change.

In Europe, the European Environment Agency (2004) has identified changes in marine phytoplankton growing season, a northward shift in zooplankton species by 1,000 km and a decrease in the size of plant populations in southern and northern Europe, all attributed to climate change. Plant diversity has increased in north-western Europe due to the northern movement of thermophilic species. The average annual growing season in Europe lengthened by about 10 days between 1962 and 1995, with an increase in 'greenness' by 12%. The survival rate of different bird species wintering in Europe has increased over the past decade, but the impact on populations has yet to be elucidated (EEA 2004). A summary of the predicted long-term ecological and socio-economic impacts of climate change (based on EEA 2004) follows:

- Increase in global temperatures
- Projected ice-free Arctic Ocean in summer 2100
- Projected rate of sea level rise between 1990 and 2100 of 2.2 - 4.4 times the rate occurring in the twentieth century
- Biodiversity changes in distribution, with species' distributions substantially affected by 2050. Changes in the rate of ecosystem functioning
- The projected increase in temperature may lead to a reduced capacity by European ecosystems in fixing further carbon.

3.2 AQUATIC SYSTEMS

Freshwater ecosystems affected by contamination from transport include groundwater systems, small or large wetlands, including coastal and estuarine swamps, ponds, lakes and rivers. Transport-related contaminants originate from road and air traffic, as well as boats and ships.

Through groundwater and riverine run-off, all contaminants released into terrestrial or freshwater systems potentially reach coastal and marine environments. Hence, coastal marine systems represent the largest sink for contaminants, and are additionally under threat from direct human pressures through intense use for human settlement, industry and ports. The commonly elevated contaminant levels in the coastal marine environment can be attributed to riverine, atmospheric and direct coastal inputs. Reducing contaminant levels in only one of these, although to be welcomed, will not necessarily result in a cleaner marine environment; an integrated approach has to be sought if attempts to protect the environment are to be taken seriously.

3.2.1 Runoff

The biological impact of runoff from roads, railways and runways is most likely to be greatest on freshwater ecosystems, as materials are washed into streams and rivers. The potential toxicity depends on the concentration and the physical and chemical properties of each contaminant present. The impact also depends on the sensitivity of organisms present to the constituents of runoff and the ability of the ecosystem to assimilate a given contaminant or mixture of contaminants (Buckler & Grenato 1999). Benthic macro-invertebrates living in contaminated sediments receive prolonged exposure to contaminants via gill cell osmosis and ingestion (Beasley & Kneale 2002). As macro-invertebrates are an important food source for many aquatic consumers, this can then result in the biomagnification of pollutants through the food chain.

Research on the effect of motorway runoff on freshwater ecosystems in England has investigated seven streams receiving runoff from the M1 motorway (Maltby et al. 1995a,b; Boxall & Maltby 1997). The main metals detected in sediments from downstream sites were zinc, cadmium, chromium and lead. At the site with the highest sediment concentrations, the most abundant PAHs were phenanthrene, pyrene and fluoranthene, the latter two being major components of crankcase oil (Maltby et al. 1995a). The results indicate that both stream water and sediment quality were altered only within a short distance (100m) of the discharge point, with only modest biological impacts.

In Canada toxicity of urban runoff from fourteen sites was investigated, including two sites receiving runoff from roads with traffic volumes of over 100,000 vehicles per day (Marsalek et al. 1999). A battery of tests was used to assess the toxicity of the runoff including Daphnia (*Daphnia magna*), Microtox and the SOS Chromotest for gentoxicity. The results show that 20% of road runoff samples were severely toxic compared with 1% of urban storm water samples. For one of these road sites the high toxicity responses were observed mainly in winter months.

High loads of suspended solids from maintenance and construction work can also cause aquatic habitat disturbance and cause a decrease in aquatic plant and animal populations.

3.2.2 Heavy metals

Heavy metals that are bioavailable are accumulated by primary producers, both algae and higher plants, as well as filter and deposit feeders and organisms at higher trophic levels. Although most metals (e.g. copper, cobalt, iron, manganese, vanadium, zinc) represent essential micronutrients essential in enzyme reactions, photosynthesis, respiration and nutrient uptake, at higher concentrations they become toxic. Other metals, in particular those derived from transport sources, such as lead, mercury, chromium (as well as organo-metal compounds, e.g. TBT) are usually not required by plants or animals and can cause toxic effects even at low concentrations.

Algae absorb metals mainly in their ionic form directly from the water; rooted plants take up metals from the sediments, and animals accumulate metals bound to particles or through consumption of algae and plants. Active uptake is linked to metabolic function, and is hence affected by ambient environmental conditions that could adversely or positively be influenced by other pollutants. For example, metal uptake is linked to growth rates, in turn affected by light penetration and nutrient loading (degree of eutrophication). Metals, including those derived from transport sources are toxic to most algae tested, but bioavailability of metals in freshwater and seawater can critically affect toxicity. Abiotic factors such as salinity, pH, the presence of inorganic and organic chelators determine the metal species presence, and hence the availability for uptake, and finally, the level of toxicity. Toxicity is metal and species specific, usually affecting germination, photosynthesis and reproduction.

In addition to metals taken up through the consumption of grazers, metals are also actively bioacummulated by deposit and filter-feeding invertebrates through consumption of particles.

About 4 million tonnes of lead annually are still used in petrol, mainly initially lost to the atmosphere but eventually deposited through precipitation. Lead is accumulated in plants and animals, and small sub lethal effects on growth rates have been observed in shellfish and birds. Mercury enters the atmosphere from combustion of fossil fuels. Levels of dissolved mercury in seawater are generally low, but sediments can be severely contaminated. However, most of this mercury however is not thought to originate from transport, but mainly from industrial outputs. Mercury is also used in antifouling paints, and elevated concentrations may exist in aquatic organisms at higher trophic levels (birds, mammals).

TBT is toxic to marine and freshwater organisms, in particular molluscs, even at concentrations as low as $0.1 \mu g/L$. The effects of TBT on shellfish has received much attention in the past after it was readily applied as an antifouling agent on boats and marine structures including fish and shellfish farms during the 1970s. Early evidence of its toxicity came from Britain and France where it was extensively used on oyster farms, and oysters and other shellfish exhibited shell deformation and reproductive abnormalities. Natural and farmed populations of at least 70 species of gastropods are now known to be affected by what is known as imposex/intersex (Smith 1981; Langston 1996), and female sterility caused severe damage to mollusc populations near ports and marinas, as well as fish and oyster farms. Although TBT was banned from boats sized <25m in 1987, it is still used on larger ocean-going ships. Effects on other marine animals are less obvious but it is clear that organotins are accumulated in fatty tissues of marine mammals (Law *et al.* 1998). Effects on fish, birds and humans appear less severe as TBT is degraded in the body (Langston1996).

Copper released from fouling paints has similar effects to TBT in that copper is readily taken up by algae and invertebrates and accumulates in the food chain. Copper is an essential element to most organisms, algae, plants and animals, but severe copper contamination has been observed in all freshwater and marine organisms including representatives along the whole food chain, ranging from phytoplankton to mammals. Despite severe local copper pollution, there is currently no evidence of species loss or community damage. Despite the high levels of copper reported for edible shellfish species, copper poisoning of humans through the consumption of contaminated shellfish is unlikely.

Cadmium from vehicle tyres and engine exhausts enters aquatic systems through water run-off into streams and rivers and atmospheric inputs. Cadmium is readily taken up by phytoplankton, and uptake is strongly dependent on water pH. Effects on phytoplankton productivity appear contradictory but similar to observations for other metals, both uptake and toxicity are affected by a range of environmental parameters (Genter 1996). Accumulation of cadmium in the food chain appears to be significant, but there are no reports of acute or chronic cadmium poisoning of aquatic organisms or any direct ecological damage.

Generally, despite sufficient evidence of heavy metal contamination in aquatic systems, reports of metal uptake by algae and subsequent accumulation in the food chain, there are only a few instances where metal contamination can directly be linked to community changes, reduction in biodiversity, changes in species composition or biomass reduction. Examples where this was evident include the elimination of gastropods from bays with high TBT concentrations, and changes in the population structure and dynamics affecting the algal-grazer relationship in polluted areas.

3.2.3 *Oxides of nitrogen and sulphur*

Effects of NO_x emissions depend on the system in question, the degree of the pollution as well as the presence of other chemicals and contaminants.

Emissions of NO_x on land may lead to local nitrogen enrichment on road-side verges leading to an increase in plant growth (Cape *et al.* 2004). Both SO_x and NO_x inputs from vehicle exhaust gases enter soils and water bodies, including streams, rivers, ponds and lakes via air (acid rain) and run-off. They can cause acidification of water bodies, as both NO_x and SO are precursors for strong acids. Acidification of rivers and lakes was first recognised in the 1980s and evident in regions polluted by industry, in particular in Eastern Europe, but also in Scandinavia, Switzerland and the US. Acidic water bodies are characterised by pH below 5.5 and acid neutralising capacity (ANC) near or below 0μeq/L (Planas 1996). Low pH leads to physical and chemical alterations of the water body such as the availability of nutrient and metals in the water body, and increase light penetration and temperature. All of these affect aquatic species. Changes in pH also directly affect species and communities by reducing photosynthesis, respiration and growth of most organisms. As individual species have different tolerances it is impossible to generalise about effects of acidification on primary producers. However, important community changes have been observed, such as decreases in species richness of both algal and macrophytic flora, and general decreases in biomass. Reports on primary productivity appear contradictory, with different effects observed for different zones of water bodies. These are to some extent due to the changes in species composition caused by physical changes in light and temperature climate. Overall, productivity is reduced which has knock-on effects on the aquatic food chain, eventually decreasing fish stocks including those of potential commercial value. The complete loss of fish populations as a direct result of severe acidification has also been reported. In addition, herbivorous and carnivorous invertebrates are directly influenced by changes in the chemical composition of the water. For example, low pH reduces calcium availability and increases heavy metal solubility, often leading to aluminium toxicity. Effects of water acidification are complex, but generally cause a shift in species abundance and result in reduced standing crop of both primary producers and consumers.

3.2.4 *Oil spills – direct effects and PAHs*

The environmental effects of oil spills depend on a number of factors including the type of oil, spread and destination, management procedures and treatment, and the intrinsic environmental and biotic factors of the ecosystem concerned. The spread and toxicity of the oil depends on the nature of the oil, e.g. light or crude oil, and the composition of different hydrocarbons (including the proportion and types of PAHs present) which is determined by the origin and state of processing of the oil. Wind speed and weather conditions can delay physical removal from the sea surface, but hot weather may increase evaporation and thus reduce the volume of oil. However, one should remember that hydrocarbons evaporated from the sea surface do not just disappear, but enter the atmosphere, and while this may reduce the immediate severity of oil pollution on nearby shores, the ecological effects are shifted to a different level and location.

Recovery of affected shores is often delayed by human intervention, in particular the addition of detergents to disperse the oil. Oil is further driven into sediments that then act as long-term sources of hydrocarbon pollution, long after the surface layers may have recovered. Dispersants are not always effective in removing the oil, and can be toxic to biota. Physical removal of oil from the sea surface appears the most environmentally favoured procedure, but this may not be possible due to bad weather or the sheer volume of the spill. In consequence, not all oil will be removed from the sea or the shore, no matter which methods, or combination of methods, are used. Natural bacterial decomposition of small amounts, in particular small droplets is possible but slow, and can be speeded up by the addition of particular strains of oil-degrading bacteria.

Chronic exposure to oil appears to change the flora in favour of tolerant species (Straughan 1972), but a number of factors determine the effects of oil spills on intertidal fauna and flora in acute instances. The physical coverage of shorelines by oil has several effects on plants and animals. Light is limited for photosynthesis, and this will reduce productivity. Also, bacterial decomposition of oil has a high biological oxygen demand (BOD) and can further reduce O_2-availability for sessile animals. Bleaching can occur and if covered by heavy oil deposits, plants are physically damaged and likely to break.

Toxicity of different components of oils on plants and algae generally increases in the order of paraffins, naphthenes, alkenes to aromatics and usually smaller hydrocarbon molecules within a class are less toxic (O'Brien & Dixon 1976). Toxicity of oil to phytoplankton varies amongst species; some exhibit enhanced growth rates, other reduced reproduction, so that it is difficult to predict oil pollution effects as environmental factors also affect toxicity. Further, the effects of oil on phytoplankton species are not always distinguishable from those caused by dispersants. In an isolated experimental ecosystem, the addition of 100ml crude oil (Kuwait) to 300L vessels decreased primary production to 50% within 3 days, and to near 0% within 7 days, some recovery was observed after 10 days (Lacaze 1974).

Recovery of exposed shores is usually faster than that of sheltered shores, due to the nature of the substratum which prevents penetration into deeper sediment layers where the oil would persist, but also directly because of enhanced physical breakdown under oxic conditions. Long-term effects are greatest in fine-grained sediments which naturally support high productivity and diverse infauna, and are critical systems at the base of the marine food web.

Toxic effects of oils on microbes and animals depend on the composition of the oil, evaporation rate and the degree of exposure. Hydrocarbons, but also dispersants, can cause fish and shellfish mortality. While some aquatic organisms can detoxify PAHs, many PAHs are carcinogenic and mutagenic to most animals. In particular larvae and juveniles are sensitive to toxins, and abnormal developments and tumours have been observed in a range of organisms as a result of PAH pollution in aquatic systems.

Humans may directly be affected by oil spills when aquaculture installations and local fish stocks are under threat. Oil spilled in freshwater systems, or indeed on land, can leach into the ground and drinking water sources, and thus represents a direct threat to human health through the intake of PAHs.

3.2.5 PCBs

Mainly derived from herbicides and pesticides as used along railway tracks, PCBs enter freshwater and marine systems either through agricultural run-off, railway drainage or through large-scale spraying from planes. As they are not usually degradable, they are considered conservative pollutants and once entered in a system, are very persistent so that today they are widespread both in freshwater and marine systems. They are highly toxic to algae and plants (used as weed killers) and animals (used as pesticides). Effects of PCBs occur at individual, population and community levels. As they also affect grazing invertebrates and other consumers, their ecological effects are very complex and cannot easily be predicted but all trophic levels are affected, at least to some extent, and interactions with the biotic, physical and chemical environment must be considered (Hoagland et al. 1996). The duration and intensity of the exposure to the toxins is important in assessing toxicity effects. PCBs inhibit photosynthesis and hence productivity. As tolerances are species-specific, changes in dominant species composition may also be observed.

PCBs are generally not water-soluble but accumulate in fatty tissue of animals; they are passed through the food chain and biomagnification occurs at higher trophic levels. PCBs have been held responsible for local fish kills, death of shrimps, lobsters, seabirds, pinnipeds and cetaceans.

3.2.6 *Contaminants in off-shore environments*
Effects of contaminants on off-shore regions have rarely been described. This is partly due to the lack of research conducted, and partly due to the dilution of estuarine and coastal contaminants by the time they reach off-shore waters. However, recent evidence suggests that transport-originating contaminants, such as wastes, oils and litter are present in substantial amounts along trans-oceanic shipping routes.
Plastics are not degradable and may act as substrata for settlement of marine organisms. Litter and plastics derived from ships, may be ingested by fish, marine mammals, turtles or seabirds and this may lead to choking or indeed malnutrition or starvation.
Organic contaminants such as PCBs and PAHs have been recorded in pelagic fish, marine mammals, seabirds and turtles, but no direct link between transport and such levels has as yet been established as these contaminants also may originate from marine and land-based industries or agriculture.

A summary of the long-term ecological and socio-economic impacts of pollutants in aquatic systems follows:
- Biodiversity reduction or loss
- Change in standing crop and species composition
- Loss of key species resulting in decrease in ecosystem stability and function
- Reduced amenity value of coastal systems for human settlement and tourism
- Losses for the fishing and shellfish industry affecting income for local communities.

4. Potential mitigation and solutions

4.1 OVERALL REDUCTION OR ELIMINATION OF EMISSIONS

Climate change is a problem with unique characteristics. It is global, long-term and involves complex interactions between climatic, environmental, economic, political, institutional, social and technological processes. This has significant international and intergenerational implications in the context of broader societal goals such as equity and sustainable development. Developing a response to climate change is characterised by decision-making under uncertainty and risk, including the possibility of non-linear and/or irreversible changes (IPCC 2001b).
The IPCC working group suggest that the type, magnitude, timing and costs of mitigation depend on different national circumstances, socio-economic and technological development paths and the desired level of greenhouse gas concentration stabilisation in the atmosphere. Development paths leading to low emissions depend on a wide range of policy choices and require major policy changes in areas other than climate change. Climate mitigation policies may promote sustainable development when they are consistent with such broader societal objectives. Some mitigation actions may yield extensive benefits in areas outside of climate change: for example, they may reduce health problems; increase employment; reduce negative environmental impacts (like air pollution); protect and enhance forests, soils and watersheds; reduce those subsidies and taxes which enhance greenhouse gas emissions; and induce technological change and diffusion, contributing to wider goals of sustainable development. Similarly, development paths that meet sustainable development objectives may result in lower levels of greenhouse gas emissions (IPCC 2001b).

Lower emissions scenarios require different patterns of energy resource development and technological change. Transport modalities will have to look at ways to reduce emission as has been done through the introduction of catalytic converters and through the reduction in emissions from most forms of transportation. In the same way alternative energy sources for transportation will be required and further development of low impact fuels and 'greener' transportation are required.

Emissions from railways are generally lower than from other modes of transport, but diesel engines, and in particular older diesel engines, can significantly contribute to air pollution through emission of particles. Modern, more efficient engines are usually fitted with catalytic converters. The use of older engines should strongly be discouraged, and ideally electric trains be used which are based on more efficient energy production at modern (and 'cleaner') power stations.

Marine vessels emit 14% of NO_x and 5% of SO_x of all fossil fuels globally (Corbett & Fischbeck 2001), and in the US the contribution of marine shipping to total emissions of particulate matter (PM) and NO_x is expected to double by 2020 due to cleaner land transport, but also increased shipping activity in the US (USEPA 2003).

The deposition of sulphuric and nitric acids, mainly derived from engine emission, has caused large-scale damage to forest and aquatic systems. The recent Gothenborg Protocol (UNECE 1999) to reduce acidification, eutrophication and low-level ozone will reduce sulphur and nitrogenous emissions from land. However, emissions from ships are not included, even though recent evidence suggests that these emissions are a major contributor to acidification. In Europe, such changes are currently discussed on the basis of cost-effectiveness, as a reduction in sulphur emissions is seen in the context of reducing overall SO_2, NO_x and NH_3 emissions in each country. The 1997 protocol to MARPOL 73/78 (Annex VI) which regulates the emission of SO_x and NO_x from ship exhausts came into operation in May 2005, it is to be monitored by the International Maritime Organisation (IMO). Reduction in SOx emissions can be achieved by the use of low-sulphur fuel or the use of exhaust gas-cleaning systems. The incineration of contaminated packaging and PCB-releasing materials on board ships is also banned under the new Annex.

In Europe, most efforts to reduce sulphur emissions from ships focus on switching to low-sulphur bunker fuel, which is available at an estimated 95-99% of EU ports (however, quantities and prices vary between the EU countries) (European Commission 2002). Availability is inconsistent elsewhere in the world although it can be found in many ports in both North and Central America. Multi-storage facilities on ships may be required when ships need to refuel outside the EU with high-sulphur fuels, but the burning of low-sulphur fuels is required on entry into EU territorial waters because of current EU legislation (Directive 1999/32/EC). Engineers, shipbuilders and repairers are challenged to address issues of cost-effective installations and conversions of ships to allow dual-fuel usage, which appears to be a realistic option for the future (European Commission 2002). In Sweden, fairway dues have been reduced for low NO_x and SO_x emitting ferries or cargo ships since 1998 (Anon 2002); this could be used as a model for EU or indeed global practice in the future. A recent proposal by the USEPA to introduce stricter standards on emissions from ships in US ports (95% of which are under a foreign flag) was not accepted by the US government since it was anticipated that international trading could be adversely affected.

Generally, reductions in emissions have direct effects on sensitive coastal ecosystems and, while the cost-effectiveness of controlling ship emissions decreases with distance from the shore, air pollution in ports and local environments could be reduced significantly (Colville et al. 2001). As far as European waters are concerned, a reduction in emissions from ships as a means to reduce acidification is probably less effective than reducing emissions from land transport, possibly because of the large proportion of long-distance transport involved in commercial shipping, and off-shore emissions of NO_x and SO_x having less localised effects than emissions from land transport.

According to Bailey and Solomon (2004), in addition to the stricter emission controls from ships and the introduction of cleaner fuels or biodiesels, pollution prevention at ports should also include idling restrictions and shore-side electric power (instead of the use of auxiliary engines while in port). They also emphasise environmental and health considerations in the selection of sites for new terminals and ports away from residential areas.

4.2 TREATMENT OF RUNOFF FROM ROAD TRANSPORTATION

Treatment of road runoff is one of the main methods used to remove the emission from land-based transport system (O'Reilly *et al.* 2004). A variety of road drainage systems are used to treat runoff and these can vary in their effectiveness. The main systems consist of infiltration technologies, filtering systems, detention, retention and vegetated practices and porous pavements (FHWA Environmental Technology Brief 1999). The principle behind most systems is to collect runoff as it drains from the road surface, then either filter or infiltrate it through soil, removing particulate and soluble matter before it is discharged into waterways.

4.2.1 *Infiltration technologies as an approach*
Infiltration technologies consist of bioretention areas, infiltration basins and infiltration trenches (FHWA Ultra-Urban Best Management Practices 2003). These systems operate by temporarily trapping runoff so it can infiltrate through soil. Infiltration trenches are mainly used to treat runoff from small drainage areas and are often present as narrow strips by the roadside. Infiltration basins are used for treating larger volumes of runoff and consist of a shallow depression, which holds runoff, until it infiltrates into the soil. A study in Switzerland on the retention of pollutants by infiltration systems that were 12 - 45years old (Mikkelson *et al.* 1996) found that there was a significant build-up of copper, zinc, cadmium, lead, PAHs and organically-bound halogens in the upper soil and sludge layers, but these contaminants decreased rapidly to background levels within depths of less than 1.5m. Infiltration systems appear to be effective pollutant traps for these types of contaminants, but their potential for reducing groundwater contamination was limited.

4.2.2 *Porous pavement*
Porous pavement is a modification of traditional asphalt road, designed to treat runoff, consisting of a top layer of asphalt or concrete with a higher than normal percentage of voids (FHWA Ultra-Urban Best Management Practices 2003). Runoff is trapped in the pores in the asphalt and percolates down to the bottom layer, which consists of a stone reservoir. Runoff is trapped here until it infiltrates down through the soil beneath.

The pavement type also influences the quality of road runoff. In Nantes, France, porous pavement reduced the load of suspended solids by 87%, total hydrocarbons by up to 90% and heavy metal loads discharged in runoff were reduced by 20% for copper and up to 74% for lead (Pagotto *et al.* 2000).

A drainage design has been created which combines the characteristics of an infiltration trench and porous pavement. The system was named PET (partial exfiltration trench), and consists of a perforated under-drain surrounded by an engineered medium and capped with a porous pavement, ideal for placing on narrow road verges (Sansalone 1999a). Silica sand coated with iron oxide (OCS) was found to be a more effective filter medium than uncoated silica sand. The *in situ* performance of an instrumented field scale PET indicated a mass removal efficiency generally greater than 80% after one year of runoff loadings (Sansalone 1999b).

However, the percentage removal efficiency was somewhat lower than predicted from a bench scale model, especially for particulate bound metals.

4.2.3 Stormwater ponds

Stormwater ponds are designed to detain runoff so that infiltration and settling of suspended solids can occur. Although these systems require a certain amount of land space, they have an effective life span of 20-50 years. However sediment accumulation can cause a decrease in storage capacity in the pond of up to 20% over 10 years, from which sediment may need to be removed every 5-10 years (FHWA Ultra-Urban Best Management Practices 2003).

The efficiency of heavy metal removal from runoff in a wet biofiltration pond at a site with an annual average daily traffic (AADT) of 140,000 vehicles per day and from a dry detention pond with similar traffic density was compared (Hares & Ward 1999). The system consisted of a wet bioinfiltration pond lined with a relatively impermeable clay base and a bed of reed mace (*Typha latifolia*). Runoff drained from this into a second basin, which was a sedimentation tank, with partitioning walls to increase sedimentation time. The treated runoff then drained into a nearby river. The dry pond system was at a site with an AADT of 120,000 vehicles per day. Runoff passed through a grit trap, an oil interceptor and a silt trap before flowing into a dry detention pond lined with a clay base. Although the site with the higher traffic volume had a higher level of metal concentration at the inlet point, the wet biofiltration pond was found to be more effective than the dry pond in the removal of metal contamination from runoff. The clay base and reed bed of the wet biofiltration pond provided a source of absorption surfaces for the binding and immobilisation of metals.

Other studies have shown that detention ponds can remove up to 90% of particulates, but are not as effective at removing soluble pollutants (FHWA Ultra-Urban Best Management Practices 2003). Four wet ponds investigated in Minnesota, USA were found to be moderately effective in removing pollutants during non-winter months but there was a marked decrease in the performance of stormwater ponds treating snowmelt runoff (Oberts 1994).

4.2.4 Wetland systems

Naturally occurring biological and chemical processes in wetland/shallow marsh systems remove pollutants (FHWA, Ultra-Urban Best Management Practices 2003). Constructed wetlands have been used in the US and parts of Europe to treat runoff. A wetland system can use a number of detention or retention ponds, which temporarily store runoff until infiltration into the soil can occur. Plants such as the common reed (*Phragmites australis*) and reed mace can take up pollutants such as hydrocarbons and heavy metals. Also, plants supply oxygen to bacteria in surrounding sediments, which encourages decomposition of organic compounds and converts ammonia to nitrates (Sriyaraj & Shutes 2001). Runoff is then released into an adjacent water body. For efficient removal of particulates, constructed wetlands should have a minimum retention time of 30 minutes and maximum retention time of 24 hours (Shutes *et al.* 1999).

The effect of runoff from a section of the M25 London motorway on a pond, wetland and stream in a nature reserve was investigated (Sriyaraj & Shutes 2001) by analysing water, sediment, plant and macro-invertebrates samples. In the reed sweet-grass (*Glyceria maxima*), and reed mace, metal concentration decreased in the order of roots>rhizomes>leaves, and reed sweet-grass had a higher uptake ability for cadmium, copper and zinc than reed mace. All metals detected in water samples were below the British water quality standards (Sriyaraj & Shutes 2001). It was concluded that filtration of runoff through pond and wetland systems decreases the concentration of metals and nutrients to acceptable levels. The use of natural wetland is inadvisable due to metal accumulation in sediment and constructed wetlands with baseflow water should be considered as an alternative treatment approach. However, to create a wetland system requires frequent maintenance. Constructed wetlands require a period of 1–3 years to mature (Shutes *et al.* 1999).

4.2.5 *Oil/grit separator*
In this system, a conventional oil/grit separator is located underground and consists of three or four concrete chambers. Runoff drains into the first chamber which is designed to settle sediment and large particles. As runoff passes into the second chamber (oil chamber) it is passed through a screen, which filters debris and traps surface oils and grease. The third chamber contains an overflow pipe to drain the treated runoff to the nearest waterway. The system is flexible and can be designed to fit the space available but frequent maintenance is needed.

4.2.6 *Filtration systems*
Filtration practices consist of underground chambers containing filters composed of sand, peat or compost filter media. Sand is the medium mostly used as it has the longest performance record and can trap 90% of the small particles in runoff (FHWA Ultra-Urban Best Management Practices 2003). When a heavy rainfall event occurs, runoff is collected into one of these chambers. Pre-treatment by settling occurs, then runoff passes through the filter media before being discharged into a storm drain. Filtration practices are useful in sites with limited land space, but have high installation costs and need subsequent maintenance. A maintained sand filter has a lifespan of 5-20 years.

4.2.7 *Vegetated swales*
Vegetated swales primarily consist of a simple grassed channel. Dry swales have a highly permeable layer of soil at the base of the channel to allow infiltration to occur. An under-drain system in the layer of soil collects the runoff. Wet swales have a layer of water and vegetation at the base of the channel to create an elongated wetland treatment system. Installation costs are minimal and maintenance is low, although sediment build up at the base of the channel needs to be removed. Vegetated swales can be installed on narrow strips of roads. Seasonal effects such as decreased vegetation in winter months can cause a decrease in pollutant uptake (FHWA Ultra-Urban Best Management Practices 2003).

4.2.8 *Filter strips*
Filter strips consist of a vegetated surface sloping downwards from the roadside. Infiltration occurs as runoff flows down along this slope. This system requires a large amount of land space and is not suitable for high-velocity flows. Generally, the level of pollutant removal is low (FHWA Ultra-Urban Best Management Practices 2003). Little maintenance is involved other than promoting plant growth; the typical lifespan of the system is 10-20 years.

4.3 RUNOFF FROM RAIL TRANSPORT

4.3.1 *Fuel and oil spillages*
Fuel and oil spillages and leakages should be reduced, ideally avoided, which may be possible if more electric and fewer diesel-fuelled trains were in operation, and older diesel engines replaced. Recent research has also improved technologies by using less corrosive and more reliable long-lasting materials. Appropriate maintenance of engines, fuel tanks and pipelines is crucial, and regular checks are now conducted in most countries, due to pressures to improve passenger and freight safety, cost-effectiveness, as well as environmental protection. Fuel consumption by the US and Canadian railways is about 16.5 billion litres of diesel annually

(Association of the American Railroads 2002), and significant spillages and leakages currently occur that potentially contaminate soils and water bodies. However, an improved design of re-fuelling systems can significantly reduce or at least contain spillages, and thereby reduce the risk of environmental contamination (Barkan 2004).

4.3.2 Herbicides

One of the major sources of pollution from railways is the use of herbicides applied along tracks, and several measures have been introduced in most countries (at least in the developed world) to avoid the application of toxic persistent organic herbicides that accumulate in soils, water and biota in the long-term and to reduce the amounts of herbicides used.

Improved management and maintenance of tracks includes a move away from persistent soil-acting herbicides, the application of smaller quantities at lower concentrations, and research on, and trials with, alternative methods for weed removal are underway. In the past mainly atrazine, diuron and other organic long-lasting herbicides were used along extensive stretches of tracks, but these have largely been replaced by more environmentally friendly glyphosphate- or imazapyr-containing substances. All tracks were sprayed with residual herbicides, resulting in heavily contaminated drainage, but spraying is now largely conducted with non-residual herbicides. The application of soil herbicides generally should be avoided, and foliar herbicides used instead. Herbicides should have a half-life of no more than 2-6 months and be totally degradable within a year of application.

Measures to reduce the need for large-scale application of herbicides include a restriction of herbicide application during adverse weather conditions, e.g. avoiding extensive spraying in rain (inefficient due to dilution) or wind (inefficient due to drift), so that larger quantities need to be applied to achieve the same effects. Generally, spraying in environmentally sensitive areas should be avoided, if possible, in particular in protected groundwater catchment regions, and alternative weed removal measures should be sought.

Such alternative methods for clearing of vegetation (instead of herbicide use) could include the removal of weeds by hot steam, but this appeared not to be sufficient in the long-term in Sweden (Anon 2004). Other methods include the use of a sealing layer, for example made of cloth, to prevent weeds from growing on embankments, but their application is obviously restricted to newly-constructed embankments. Also, mechanical clearance has been suggested in environmentally-sensitive areas, however, this is slow and labour intensive.

Clearly, through optimal dosage and accurate spraying only where and when needed, the application even of less toxic herbicides can, and should, be reduced. For example, in Banverkct, Sweden there are specially designed small trains that allow efficient spraying with computer-controlled dosing (Anon 2004).

Even if contaminant release is reduced today because of increased environmental awareness and improved technology, large quantities of persistent pollutants have already leached into nearby soils and sediments of adjacent water bodies, so that while a reduction in the use of herbicides should obviously be sought, soils are already contaminated. To mitigate environmental contamination in such cases, the only option is to remove contaminated soils and materials, which is costly and causes disruption to rail traffic, and hence in many cases is not considered a realistic option. An increasing demand for environmental mitigation procedures has lead to the existence of a large range of consultancies specialising in environmental solutions and improved mitigation technologies in environmental engineering and management. The removal of persistent herbicides from water bodies can be undertaken, for example by applying ozone, with or without peroxide, but it is costly. Also, while persistent herbicides may be removed from soils in this way, other ecological damage and physical destruction of habitats may occur instead.

4.3.3 *Other contaminants*

Making railways 'greener' also involves the use of environmentally friendlier materials (and their recycling) in track construction, train component materials and better route planning. Recently, more emphasis has been placed on environmentally-friendly construction works, with stricter guidelines on emissions from diesel vehicles, contaminant runoff and spills, including oil spills.

In the past, wastes and hazardous materials including mine spoil were used in the construction of embankments; although this practice has now been abandoned, contaminants can be released during maintenance and reconstruction of such embankments, and their replacement with less hazardous materials should be considered.

Mitigation measures range from the removal of soils and detoxification of materials deliberately used in the construction, to soils accidentally contaminated by herbicide spraying, fuel leakage, for example during maintenance or accidents, and freight spillage. Technologies required include specific soil excavation, transport and disposal of hazardous materials, and often the reconstruction of embankments and tracks, and potentially necessary soil bioremediation. All of these are expensive.

Similarly, environmentally-friendly wood preservatives are available today, and building of new railway lines, or the replacement of old sleepers and other timber products should be considered, however keeping a watchful eye on the destination of recycled old materials. For example, old railway sleepers have in the past been used as borders for sand pits in children's playgrounds, despite having been treated with creosote. Contaminants leached into the sand and thus represented a serious health hazard, with some playing sands containing phenol concentrations of up to 100 mg/kg. The use of more environmentally-friendly materials will in the future allow more efficient recycling or reusing. For example, crushed concrete sleepers or timber poles treated with modern wood preservatives could be reused in building without causing environmental harm to biota or humans. Also, sites and facilities for wood preservation used in railway construction are often heavily contaminated, with contaminants leached into soils and groundwater for decades; removal of soils is possible, but access to groundwater must be blocked off.

Solid and liquid wastes from trains need to be collected and appropriately disposed of at railway depots, instead of being directly disposed of along tracks; this includes sewerage collection and management from passenger traffic as is now the case in modern trains at least in Europe, with tank systems that can be serviced similarly to those on aircraft. In addition, modern servicing of trains and railway stations should be considered, and planning considerations include the accessibility of stations by environmentally friendly means such as bicycles or buses, rather than cars.

4.4. AQUATIC TRANSPORT

In comparison with terrestrial systems and the problems associated with road or runway runoff, the sources of pollutants in aquatic systems are often more difficult to identify. A range of other pollutants that threaten aquatic, and in particular coastal, systems, are derived from land or air-transport through riverine or atmospheric inputs, and reduced inputs into groundwater, streams and rivers will also improve water quality and environmental health of estuarine, coastal and marine systems.

Specific measures to prevent or mitigate pollution in aquatic environments have concentrated on oil pollution through improvements in tanker operations, movement and cleaning procedures, ballast water and antifouling measures, and a reduction in emissions from ship engines. In freshwater systems, measures to prevent or mitigate eutrophication and acidification have also received much attention.

4.4.1 *Oil pollution, preventative measures and mitigation of impact*

As most oils enters aquatic environments through accidental and deliberate oil spills from tankers, recent legislation and an improved practises have concentrated on these. Intentional spillages of oil occur during normal operation when tankers, after discharging the oil, fill cargo tanks with ballast water for stability and cleaning, and then release dirty ballast water (a mixture of oil and water) into the sea.

Such spillage during ballast water exchange or cleaning operations has been considerably reduced since the introduction of separators in hulls of modern tankers where ballast water is kept separate from oil compartments, and can be released into the water without having been in contact with oil. Other methods include the consecutive cleaning of different compartments with high-pressure jets and maintenance of water in compartments until the oil floats on top of the water. Instead of being released into the sea, the dirty ballast water is pumped into separate tanks and stored until the oil forms a layer on top; the oil-free water underneath can be pumped out into separate tanks and eventually be released into the sea. The oil can be recovered and transferred into a slop tank, and fresh oil is loaded on top ('Load on Top' method). Even more effective in preventing oil release into the sea is the 'crude-oil washing' developed in the 1970s which involves the cleaning of cargo compartments with oil (instead of water) which makes the process more effective and reduces the amount of oil-water mixtures produced. Crude oil washing was made mandatory for new tankers in 1978 by the MARPOL Convention protocol, and revised specifications regarding design, operation and controls were adopted by IMO in 1999.

Changes in policies and their consequences for tanker route management and operation have been reviewed and analysed by Ketkar and Babu (1997), Owen (1999), Talley *et al.* (2001) and Ketkar (2002). The Oil Pollution Act of 1990 (OPA-90) has established accountability for vessel oil spills, and requires single hull tankers travelling in US waters to be phased out by 2015 with the assumption that oil spills from double-hull vessels would be less severe. Similarly, the European Union has called for a worldwide phasing out of single-hulled tankers by 2012. Oil spills are related to size and width of the water body, traffic density and rivers, but Meade *et al.* (1983) showed that 75% of the variance of accidents could be explained by the size, age and flag of the vessel. Ketkar and Babu (1997) show that oil spillage size is significantly affected by vessel type or age, but spills are not directly related to oil imports but rather domestic movements (in the US), rough weather and hydrographic conditions and traffic density. In the UK, analysis of tanker route management suggests that local ports are under severe pressure as they are largely responsibility for tanker routes and managing a high traffic density in many environmentally sensitive and protected marine areas. This could significantly be improved by further national or international support (Owen 1999).

Comparisons of recent (post OPA-90) oil spillage data indicate that, despite recently introduced legislation, spillages from tank barges are still considerably greater than from non-oil cargo vessels, but spillages from tankers are not greater than from non-oil cargo vessels except for freight ships (Talley *et al.* 2001). While the OPA-90 is considered a deterrent to accidental spills, the frequency of oil spills is expected to increase, and estimated costs of adhering to major regulations currently exceed estimated benefits (Ketkar 2002).

Despite legislation, oil spills are becoming more frequent as tanker traffic and shipping density on busy routes increases. In addition to the phasing out of aged and single-hull tankers, stricter regulation such as single traffic routes along busy tanker routes and navigational communication could reduce the frequency of accidents. The greater use of tug boats (including emergency tug boats) is recommended, and increased environmental consideration by port authorities should be enforced by stricter regulations and implemented by national governments. Other preventative measures to improve tanker safety, and ultimately reduce the number and size of oil spills, include better training of pilots and crew; it is thought that 80% of accidents are due to human error caused by poor crew performance (Owen 1999).

While the IMO have established sound regulations and safety standards, the problem remains that international legislation is slow to be introduced and difficult to be implemented by the IMO or local port authorities that are responsible for safe navigation, yet are without the powers to enforce adherence to regulations.

The different procedures of dealing with oil spills described above have different environmental consequences. According to the size and extent of the oil spill, the type and origin of oil, age of the spill, the equipment available, as well as the location and local meteorological conditions, different methods and combinations thereof are used, with the aim of minimising the detrimental impacts of the spillage on nearby coastal ecosystems or fisheries, as well as general global impacts through evaporation or burning of large volumes of oil.

All oil spills are detrimental to the environment, and emphasis should be placed not on their mitigation, but on improved preventative procedures as discussed above. Measures should include the better design, construction, technical condition and maintenance of ships, stricter safety regulations that are better policed, as well as the more efficient implementation of management plans.

The management of oil spills is crucial, and the International Convention on Oil Pollution Preparedness, Response and Co-operation (OPRC) that was adopted in 1990 supports and facilitates international co-operation and mutual assistance of governments. Specific information, training and emergency response centres have been set up to allow more effective prevention and response to oil spills in different seas. The OPRC also supports oil spillage-related research and development, and provides guidelines for governments and individuals involved in oil transfer and transport, and those involved in the planning, conduct and administration of salvage operations.

4.4.2 Antifouling paints

In 2001 the IMO adopted an International Convention on the control of harmful anti-fouling systems on ships and platforms and other structures which bans the use of organo-tin containing compounds as biocides on ships, in particular TBT, due to its known toxicity at species, habitat and ecosystem level, as well as potential impacts on human health through the consumption of seafood. However, the Convention does not cover naval, or other government non-commercial ships, and not all countries have signed it.

Despite the phasing-out of TBT-based paints over recent years, copper-based paints are still used. They also leach into freshwater or marine systems and are toxic to organisms, but are thought to be less bioavailable as copper readily forms organic complexes. However, copper accumulation in algae and other aquatic biota has occurred and copper contamination near harbours and marinas has been observed (Stengel & Dring 2000).

Other, new antifouling paints that are currently being tested are still metal-based, but their active ingredients (e.g. 4,5-dichlor-2-n-octyl-4-isothiazolin-3-on (DCOI) in Sea-Nine 211, and zinc pyrithion) are thought to be biodegradable. Although they are very toxic to marine biota, they break down into less toxic substances rapidly and hence are unlikely to bioaccumulate in marine environment, and are thus considered more environmentally friendly in the long-term.

Current research focuses on non-stick foulant release substances including silicon- and ceramic-epoxy coatings. Non-adhesive surfaces can deter microbial and algal settlement, which is usually the first stage in biofouling succession. Other methods include electric sterilisation of ships' hulls or the use of non-adhesive surface structures such as microscopic prickles. Other new alternatives are coatings containing natural antifouling compounds; some of these are derived from marine bacteria, algae, corals and sponges. Natural biocides may be enzyme-based that prevent the settlement of bacteria by breaking down attachment mucilages. Routine mechanical removal of fouling organisms from large ships and aquatic surfaces is considered impractical as it is inefficient and labour intensive because it involves the physical removal by divers or extended dry-docking. However, mechanical removal often precedes renewal of antifouling coatings.

4.4.3 *Waste*

Waste from ships should be collected at ports and ferry terminals, in similar fashion to procedures at airports and the servicing of planes. Several international conventions have banned dumping of oils, waste and rubbish (e.g. MARPOL Conventions of 1973, 1978 and 1992), but implementation in off-shore waters remains a problem. As of August 2005, Annex IV of MARPOL 73/78 prohibits the release of raw sewage within 12 nautical miles of land, and of treated sewage within 3 nautical miles of land. However, this only applies to new ships of a certain size that carry more than 15 passengers; older ships are required to comply with regulations within the following 5 years. The dumping of solid waste including waste and harmful substances is also restricted under MARPOL 73/78 (Annex V and III, respectively). The disposal of plastics is prohibited anywhere in the sea, while dumping other rubbish is restricted only in coastal and protected areas, and record books logging waste disposed and incinerated are mandatory for large ships. Governments are required to provide reception facilities for sewage and waste at ports and terminals. However, despite the early legislation, the continued evidence of wastes, including plastics, washed up on shores worldwide suggests that, in addition to the regulations, better education of crew and passengers is required.

4.4.4 *Remediation of fresh water systems*

To decrease pH in acidified water bodies, lime ($CaCO_3$) can be added to lakes, streams and or whole catchments (Donnelly *et al.* 2003). In particular, in Scandinavian lakes, large-scale additions of lime from boats, planes and helicopters were undertaken to neutralise waters acidified by dry and wet deposition of sulphuric and nitric acids. Lime is relatively inexpensive, it dissolves readily in water, at low concentrations is non-toxic to biota, and also reduces the toxicity of heavy metals, in particular aluminium, copper, cadmium, lead and zinc. The effectiveness of liming of water bodies as a means to reduce water pH in the long-term largely depends on bedrock characteristics and mineral composition in the water, as well as the method, dosage and intensity of application (Donnelly *et al.* 2003).

Nitrogen can be removed chemically by ion-exchange resins, these can effectively remove 99% of all NH_3, but this method is costly (Farmer 1997). Other methods include the introduction of microbes that that oxidise NH_3 to NO_3 (by the use of aerated gravel beds or activated sludge), followed by denitrification of NO_3 to N_2 gas which can be emitted (Farmer 1997). Alternatively, fast-growing large (floating) plants can be introduced that take up nitrogen quickly and can easily be removed from the water. The construction of artificial reed beds (mainly *Phragmites australis*) to remove and soak up nutrients (or metals from runoff) has been discussed above. The use of (mainly) barley straw in eutrophic lakes and rivers to successfully reduce algal growth has also received much attention recently (Everall & Lees 1996).

4.4.5 *Remediation of toxic sediments*

Coastal and estuarine sediments are continuous sinks for contamination. Toxic sediments can be physically removed and dumped elsewhere. However, both the method of the removal and the destination of toxic sediments are critical. Physical disturbance of the sediments and re-suspension and oxygenation of toxins may increase their bioavailability to biota. Dumping of toxic marine sediments on land has occurred in the past, leading to the leaching of contaminants into groundwater. Regular dredging of sediments which may or may not be toxic occurs near entrances to ports, and sediments chronically polluted with high concentrations of PAHs and PAH-derivatives have often been dumped in designated marine (coastal or off-shore) dump sites. Although these are generally away from busy shipping routes, the actually dredging causes disturbance of (toxic) sediments, and stimulates further release of nutrients or toxins into the marine systems where they may become available for uptake by marine biota. These biota should be assessed using tiered ecotoxicity assessment approaches (Hartl *et al.* 2005; Davoren *et al.* 2005). Seaweeds that naturally absorb nutrients, either as particulates or in a dissolved form, from the water may be cultured in the disturbed sediments. Seaweed biomass can be processed into an energy-rich, valuable biofuel (Schramm 1991).

5. Future perspectives

As the contribution of transport-derived contaminants to air, water and land pollution cannot currently be estimated reliably, their emission should be reduced to a minimum. In addition, mitigation procedures such as described above, should be implemented wherever feasible.

Although many of the above solutions are directed at reducing the impact of run-off from existing types of combustion engines, there is an increasing trend to explore alternative technologies and fuel sources for land-based transportation.

With respect to emissions, electric engines are probably the cleanest transport mode (assuming that the energy used is produced without causing environmental damage), but as diesel locomotives are still used, even railways, often considered the 'green option' are still major emitters of air pollutants. Also, regarding organic pollution, the tradition of using large quantities of weed killers along tracks has created a contamination legacy for railways that needs to be addressed, in particular where organic residual herbicides have accumulated in soils and groundwater over decades.

In the case of individual vehicle ownership there has been a sustained drive towards higher performance and reduced emissions for several decades. It is likely that we are approaching the limits of the technology here in terms of reducing emission, with existing fuels. There has been little development in the use of biofuels, although with increasing fossil fuel cost this may change. It is likely that additional benefits are likely to come either from hybrid vehicles that combine fossil fuel and electric technologies, electric vehicles or the introduction of fuel-cell technology.

However, electric or fuel-cell vehicles still rely on energy use, which will increase significantly over the coming decades, so debate in the future may focus even more on the need for 'clean' energy production, as well as its economic consequences.

In the case of aviation, it is difficult to see how the impact of emissions will be managed. In the foreseeable future it is unlikely that aviation will use fuels other than kerosene. It is likely however, that land-based activities associated with air transport will explore the use of non-toxic de-icing agents, manage waste better and provide for management of runway run-off in sustainable ways.

Concerning aquatic transport, there are two main issues that need to be addressed. Firstly, emissions from ships need to be regulated and included in global emission restrictions and guidelines. The implementation of this, possibly through the introduction of taxes, will prove difficult, especially in international waters. The second issue is the ever-increasing risk of major oil spills from tanker incidents, and the release of oil during illegal (or even legal) tanker operations in international waters. Both issues need to be addressed through further international legislation, and equally importantly, effective implementation and policing with international agreement.

Clearly, measures to avoid or minimise environmental contamination first and foremost should include protective measures to physically avoid or reduce, or treat, runoff from into soils and groundwater, but also appropriate planning of new routes, in particular to protect sensitive habitats or groundwater catchment areas. However, even if direct run-off of pollutants can be avoided, transport-derived pollutants can still reach drinking water catchments or other aquatic systems through atmospheric deposition. This emphasises the fact that environmental pollution is not a local but a regional, if not global, issue. As has become evident from the discussion above, an integrated approach to environmental management and mitigation procedures must be taken. However, this should not be considered an excuse not to change individual behaviour on a small scale (e.g. personal choice of transport mode), community (e.g. choice of local transport system planning), regional or even country-wide scale (e.g. transport investment, planning and management policies).

There is a growing demand for the development of new technologies that will improve environmental performance of transport systems and the infrastructure associated with them. Today, biotic and abiotic cycles pathways of contaminants and potential pollutants are understood better than a decade ago, but long-distance transport and long-term accumulation of contaminants, in particular the effects and mechanisms of bio-accumulation, remain poorly understood. Current research focuses on understanding the mechanisms of environmental contamination on ecosystems change, as well as mitigation measures.

Even today, after half a century of study, toxic effects of substances previously considered harmless are still discovered. In many cases where a substance is suspected to be hazardous, there is not enough hard scientific evidence to prove it. Safety regulations and standards should therefore be based on precautionary principles rather than toxicity evidence. In other words new substances (e.g. new antifouling paints on ships) should be treated with caution until proven harmless.

Also, long-term effects of pollutants require more attention, as short-term observations cannot predict long-term environmental change. In particular, ecosystem responses at the biodiversity level and ecosystem function are poorly understood. Also, feedback mechanisms of global climate change in combination with environmental pollution are currently not predictable.

There is no doubt that, in parallel with general reductions in transport emissions and developments in modern approaches to transportation, sensitive habitats and ecosystems should be protected as soon as possible. Such habitats include groundwater catchment areas and coastal marine systems which act as typical sinks for pollutants and require special protection. Although it is clear that environmental pollution is a global problem, local environment protection schemes have an important role to play in securing drinking water supplies, species diversity, agricultural crops and improving local amenity and, ultimately, preserving human health.

Recently, several attempts have been made to compare the value of different transport systems in relation to environmental pollution and other costs (e.g. Hall *et al.* 1992; Small & Kazimi 1995; Fankhauser & Tol 1996; Romilly 1999; Nilsson & Küller 2000; Hiscock *et al.* 2002). Costs of different transport systems can be estimated in terms of emissions, actual running costs (including construction and maintenance of vehicles, train and planes or road, track and airports), mortalities (human or other) that occur due to pollution, habitat loss and loss in biodiversity, as well as mitigation and compensation procedures. The difficulty lies in placing comparable monetary values on emissions, atmospheric or water pollution, quality of life, as well as human and ecosystem health.

A major challenge lies in persuading governments to commit to international agreements to reduce emissions such as the Kyoto 1997 agreement, and enabling environmental agencies, as well as international organisation such as the IMO, to implement the standards at an international level. Future emphasis should be placed on environmental training and education. Research should focus on improving existing, and developing new, technologies for the prevention of transport-derived pollution and providing effective mitigation measures that will undoubtedly become more critical in environmental management in the future.

References

Anon 2002; Air pollution from ships. Swedish NGO Secretariat on Acid Rain. European Environmental Bureau. European Federation for Transport and Environment.

Anon 2004; Banverket – Environment. www.banverket.se, updated December 2004 (accessed 4-01-2005).

Association of the American Railroads. 2002; Railroad Facts, 2002 Edition. Association of the American Railroads, Washington, DC.

Bailey D, Solomon G. 2004; Pollution prevention at ports: clearing the air. Environmental Impact Assessment Review 24: 749-774.

Barkan CPL. 2004; Cost effectiveness of railroad fuel spill prevention using a new locomotive refuelling system. Transportation Research Part D, 9: 251-262.

Beasley G, Kneale P. 2002; Reviewing the impact of metals and PAHs on macroinvertebrates in urban watercourses. Progress in Physical Geography 26(2): 236-270.

Bellinger DC. 2004; Lead. Paediatrics 113: 1016-1022.

Bielefedldt AR, Illangasekare T, LaPlante R. 2004; Bioclogging of sand due to biodegradation of aircraft deicing fluid. Journal of Environmental Engineering 130: 1147-1153.

Bomboi MT, Hernandez A. 1991; Hydrocarbons in Urban Runoff: Their Contribution to the Wastewaters. Water Research 25(5): 557-565.

Boxall ABA, Maltby L. 1997; The Effects of Motorway Runoff on Freshwater Ecosystems: 3. Toxicant Confirmation. Environmental Toxicology and Chemistry 33: 9-16.

Broster M, Pritchard C, Smith DA. 1974; Wheel/rail adhesion: its relation to rail contamination on British railways. Wear 29: 309-321.

Buckler DR, Grenato GE. 1999; Assessing Biological Effects from Highway Runoff Constituents. US Department of Transportation. Open file report 99-240.

Cape JN, Tang YS, van Dijk N, Love L, Sutton MA, Palmer SCF. 2004; Concentrations of ammonia and nitrogen dioxide at roadside verges, and their contribution to nitrogen deposition. Environmental Pollution 132 (3): 469-478.

Carpenter TG. 1994; Pollution. In: Carpenter TG (ed) The environmental impact of railways, 165-183. Wiley & Sons Ltd, Chichester, UK.

Clark RB 2003; Marine Pollution. Oxford University Press, Oxford, 237pp.

Colvile RN, Hutchingson EJ, Mindell JS, Warren RF. 2001; The transport sector as a source of air pollution. Atmospheric Environment 35: 1537-1565.

Cooper DA. 2003; Exhaust emissions from ships at berth. Atmospheric Environment 37: 3817-3830.

Corbett J, Fischbeck P. 2001; Sources and transport of air pollution from ships: current understanding, implications and trends. Conference on marine vessels and air quality, EPA Region 9, Feb 2001, San Francisco, California; http://www.epa.gov/region09/air/marinevessel/ (accessed 10-5-2005).

Crick HQP. 2004; The impact of climate change on birds. Ibis 146: 48-56.

Crick HQP, Dudley C, Glue DE, Thomson DL. 1997; UK birds are laying eggs earlier. Nature 388 (6642): 526-52.

Davoren M, Shuilleabhain SN, O'Halloran J, Hartl MGJ, Sheehan D, O'Brien NM, van Pelt FNAM, Mothersill C. 2005; A Test Battery Approach for the Ecotoxicological Evaluation of Estuarine Sediments. Ecotoxicology 14:741-755.

Donnelly A, Jennings E, Allot N. 2003; A review of liming options for afforested catchments in Ireland. Biology and Environment: Proceedings of the Royal Irish Academy103B: 91-105.

EEA. 2001; European Environmental Agency. EMEP/CORINAIR Emissions Inventory Guidebook, 3rd Edition. EEA, Group 08.http://reports.eea.eu.int/EMEPCORINAIR3.

European Commission. 2002; Quantification of emissions from ships associated with ship movements between ports in the European Community. Entec UK Limited, July 2002. www.europa.eu.int/comm/environment/air/background.htm#transport Accessed 10-05-2005.

European Environmental Agency. 2002; Annual European Community CLRTAP emission inventory 1990-2000. Technical Report 91. http://reports.eea.eu.int/technical_report_2002_73/en. (accessed 10-05-2005).

European Environment Agency. 2004; Impacts of Europe's changing climate. An indicator based assessment. EEA Report No 2/2004.

Fankhauser S, Tol RSJ. 1996; Climate change costs recent advancements in the economic assessment. Energy Policy 24: 665-673.

FHWA (Federal Highway Administration). 2003; Ultra-Urban Best Management Practices, US Department of Transportation.

FHWA (Federal Highway Administration). 1999; Environmental Technology Brief Office of Infastructure Research and Development, Turner-Fairbank Highway Research Centre, 6300 Georgetown Pike, Mc Lean, VA 22101.

Genter, RB. 1996; Ecotoxicology of inorganic stress to algae. In: Stevenson RJ, Bothwell ML & Lowe RL (eds) Algal ecology – Freshwater Benthic Ecosystems. Academic Press, San Diego, California.

Green JE. 2003; Civil aviation and the environmental challenge. Aeronautical Journal 107 (1072): 281-299.

Hall A, Fielding AH, Butler M. 1979; Mechanisms of copper tolerance in the marine fouling alga Ectocarpus siliculosus- evidence for an exclusion mechanism. Marine Biology 54: 195-199.

Hall JV, Winder AM, Kleinman MT, Lurmann FW, Brajer V, Colome SD. 1992; Valuing the health benefits of clean air. Science, 255: 812-817.

Hares RJ, Ward NI. 1999; Comparison of the heavy metal content of motorway stormwater following discharge into wet bioinfiltration and dry detention ponds along the London Orbital (M25) Motorway. The Science of the Total Environment. 235: 169-178.

Harrison PA, Berry PM, Dawson TO. (eds) 2001; Climate Change and Nature Conservation in Britain and Ireland: Modelling natural resource responses to climate change (the Monarch project). UKCIP Technical report Oxford.

Hartl M, van Pelt FNM, O'Halloran J. 2005; Development and Application of sediment toxicity tests for regulatory purposes. In: Wiley Encyclopaedia of Water Pollution.

Hewitt CN, Rashed MB. 1992; Removal rates of selected pollutants in the runoff waters from a major rural highway. Water Research 26(3): 311-319.

Hoagland KD, Carder JP, Spawn RL. 1996; Effects of organic toxic substances. In: Stevenson RJ, Bothwell ML & Lowe RL (eds) Algal ecology – Freshwater Benthic Ecosystems. Academic Press, San Diego, California.

International Maritime Organisation. 1998; Annex VI to MARPOL 73/78 regulations for the prevention of air pollution from ships and NO$_x$ technical code. Report IMO-664E.

IPCC 2001a; Intergovernmental Panel on Climate Change. Climate Change 2001: the scientific basis. Contributions of working group I to the Third Assessment report of the Intergovernmental Panel on Climate Change. (J. Houghton et al., eds). Cambridge University Press, Cambridge.

IPCC 2001b; Intergovernmental Panel on Climate Change. Climate Change 2001: Mitigation. Contributions of working group III to the Third Assessment report of the Intergovernmental Panel on Climate Change. Cambridge University Press, Cambridge.

Isakson J, Persson TA, Selin Lindrgren E, 2001; Identification and assessment of ship emissions and their effects in the harbour of Göteborg, Sweden. Atmospheric Environment, 35: 3659-3666.

Ketkar KW. 2002; The Oil Pollution Act of 1990: a decade later. Spill Science and Technology Bulletin 7: 45-52.

Ketkar KW, Babu AJG. 1997; An analysis of oil spills from vessel traffic accidents. Transportation Research. Part D, 2: 35-41.

Lacaze JC. 1974; Ecotoxicology of crude oils and the use of experimental marine ecosystems. Marine Pollution Bulletin 5: 153-156.

Langston WJ. 1996; Recent developments in TBT ecotoxicology. Toxicology and Ecotoxicology News 3: 179-187.

Law RJ, Blake SJ, Jones BR, Rogan E. 1998; Organotin compounds in liver tissue of harbour porpoises (Phocena phocena) and grey seals (Halchoerus grypus) from the coastal waters of England and Wales. Marine Pollution Bulletin 36: 241-247.

Legret M, Pagotto C. 1999; Evaluation of pollutant loadings in runoff waters from a major rural highway. The Science of the Total Environment 235: 143-150.

Maltby L, Boxall ABA, Forrow DM, Calow P, Betton CI. 1995b; The Effects of Motorway Runoff on Freshwater Ecosystems: 2. Identifying the Major Toxicants. Environmental Toxicology and Chemistry 14 (6): 1093-110.

Maltby L, Forrow DM, Boxall ABA, Calow P, Betton CI. 1995a. The Effects of Motorway Runoff on Freshwater Ecosystems: 1 Field Study. Environmental Toxicology and Chemistry 14(6): 1079-1092.

Marsalek J, Rochfort Q, Brownlee B, Mayer T, Servos M. 1999; An exploratory study of urban runoff toxicity. Water Science and Technology 39(12): 33-39.

Meade N, Lapointe T, Anderson R. 1983; Multivariate analysis of worldwide tanker accidents. National Oceanographic and Atmospheric Administration, Washington, DC, March.

Mikkelson PS, Hafliger M, Ochs M, Tjell JC, Jacobsen P, Boller M. 1996: Experimental Assessment of soil and groundwater contamination from two old infiltration systems for road runoff in Switzerland. The Science of the Total Environment 189/190: 341-347.

Nilsson M, Küller R. 2000; Travel behaviour and environmental concern. Transportation Research. Part D. 5: 211-234.

Oberts G. 1994; Performance of Stormwater Ponds and Wetlands in Winter. Watershed Protection Techniques. 1(2): 64-68.

O'Brien PY, Dixon PS 1976; The effects of oils and oil components on algae: a review. British Phycological Journal 11: 115-142.

O'Reilly S, Morgan G, Heffron JJA, O'Halloran J. 2004; A review of the compostion, treatment and ecological effects of road runoff. In: Davenport J & Davenport JL (eds) The effects of human transsport on Ecosystems: Cars and Planes, Boats and Trains, 63-82. Royal Irish Academy, Dublin.

Owen J. 1999; The environmental management of oil tanker routes in UK waters. Marine Policy 23: 289-306.

INDEX